FRANCIS GALTON AT CAMBRIDGE

Francis Galton at Cambridge

Letters and Diaries, 1840-1844

Gavan Tredoux

galton.org

ISBN: 978-1-963225-14-3
Not for sale in the United Kingdom.

Francis Galton: a Lifetime of Exploration

1. Francis Galton's Nature and Nurture: 1822–1865.
2. Francis Galton's Genius: 1865–1911.

Supplements:

1. Sir Francis Sacheverel Darwin.
2. Francis Galton at Cambridge: Letters and Diaries, 1840–1844.
3. The Diary of Charles John Andersson: 1850–1851.
4. Francis Galton on Mars: the Discontinuous Variation Notebook.
5. Francis Galton and Alphonse de Candolle: Notes and Correspondence.
6. Francis Galton and Alfred Binet: Correspondence.
7. Francis Galton: Selected Papers, Notebooks and Diaries.

Figure i: Galton by Octavius Oakley, January 1841.

Contents

Preface

Francis Galton went up to Trinity College at Cambridge in October 1840, and left on the death of his father in 1844, by which time he had already graduated but was still attending medical lectures. His original intention was to study mathematics, a subject that promised to be a welcome relief from medicine—which he had been training for in London but had lost most of his interest in, despite the wishes of his family. He was by inclination a "reading man", rather than one of the rowdies known as "rowing men", but the additional obligation to read classics was a nuisance.

Although the Galtons were wealthy, they were always thrifty, so he enrolled as an ordinary student, or "pensioner", rather than as a more luxurious "fellow commoner". But he did pay dearly to be coached by the noted private tutor William Hopkins, famous for producing leading "wranglers" like Arthur Cayley, William Thomson, James Clerk Maxwell and George Stokes. The University provided little instruction itself, and had only been reluctantly persuaded to modernize its curriculum, to catch up with continental advances.

However, Galton's peculiarly obsessive mentality led to grim and monotonous overwork and a severely debilitating breakdown in his second year. He had to withdraw from the highly competitive honours examinations used to rank the "wranglers" and settle for an ordinary degree instead. After this he went through the motions of studying for a medical degree, until the traumatic death of his father removed all incentive to continue. It would take him many years to recover and eventually launch the innovative scientific career that he is now remembered for.

During his time at Cambridge, Galton kept up an extensive correspondence with his family, particularly with his father, who had retired to Leamington Spa some years earlier, but was keenly interested in the progress of his son, and preserved most of the material that is still extant. A generous selection of the correspondence from this formative period was included in Karl Pearson's monumental *Life, Letters and Labours of Francis Galton*[1]—but often only in partial form, with important omissions, numerous redactions, editorial interventions, and inevitable errors.

This collection restores the original text of the letters, transcribed directly from the manuscripts to modern standards. Additional letters completely omitted by Pearson are included, together with the sketches and doodles for all the letters, which are essential for understanding them.

A slew of mechanical inventions and contrivances dominate the earlier letters, but fade away as compulsive overwork and stress wear on Galton. He was increasingly fixated, then dizzy and palpitating. His own Diary entries survive for 1843 and show us a glimpse of his college and home life; together with the letters, they allow us to trace the development of his ongoing mental breakdown and his strategies for coping with it. Entries from the shorthand diary kept by his doting father throughout this period are added to all this material, in chronological order, and fill in many gaps.

Many obscurities in the texts have been clarified in the footnotes, but Galton's original punctuation and spelling have, as far as possible, been left undisturbed. A register with biographical details of the more important figures has also been provided.

The printed editions also include facsimiles of all the original manuscripts. As they reflect Galton's personality and (at times, wine-stained) circumstances in a way that transcriptions cannot capture; the original documents form an important record. The reader may then correct and clarify any errors and obscurities that remain.

Gavan Tredoux
`http://galton.org`
December 2017.

[1] Pearson 1914.

Correspondents

FG	Francis Galton	.
STG	Samuel Tertius Galton	Father
FVG	Frances Violetta Galton	Mother
EAG	Elizabeth Anne Galton	Sister
AG	Adèle Galton	Sister
ESG	Emma Sophia Galton	Sister

Chapter 1

1840

1840/03/31. Joseph W. Blakesley to Samuel Tertius Galton

Trin. Coll.
Mar. 31st / 40.

Dear Sir,

I have found by experience that, except in cases of deficiency, it is extremely unfavourable to the mental development of a young man that he should be put into the hands of a private tutor immediately on his arrival in the University. I have therefore always strongly recommended my pupils to pass at least the first term without any other assistance than the College lectures, which are amply sufficient for the purpose of directing their attention to the proper course of reading. If a young man takes a private tutor before he has had time to feel what his own strength is, and what his peculiar defects are, he frequently makes an injudicious selection, and even his University prospects suffer in consequence. I would much rather therefore, if you have no objection, not make any arrangement of this kind for your son, especially as by your account he appears to have talents & taste for the study of mathematics. In every point of view he will be benefited by passing his first term without a private tutor: indeed with Mr. Hopkins[1] I should recommend that he does not begin reading until after the Examination in the Easter Term of 1841. If however you should view the matter in a different light, I will certainly make an application to that gentleman. I remain, Dear Sir,

[1] Hopkins, William. See page 484.

Faithfully yours
J. W. Blakesley.[2]

1840/10/13. Diary of Samuel Tertius Galton

October 13th. Tuesday. Leamington.
I started with Francis at 7:15 by the Rising Sun Coach for Cambridge distant 86 miles which we reached about 6:30. At Weadon we were joined by many students from the train among whom was George Hughes. We lodge at the Hoop Inn at which the coach stopped. The best Inns are the Eagle, the Hoop, and the Lion. A beautiful, clear day.

1840/10/14. Diary of Samuel Tertius Galton

October 14th. Wednesday. Cambridge.
Fine Day. After breakfast Francis and myself went over St. John's, Trinity, Kings College Chapel etc. We visited the County Gaol, Castle Hill, called upon Mr. [Blakesley], who showed us Francis's room in the New Court B, 2nd floor. Francis bought his cap, gown etc. and dined in the Hall and I dined with the Fellows with Mr. [Blakesley] who introduced me to his colleagues. Mr. Percy and a Mr. Snow and Mr. Oliver who lives near the Wedgwoods and both of whom have sons freshmen. I attended Trinity College chapel in the evening. I was informed at the County Gaol that the silent and separated systems were tried with good effect …

1840/10/15. Diary of Samuel Tertius Galton

October 15th. Thursday. Cambridge.
Weather continues fine. Francis examined this morning and took possession of his rooms. I visited Downing and several other colleges and the Public Library, also Mr. Hughes introduced me to the Fitzwilliam and Mesman museums. Francis dined in the Hall and called upon me at the Hoop Inn and left me for Chapel, after which I went to his rooms and drank tea with him and stayed with him whilst he was destroying some old letters and papers. I left him about 9 p.m. near the entrance of the guard court where

2 Blakesley, Joseph Williams. See page 480.

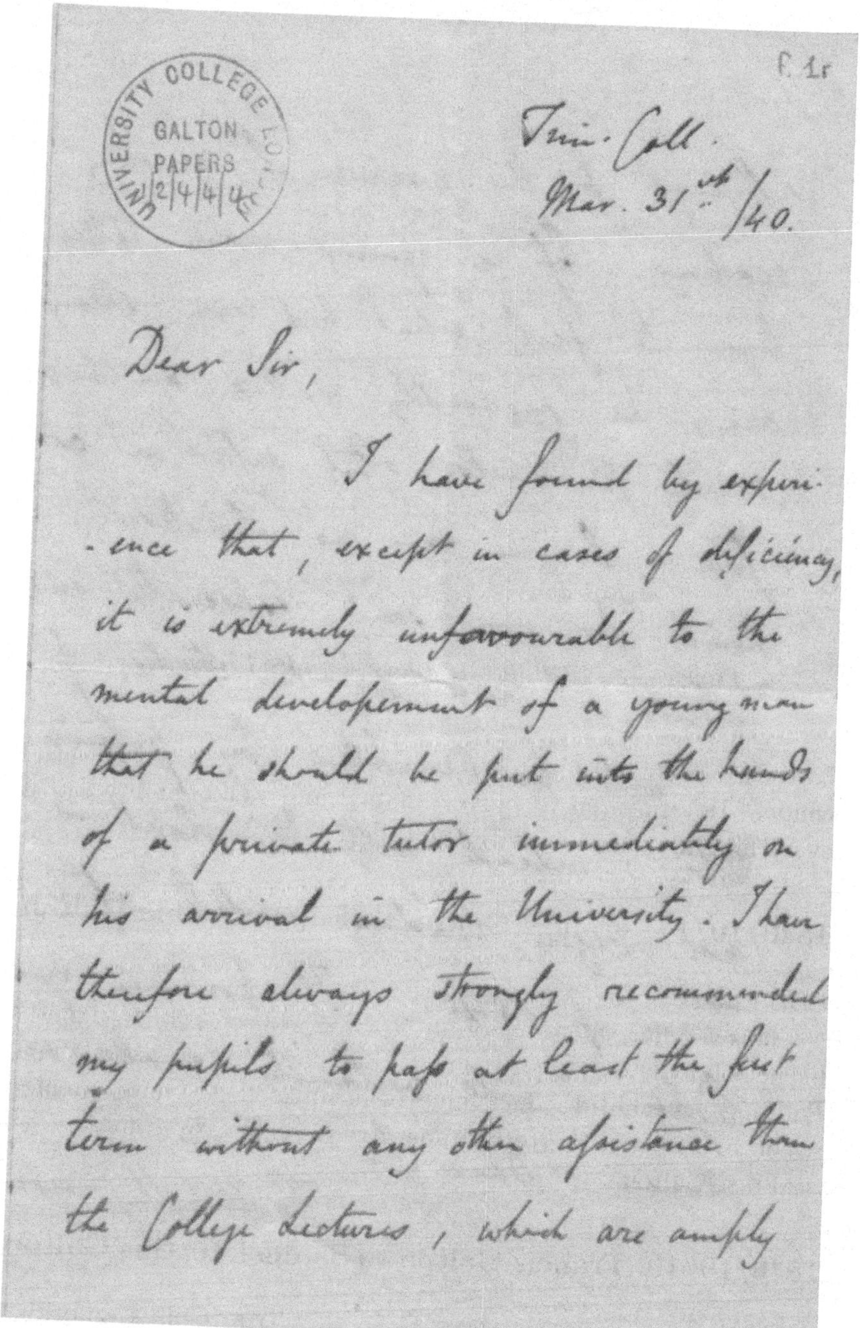

f. 1r

Trin. Coll.
Mar. 31st /40.

Dear Sir,

I have found by experi-ence that, except in cases of deficiency, it is extremely unfavourable to the mental developement of a young man that he should be put into the hands of a private tutor immediately on his arrival in the University. I have therefore always strongly recommended my pupils to pass at least the first term without any other assistance than the College Lectures, which are amply

Figure 1.1: 1840-03-31. Blakesley to STG page 1.

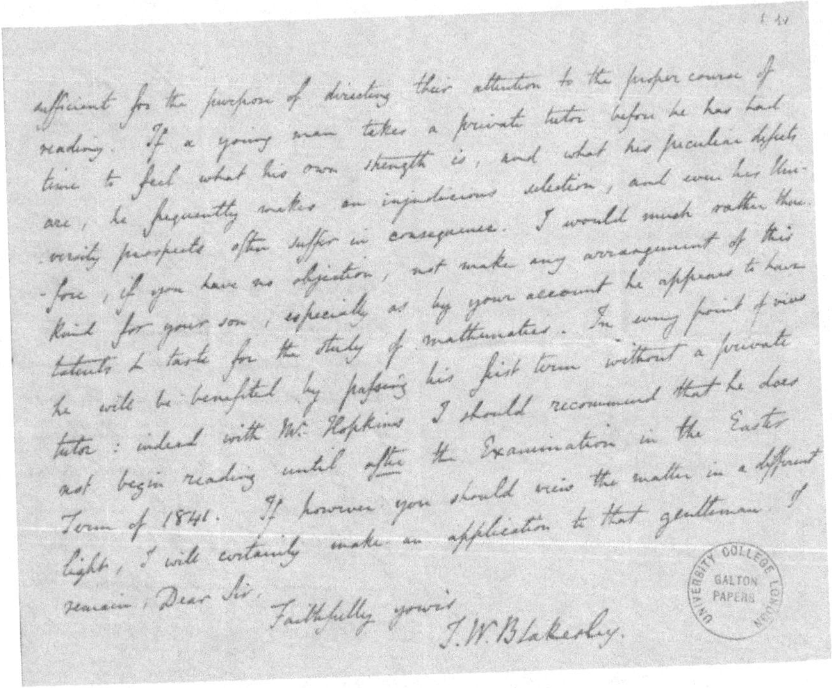
sufficient for the purpose of directing their attention to the proper course of reading. If a young man takes a private tutor before he has had time to feel what his own strength is, and what his peculiar defects are, he frequently makes an injudicious selection, and even his University prospects often suffer in consequence. I would much rather therefore, if you have no objection, not make any arrangement of this kind for your son, especially as by your account he appears to have talents & taste for the study of mathematics. In every point of view he will be benefited by passing his first term without a private tutor: indeed with Mr Hopkins I should recommend that he does not begin reading until after the Examination in the Easter Term of 1841. If however you should view the matter in a different light, I will certainly make an application to that gentleman. I remain, Dear Sir,

Faithfully yours

J. W. Blakesley.

Figure 1.2: 1840-03-31. Blakesley to STG page 2.

he turned back and I watched him as he passed the toll near the chapel. He appeared to be very happy and well contented with everything in connexion with the college.

1840/10/17. Francis Galton to Emma Sophia Galton

Dear Emma

This is a hurried scetch of my room. I will send you a vignette in each of my letters to pay you "in kind" for your two water colour Paintings—Cambridge has some beautiful college views, greatly set off by rows of elm trees. I must try & scetch you some.

Fras. Galton.

1840/10/19. Francis Galton to Samuel Tertius Galton

Trin: Coll: Cambridge
Monday Oct. 1840.

My dear Father

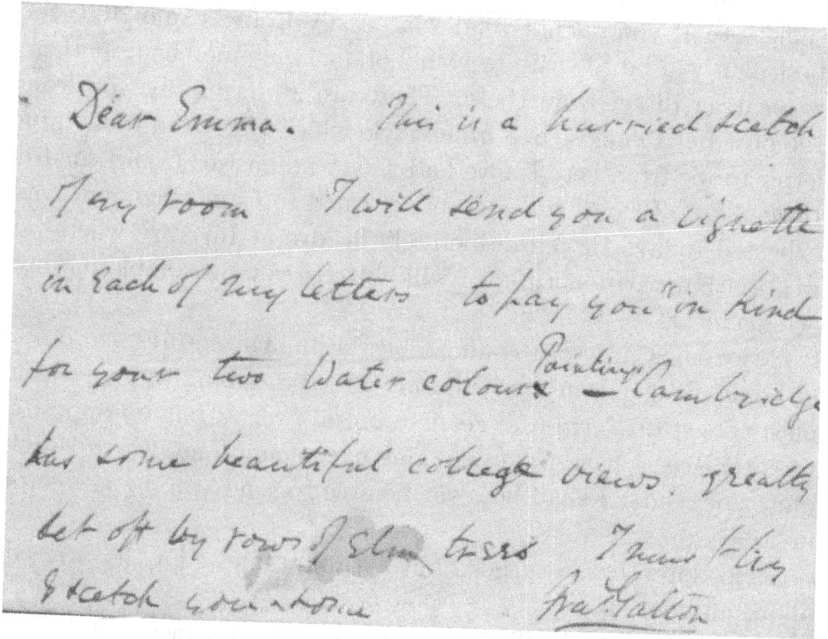
Dear Emma. This is a hurried sketch of my room I will send you a vignette in each of my letters to pay you in kind for your two Water colour Paintings — Cambridge has some beautiful college views, greatly set off by rows of Elm trees I must beg [illegible]

Fra. Galton

Figure 1.3: 1840-10-17. FG to ESG page 1.

Figure 1.4: 1840-10-17. Sketch of his Trinity Rooms by FG.

Thank you for your letter. Six silver teaspoons will be amply sufficient. If you cannot send wine easily from Leamington, the best plan will be to write to your London wine-merchant as there is a carrier direct from there. There are no letters for you from the post but I enclose one from Adèle which I received today in a letter to me from her. I have had as yet no answer from Leonard Horner.[3] O'Brien[4] has not yet returned to Cambridge, but was expected today. He fell desperately in love at Inverary where he went with a party as tutor. I will write to you on completing my arrangements.

My rooms are very comfortable. Emma's pictures are quite at home, as usual, in my bedroom, and I am going to invest in a plaster bust of Newton & get it bronzed over & put up opposite the fireplace. I have got everything except my linen which is not quite got ready. I shall however be able to send you my accounts in 2 days.

Theodore[5] has returned, but I cannot find him, he arrived about an hour since.

Perry[6] gave us his first lecture today, what a pleasure it is to hear a real Senior Wrangler speak. My organ of veneration is so very strong, that I doubt when I shall dare to address him. What a fine sight a surplice night is, the bright light of the wax candles & the white dresses so well contrasting with the dark panelled oak behind gives no slight resemblance to a scene in Revelations.

I am as happy as possible & am preparing for a long & strong pull at reading. Love to Mater & all.

Your affectionate Son
Fra^{s}. Galton

1840/10/23. Samuel Tertius Galton to Francis Galton

Claverdon, 23 Oct
1840

My dear Mr. Freshman,

3 Horner, Leonard. See page 484.

4 O'Brien, Matthew. See page 487.

5 His cousin Theodore Howard Galton (1820-1881), who had gone up to Trinity in 1838. He was the son of John Howard Galton of Hadzor. He would later practice at the Bar.

6 Perry, Charles. See page 490.

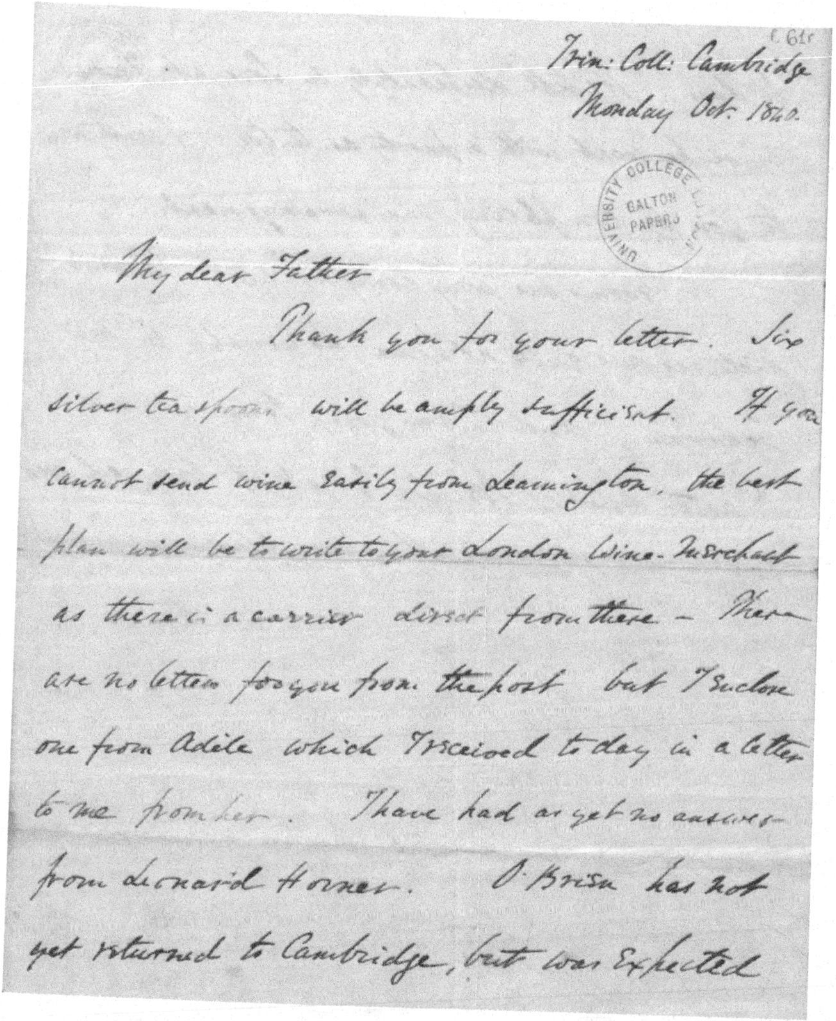

f. 61r

Trin: Coll: Cambridge
Monday Oct: 1840.

My dear Father

Thank you for your letter. Six silver tea spoons will be amply sufficient. If you cannot send wine easily from Leamington, the best plan will be to write to your London Wine-Merchant as there is a carrier direct from there — There are no letters for you from the post but I enclose one from Adele which I received to day in a letter to me from her. I have had as yet no answer from Leonard Horner. O'Brien has not yet returned to Cambridge, but was expected

Figure 1.5: 1840-10-19. FG to STG page 1.

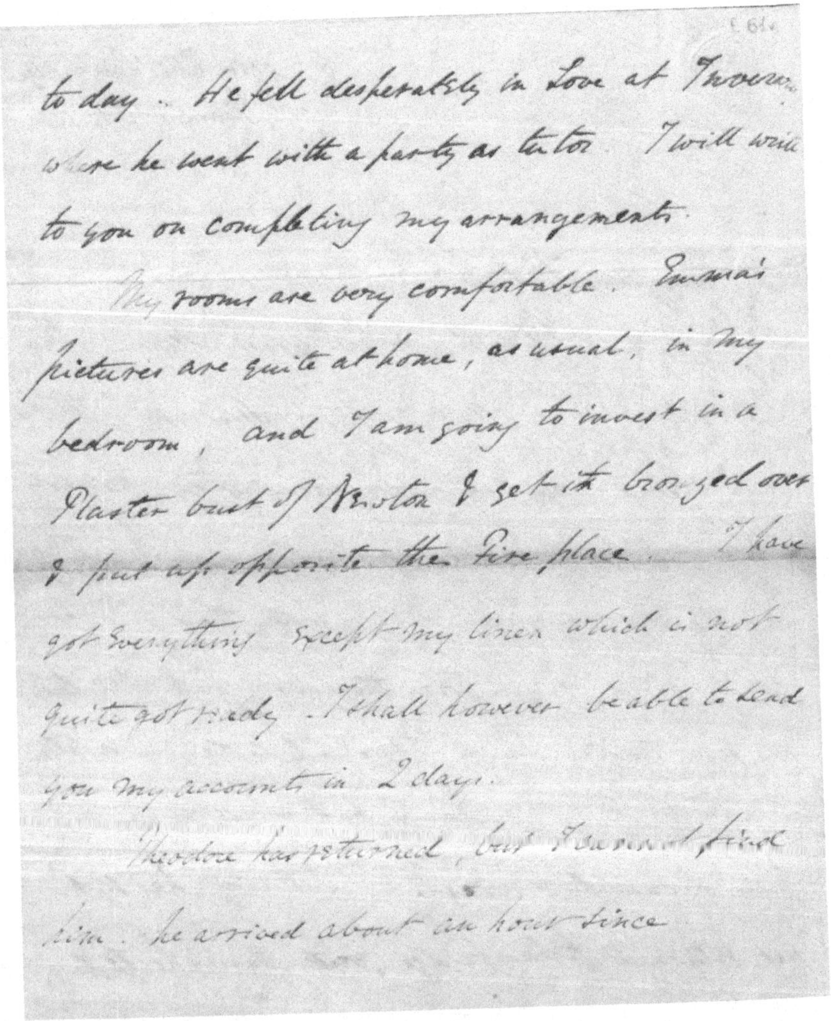

to day .. He fell desperately in Love at Inverness where he went with a party as tutor. I will write to you on completing my arrangements.

My rooms are very comfortable. Emma's pictures are quite at home, as usual, in my bedroom, and I am going to invest in a Plaster bust of Newton & get it bronzed over & put up opposite the Fire place. I have got Everything except my linen which is not quite got ready. I shall however be able to send you my accounts in 2 days.

Theodore has returned, but I cannot find him. he arrived about an hour since

Figure 1.6: 1840-10-19. FG to STG page 2.

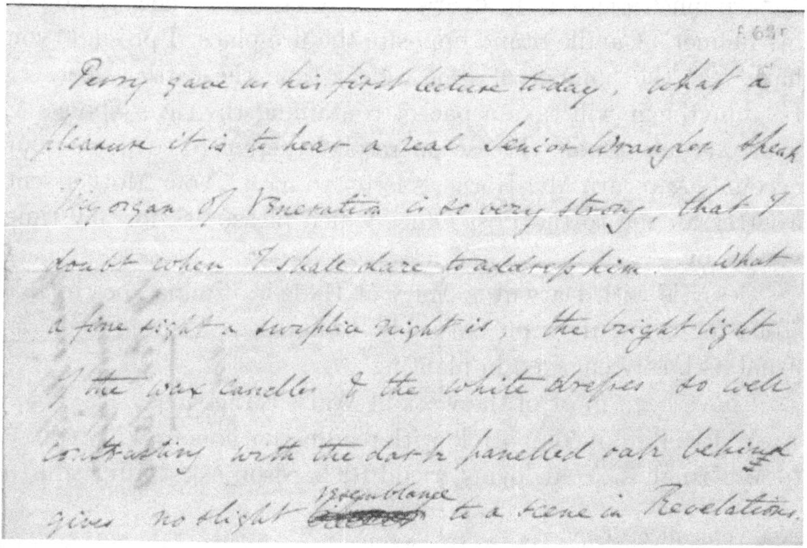

f 63r

Perry gave us his first lecture today, what a
pleasure it is to hear a real Senior Wrangler speak
My organ of Veneration is so very strong, that I
doubt when I shall dare to address him. What
a fine sight a surplice night is, the bright light
of the wax candles & the white dresses so well
contrasting with the dark panelled oak behind
gives no slight resemblance to a scene in Revelations.

Figure 1.7: 1840-10-19. FG to STG page 3.

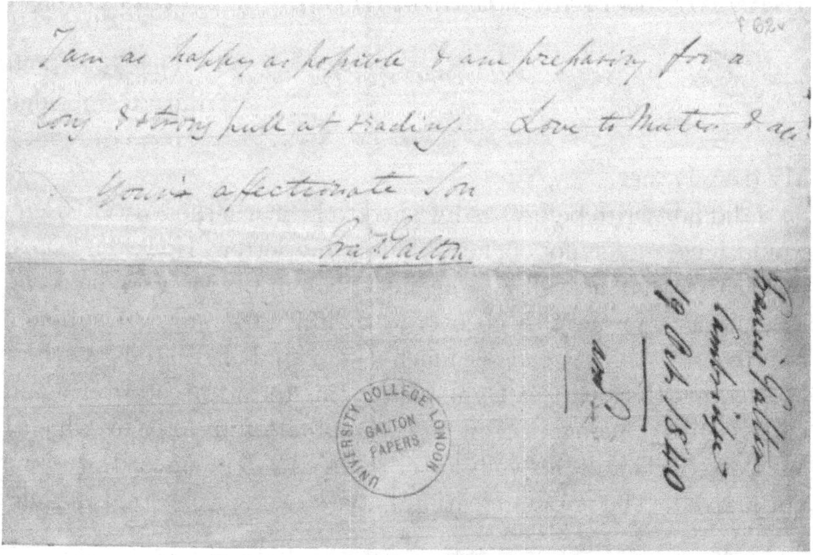

f 62v

I am as happy as possible & am preparing for a
long & strong pull at reading. Love to Mater & all
Your affectionate Son
Fra^s Galton

Francis Galton
Cambridge
19 Oct 1840
ans^d

UNIVERSITY COLLEGE LONDON
GALTON PAPERS

Figure 1.8: 1840-10-19. FG to STG page 4.

I am delighted that you are delighted with your College Life & take much pleasure in figuring to myself the student seated at his Palmer's Candle Stand opposite the fire place. I presume you have rec^{d} the hamper of wine. Next time the carriage goes to Leamington it will take a packet containing the silver spoons & your Account Book. Please on receipt thereof to send me your accounts & return Mr. Horner's letter to me.[7] Your Mother sent a letter for me to the Post Office which please ask for any time you go by.

Bessy is with Darwin & Mary at Hadzor. Emma goes to Edstone for a few days on Saturday, tomorrow. Your Mother, as usual, is busily engaged in planting.

I have been most of the week at Warwick Sessions.

Write often, if compatible with reading for honours. My comp^{s} to Mr. Hughes. Your loving friend Richardson asked after you.

Your affect^{e} Govr
STG

NB. Emma much pleased with y^{r} sketch of Eugene Cham's apartment.[8]

1840/10/23. Francis Galton to Samuel Tertius Galton

Oct. 23d 1840
Trinity Cambridge

My dear Father

I did not write before, as I wished to hear whether Mr O'Brien[9] would have me or not, before sending my letter. However I have just learnt that he will & I begin with him on Monday next, he recommends me not to go over with him now, my old subjects but to start off & read as much I can of Differentials & their application to Statics & Dynamics & this term to read over again what I have done against my first examination in May by which I am classed. Thanks for your letter received 2 days since. Port wine not arrived. The communication between the intellectual nucleus

7 Horner, Leonard. See page 484.

8 The letter appends a short "Essay on Book-keeping" and a statement of Francis' accounts drawn up by his father, titled "Dr. Francis Galton Esq. in account with his Treasurer".

9 O'Brien, Matthew. See page 487.

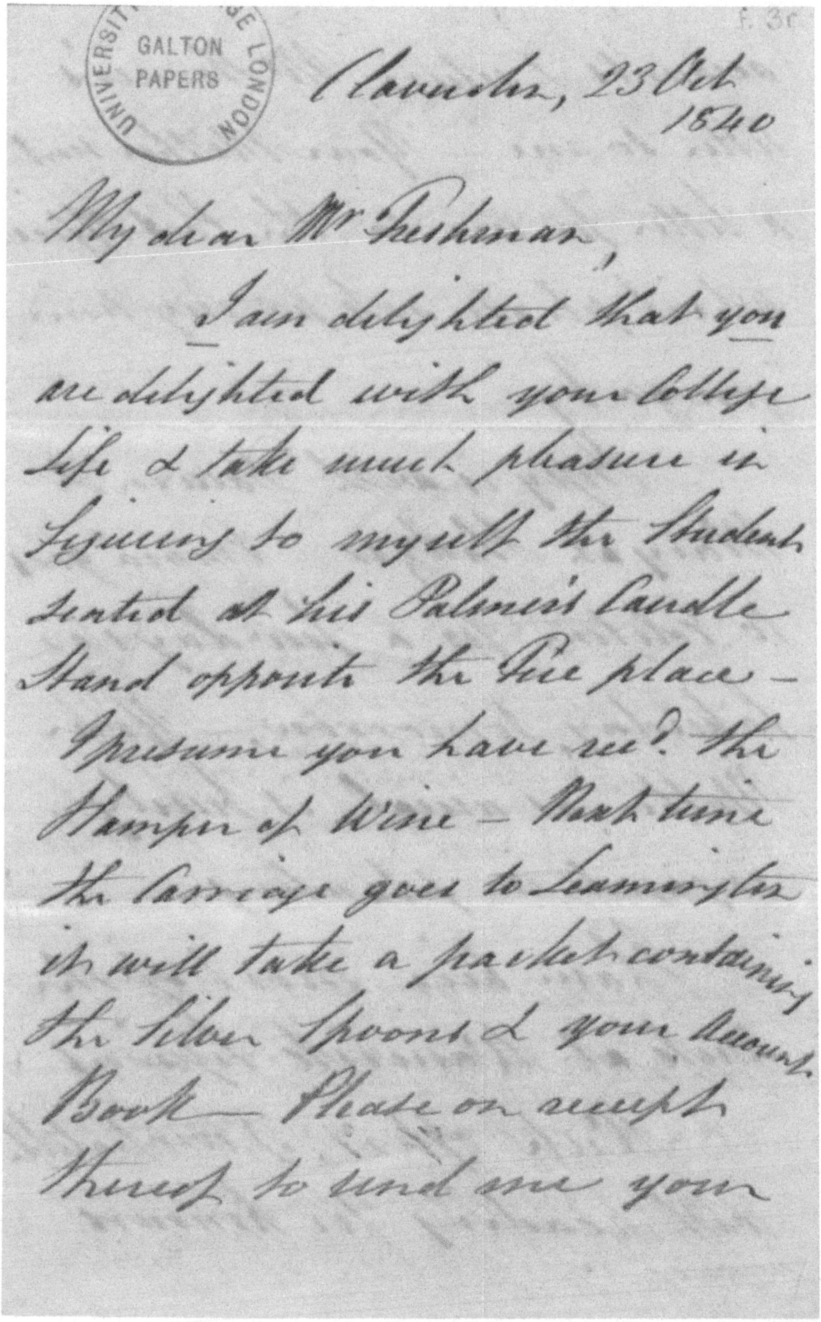

Claverdon, 23 Oct
1840

My dear Mr Freshman,

I am delighted that you are delighted with your College life & take much pleasure in figuring to myself the Student seated at his Palmer's Candle Stand opposite the fire place — I presume you have rec^d. the Hamper of Wine — Next time the Carriage goes to Leamington it will take a packet containing the Silver Spoons & your Account Book — Please on receipt thereof to send me your

Figure 1.9: 1840-10-23. FG to STG page 1.

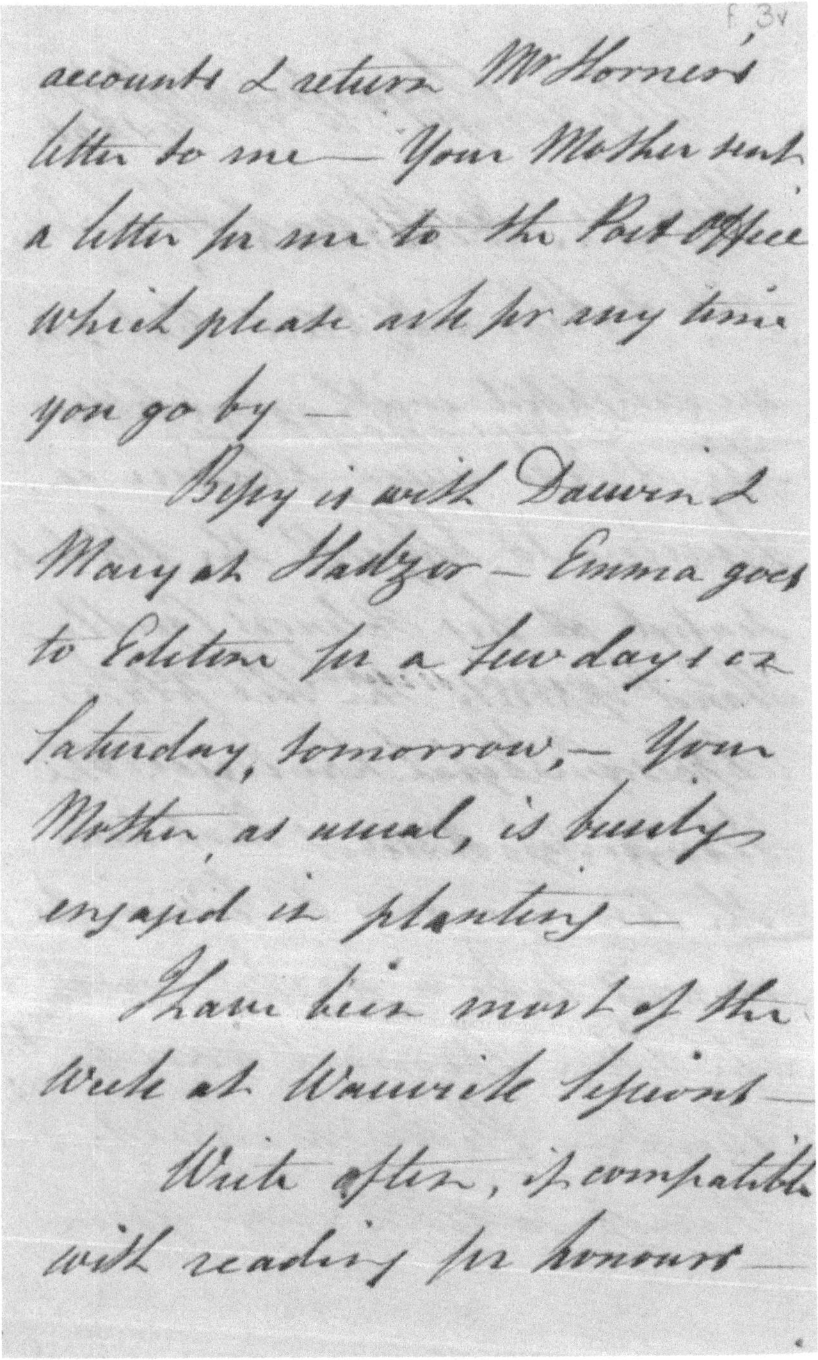

f. 3v

accounts & return Mr Horner's
letter to me — Your Mother sent
a letter for me to the Post Office
which please ask for any time
you go by —
Bessy is with Darwin &
Mary at Hadzor — Emma goes
to Edstone for a few days on
Saturday, tomorrow, — Your
Mother, as usual, is busily
engaged in planting —
I have been most of the
week at Warwick Sessions —
Write often, if compatible
with reading for honours —

Figure 1.10: 1840-10-23. FG to STG page 2.

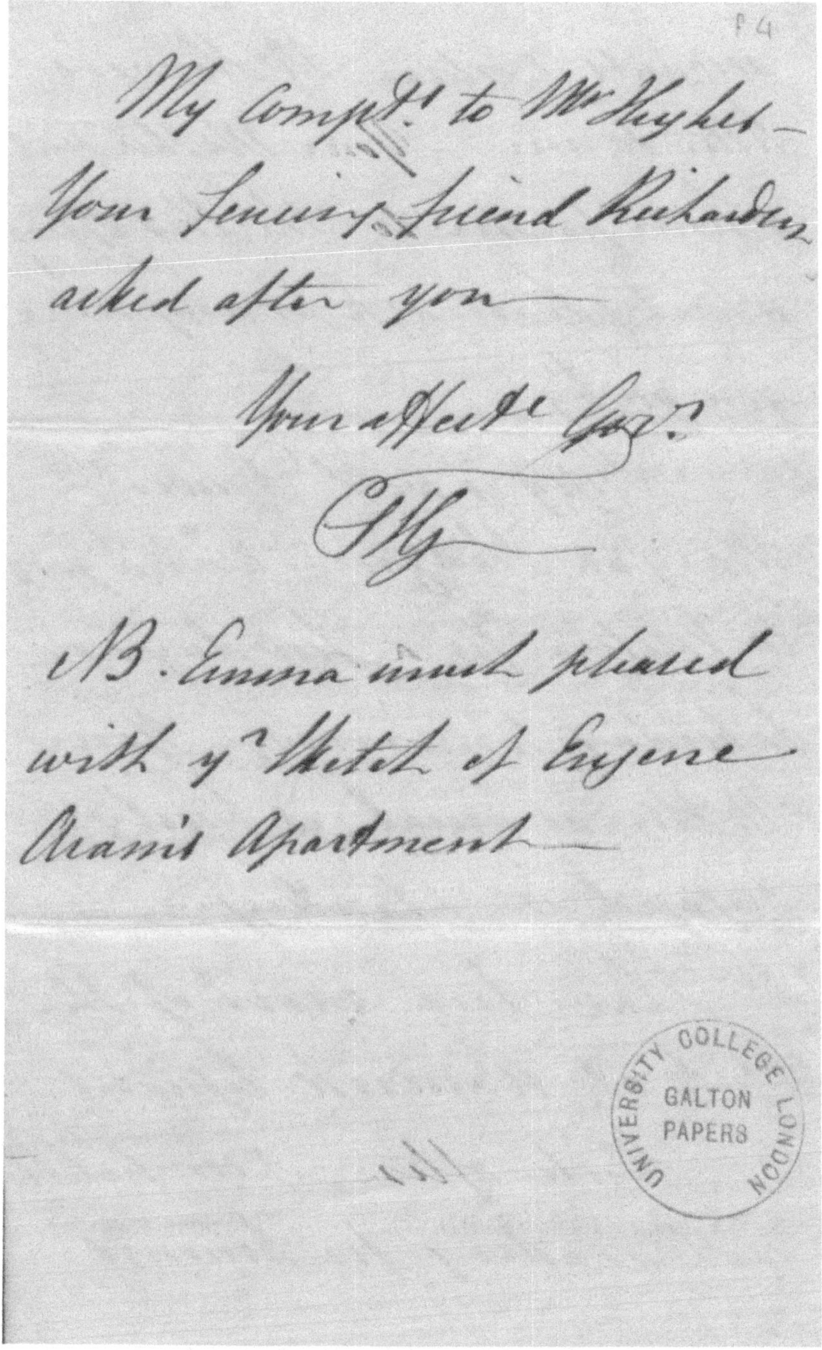
P 4

My Compts. to Mr Hughes —
Your Fencing friend Richardson
asked after you —
Your affecte Bror.
FG
N.B. Emma much pleased
with yr Sketch of Eugene
Aram's Apartment —

Figure 1.11: 1840-10-23. FG to STG page 3.

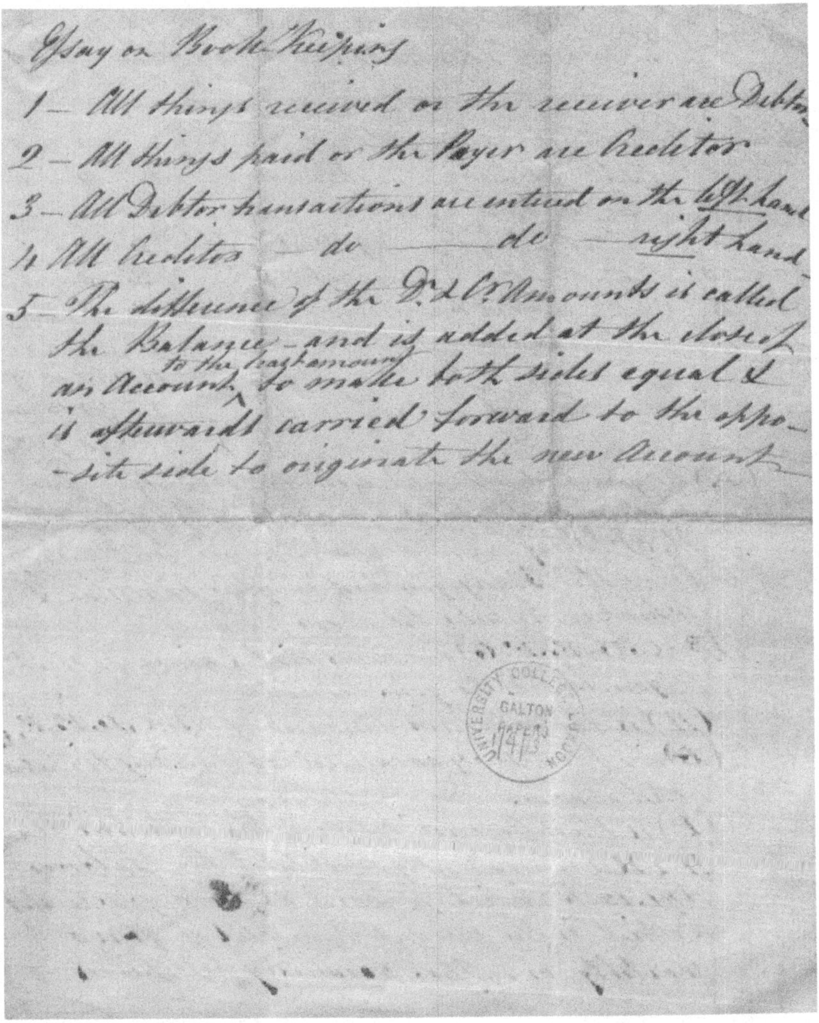

Essay on Book Keeping

1— All things received or the receiver are Debtor

2— All things paid or the Payer are Creditor

3— All Debtor transactions are entered on the left hand

4 All Creditor — do — do — right hand

5— The difference of the Dr & Cr Amounts is called the Balance — and is added at the close of an Account to the least amount to make both sides equal & is afterwards carried forward to the opposite side to originate the new Account

Figure 1.12: 1840-10-23. FG to STG page 4.

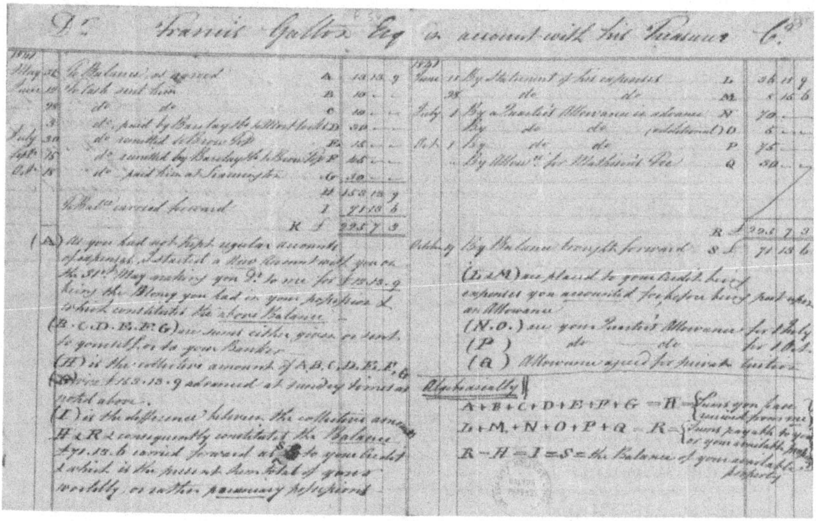

Figure 1.13: 1840-10-23. FG to STG page 5.

of Cambridge & the Boeotian town of Leamington is excessively tardy. I really think that our present economising Government must have made a contract with the Carrier for the transmission of the mail-bag, as, if the postmaster at Warwick was not seized with an apoplexy & thereby occasioned a delay, your letters take 3 days to get here. Yesterday I had a letter directed to you in my Mother's handwriting the enclosed letter sent me which must have slept on the road many days. I for a wager any day would undertake to be on the top of the Drachenfels by Bonn before a letter put in the post at Cambridge at midday would reach you in Leamington. I waste paper fearfully i.e. scribble over both sides of it innumerable x' y' & funny looking triangles. Mrs Hoppit says that "it's a great comfort to her to have a reading gentleman because there is then always plenty of stuff to light the fires with." Theodore looks blooming he hangs up pictures of Cerito in his rooms & talks of the Ō/pey/ra.[10] I tea with Boulton[11] to night, he is not much altered but very shy. Talking of tea please send me some soon, as there are many sloe leaves in the Cambridge. I have proved this by microscopical observation. Hughes is very goodnatured & has introduced me to some nice men, he was officious at first wanting I think to make me as dependent on him as Theodore is, but there

[10] Francesca "Fanny" Cerito (1817-1909) the Italian La Scala ballerina.
[11] Boulton, Matthew Piers Watt. See page 480.

was then a difference of opinion between us, & now we are great allies.

Good bye Your affectionate Son
Fras. Galton.

Cap and Gown	3	16	
Stationery	3	19	6
addendum		13	6
Ironmongers bill	6	5	
Linen &c.	6	17	6
Looking Glass		4	6
& many other little things which I have properly entered		10	3
Making in whole	30	17	9

Oct 14th Received from you	2
*from Mortlock's bank	50
	52

(* I have only drawn £30)

I am now quite set as I have left more than £27.

1840/11/03. Francis Galton to Samuel Tertius Galton

My dear Father

I should have sent a letter to you yesterday if it had not been that the one that I had written was spoilt by an accident in my Gumption reviver Machine which covered it with water, which machine as it has perhaps come into use since your time I will describe to you.[12]

A large funnel is supported on a double stand about 6 ft. high, it has a graduated stop cock at the bottom by which the size of the aperture can be regulated, this as you read is placed above your head & filled with water. Round the head a napkin is tied dependant on one side where the bow & end is so that the water may drop off. [See Figure 1.21, p. 23.]

Now it is calculated that as the number of hours of study increases in an arithmetic ratio, so will the weariness consequent on

[12] Samuel Tertius Galton had been admitted to Trinity as a pensioner on October 23, 1799, but left without taking a degree.

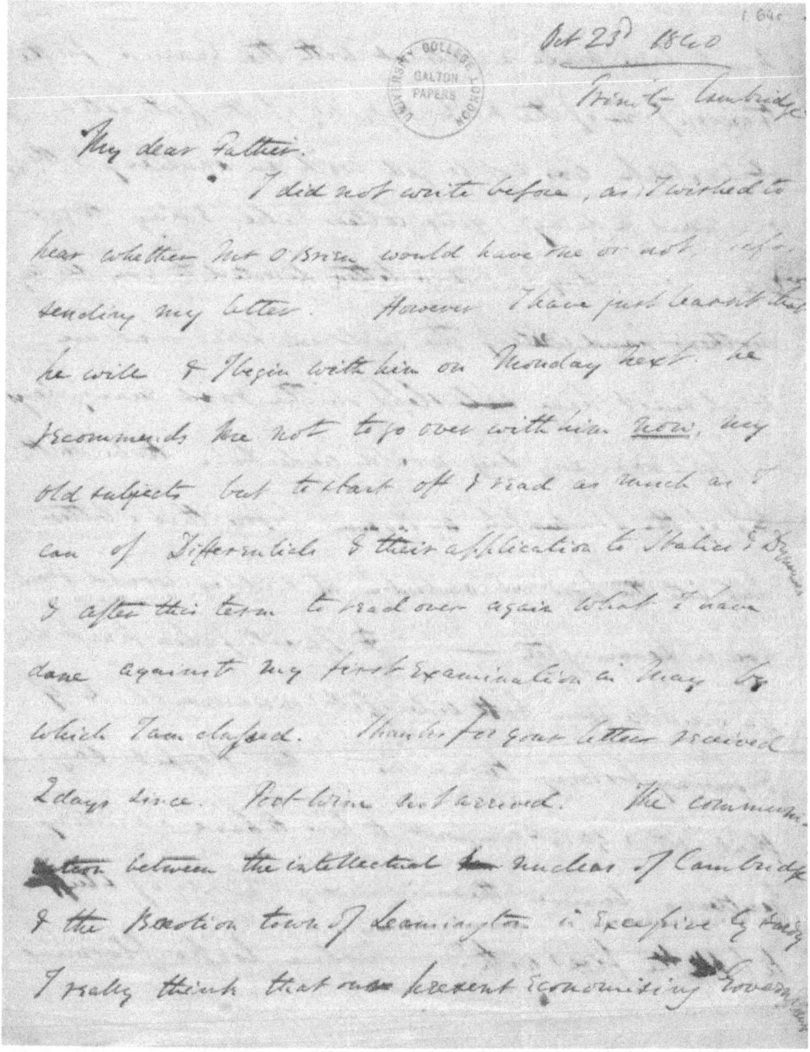

Oct 23d 1840

Trinity Cambridge

My dear Father,

I did not write before, as I wished to hear whether Mr O'Brien would have me or not, before sending my letter. However I have just learnt that he will & I begin with him on Monday next. He recommends me not to go over with him now, my old subjects but to start off & read as much as I can of Differentials & their application to Statics & Dynamics & after this term to read over again what I have done against my first Examination in May by which I am classed. Thanks for your letter received 2 days since. Port Wine not arrived. The communication between the intellectual nucleus of Cambridge & the Boeotian town of Leamington is excessively slow. I really think that our present Economising Government

Figure 1.14: 1840-10-23. FG to STG page 1.

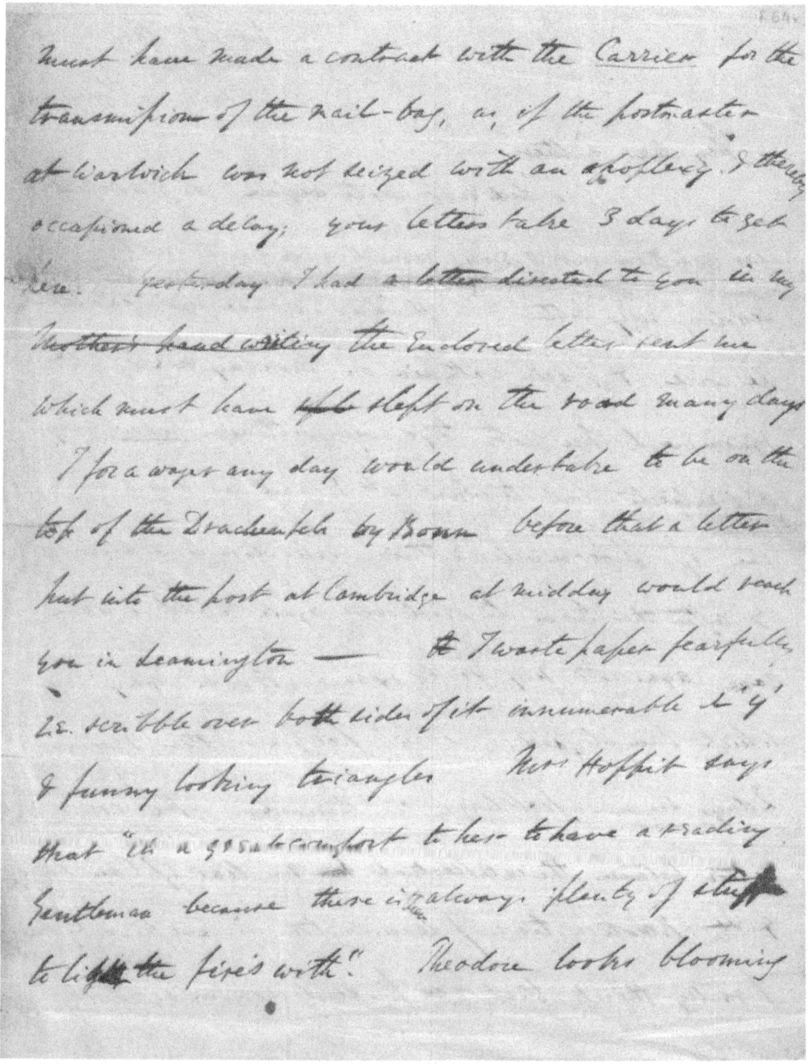

must have made a contract with the Carrier for the
transmission of the mail-bag, as, if the postmaster
at Warwick was not seized with an apoplexy & thereby
occasioned a delay; your letters take 3 days to get
here. Yesterday I had a letter directed to you in my
Mother's hand writing the Enclosed letter sent me
which must have slept on the road many days
I for a wager any day would undertake to be on the
top of the Drachenfels by Bonn before that a letter
put into the post at Cambridge at midday would reach
you in Leamington — I waste paper fearfully
i.e. scribble over both sides of it innumerable x's y's
& funny looking triangles Mrs Hoppit says
that "it is a great comfort to her to have a reading
Gentleman because there is always plenty of stuff
to light the fires with." Theodore looks blooming

Figure 1.15: 1840-10-23. FG to STG page 2.

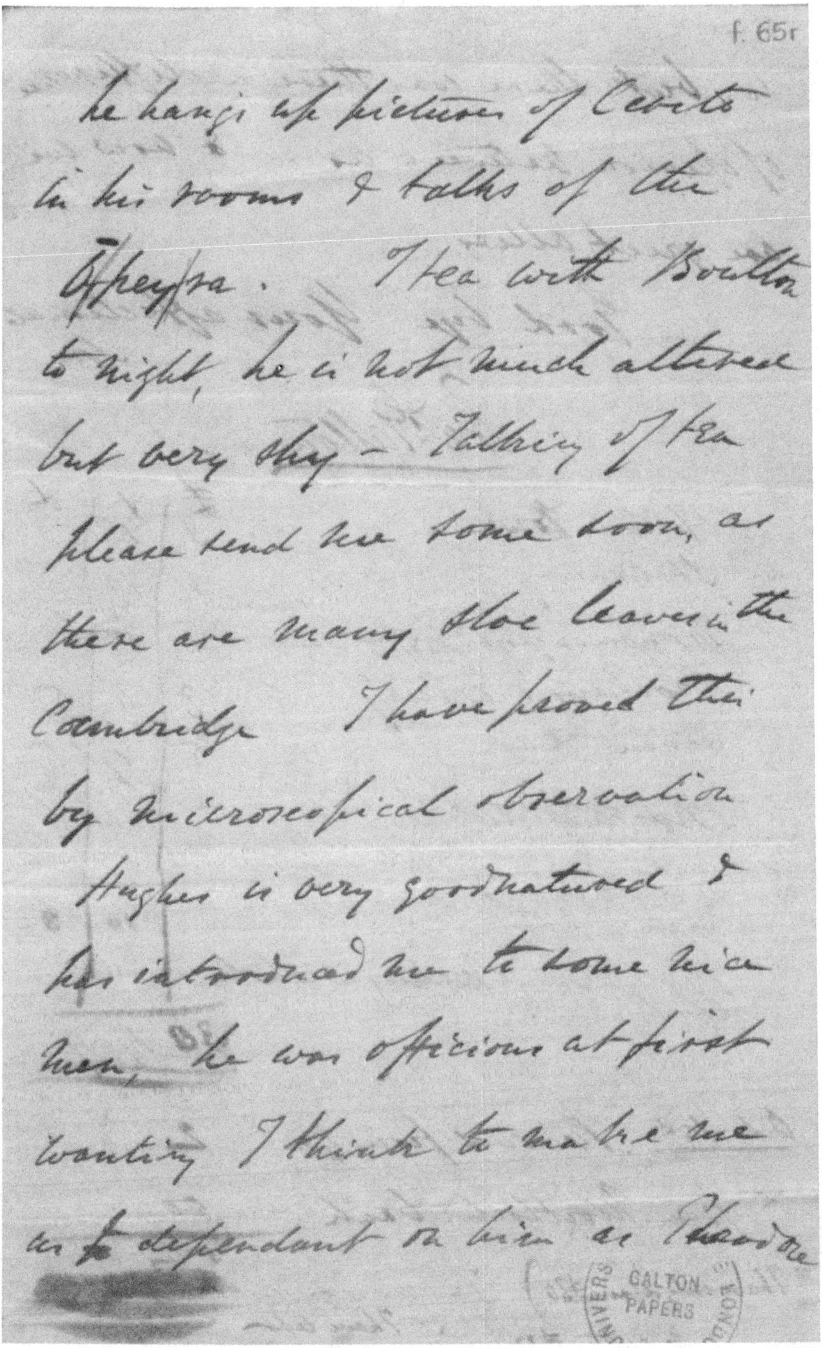
f. 65r

he hangs up pictures of Cevito
in his rooms & talks of the
Oppeyfra. I tea with Boulton
to night, he is not much altered
but very shy — Talking of tea
please send me some soon, as
there are many sloe leaves in the
Cambridge I have proved this
by microscopical observation
Hughes is very goodnatured &
has introduced me to some nice
men, he was officious at first
wanting I think to make me
as dependant on him as Theodore

Figure 1.16: 1840-10-23. FG to STG page 4.

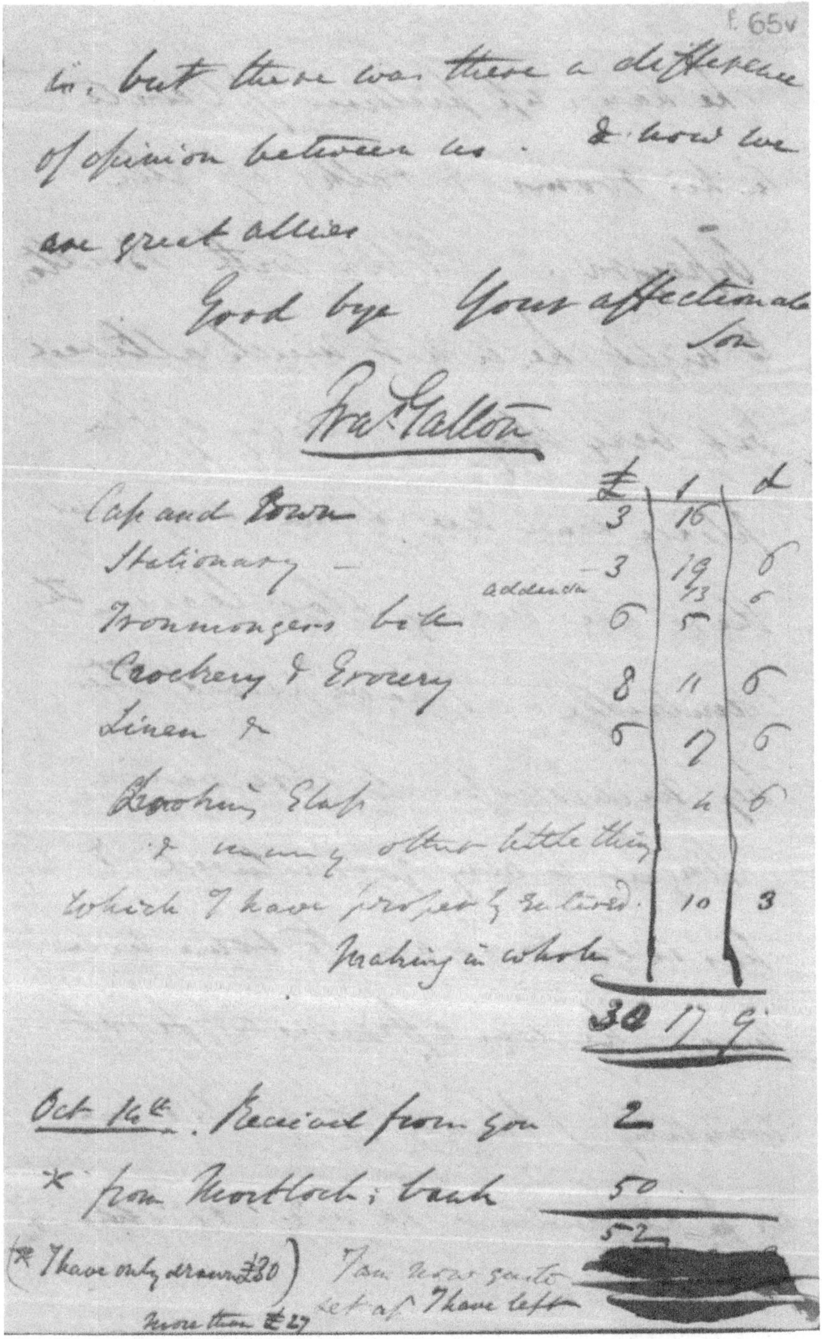

in, but there was there a difference of opinion between us. & now we are great allies

Good bye Your affectionate son

Fra^s Galton

	£	s	d
Cap and Gown	3	16	
Stationary —	3	19	6
addenda		13	6
Ironmongers bill	6	5	
Crockery & Grocery	8	11	6
Linen &	5	7	6
Looking Glass		4	6
& many other little things which I have properly entered.		10	3
Making in whole	30	17	9

Oct 16th. Received from you 2

* from Mortlock's bank 50

52

(* I have only drawn £30 more than £27) I am now quite set up I have left [redacted]

Figure 1.17: 1840-10-23. FG to STG page 5.

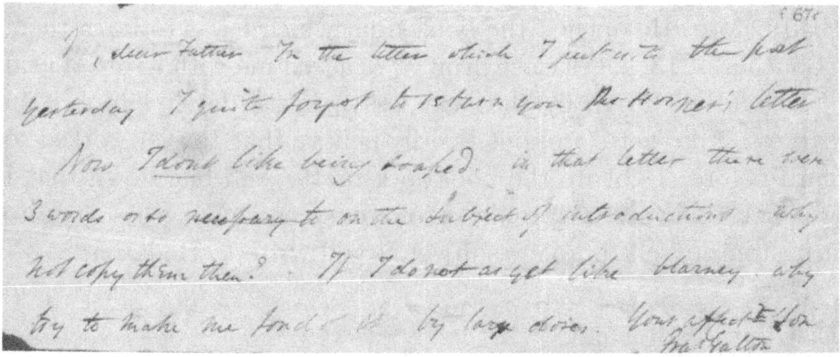

My dear Father In the letter which I put into the post
yesterday I quite forgot to return you Mr Horner's letter
Now I dont like being soaped. in that letter there were
3 words or so necessary to on the subject of introductions. why
not copy them then? If I do not as yet like blarney why
try to make me fond of it by large doses. Your affecte Son
Fras Galton

Figure 1.18: 1840-10-23. FG to STG page 6.

it increase in a geometrical ratio, & the ~~supply~~ stream of water must in that ratio be increased. The geometric ratio used in the 1st year i.e. for freshmen is 2 in the 2nd year 3, & in the term before taking the degree 5. At that time the Gyp[13] has to call every quarter of an hour to refill the funnel, the clothes are then also not protected as damp shirts do not invite repose. We generally begin to use this machine about 10 at night & continue it till 1 or 2, it is very useful. [See Figure 1.19, p. 22, and Figure 1.20, p. 23.] My private tutor recommended it to me as the first thing, it is in fact quite indispensable to a high wrangler. I have received wine spoons & tea for all of which thank you. So Fanny Broadley is going to be spliced.[14] I congratulate her heartily. Mr. Burrows[15] may think himself uncommonly lucky for I think she was the prettiest girl I almost ever saw.

You mention the case of poor Mr. Holman of Catharine Hall.[16] I hear that he had worked himself almost to madness, but was quite unable to succeed on account of his natural powers, poor fellow he did the best thing that he could do though. As to who were plucked nobody knows except the pluckers & the plucked it is done very quietly. I have been reading very hard and am accordingly rather dull. On going to O'Brien[17] my private tutor (a 3rd Wrangler) he set me about Conic Sections, which I had not

---

[13] A college servant.

[14] Broadley, Anne 'Fanny'. See page 482.

[15] Henry Nicholson Burrows (1809-1859).

[16] John Anthony Holman of St Catharine's Hall committed suicide on October 20th 1840 in his room, by taking laudanum, after being "plucked", i.e. failed. A verdict of "temporary insanity" was returned.

[17] O'Brien, Matthew. See page 487.

read before. He opened the book from which I was to learn them (Boucharlat IX pages close print)[18] & asked me with a sort of grin if I could get it up by the next lesson in $2\frac{1}{2}$ days. I took it in earnest & did get it up, but I verily believe that I never worked so hard before. I got up the bookwork pretty well but I own that I was not able to work problems, of course not having had time to accustom myself to a new subject. I will write to Hodgson.[19]

Distribution of Time.

Up at Chapel at 7
ditto to 7 2
Reading & breakfast to 9
Lectures to 11
Reading by myself & with O'Brien to 2
walk to 4——a 4 mile walk——
Hall to 4.20
Read including tea $10\frac{1}{2}$.

Lectures 2 hours a day Reading (full tide) $10\frac{1}{2}$ hours.

I shall cut this down to 6 as it is really too much. Tell Bessy that there is the most extraordinary possible change in my complexion the tan having quite disappeared. Breadth of phiz ~~decreasing~~ on the wane. Loves universally.

Fras. Galton.

Figure 1.19: The Stand.

[18] Jean-Louis Boucharlat (1775-1848).
[19] Hodgson, Joseph. See page 483.

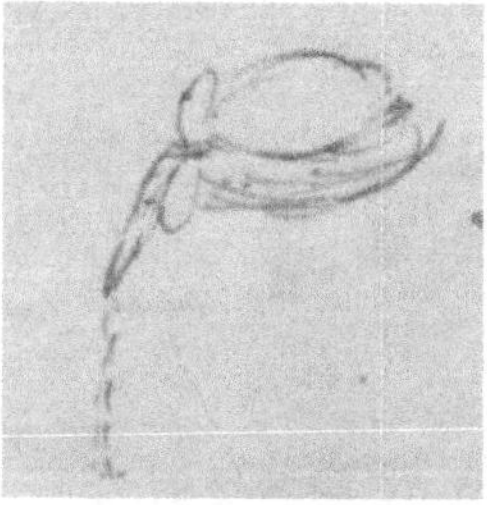

Figure 1.20: The Dripping Tourniquet.

Figure 1.21: The Gumption-reviver Machine.

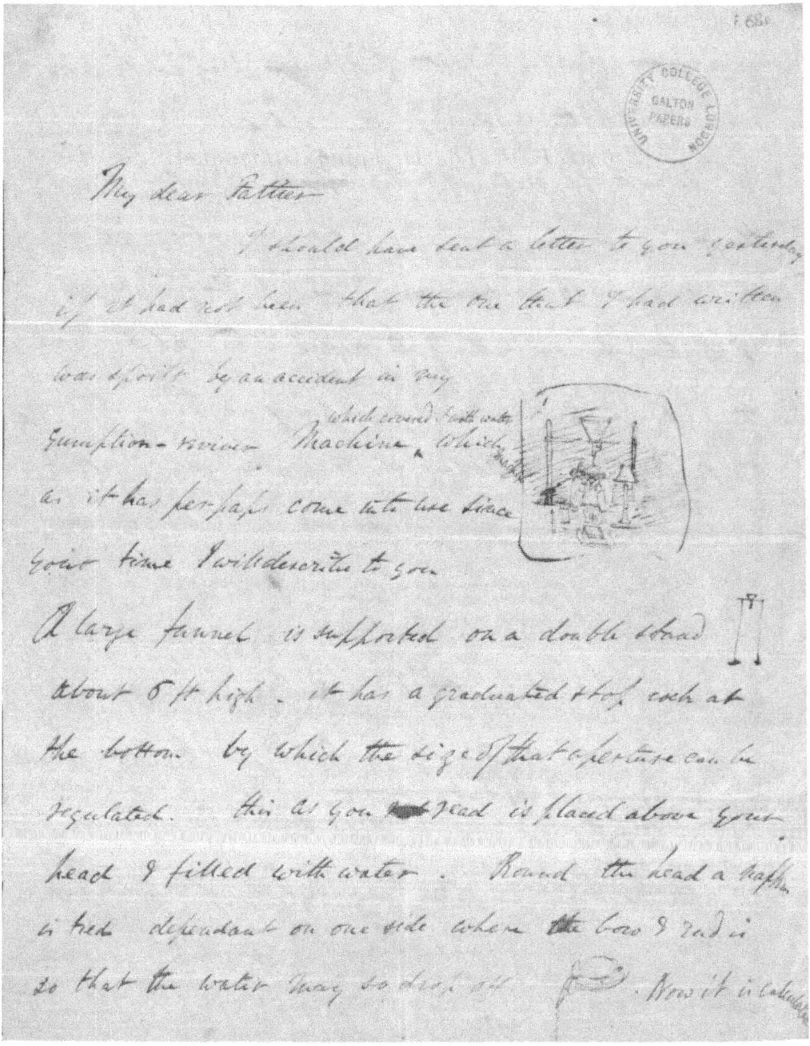

My dear Father

I should have sent a letter to you yesterday if it had not been that the one that I had written was spoilt by an accident in my Gumption-reviver Machine, which covered it with water, which as it has perhaps come into use since your time I will describe to you.

A large funnel is supported on a double stand about 6 ft high. it has a graduated stop cock at the bottom by which the size of that aperture can be regulated. this as you read is placed above your head & filled with water. Round the head a napkin is tied dependant on one side where the bow & end is so that the water may so drop off. Now it is calcula

Figure 1.22: 1840-11-03. FG to STG page 1.

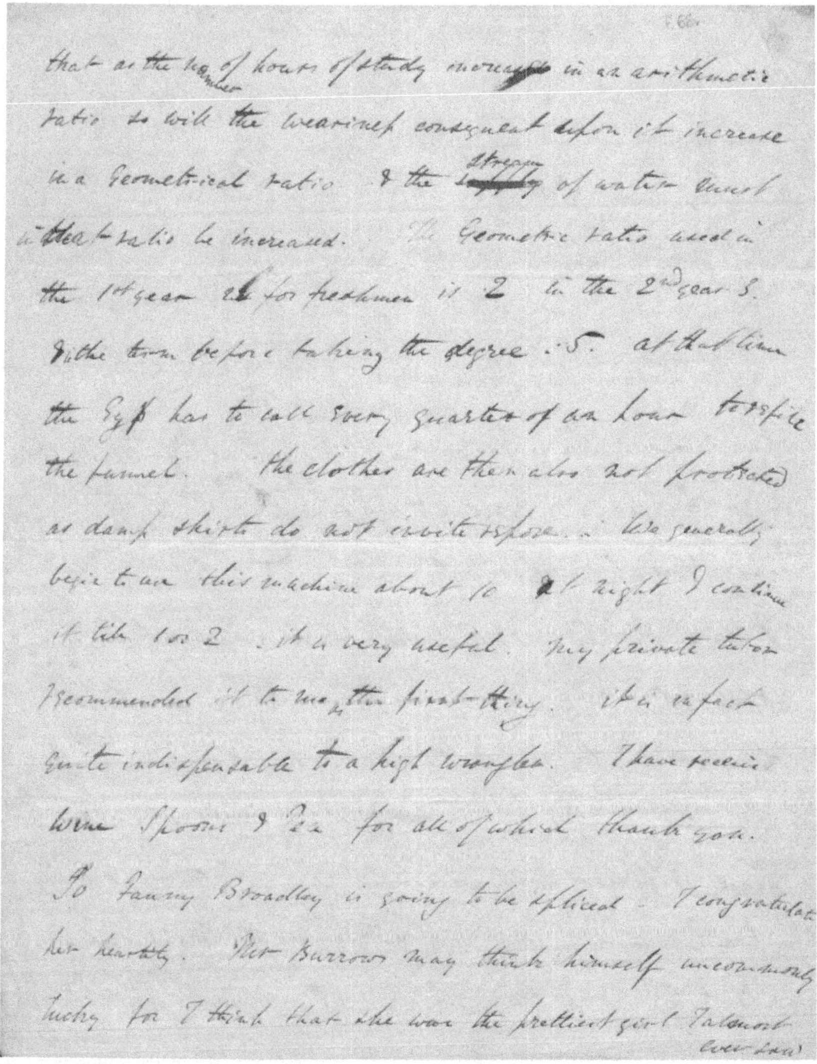

that as the number of hours of study increases in an arithmetic
ratio so will the weariness consequent upon it increase
in a Geometrical ratio & the stream of water must
in that ratio be increased. The Geometric ratio used in
the 1st year &c for freshmen is 2 in the 2nd year 3.
& the term before taking the degree ∴ 5. at that time
the Gyp has to call every quarter of an hour to refill
the funnel. the clothes are then also not protected
as damp shirts do not invite repose. We generally
begin to use this machine about 10 at night & continue
it till 1 or 2. it is very useful. my private tutor
recommended it to me the first thing. it is in fact
quite indispensable to a high wrangler. I have received
Wine Spoons & Tea for all of which thank you.
So Fanny Broadley is going to be spliced. I congratulate
her heartily. Mr Burrows may think himself uncommonly
lucky for I think that she was the prettiest girl I almost
ever saw

Figure 1.23: 1840-11-03. FG to STG page 2.

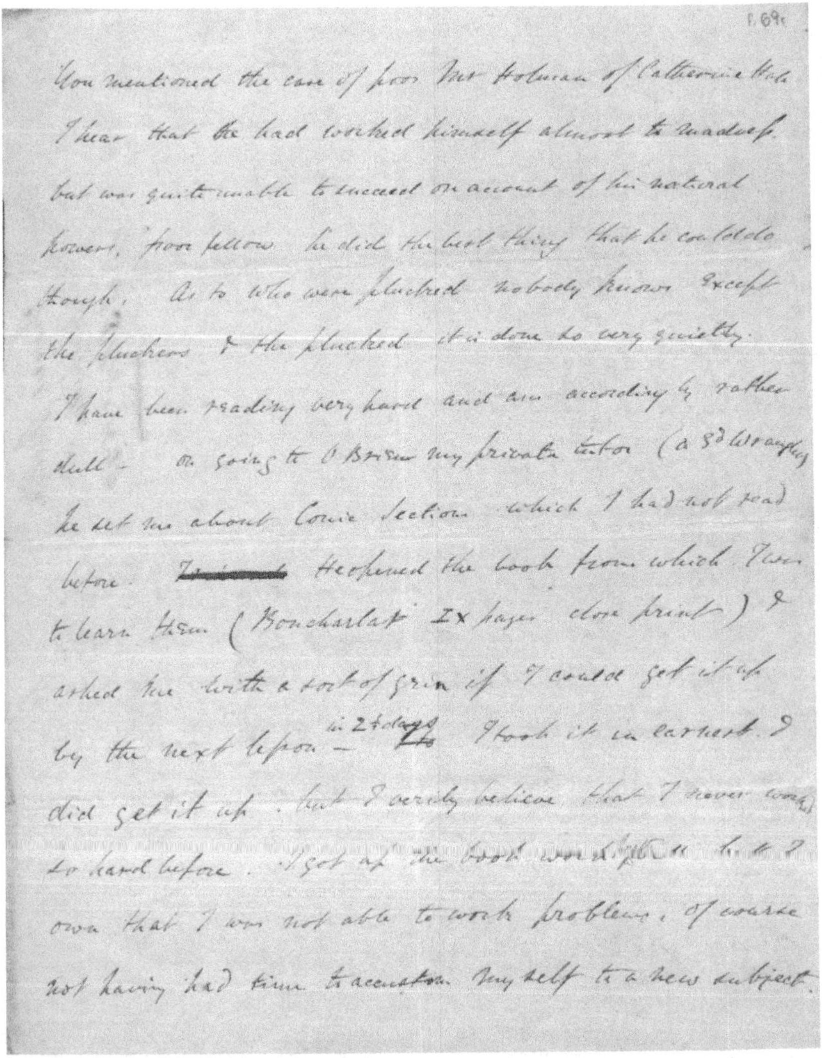
You mentioned the case of poor Mr Holman of Catherine Hall
I hear that he had worked himself almost to madness
but was quite unable to succeed on account of his natural
powers, poor fellow he did the best thing that he could do
though. As to who were plucked nobody knows except
the pluckers & the plucked it is done so very quietly.
I have been reading very hard and am accordingly rather
dull - on going to O Brien my private tutor (a 3d Wrangler)
he set me about Conic Sections which I had not read
before. He opened the book from which I was
to learn them (Bouchorlat IX pages close print) &
asked me with a sort of grin if I could get it up
by the next lesson - in 2 days I took it in earnest &
did get it up. but I verily believe that I never worked
so hard before. I got up the book work [illegible] but I
own that I was not able to work problems, of course
not having had time to accustom myself to a new subject.

Figure 1.24: 1840-11-03. FG to STG page 3.

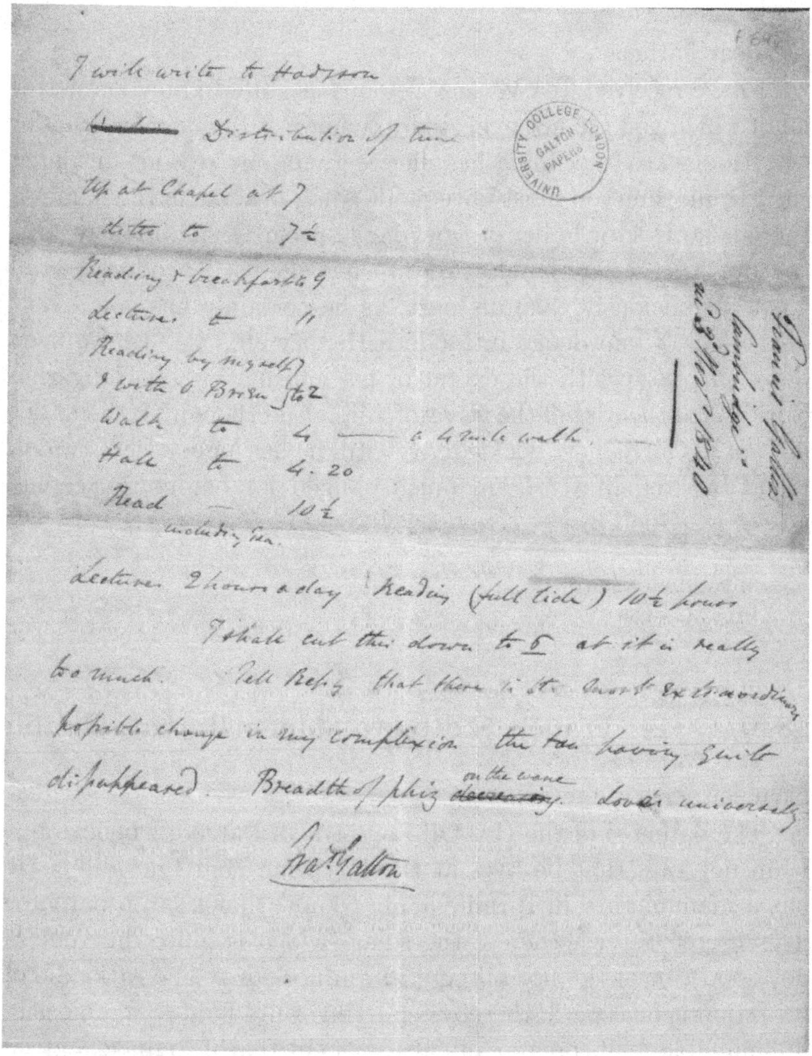

I will write to Hodgson

~~Hudson~~ Distribution of time

Up at Chapel at 7
ditto to 7½
Reading & breakfast to 9
Lectures to 11
Reading by myself } & with O Brien } to 2
Walk to 4 — a 4 mile walk.
Hall to 4.20
Read — 10½
including tea.

Lectures 2 hours a day Reading (full tide) 10½ hours

I shall cut this down to 8 as it is really too much. Tell Bessy that there is the most extraordinary possible change in my complexion the tan having quite disappeared. Breadth of phiz ~~decreasing~~ on the wane. Love is universally

Fra^s Galton

Francis Galton
Cambridge
3rd Nov 1840

Figure 1.25: 1840-11-03. FG to STG page 4.

1840/11/05. Francis Galton to Samuel Tertius Galton

Trinity College
Cambridge
5th Novr 1840

My dear Father

I have received Tea Wine & Spoons as stated in my last letter all of which you are not expected to believe & I am now writing to Mr. Hodgson[20] declining his offer & giving my reasons for doing so. I think that you must have made some mistake in the name as there is no Doctor Fuller of Downing. I progress salubriously. Bye the bye in case any laughs are directed against Theodore I shall most pugnaciously take his part, as he certainly has got a very great deal of knowledge in Modern History &c. &c. I have been quite surprised with the extent of his information on Hungarian Turkish & other out of the way worthies. And though Mathematics are not his forte most decidedly yet still he has a great deal in him. Good bye tell Mater I am much obliged for her house-keeping advice. & Good bye

affectionate son
Fras. Galton

1840/11/11. Francis Galton to Samuel Tertius Galton

Dear Father

The patentee of the Hot Oil Lamps is J. Parker. I believe but I am not sure that he lives at the Egyptian Hall Piccadilly. He has a manufactury in Birmingham. I know that a lamp patentee lives there, but whether of the solar (which is quite different & not good), or of the hot oil lamp, I am not sure. The rationale of its action is heating & thereby rendering quite fluid & more easily inflammable, half concrete oil. It is evident that if to make a given oil burn with a given brilliancy (i.e. a given qty consumed in a known time) it requires a heat (h) & if the heat of a flame raise oil from 60°to (h − x); [then] where the oil is in the first instance at the temperature of (60 + x), the same flame will raise it to (h). [See Figure 1.29, p. 32.]

20 Hodgson, Joseph. See page 483.

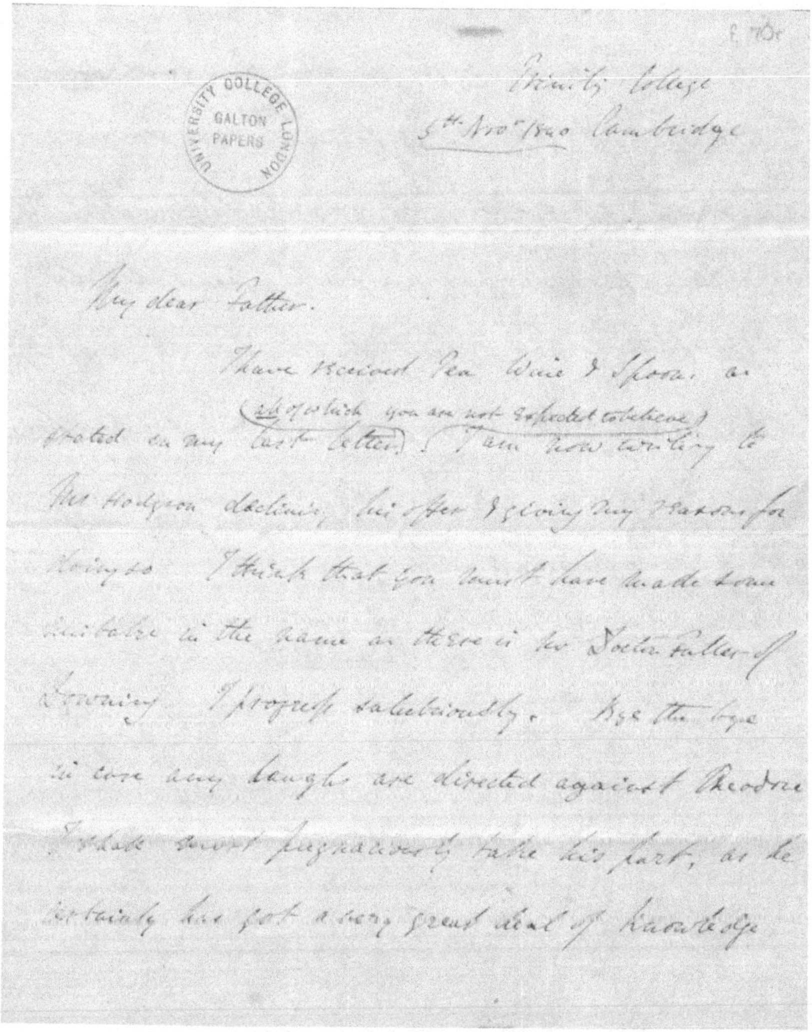

f. 70r

UNIVERSITY COLLEGE LONDON GALTON PAPERS

Trinity College
5th Nov 1840 Cambridge

My dear Father.

I have received Tea Wine & Spoons as (all of which you are not expected to believe) stated in my last letter. I am now writing to Mr Hodgson declining his offer & giving my reasons for doing so. I think that you must have made some mistake in the name as there is no Doctor Fuller of Downing. I progress salubriously. Bye the bye in case any laughs are directed against Theodore I shall most pugnaciously take his part, as he certainly has got a very great deal of knowledge

Figure 1.26: 1840-11-05. FG to STG page 1.

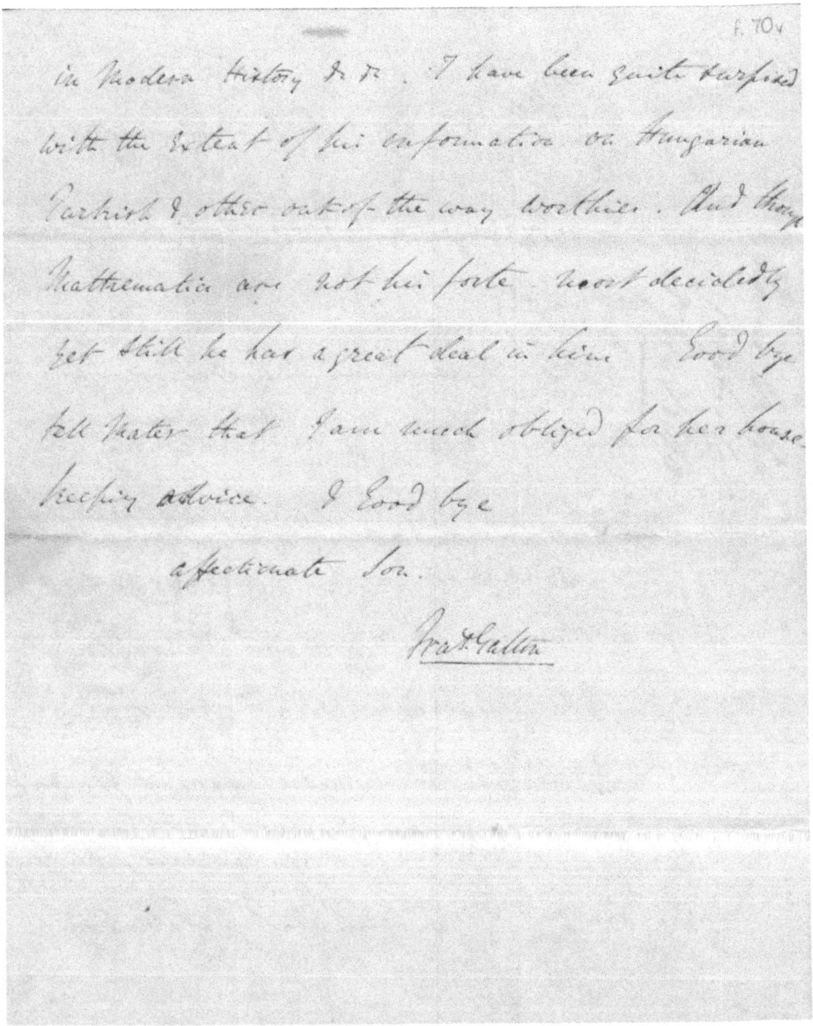
f. 70v

in Modern History &c &c. I have been quite surprised with the extent of his information on Hungarian Patriots & other out of the way worthies. And though Mathematics are not his forte most decidedly yet still he has a great deal in him. Good bye tell Mater that I am much obliged for her housekeeping advice. & Good bye

affectionate Son.

Fras Galton

Figure 1.27: 1840-11-05. FG to STG page 2.

Figure 1.28: 1840-11-05. FG to STG page 3.

It gives an exceedingly bright light but the lamp looks ugly; it is more adapted for a hall light. You ought to choose one yourself. I do not know whether a common lamp can easily be changed to an hot oil.[21]

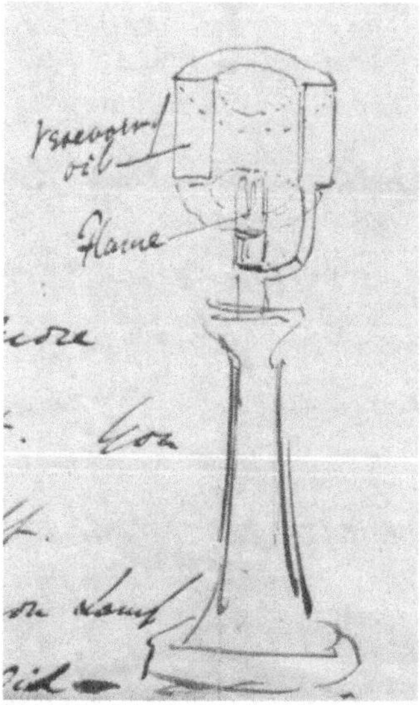

Figure 1.29: Hot Oil Lamp.

1840/11/26. Francis Galton to Samuel Tertius Galton

My dear Father

Thanks for letter received yesterday. I am much obliged to you about getting me nominated to the Athenaeum, please thank Uncle Howard for me. As he explained to me on a former occasion it will be much better to make use of his assistance than for me to get Daniell or Partridge[22] to nominate me, in which latter case I should be sure of a professional opposition.

I hope Bessy will get better soon.

The reason why I write in pencil is as I am lying on my back, I can't get a pen to write (I have been confined to my bed for some days, rheumatism not over reading but will shortly be released. It has put a pro tempore dead stop to Maths.)

I have just received a letter from Horner,[23] he ~~says~~ offers to get

[21] Missing pages.
[22] Partridge, Richard. See page 490.
[23] Horner, Leonard. See page 484.

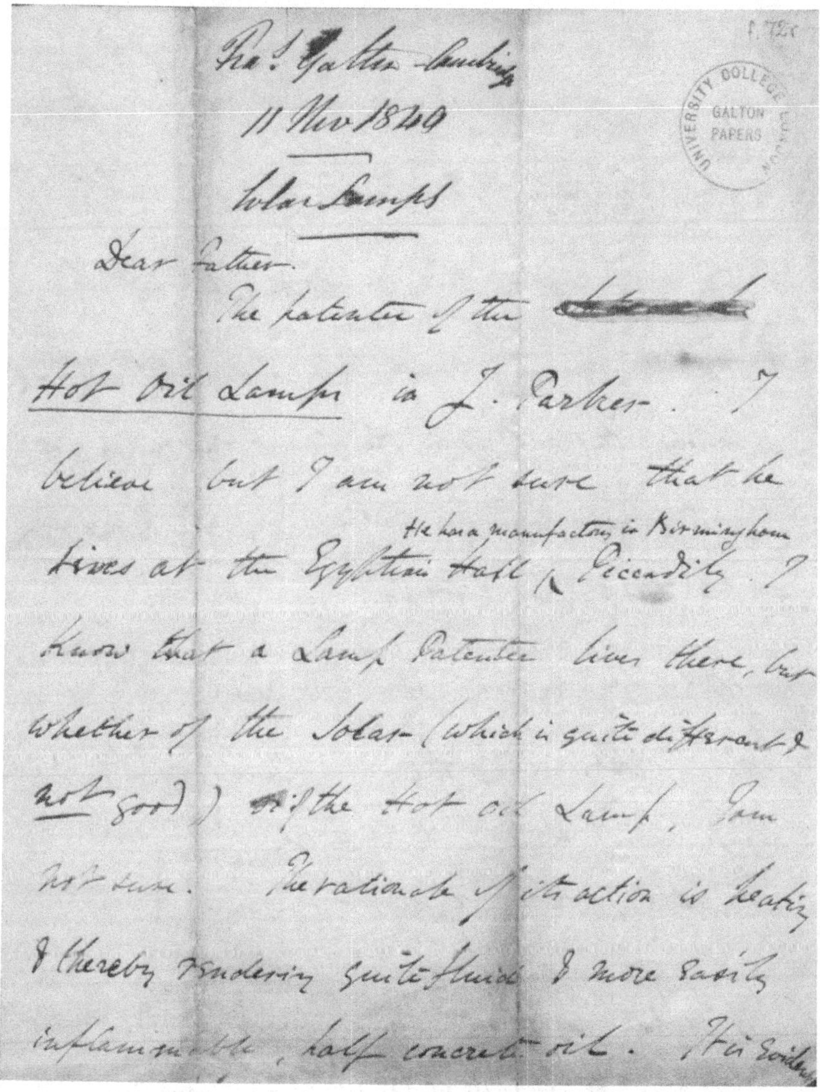

Fra.s Galton Cambridge
11 Nov 1840
Solar Lamps

Dear Father.

The patentee of the ~~[illegible]~~ Hot Oil Lamps is J. Parker. I believe but I am not sure that he lives at the Egyptian Hall, Piccadilly. [He has a manufactory in Birmingham] I know that a Lamp Patentee lives there, but whether of the Solar (which is quite different & not good) or of the Hot oil Lamp, I am not sure. The rationale of its action is heating & thereby rendering quite fluid & more easily inflammable, half concrete oil. It is evident

Figure 1.30: 1840-11-11. FG to STG page 1.

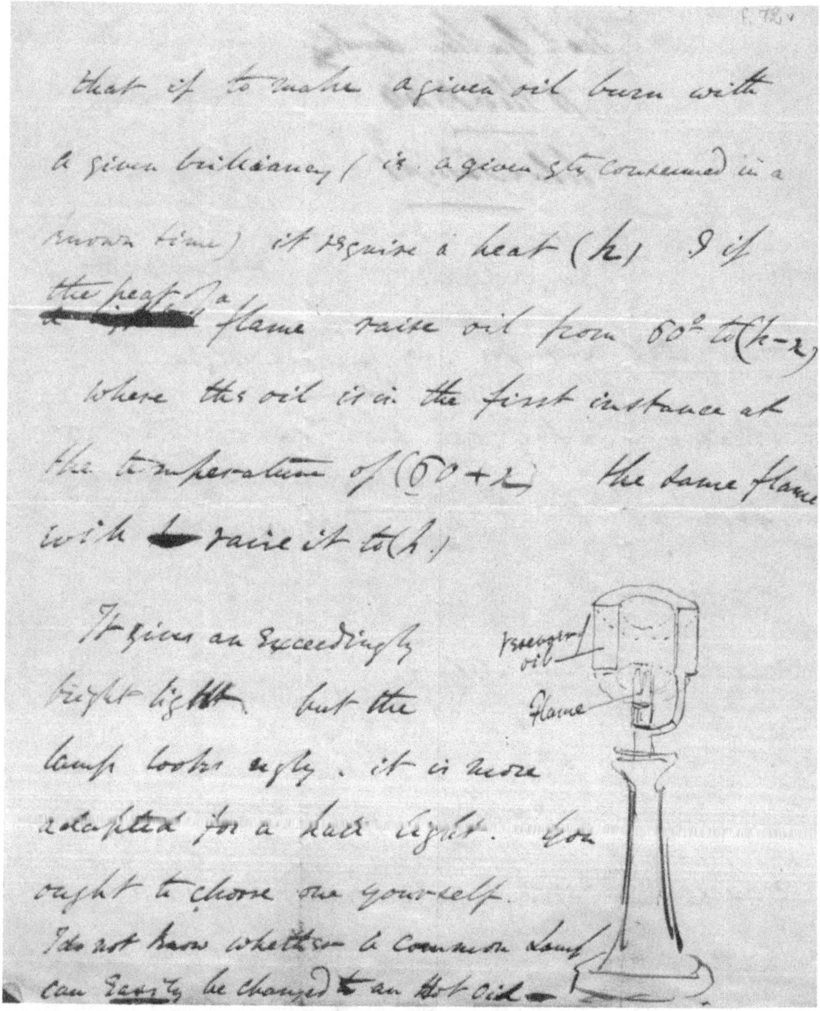
that if to make a given oil burn with
a given brilliancy (i.e. a given qty consumed in a
known time) it require a heat (h) & if
the heat of a flame raise oil from 60° to (h−x)
where the oil is in the first instance at
the temperature of (60+x) the same flame
will raise it to (h.)

It gives an exceedingly
bright light but the
lamp looks ugly. it is more
adapted for a hall light. You
ought to choose one yourself.
I do not know whether a common lamp
can easily be changed to an Hot Oil

Figure 1.31: 1840-11-11. FG to STG page 2.

me an introduction to any man I may like, said he did not write before to give me time to settle & to find out the most desirable quarters for introductions. I must make many enquiries before answering his letter.

Good bye affect. Son
Fras. Galton.

Hurra for the Queen's Kinchin![24] I have ordered 3 dozen of "audit ale" on a venture for you at Leamington, but am afraid that it will not be ready for X^{tmas}. I shall be with you certainly not ~~after~~ later than the 18th Decr.

1840/12/03. Francis Galton to Samuel Tertius Galton

My dear Father

Would you please send me by return of post some money as I do not know what my doctor's bill will amount to, was released part of Monday from bed (the 12th day of my rigid confinement thereto). As I had been extremely ill the Doctor came 4 times in two days (fever, touch of delirium). I am in a great hurry for the post. Shall not be with you before the 18th as I have some things to do in London, we are free for 5-7 weeks beginning on the 12th.

Shall begin lectures again on Monday next.

Fras. Galton

1840/12/04. Diary of Samuel Tertius Galton

December 4th. Friday. Leamington.

.. I received a letter from my dear son Francis mentioning his having been ill and I wrote to Theodore to enquire how he was going on. ...

1840/12/06. Diary of Samuel Tertius Galton

December 6th. Sunday. Cambridge.

... arrived at 9 o'clock at Francis' rooms, b. staircase, New Court, where I found him much better than I expected but much reduced in size. I saw his medical attendant Dr. Fisher of Downing who gave me a good account of him. Theodore Galton called and

[24] Child.

Figure 1.32: 1840-11-26. FG to STG page 1.

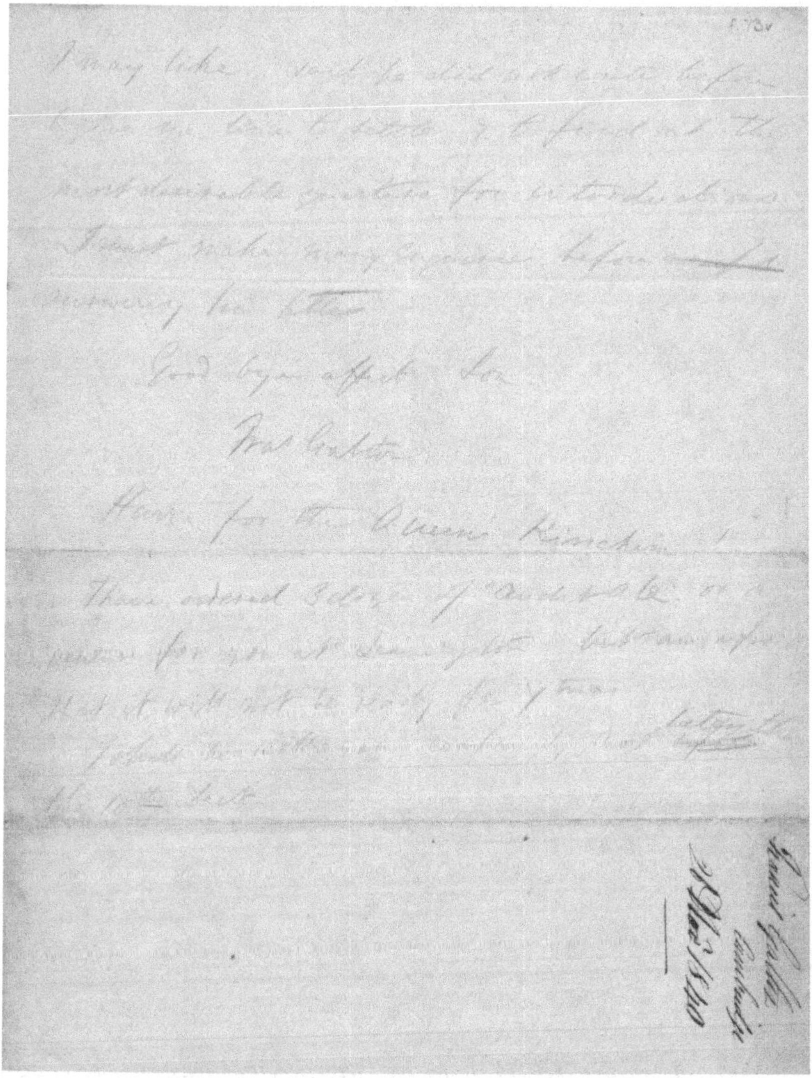

Figure 1.33: 1840-11-26. FG to STG page 2.

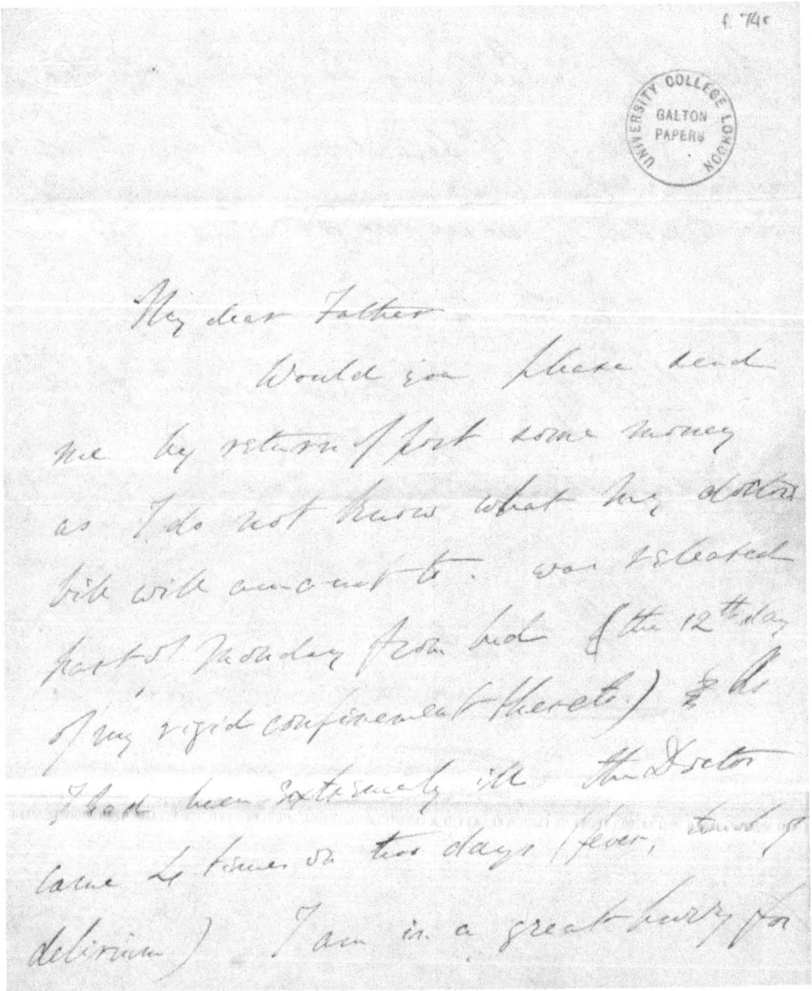
f. 74r

My dear Father

Would you please send
me by return of post some money
as I do not know what my doctor's
bill will amount to. I was released
last Monday from bed (the 12th day
of my rigid confinement thereto) & as
I had been extremely ill. the Doctor
came 4 times on two days (fever, tending to
delirium) I am in a great hurry for

Figure 1.34: 1840-12-03. FG to STG page 1.

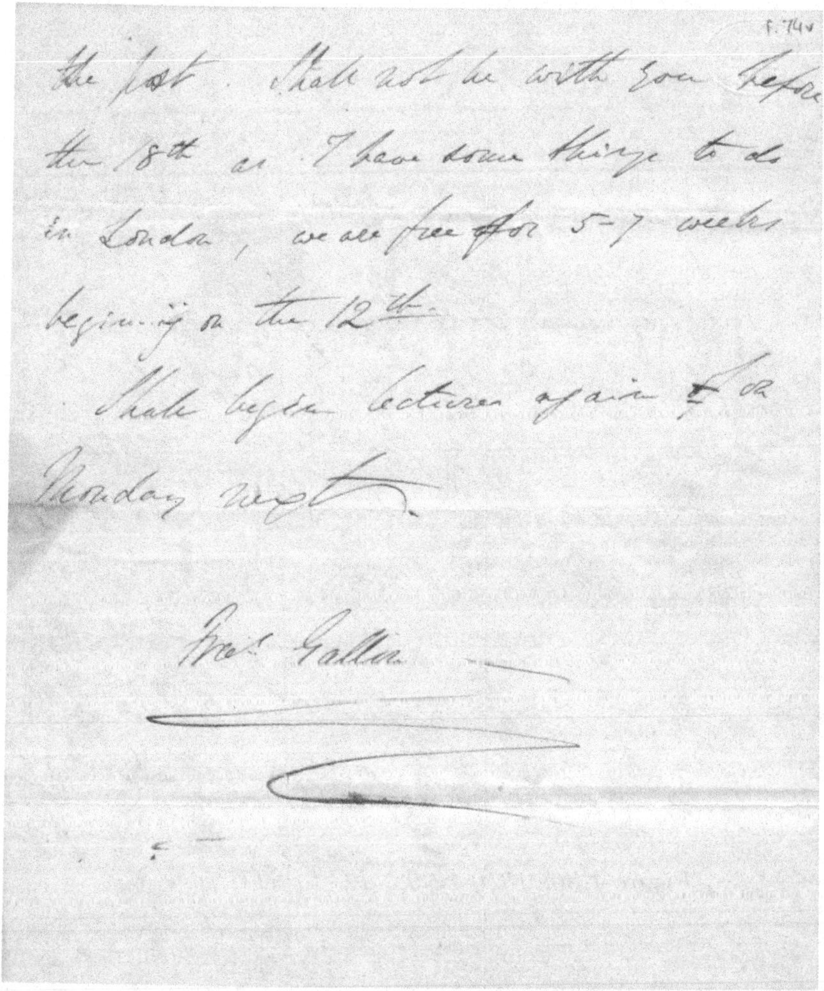

f. 74v

the post. Shall not be with you before the 18th as I have some things to do in London, we are free for 5-7 weeks beginning on the 12th.

Shall begin lectures again [illegible] on Monday next.

Fras. Galton

Figure 1.35: 1840-12-03. FG to STG page 2.

Figure 1.36: 1840-12-03. FG to STG page 3.

sat with us both morning and afternoon. I also met Matthew Boulton.[25] I dined at the Eagle and returned to Francis after dinner. He had a variety of projects for improving lamps, applying galvinism as a motive power etc., some of which he intends to get executed in London. I left him soon after 8 p.m. I was glad to hear from him that Perry[26] had conversed with him and spoken of him in favourable terms.

1840/12/08. Francis Galton to Samuel Tertius Galton

Tuesday 8th Decr. 1840
Trin: Coll Cambridge

My dear Father

On reconsideration concerning O'Brien[27] I think that it will be better for me to go on with him for one term more. For during the vacation I shall have plenty of time to digest well what I have already gone over with him, and to read as much further as I easily can without explanations. Next term he tells me that I had better go over the easy parts of Maths with him, where he would certainly be of the greatest use to me with reference to my approaching examination. For though I believe that I know those subjects very well, in the way that I was taught them, yet a Cambridge gloss makes much difference in the marks. The term is short & I think that it would be a pity to lose O'Brien by a hasty decision. He called on me this morning to see how I was, I said nothing however to him about leaving off.

On calling on Perry yesterday he mentioned Mathison[28] & said that a tutor would be a decidedly good thing in the long vacation. Of course O'Brien would be better.

I may as well add that my present decision has been made quite independently of other men's advice.

As you know all the news I have nothing else to say but that I send a liberal supply of loves to be equally divided.

Good bye Your affectionate Son
Fras. Galton.

---

[25] Boulton, Matthew Piers Watt. See page 480.
[26] Perry, Charles. See page 490.
[27] O'Brien, Matthew. See page 487.
[28] Mathison, William Collings. See page 486.

P.S. I forgot to say that I am getting on well, shall not I think dissect in London but give up my time to Maths. and Classics.

1840/12/09. Francis Galton to Samuel Tertius Galton

Wednesday 9th 1840
Trin. Coll: Cambridge.

Please, bed made warming pan in trim plenty of hot & cold water by seven & a half o'clock. Saturday Evening 12th.

Too ill for London—in bed again, cold in lecture room this morning get out again tomorrow.

Fras. Galton.

1840/12/12-26. Diary of Samuel Tertius Galton

December 12th. Saturday. Leamington.

... Francis arrived at 6 p.m. looking much better than when I left him at Cambridge.

December 16th. Tuesday. Leamington.

... Emma, Francis and Charlotte went to the Ball.

December 17th. Thursday. Leamington.

... Francis complained of not being very well but got some refreshing sleep.

December 18th. Friday. Leamington.

.. Francis much better.

December 22nd. Tuesday. Leamington.

Violetta, Francis, Charlotte Wood and myself went to Claverdon and from thence to Edstone but found Mary rather poorly. ... Francis and myself dined at the Miss Phillips ... Francis, Emma and Charlotte went to Mr. Thompkinson's.

December 26th. Saturday. Leamington.

Francis started for Birmingham at 7:45 a.m.

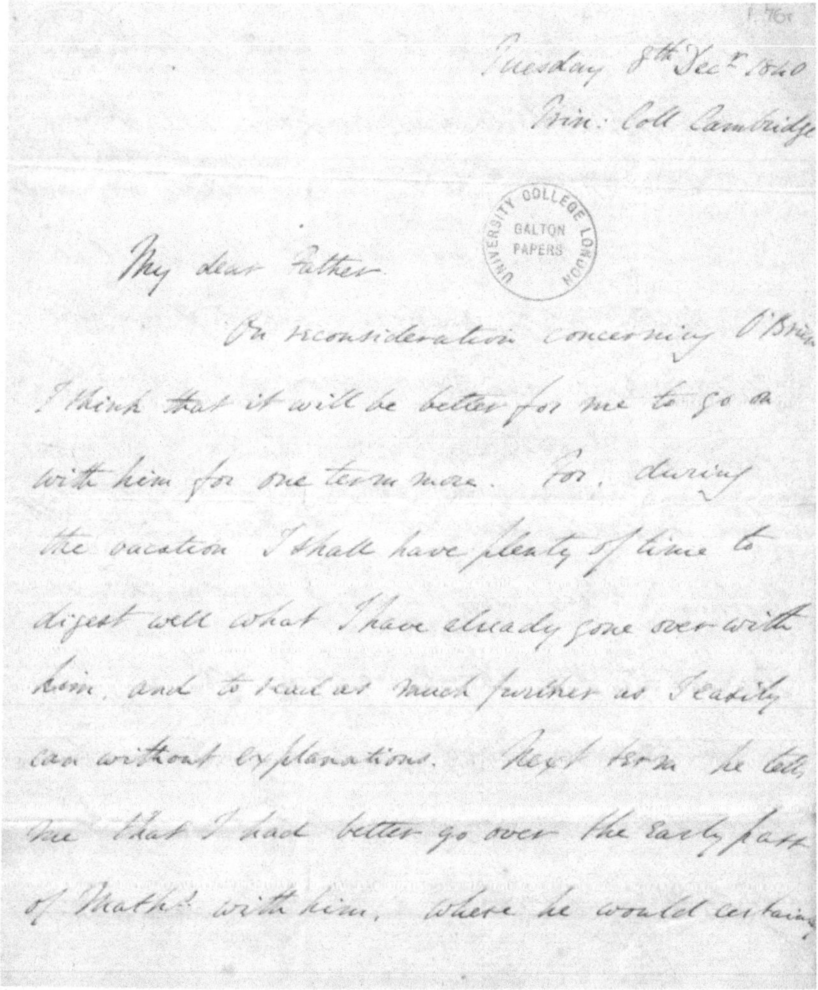

f. 76r

Tuesday 8th Decr 1840
Trin: Coll Cambridge

My dear Father

On reconsideration concerning O'Brien I think that it will be better for me to go on with him for one term more. For, during the vacation I shall have plenty of time to digest well what I have already gone over with him, and to read as much further as I easily can without explanations. Next term he tells me that I had better go over the early part of Math^{s} with him, where he would certainly

Figure 1.37: 1840-12-08. FG to STG page 1.

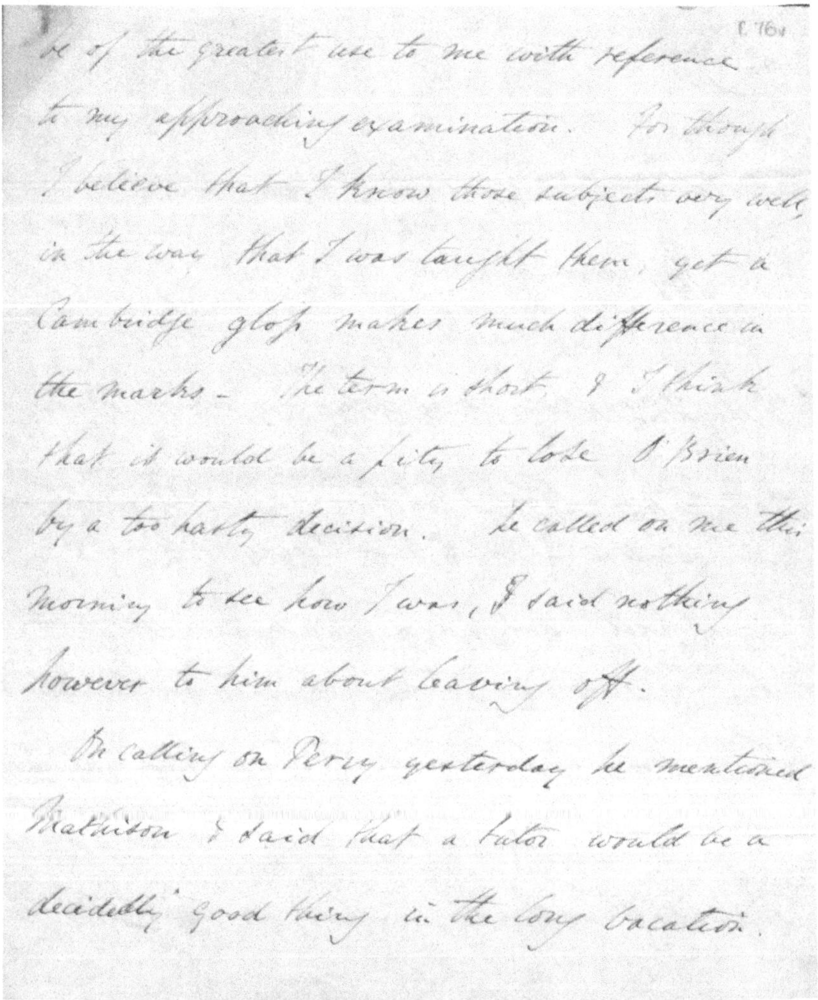

be of the greatest use to me with reference
to my approaching examination. For though
I believe that I know those subjects very well,
in the way that I was taught them, yet a
Cambridge gloss makes much difference in
the marks - The term is short, & I think
that it would be a pity to lose O'Brien
by a too hasty decision. He called on me this
morning to see how I was, I said nothing
however to him about leaving off.

On calling on Perry yesterday he mentioned
Matthison & said that a tutor would be a
decidedly good thing in the long vacation.

Figure 1.38: 1840-12-08. FG to STG page 2.

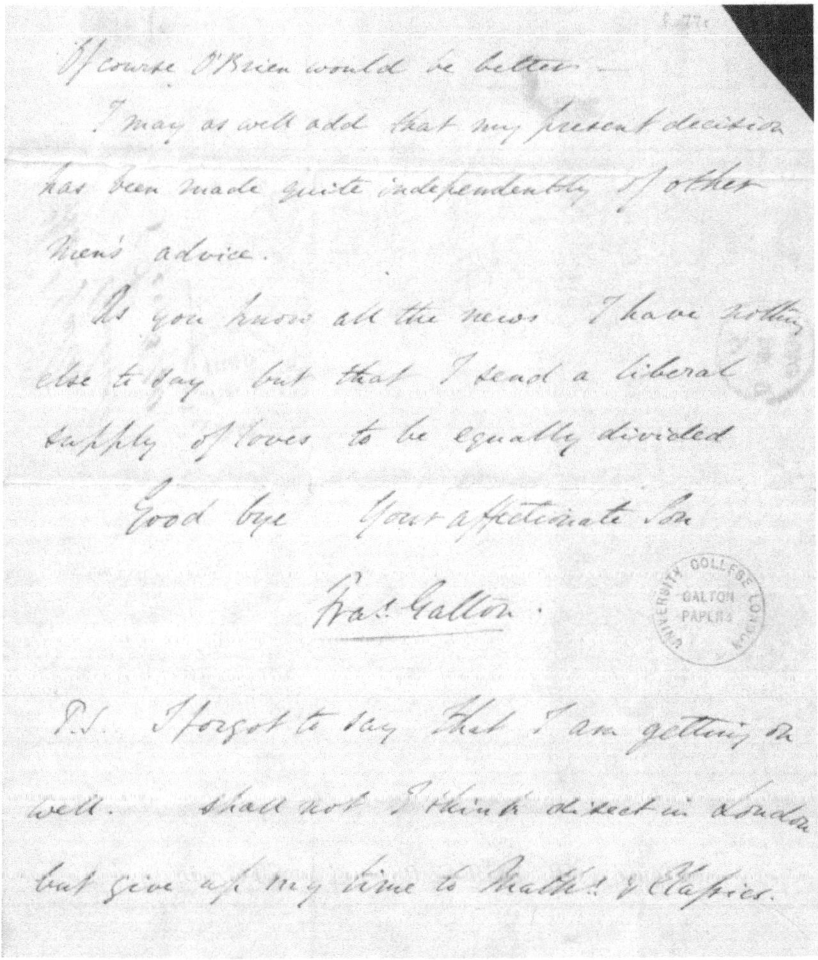

Of course O'Brien would be better —
I may as well add that my present decision
has been made quite independently of other
men's advice.
As you know all the news I have nothing
else to say but that I send a liberal
supply of loves to be equally divided
Good bye Your affectionate Son
Frat. Galton.

P.S. I forgot to say that I am getting on
well. shall not I think direct in London
but give up my time to Math! & Classics.

Figure 1.39: 1840-12-08. FG to STG page 3.

Figure 1.40: 1840-12-08. FG to STG page 4.

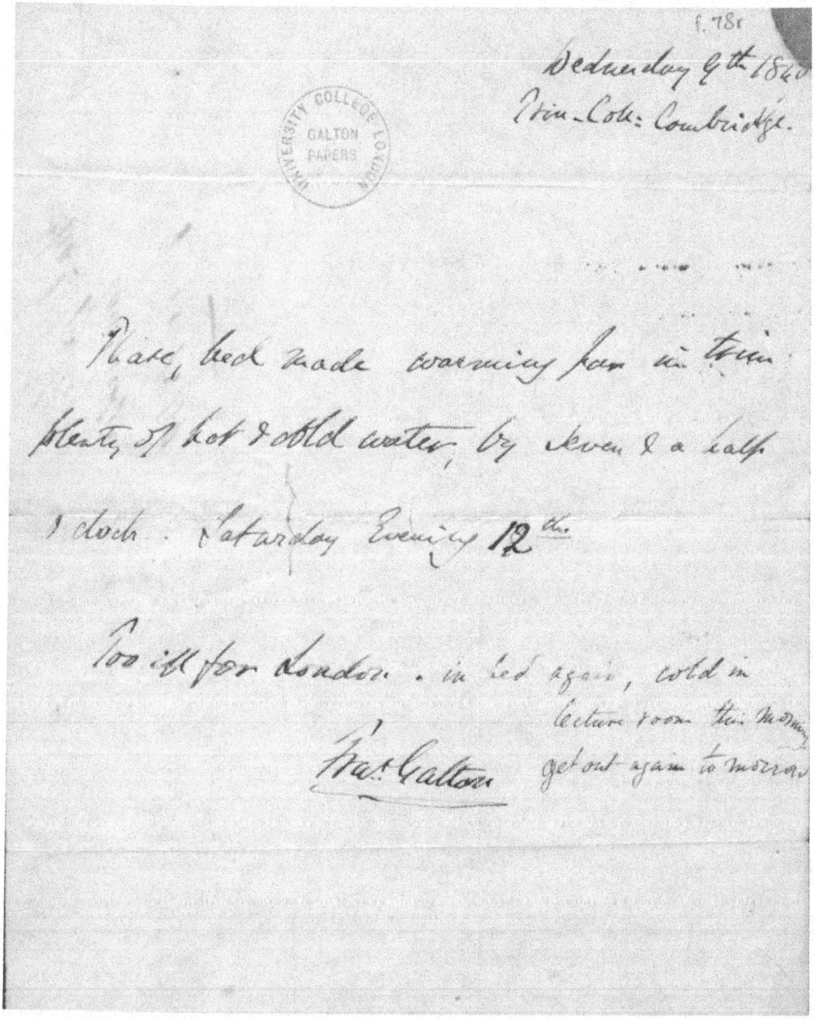

f. 78r

Wednesday 9th 1840
Trin. Coll: Cambridge.

UNIVERSITY COLLEGE LONDON GALTON PAPERS

Have, bed made warming pan in it & plenty of hot & cold water, by seven & a half o'clock. Saturday Evening 12th.

Too ill for London. in bed again, cold in lecture room this morning get out again to morrow

Fra^s. Galton

Figure 1.41: 1840-12-09. FG to STG page 1.

Figure 1.42: 1840-12-09. FG to STG page 2.

1840/12/26. Francis Galton to Samuel Tertius Galton

Dear Father,

Edwards & Ball 82 High Street[29] are the silversmiths Mr. Pritchard spoke of. On Monday evening I dine with James[30] at some club in Sutton Coldfield[31] so cannot come before Tuesday. My arrangements after that are uncertain. Hodgson[32] wants me to meet Partridge[33] at dinner with him so my plans are unsettled. I shall however be back next week.

Fras Galton

1840/12/29-31. Diary of Samuel Tertius Galton

December 29th. Tuesday. Leamington.

Francis returned home and went to the Archeological Ball at Warwick. . . .

December 31st. Thursday. Leamington.

. . . Francis was engaged inventing a new kind of [lock].[34]

---

29 Birmingham.
30 Possibly James Moilliet (1806-1878), who married Galton's sister Lucy.
31 Birmingham.
32 Hodgson, Joseph. See page 483.
33 Partridge, Richard. See page 490.
34 The shorthand transcription reads 'leg'.

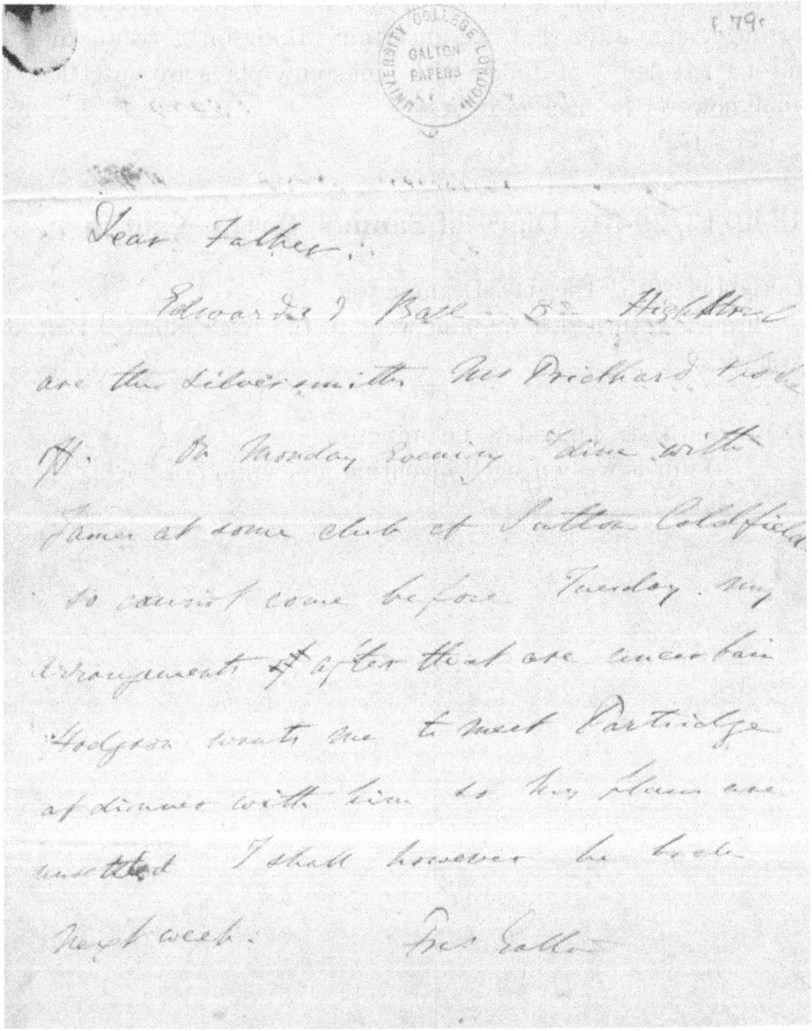

UNIVERSITY COLLEGE LONDON GALTON PAPERS

f. 79r

Dear Father,

Edwardes & Ball 82 High Street
are the silversmiths Mrs Prichard took
it. On Monday evening I dine with
James at some club at Sutton Coldfield
so cannot come before Tuesday. my
arrangements after that are uncertain
Hodgson wants me to meet Partridge
at dinner with him so my plans are
unsettled I shall however be back
next week. Fras Galton

Figure 1.43: 1840-12-26. FG to STG page 1.

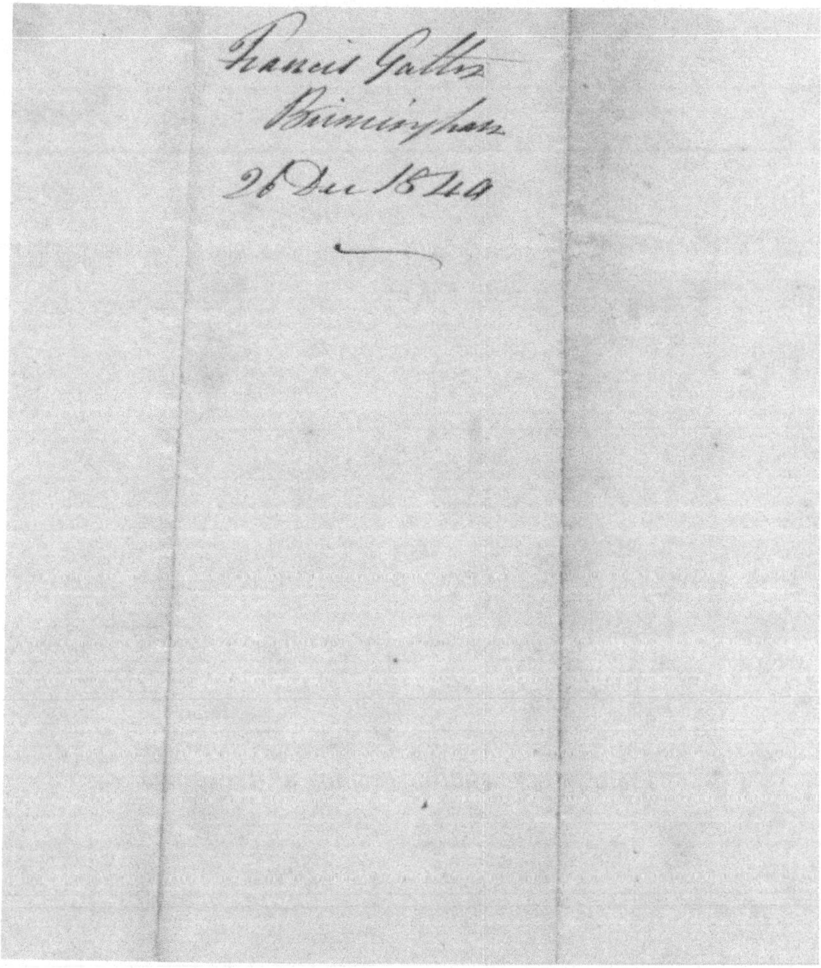

Figure 1.44: 1840-12-26. FG to STG page 2.

Figure 1.45: Galton's rooms at Trinity.

Chapter 2

1841

1841/01/01-23. Diary of Samuel Tertius Galton

January 1st. Friday. Leamington.

Francis showed me the model of a lock which he had been planning in the night of a construction which appeared to him to present more complexities than the one constructed by Chubb. He started by the Regulator coach for Birmingham at 9:45 a.m. ...

January 3rd. Sunday. Leamington.

... Francis returned from Birmingham.

January 4th. Monday. Leamington.

Snowy Day. Francis sat to Oakley for his portrait. I took a drive with Violetta.

January 7th. Thursday. Leamington.

The Horners[1] went with Francis to Warwick Castle.

January 9th. Saturday. Leamington.

... Francis and myself dined with Mr. Russell[2] where we met a party of gentlemen.

January 14th. Thursday. Leamington.

Thaw continues; several sledges in Leamington. Went to dine at Edstone with Emma and Batt where we found Francis, Alston and Mason and went to the Stratford Ball in the evening and af-

---

1 Horner, Leonard. See page 484.

2 Possibly Russell, James, see page 490.

terwards brought Francis home with us.

January 16th. Saturday. Leamington.
... Francis started for Birmingham and Loxton at 4:30 p.m.

1841/01/23. Francis Galton to Samuel Tertius Galton

My dear Father

I set off at 2 this afternoon for Paris, where I intend to stay till the end of the Vacation, that is to say I should have done, but that plaguy thing conscience prevented me. The placards about Boulogne Steamers looked very tempting.

I have just been to Bramah about my lock was more than $1\frac{1}{2}$ hour with his top-sawyer man, who was in raptures & most deferential he thinks about it today & I call again on Monday to explain anything that he may not fully understand. I enter my caveat for lamp on Monday. Now for proceedings.

Arrived at Brum[3] at $8\frac{1}{2}$ p.m., theatre & whiskey grog till $12\frac{1}{2}$. Was shut up with a frowsy fat old gentleman & a fast young gentleman in the coach whose lungs were, judging by his breath, entirely composed of full flavoured Cubas and the Cream of the Valley. The latter was not a very pleasant companion for vinous fumes ascending into his cranium displaced what reason had existed there & shewed their presence by causing him primarily to carol forth Nix my dolly pals[4]—2[ndly] to sing a very sentimental song & at last to open the window & afford me a very convincing proof that gin and cigars act as a strong emetic. Fell asleep & awoke about 10 miles from Bristol. Cross coach had had an accident waited an hour till it was mended occupying my time in eating 2 eggs, 4 slices of beef 2 plates of muffins & half a quartern loaf & then sallied forth & studied St Mary le Port.[5] Went to Cross box-seat a provincial medical man sitting behind with 2 friends we got into a dreadful quarrel about homoeopathy & as he was giving in & I was blarnying about Hippocrates, a gust of wind carried my patent gossamer hat down a steep hill into the middle of a pond (what was more natural than that beaver should take to water). The guard reclaimed it but it presented the appearance of a chemical filter as well it might for

3 Birmingham.
4 Broadside Ballad, Roud Number 12735.
5 An ancient church in the city of Bristol.

on inspection I found that its substance was composed of brown paper the day was beautiful. [See Figure 2.2, p. 56.]

Arrived at Cross, Erasmus had been in the morning to meet the first coach & had gone away again. I gigged it to Loxton. Sun shine, quite mild. Somersetshire is really the most beautiful country I have ever seen, north of the Alps (for Bessy), and of all dull pig headed stupid bipeds the Somersetshire clown stands preeminent. Arrived at Loxton the manor house commodious but not gaudy. Eras: girth visibly increased Delly[6] all smiles & lawn collar the last mentioned article being as whitewash to a sepulchre or as charity, covering a multitude of deficiencies. They really both look as happy as possible don't clash at all being separate all the morning and in the evening whilst Delly writes letters for 4 hours & reads others for 1 quarter Rassy[7] pulls Track (the dog) by his tail and ears alternately causing him to growl ferociously for 1 hour, then sleeps $3\frac{1}{4}$ & after that both adjourn to the dinner room to edify 3 maid-servants & a small boy with a learned commentary on the psalms giving the true interpretation pronunciation & critical dissertation upon the more difficult Hebrew or Chaldee words. Was knocked up next morning at 6 of course fell asleep again but was awoke at $6\frac{1}{2}$ by a cracked dinner bell in hysterics when the farming men go to their work. Got up & the small boy aforementioned brought me a pair of shoes with nails at the bottom like the teeth of the cog wheel attached to the fly of a 10-horse steam engine, this I found was truly necessary to Somersetshire walking. [See Figure 2.1, p. 55.]

I am in a great hurry, will finish tomorrow—but must say that Rassy & Delly were most kind. I enjoyed my visit greatly. Rassy works hard at his farm & evidently takes the greatest interest in it. I went to Bath to call upon my earliest flame Douglas Hunter. I have no time to write more. How is Charlotte?

Fras. Galton.

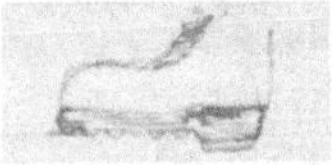

Figure 2.1: 1841-01-23 FG to STG: Boots.

[6] Adèle Galton.

[7] Erasmus Galton.

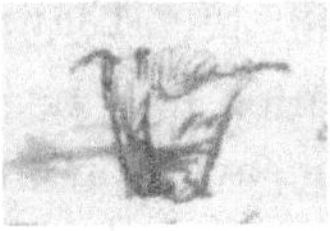

Figure 2.2: 1841-01-23 FG to STG: Filter.

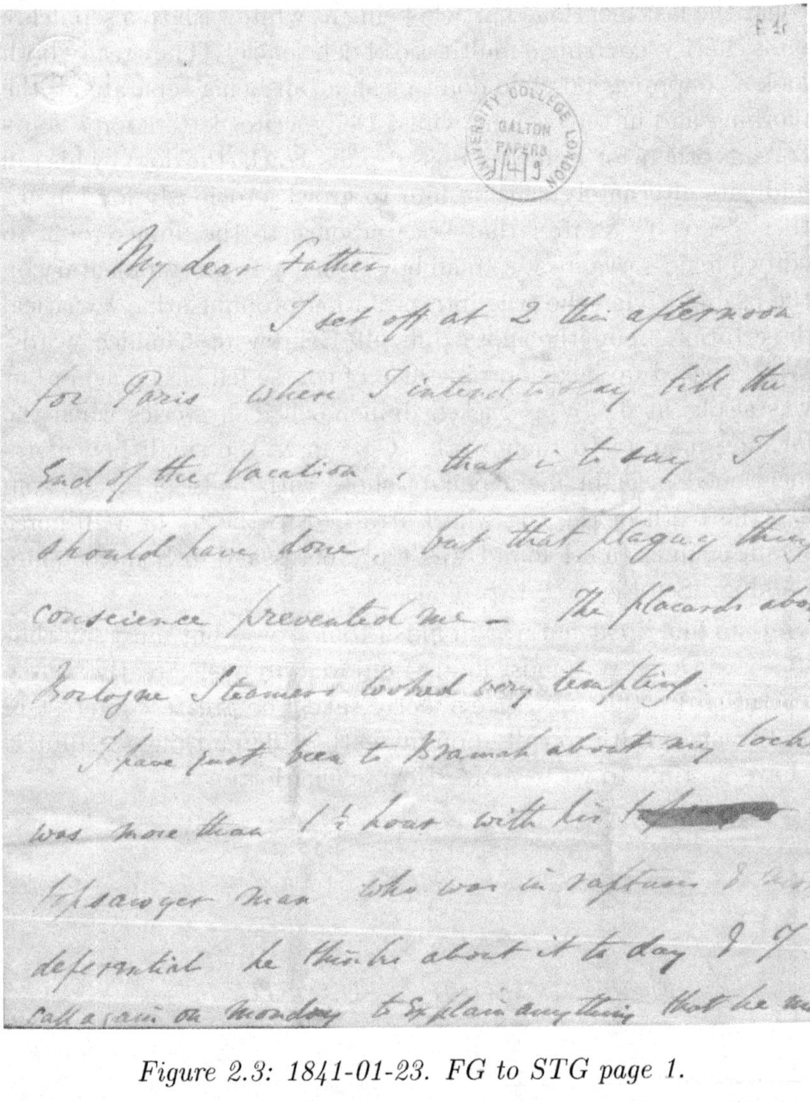

My dear Father
I set off at 2 this afternoon
for Paris where I intend to stay till the
End of the Vacation, that is to say I
should have done, but that plaguy thing
conscience prevented me — The placards abo
Boulogne Steamers looked very tempting.
I have just been to Bramah about my lock
was more than 1½ hour with him
his passenger man who was in rapture & very
deferential he thinks about it today & I
call again on monday to explain anything that he wa

Figure 2.3: 1841-01-23. FG to STG page 1.

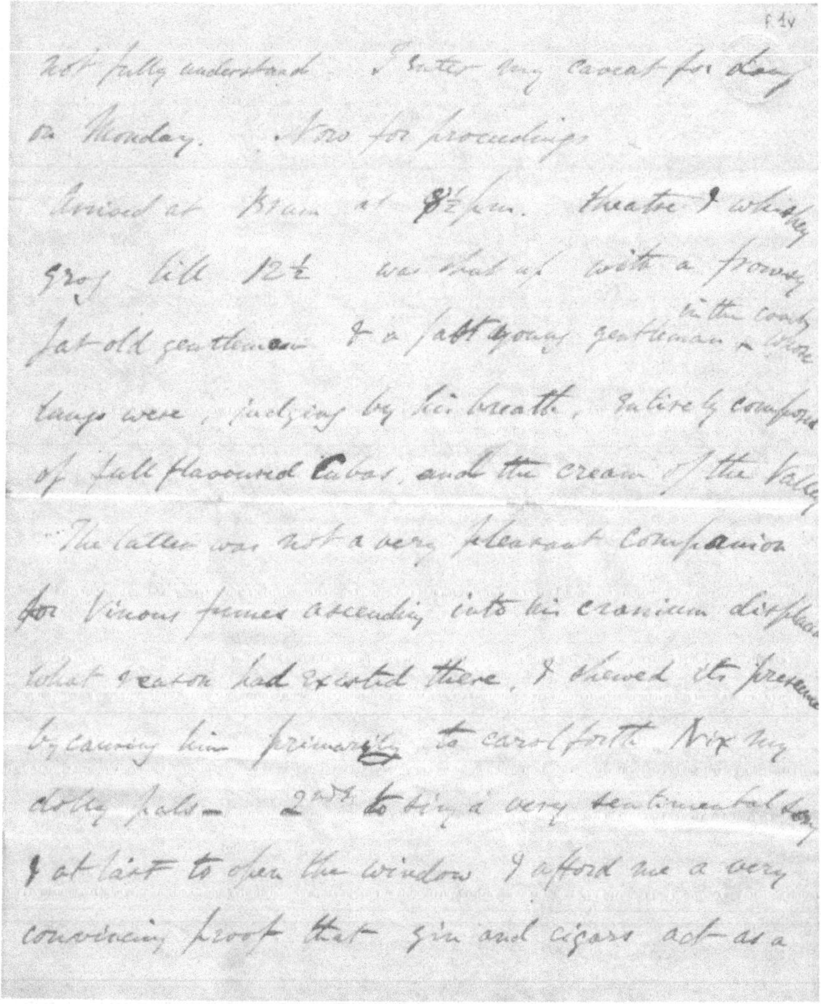
f.1v

not fully understand. I enter my caveat for Long
on Monday. Now for proceedings.
Arrived at Brum at 8½ pm. theatre & whiskey
grog till 12½ was shut up with a prosy
fat old gentleman & a fatt young gentleman in the coach whose
lungs were, judging by his breath, entirely composed
of full flavoured Cubas, and the cream of the Valley
The latter was not a very pleasant companion
for Vinous fumes ascending into his cranium displaced
what reason had existed there, & shewed its presence
by causing him primarily to carol forth Nix my
dolly pals — 2ndly to sing a very sentimental song
& at last to open the window & afford me a very
convincing proof that gin and cigars act as a

Figure 2.4: 1841-01-23. FG to STG page 2.

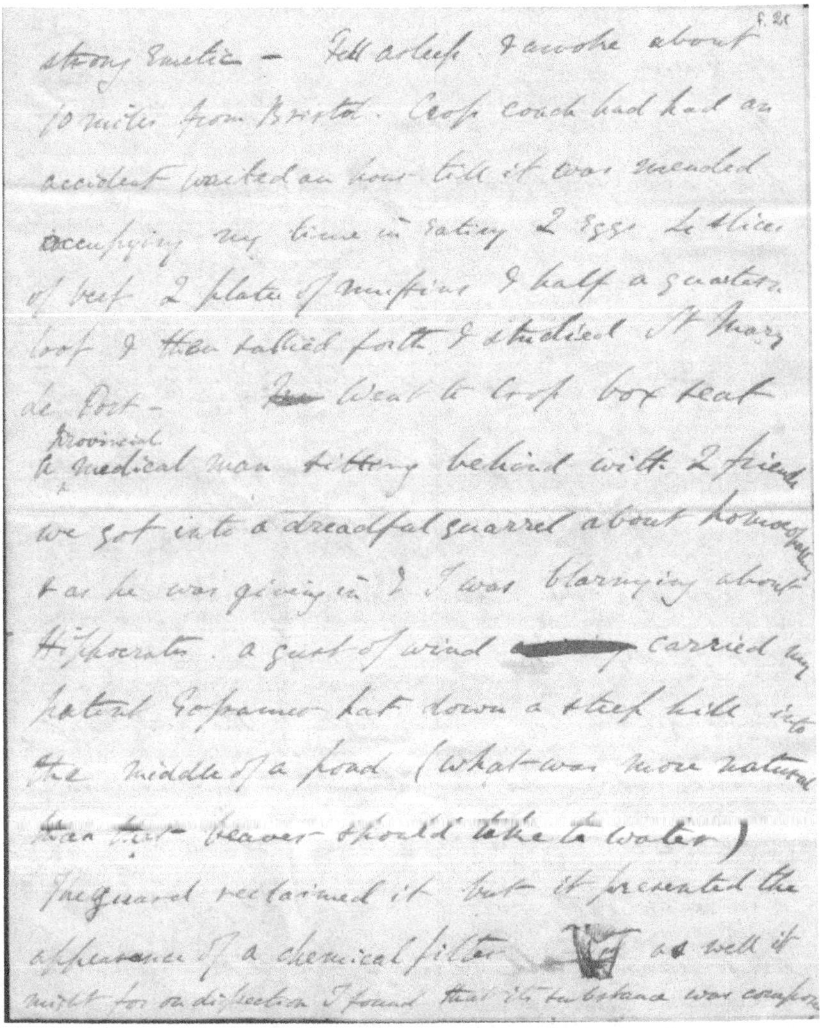

f. 21

strong Emetic — Fell asleep & awoke about
50 miles from Bristol. Cross coach had had an
accident waited an hour till it was mended
occupying my time in eating 2 eggs 5 slices
of beef 2 plates of muffins & half a quarter
loaf & then sallied forth & studied St Mary
de Port — Went to cross box seat
a provincial medical man sitting behind with 2 friends
we got into a dreadful quarrel about homoeopathy
& as he was giving in & I was blarneying about
Hippocrates. a gust of wind carried my
patent Gossamer hat down a steep hill into
the middle of a pond (what was more natural
than that beaver should take to water)
The guard reclaimed it but it presented the
appearance of a chemical filter as well it
might for on dissection I found that its substance was compos

Figure 2.5: 1841-01-23. FG to STG page 3.

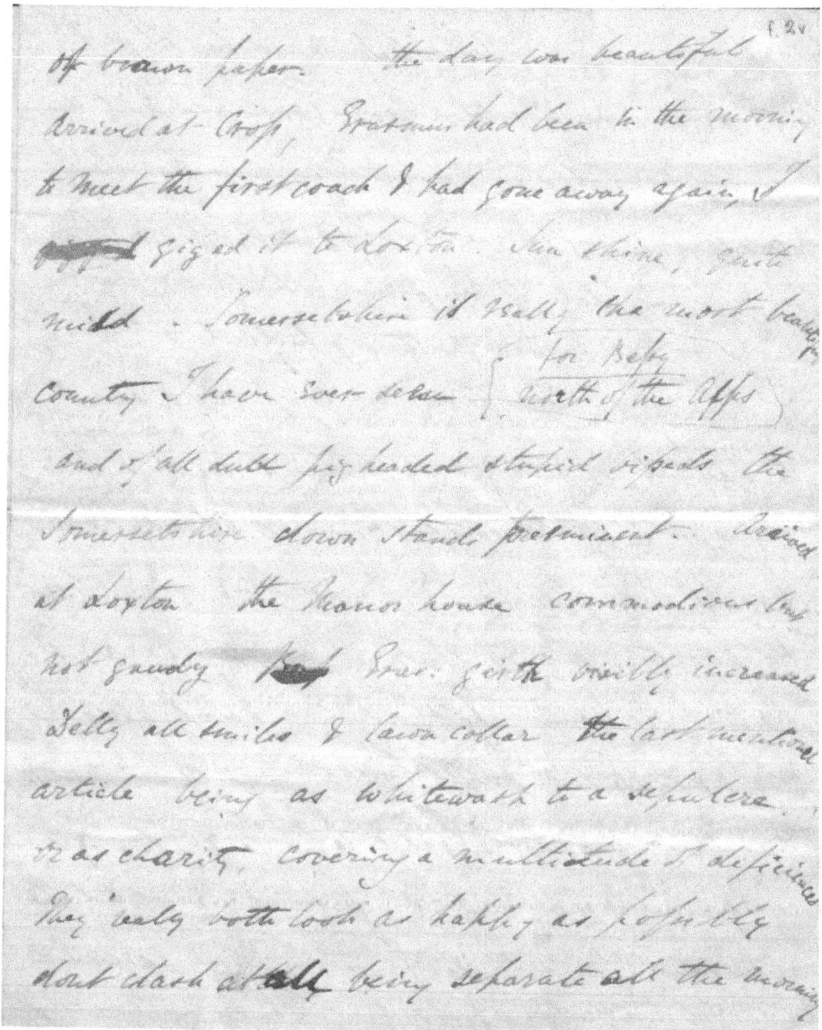
f. 2v
of brown paper. the day was beautiful
Arrived at Crops, Grannie had been in the morning
to meet the first coach & had gone away again, I
gigged it to Loxton. Sun shine, quite
mild. Somersetshire is really the most beautiful
country I have ever seen {for Betsy north of the Alps}
and of all dull pig headed stupid bipeds the
Somersetshire clown stands preeminent. Arrived
at Loxton the Manor house commodious but
not gaudy Emer. girth visibly increased
Betty all smiles & lawn collar the last mentioned
article being as whitewash to a sepulchre
or as charity covering a multitude of deficiencies
they really both look as happy as possibly
don't clash at all being separate all the morning

Figure 2.6: 1841-01-23. FG to STG page 4.

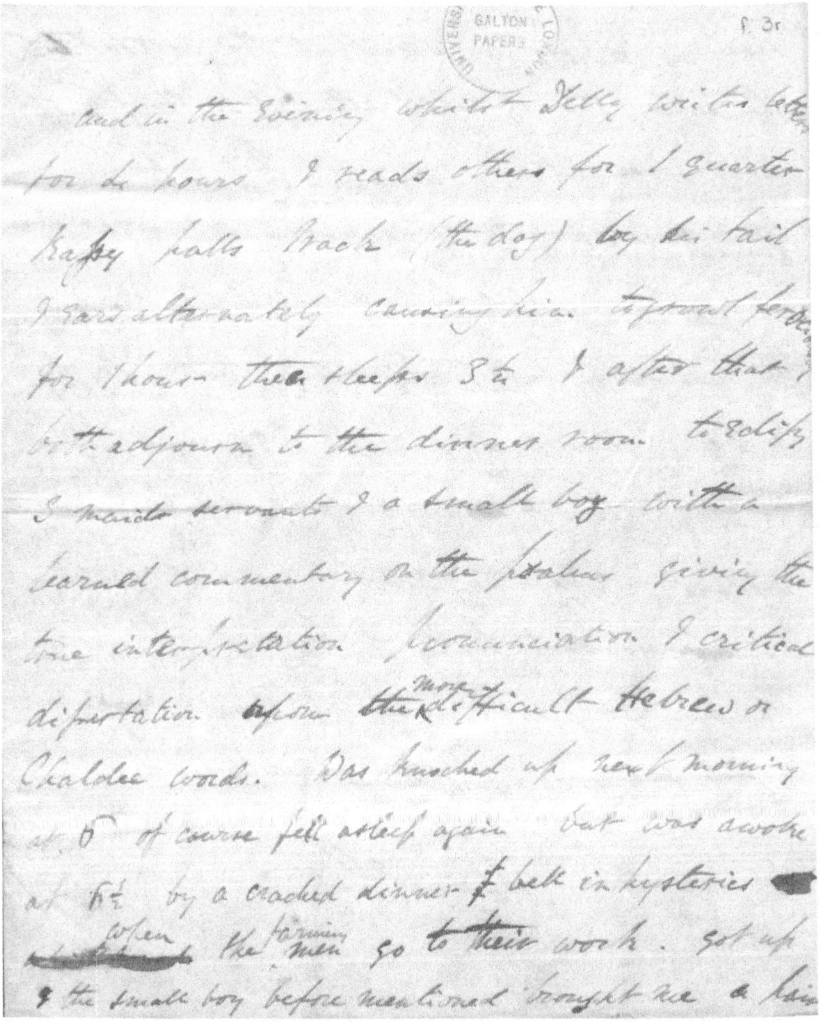

f. 3r

and in the Evening whilst Delly writes letters
for 2 hours I reads others for 1 quarter
Happy pulls Bach (the dog) by his tail
& saw alternately causing him to growl ferociously
for 1 hour — then sleeps 3/4 & after that
both adjourn to the dinner room to edify
3 maids servants & a small boy with a
learned commentary on the psalms giving the
true interpretation pronunciation & critical
dissertation upon the more difficult Hebrew or
Chaldee words. Was knocked up next morning
at 6 of course fell asleep again but was awoke
at 6½ by a cracked dinner bell in hysterics
when the farming men go to their work. got up
& the small boy before mentioned brought me a bas[in]

Figure 2.7: 1841-01-23. FG to STG page 5.

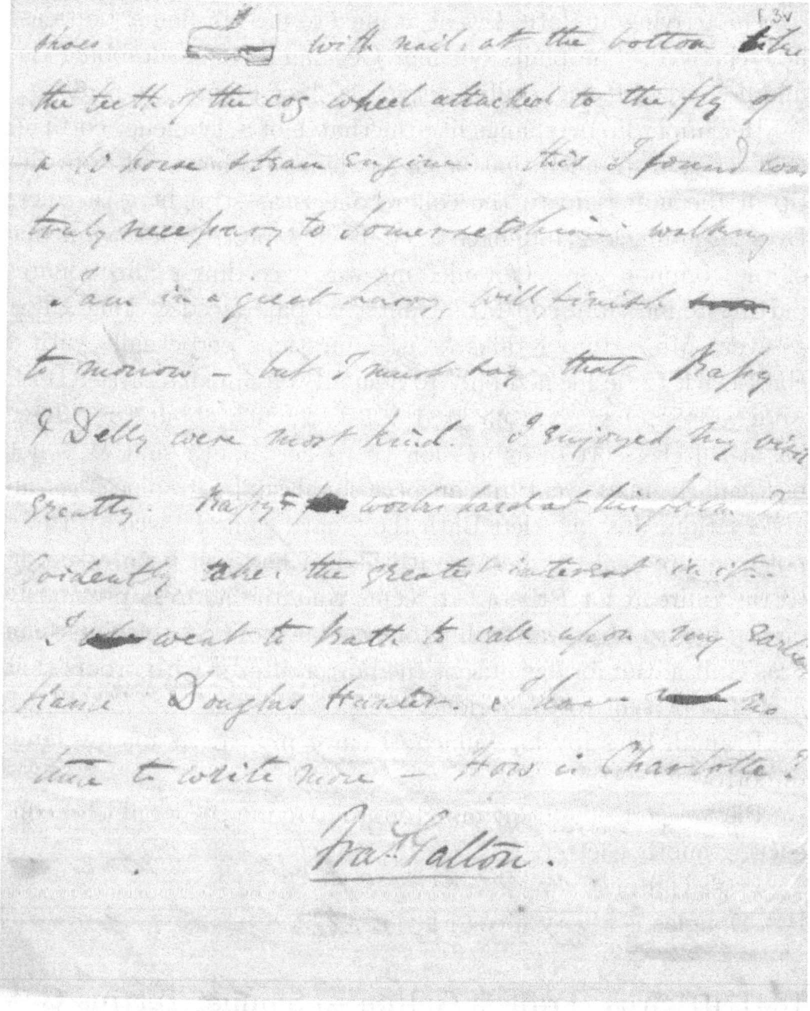
f. 3v

shoes with nails at the bottom like
the teeth of the cog wheel attached to the fly of
a 10 horse steam engine. this I found was
truly necessary to somersetshire walking
I am in a great hurry, will finish
to morrow — but I must say that Happy
& Delly were most kind. I enjoyed my visit
greatly. Happy works hard at his farm &
evidently takes the greatest interest in it.
I went to Bath to call upon my Darling
Aunt Douglas Hunter. I have no
time to write more — How is Charlotte?

Fra^s Galton.

Figure 2.8: 1841-01-23. FG to STG page 6.

1841/01/27. Francis Galton to Adèle Galton

Sunday
17 New Str, Spring Gardens

My dear Adèle

On arriving at Bath, I went at once to my fair flame, but alas, how changed. She plaits her hair German fashion scrabbing the middle part of it frightfully backwards. [See Figure 2.9, p. 63.]

Her upper lip overhangs like the thatch of a Tyrolene[8] cottage, chin retreats so much that when she holds up her head from the tip of the upper lip to the collar-bone is as straight as a ruler. Eyes large quick & intelligent. Front view pretty. Side view out of the common way. Our meeting was exceedingly affectionate, parting remarkably so. Mr. Hunter as pale faced & long gilled as ever. Mrs. Hunter talks & asks questions perpetually with a clack-clack to be likened only to a small watchman's rattle. Both were very kind & good natured & asked much about you. Jesse has a Butcher-sort of expression keeps her mouth shut & is not particularly mad, ventures an occasional smile & sometimes an observation. Just as I left Bath the Guard came & asked if I had not a square deal box because if I had, it had just been taken off to the railroad for Bristol. It seems that the address previously on the box to Miss Galton Loxton was far more conspicuous than Fras Galton Cambridge. I sent the porter after it with orders that it should be sent to Cambridge.

Bramah likes my lock much. I enter my caveat for my lamp tomorrow.

Please send this to Leamington, to save me the trouble of concocting another letter.

Fras. Galton

P.S. Douglas is a very nice girl.

1841/01 circa. Francis Galton to Samuel Tertius Galton

My dear Father

Lamp and lock both dished but have come off ~~well~~ honorably in both. Capt. Basil Hall, aided by Wheatstone had hit upon the same idea a short time since & has since been making experiments.

[8] Tyrolean.

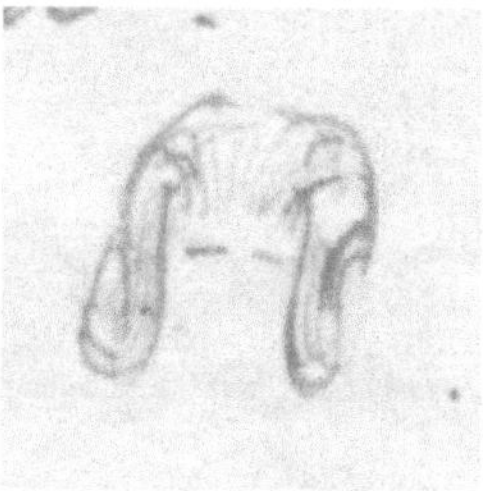

Figure 2.9: 1841-01-27 FG to AG: Plaits.

The light appears not advantageous as regards illuminating rooms though it is useful for lighthouses.

As regards my lock, Bramah complimented it, & spoke very ~~well recommen~~ sensibly about it, he said that it was certainly much more difficult to pick than any one of the same size & of a different construction, but the chances were quite great enough for security against a chance key in either the Bramah & Chub, the only thing to be feared was a model being taken of the original key when accidentally left about. Now mine being merely ~~adapted~~ a piece of bent wire could be imitated from the impression left on almost any substance—or traced on paper—whilst his (here he grew coxy) required a very careful modeller & much time to imitate—my lock would also be expensive. This was very true & I quite agree with him, but as regards the security of his lock when the key had been left in the hands of a pickpocket I offered to ~~take~~ make a false key in 5 minutes if he would leave the original key in my hands for 5 seconds—he of course stood up in defence of his own key so I got 10 knitting pins 5 large & 5 small & one wooden one which was central the others surrounding it. [See Figure 2.14, p. 68.]

On pressing the central one down the bore of the key, the other ones were variously depressed according to the teeth in the key as in the drawing the other end B of course exactly represents the key (A) the ward (C) is always the same distance from the end & could therefore be fixed to one of the pins. Knitting pins are of course clumsy but with a little contrivance a perfect picklock can be made (the breadth of the slits is of no consequence only the depth). Bramah was very fierce, I ~~asked~~ told him that I had some intention of patenting the picklock, & advertising "Important to Thieves, Housebreakers & others." I enclose a model of one of his show locks, by five pins. Bramah who was called down to see my

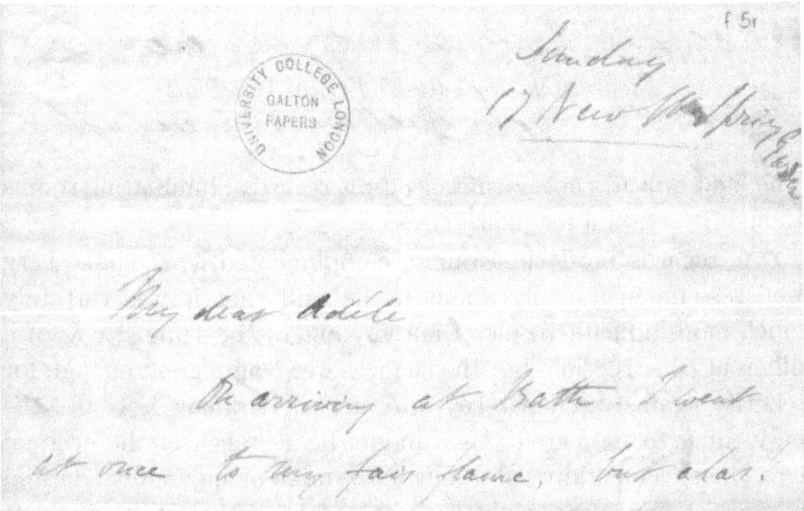

Sunday
17 New St Spring Gardens

My dear Adèle

On arriving at Bath I went at once to my fair flame, but alas! how changed. She plaits her hair German fashion scrubbing the middle part of it frightfully backwards, her upper lip overhangs like the thatch of a Tyrolese Cottage, chin retreats so much that when she holds up her head from the tip of the upper lip to the collar bone is as straight as a ruler. Eyes large quite

Figure 2.10: 1841-01-27. FG to AG page 1.

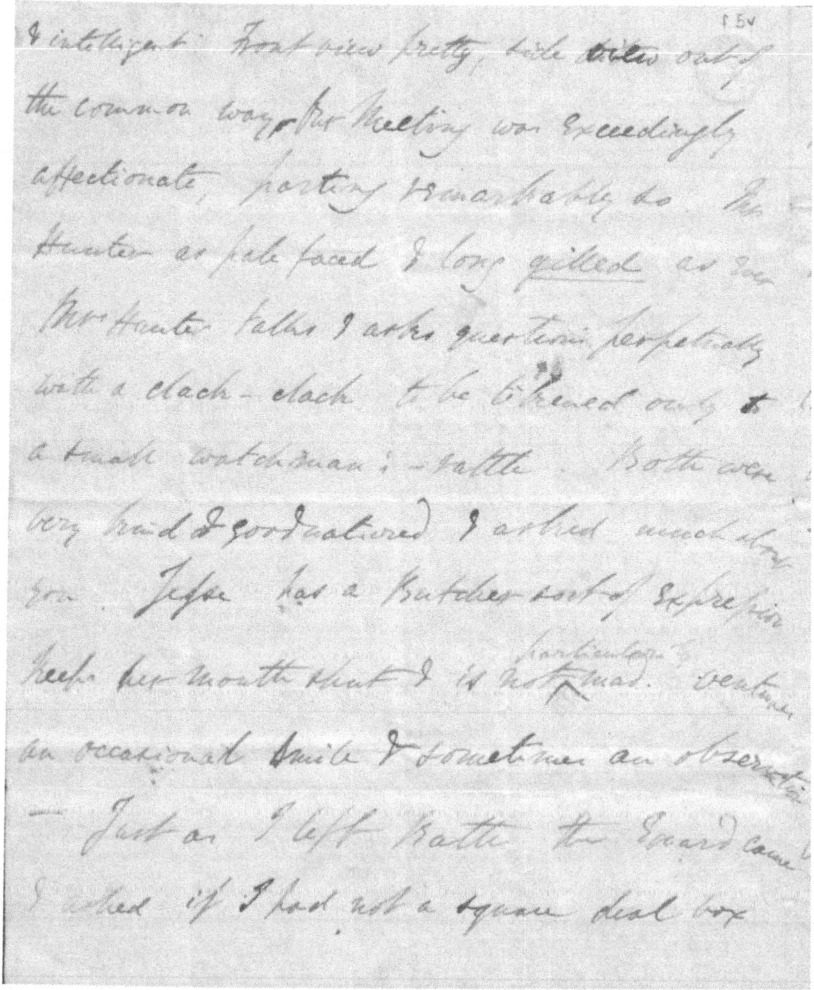

& intelligent. Front view pretty, tho' ~~rather~~ rather out of
the common way. The Meeting was exceedingly
affectionate, parting remarkably so. Mr
Hunter as pale faced & long gilled as ever
Mrs Hunter talks & asks questions perpetually
with a clack-clack to be likened only to
a small watchman's-rattle. Both were
very kind & goodnatured & asked much about
you. Jesse has a butcher sort of expression
keeps her mouth shut & is not particularly mad. ventures
an occasional smile & sometimes an observation
— Just as I left Bath the Guard came
& asked if I had not a square deal box

Figure 2.11: 1841-01-27. FG to AG page 2.

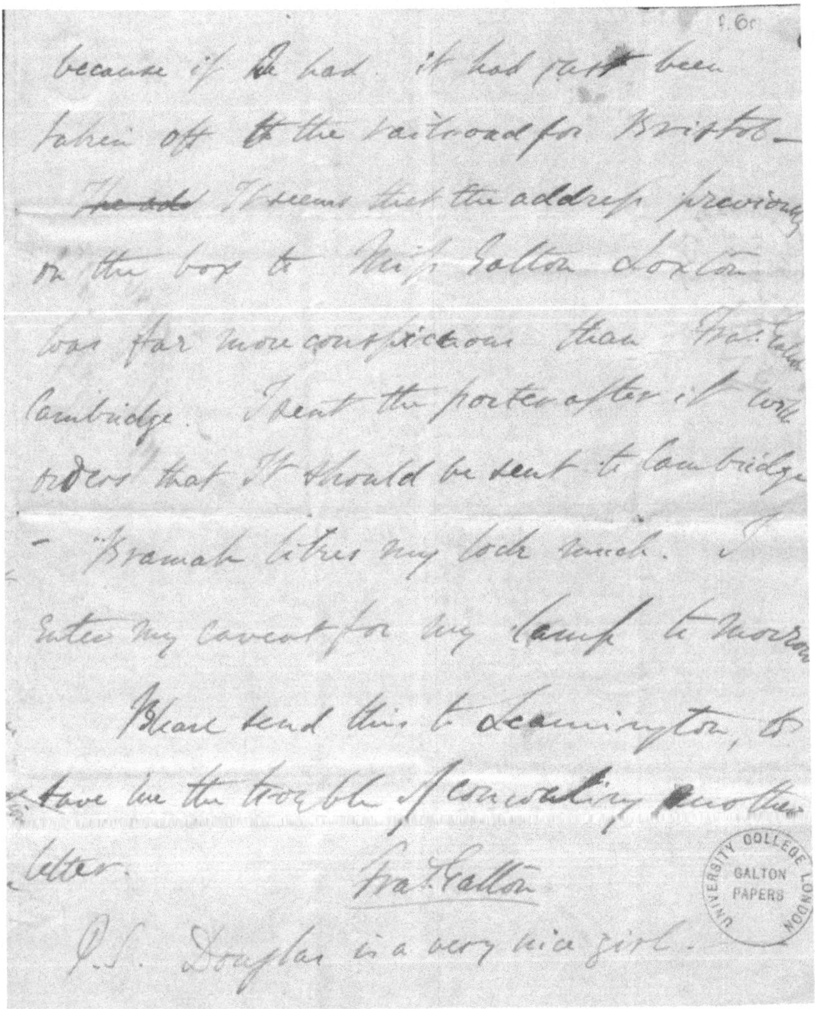

f. 6r

because if he had it had just been
taken off to the railroad for Bristol —
It seems that the address previously
on the box to Miss Galton Loxton
was far more conspicuous than Fra. Galton
Cambridge. I sent the porter after it with
orders that it should be sent to Cambridge
— Bramah likes my lock much. I
enter my caveat for my lamp to morrow

Please send this to Leamington, to
save me the trouble of concocting another
letter.

Fra. Galton

P.S. Douglas is a very nice girl.

Figure 2.12: 1841-01-27. FG to AG page 3.

Figure 2.13: 1841-01-27. FG to AG page 4.

knitting pins looked angry.

Fras. Galton.

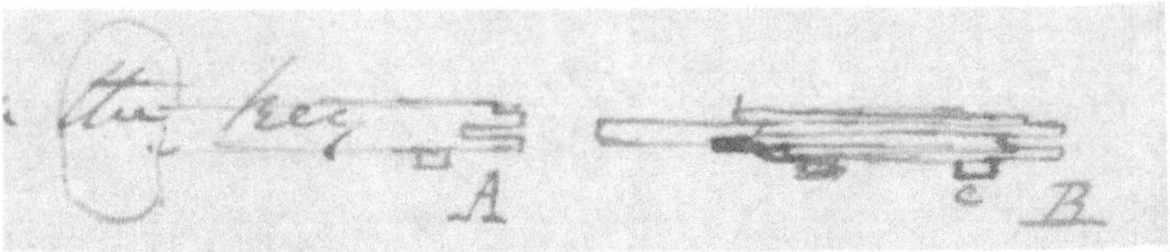

Figure 2.14: 1841-01 FG to STG: Key.

1841/01/31. Francis Galton to Samuel Tertius Galton

My dear Father

Thanks for letter. Lectures begin to-morrow (Monday) Poor Hughes! Charlotte's cousin was unable to stand the examination more than 3 days on account of health & so got an honour aegrotat.[9] I was very much vexed about my lamp but am now trying other things. I have I think a neat plan of making any balance weigh (by double weighing) to the greatest accuracy. I do it by fixing the balance near its centre to a bar of steel (magnetic) perpendicularly. The upper end of this carries a steel point which works against an iron plate, which plate can be screwed up against one of the poles of a fixed magnet, downwards, through a small space. [See Figure 2.19, p. 73.]

From this iron plate A, a rod B passes vertically downwards to which is attached C, a plate on which as in fig I the unsupported scale beam rests. Now if A be screwed up nearer and nearer the magnet at a certain point the scale beam will leave C & by its iron point will touch A and be suspended. The least possible friction can be easily ~~ascertained~~ attained. [See Figure 2.20, p. 74.]

A small horseshoe magnet that I have in my pocket will lift a heavy bunch of keys in contact with one pole. [See Figure 2.20, p. 74.] I do not doubt in getting easily a magnet that will suffice. You probably understand the process of double weighing. It is putting the substance to be weighed in one scale & some indifferent substance as sand in the other, then taking out the substance to be weighed place the weights in the same scale pan. Under these circumstances inequality of the length of the arms will have

9 George Bilsborrow Hughes (1818-1893) of Jesus College.

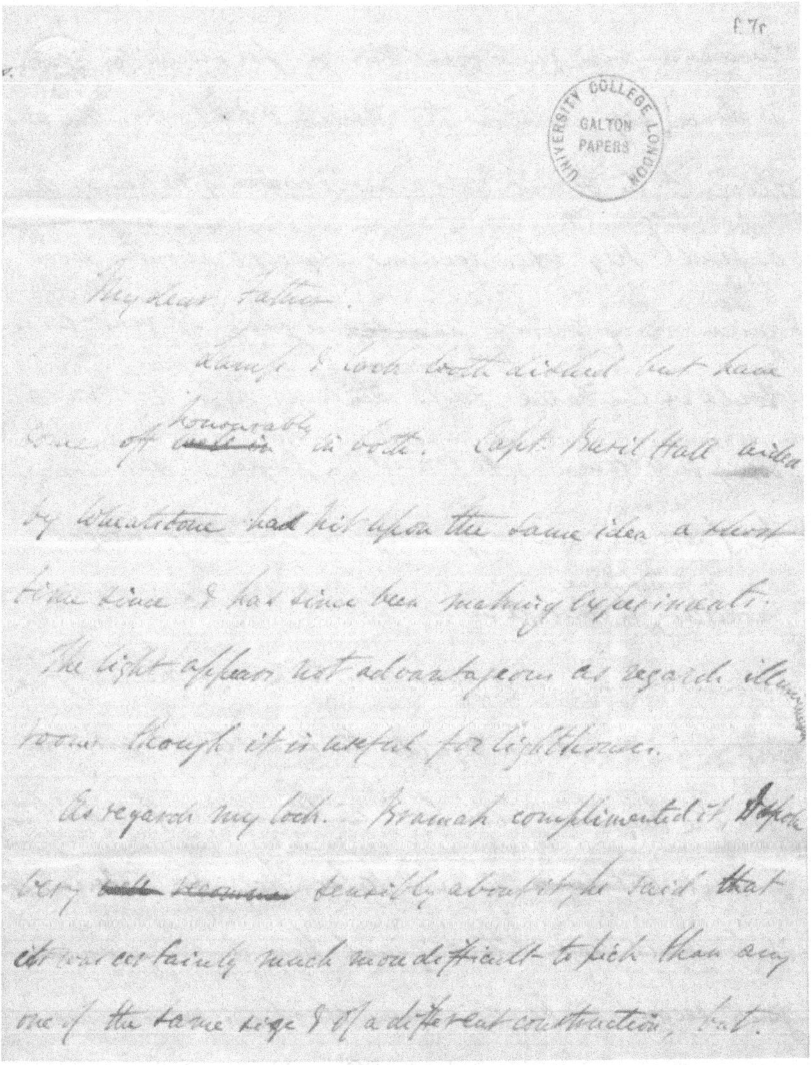

f. 7r

My dear Father.

Lamps & lock both dished but have come off ~~with in~~ honourably in both. Capt. Basil Hall aided by Wheatstone had hit upon the same idea a short time since & has since been making experiments. The light appears not advantageous as regards illumination though it is useful for lighthouses.

As regards my lock. Bramah complimented it, & spoke very ~~well~~ ~~[illegible]~~ sensibly about it, he said that it was certainly much more difficult to pick than any one of the same size & of a different construction, but.

Figure 2.15: 1841-01. FG to STG page 1.

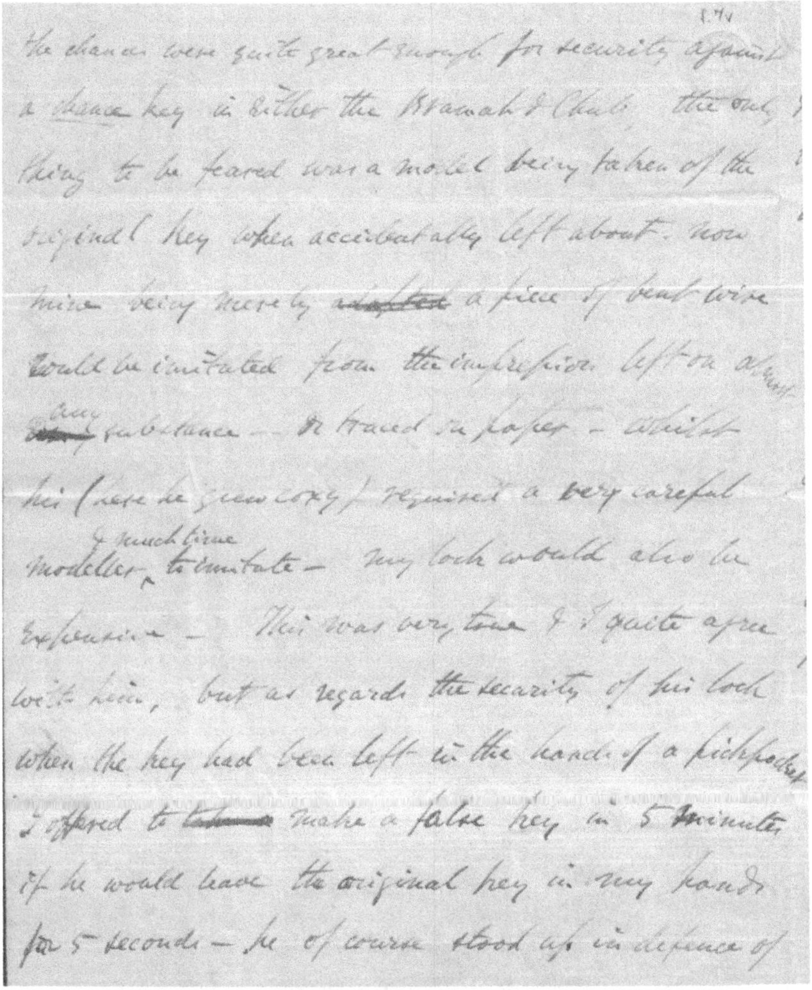

the chances were quite great enough for security against
a chance key in either the Bramah & Chub, the only
thing to be feared was a model being taken of the
original key when accidentally left about. now
mine being merely ~~[illegible]~~ a piece of bent wire
would be imitated from the impression left on any
~~[illegible]~~ any substance — or traced on paper — whilst
his (here he grew cozy) required a very careful
modeller & much time to imitate — my lock would also be
expensive — This was very true & I quite agree
with him, but as regards the security of his lock
when the key had been left in the hands of a pickpocket
I offered to ~~[illegible]~~ make a false key in 5 minutes
if he would leave the original key in my hands
for 5 seconds — he of course stood up in defence of

Figure 2.16: 1841-01. FG to STG page 2.

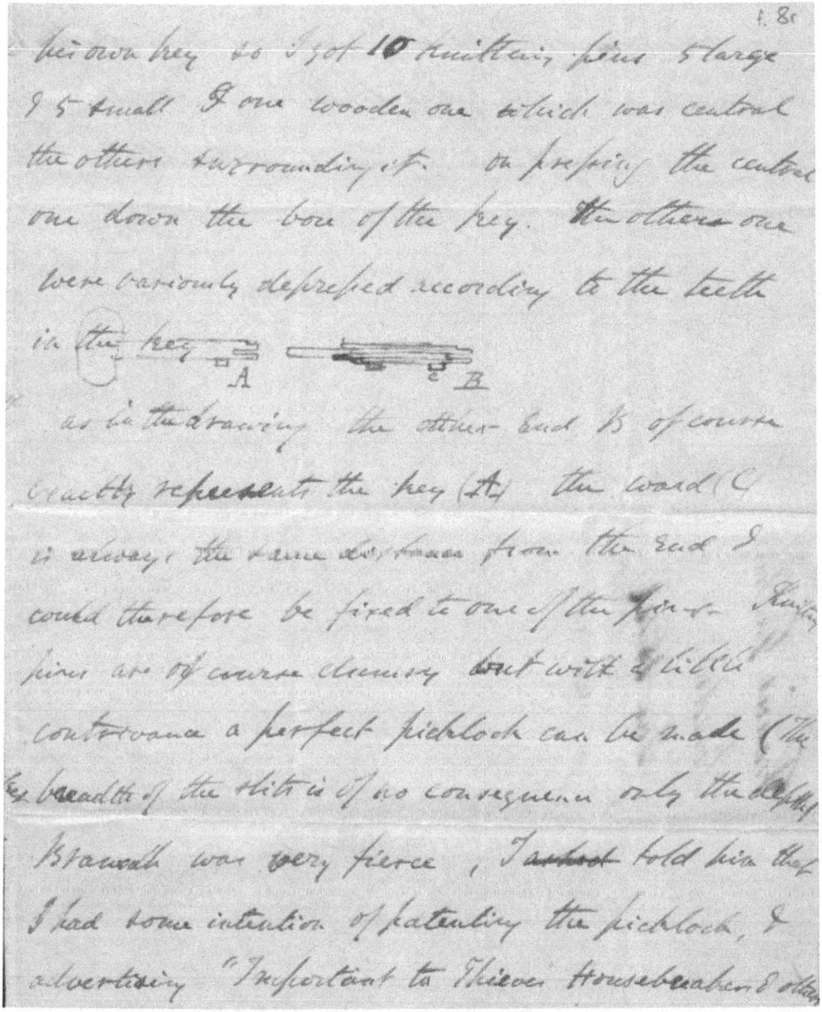

f. 8r

his own key so I got 10 knitting pins 5 large & 5 small & one wooden one which was central the others surrounding it. On pressing the central one down the bore of the key. the others one were variously depressed according to the teeth in the key

as in the drawing the other end B of course exactly represents the key (A) the ward (C is always the same distance from the end & could therefore be fixed to one of the pins. Knitting pins are of course clumsy but with a little contrivance a perfect picklock can be made (The breadth of the slits is of no consequence only the depth

Bramah was very fierce, I told him that I had some intention of patenting the picklock, & advertising "Important to Thieves Housebreakers & others

Figure 2.17: 1841-01. FG to STG page 3.

I enclose a model of one of his three locks, by [illegible]
Bramah who was called down to see my brother
piece looked angry. Fras Galton.

Francis Galton
London
Jan 1841

Figure 2.18: 1841-01. FG to STG page 4.

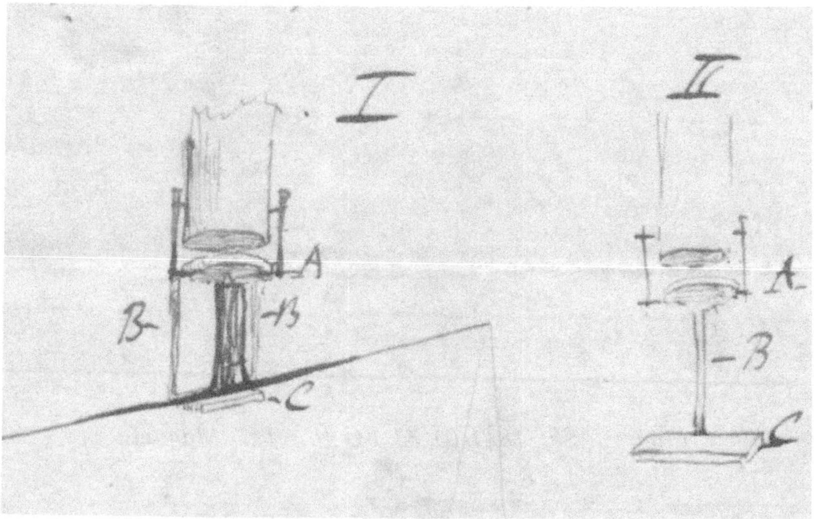

Figure 2.19: 1841-01-31 FG to STG: Scale, Figs. I and II.

no effect in giving an erroneous result (besides, in an ordinary balance, increase of friction) as both the weights & the substance to be weighed were in exactly similar situations. I think that in this way a very accurate balance might be constructed for 10 shillings which would be a desideration. I will make one.

I am having a ~~Chubb~~ Bramah picklock made. I smoke my Turkish or German pipe nightly with somebody else & give Theodore eau sucré to drink with it—bless his innocence—it comes uncommon cheap—no man can drink more than three tumblers full, or it would make him sick.

O'Brien[10] begins on Thursday.

Fras. Galton.

1841/02/03. Francis Galton to Samuel Tertius Galton

Wednesday
Trin Coll

My dear Father

Atwood[11] came down this morning & breakfasted with me & I have left him in the hands of Boulton[12] to lionise as I am invalided

[10] O'Brien, Matthew. See page 487.
[11] Atwood, H. A. S.. See page 479.
[12] Boulton, Matthew Piers Watt. See page 480.

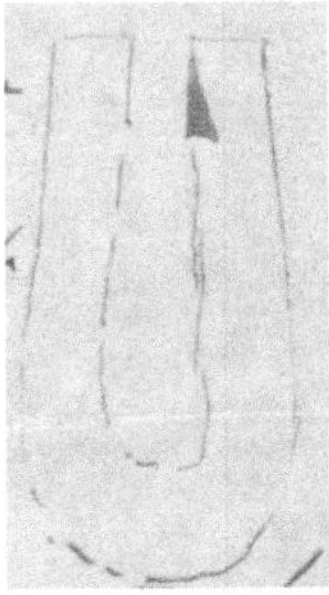

Figure 2.20: 1841-01-31 FG to STG: Magnet.

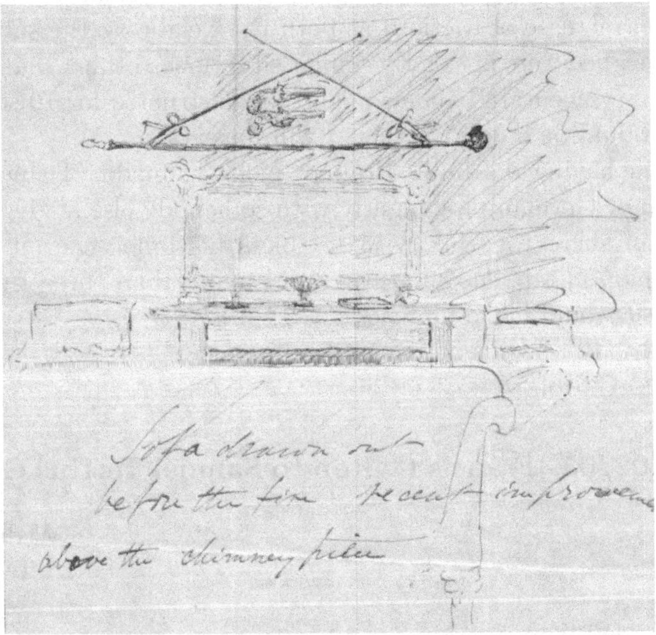

Figure 2.21: Hearth, Pistols and Swords. Sofa drawn out before the fire, recent improvements above the chimney place

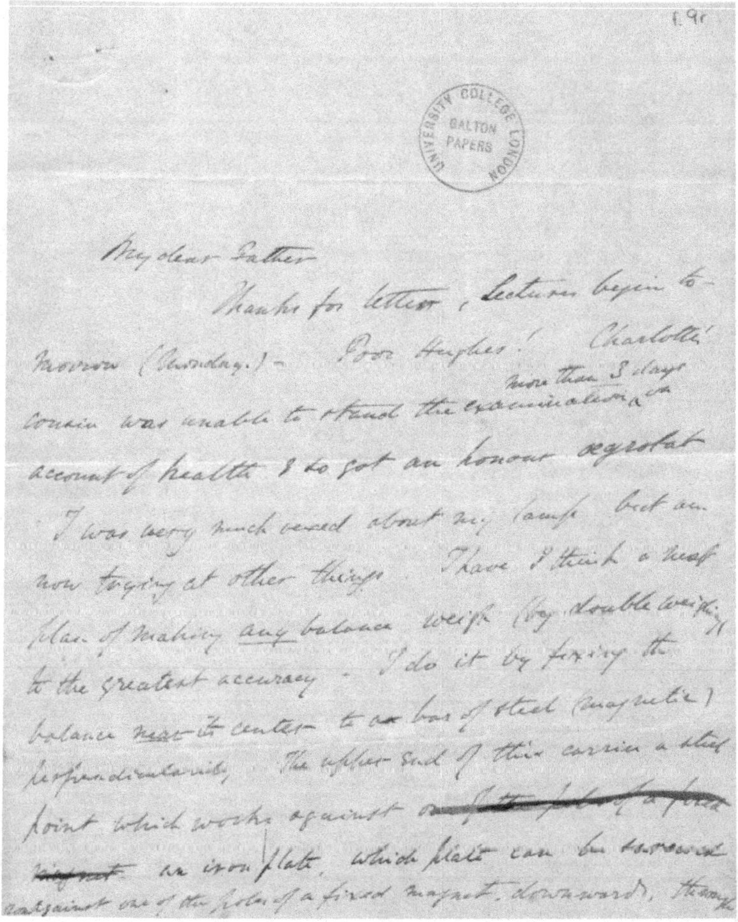

My dear Father

Thanks for letter, Lectures begin to-
morrow (Monday.) – Poor Hughes! Charlotte's
cousin was unable to stand the examination more than 3 days on
account of health & so got an honour aegrotat
I was very much vexed about my lamp but am
now trying at other things. I have I think a neat
plan of making any balance weigh (by double weighing)
to the greatest accuracy. I do it by fixing the
balance near its center to a bar of steel (magnetic)
perpendicularly. The upper end of this carries a steel
point which works against ~~[illegible]~~
~~[illegible]~~ an iron plate, which plate can be screwed
against one of the poles of a fixed magnet, downwards, the

Figure 2.22: 1841-01-31. FG to STG page 1.

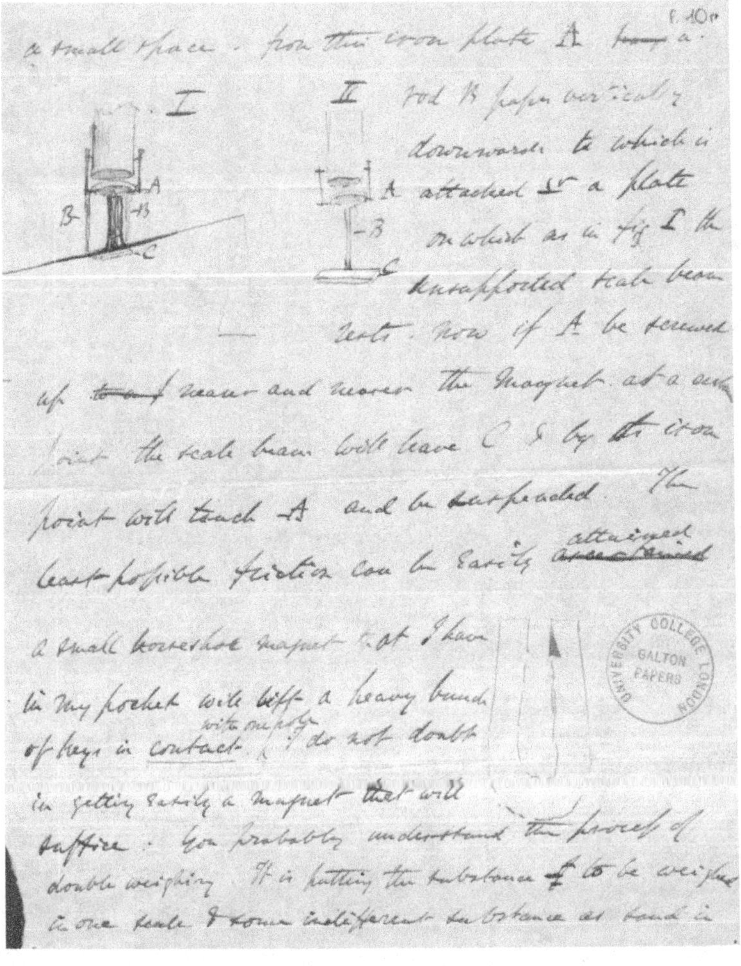

f. 10r

a small space - from this iron plate A ~~from~~ a
rod B paper vertically
downwards to which is
attached a plate
on which as in fig I the
unsupported scale beam
rests. Now if A be screwed
up ~~to~~ nearer and nearer the Magnet at a certain
point the scale beam will leave C & by the iron
point will touch A and be suspended. The
least possible friction can be easily ~~ascertained~~ attained

A small horseshoe magnet that I have
in my pocket will lift a heavy bunch
of keys in contact with one pole I do not doubt
in getting easily a magnet that will
suffice. You probably understand the process of
double weighing It is putting the substance ~~f~~ to be weighed
in one scale & some indifferent substance as sand in

Figure 2.23: 1841-01-31. FG to STG page 2.

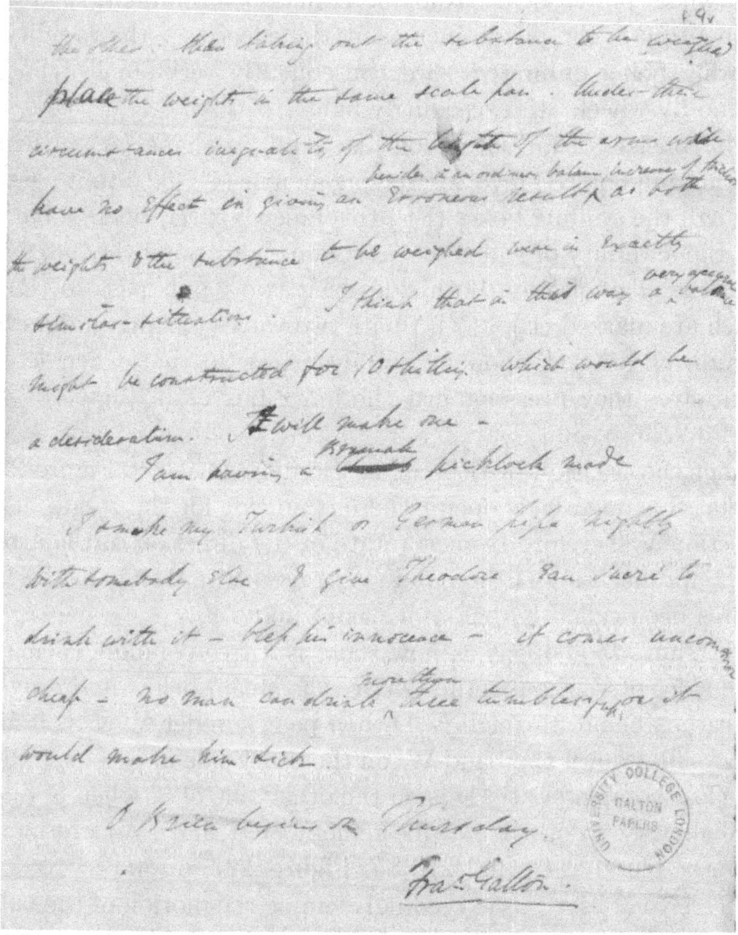

f.9v

the other. then taking out the substance to be weighed

place the weights in the same scale pan. Under these

circumstances inequalities of the length of the arms which

besides in an ordinary balance increase of friction

have no effect in giving an erroneous result, as both

the weights & the substance to be weighed were in exactly

very accurate balance

similar situation. I think that in that way a

might be constructed for 10 shillings which would be

a desideratum. I will make one –

Bramah

I am having a picklock made

I smoke my Turkish or German pipe nightly

with somebody else I give Theodore Eau sucré to

drink with it – keep his innocence – it comes uncommon

more than

cheap – no man can drink three tumblers full for it

would make him sick.

I write again on Thursday.

Fra^s Galton.

Figure 2.24: 1841-01-31. FG to STG page 3.

from a relapse of my old illness which came on on Saturday without any cause to which I can assign it. I am all but well, it has not confined me to my bed but only to my room.

Thanks for lecture per post. I am rather mad about a ~~revolving~~ rotatory Steam Engine which I have been contriving, Boulton thinks it will do. [See Figure 2.25, p. 79.] Advantages being 1st The whole power being available cranks being absent. 2nd The momentum of the piston increasing the effect & ∴ the rapidity of working being unlimited. 3rd consequently very small cylinder, 4th no fly wheel, 5th exceedingly light.

It consists of an outer box like a pill in shape (A) in which another pill box (B) revolves, having an axis (C) fixed to it passing through the stuffing boxes (FF) this inner box (B) bears a barrier (D) the externity of which moves steam tight round the outer box from the edge of Y on both sides of it two bands pass to (B) — which are marked (E) – (Y)(Y) are two valves moving in direction of radii & pressed upon by spiral springs so that when left to themselves they press against the inner box as in (fig. 3). [See Figure 2.29, p. 82.]

Let the position of B be as in the diagram. [See Figure 2.26, p. 80.] S is open E' is open. There is no exit for the steam in the direction V therefore B must rotate in the direction of the arrow. As D approaches (V) the band (W) raises V' (W has in the first figures been called E) before it passes under E'.

E' shuts & E opens as now there is a free communication on both sides of V' the [band] D drives the steam below it through E instead as before through E' D then passes under V' & as it goes on V' slides down the band W' on the other side of it, but V' does not close completely till D is on the other side of S' when S' opens & S' shut, & so on (it is to be remembered that (W) are merely bands of metal as in (fig. 2)) [See Figure 2.27, p. 80.]

This you observe is exceedingly simple the motion of the valves of S, S', E, E' can be communicated at once & the edges of the valves & vane may be as broad as you like. Now for the power.

I have power raised 33,000 lbs./foot in 1 minute. Suppose the diameter of inner circumf. of outer box = 6 inches, [the diameter of] outer [circumf. of] inner box = 2 inches, the mean radius of same would be 2 inches & its length 1 inches. [See Figure 2.28, p. 81.]

Let the height be one inch then 1 rotation of the valve = 1 square inch moving forward 2 inches since the mean radius of the

vane is 2 inches now 33,000 lbs./foot per minute = 550 lbs./feet per second = 550 revolutions under a pressure of 1 per second = 37 revolutions under pressure of 15 lbs or one atmosphere per second. Now it could easily accomplish 10 in a second ∴ with a pressure of 4 atmospheres an engine of size above described = 1 horsepower.
Fras Galton

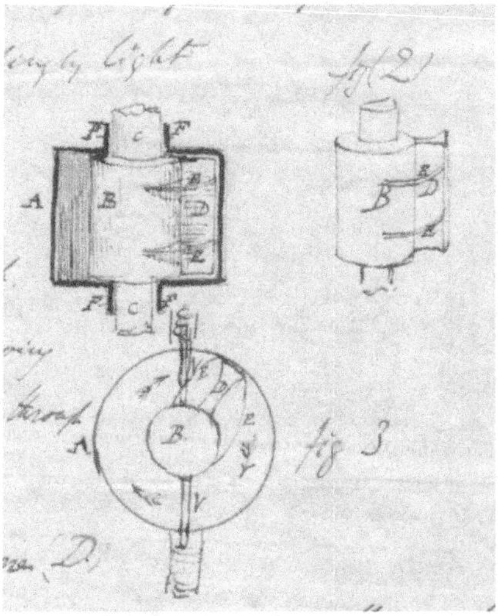

Figure 2.25: 1841-02-03 FG to STG: Steam Engine Figs. 1, 2, 3.

1841/03 circa. Francis Galton to Emma Sophia Galton

Dear Emma

I send 17 shillings worth of etchings &c. Some of them by C. [Schub] are exceedingly good. I was unable to get some outlines by Rembrandt which I was anxious to have done.

I am very sorry that my Father is so unwell, perhaps this attack will do for asthma & all.

Yesterday I made my appearance before the eyes of wondering Cantabs. Where do you think? Why right in the midst of a den containing 1 Lion 1 Lioness 1 huge Bengal Tiger & 4 Leopards in Wombwell's menagerie.[13] The Lion snarled awfully. I was a wee

[13] George Wombwell (1877-1850) ran a travelling menagerie which toured round England. Cambridge students were forbidden to attend it in later years.

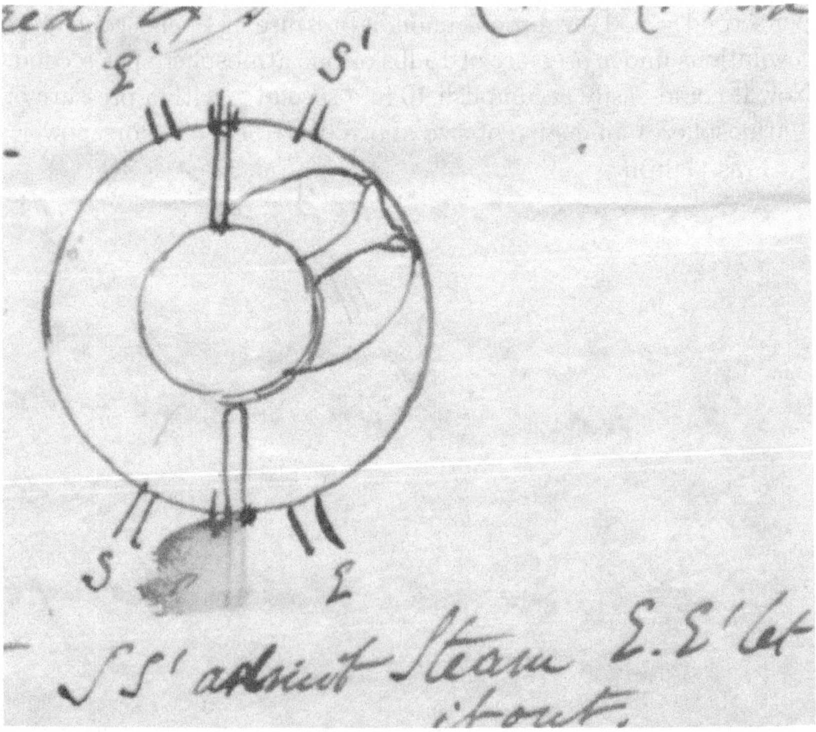

Figure 2.26: 1841-02-03 FG to STG: Steam Engine: S S' admit steam, E E' let it out.

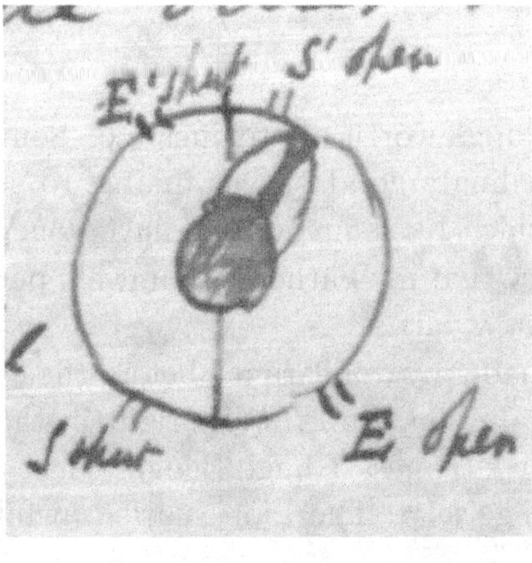

Figure 2.27: 1841-02-03 FG to STG: Valves Reverse.

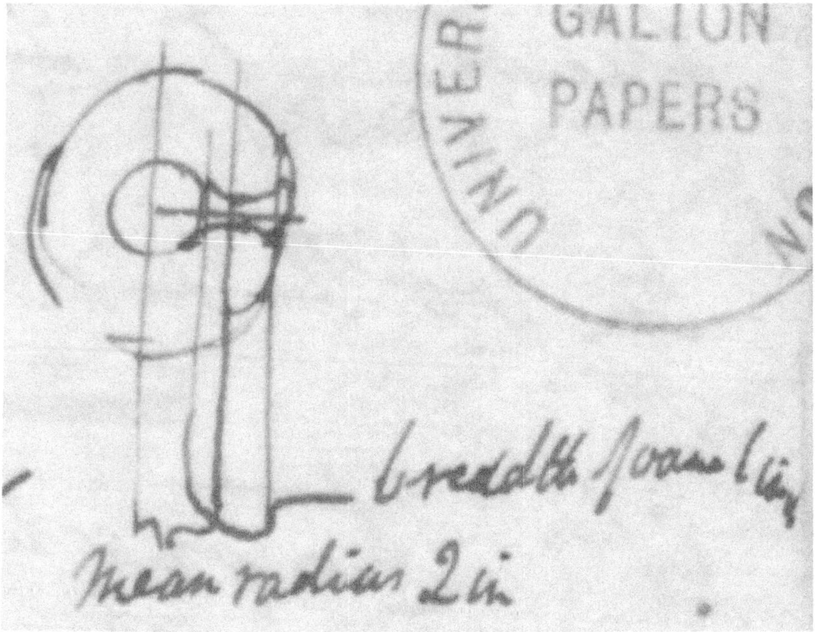

Figure 2.28: 1841-02-03 FG to STG: Steam Engine: Power.

frightened for the Brute crouched so. The keeper told me that I was only the 4th that had entered that den.

Nothing like making oneself a "Lion" at Cambridge. My Turkish tour & medical education does wonders & my late van Amburg performance promises to crown my reputation.[14]

FG

P.S. I send a view I had of a street in Smyrna thinking that it would make a very good picture a la Prout.

1841/03/21. Francis Galton to Samuel Tertius Galton

My dear Father

Thanks for your letter. Tomorrow I will see if Mortlock has received the £20. I will send my accounts. Thanks greatly about Aberystwith.

I am rather cut up by the sudden death of a College friend of mine. Poor fellow he wined with me last Tuesday, walked with me next day, complaining only of a slight headache. I heard that he

[14] Isaac van Amburgh (1811-1865), the animal trainer.

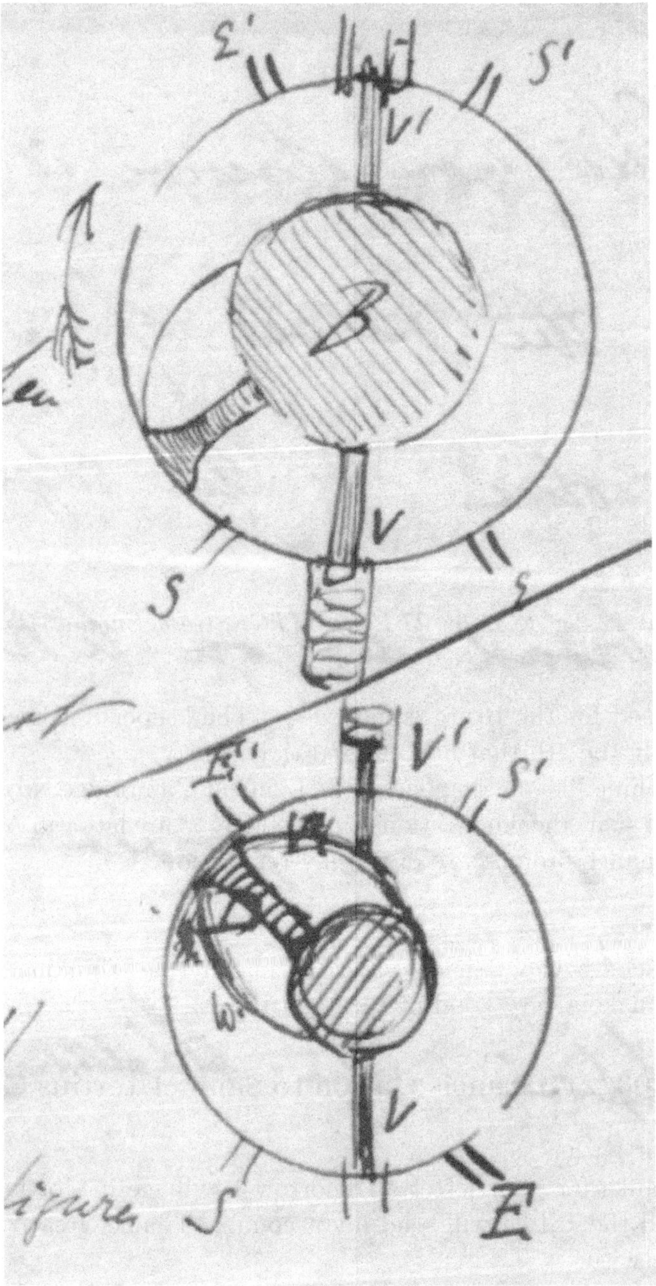

Figure 2.29: 1841-02-03 FG to STG: Steam Engine: S is open E' is open.

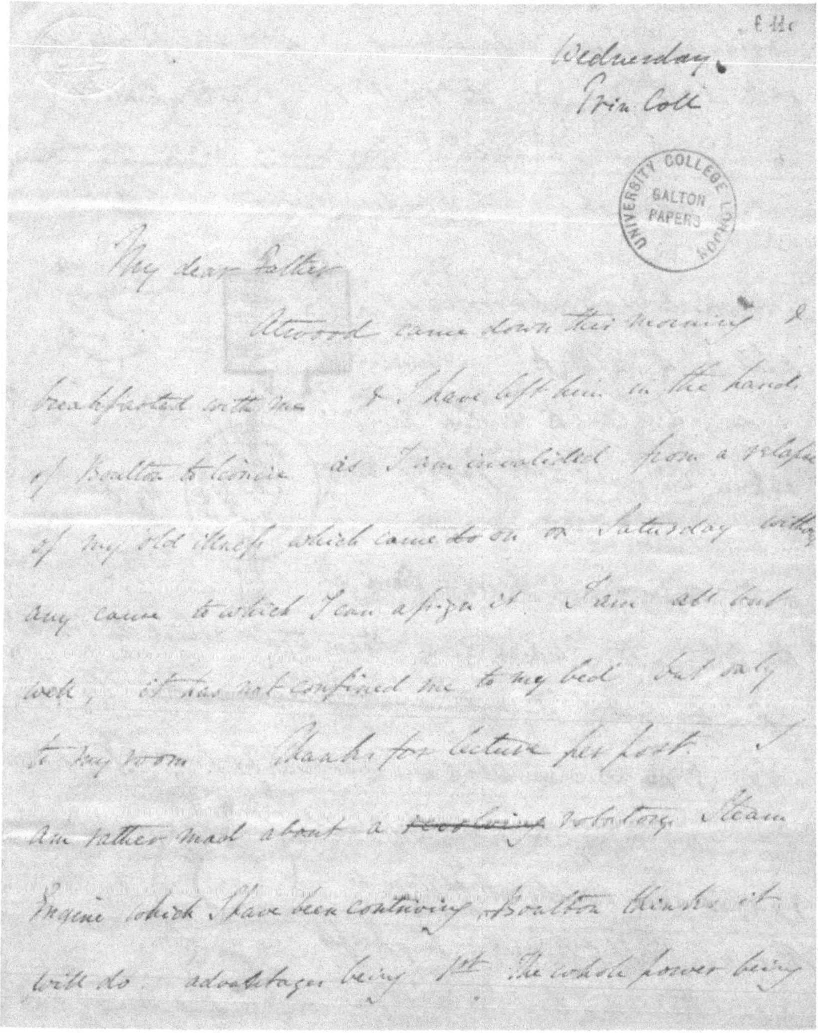

Wednesday.
Trin Coll

My dear Father

Atwood came down this morning & breakfasted with me & I have left him in the hands of Boulton to dinner as I am invalided from a relapse of my old illness which came on on Saturday without any cause to which I can assign it. I am all but well, it has not confined me to my bed but only to my room. Thanks for lecture per post. I am rather mad about a ~~revolving~~ rotatory Steam Engine which I have been contriving, Boulton thinks it will do. advantages being 1st The whole power being

Figure 2.30: 1841-02-03. FG to STG page 1.

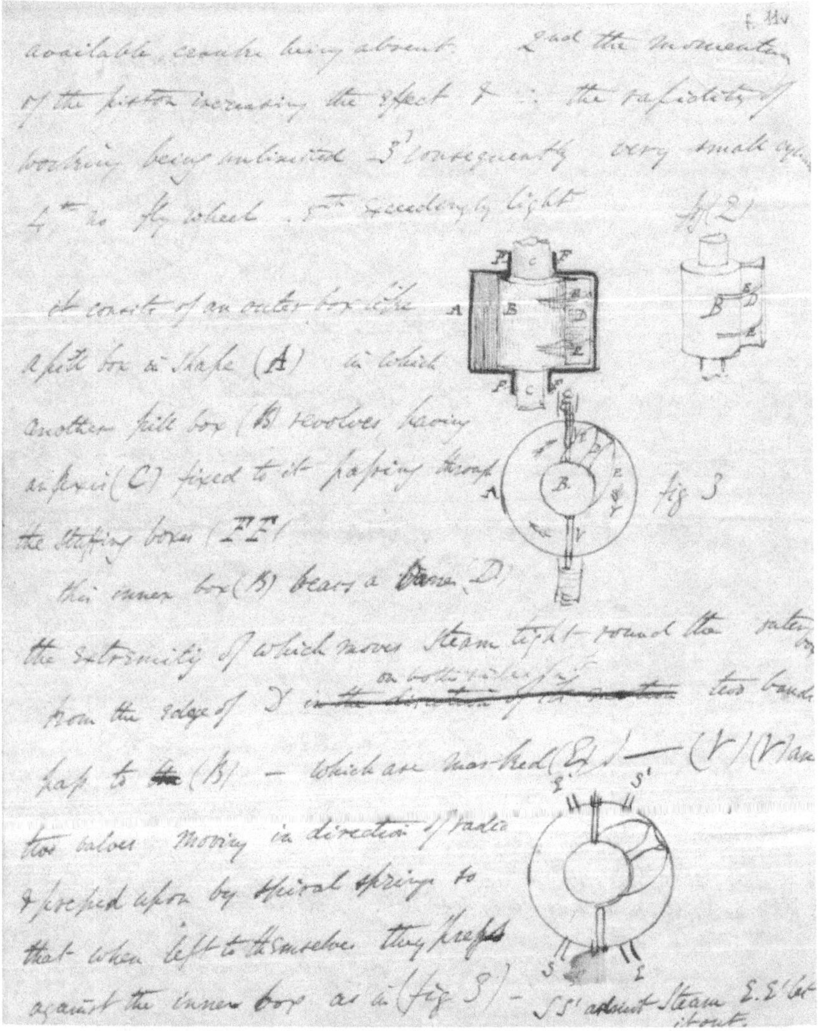
available, counter being absent. 2nd the momentum of the piston increasing the effect & ∴ the rapidity of working being unlimited 3rd consequently very small cyl... 4th no fly wheel. 5th exceedingly light

fig (2)

it consists of an outer box like a pill box in shape (A) in which another pill box (B) revolves having an axis (C) fixed to it passing through the stuffing boxes (FF)

this inner box (B) bears a ~~vane~~ (D)

the extremity of which moves steam tight round the interior of A from the edge of D on both sides of it ~~in the direction of its motion~~ two bands pass to ~~the~~ (B) — which are marked (E) — (V)(V) are two valves moving in direction of radii & pressed upon by spiral springs so that when left to themselves they press against the inner box as in (fig 3) — S S' admit steam E.E' let it out.

Figure 2.31: 1841-02-03. FG to STG page 2.

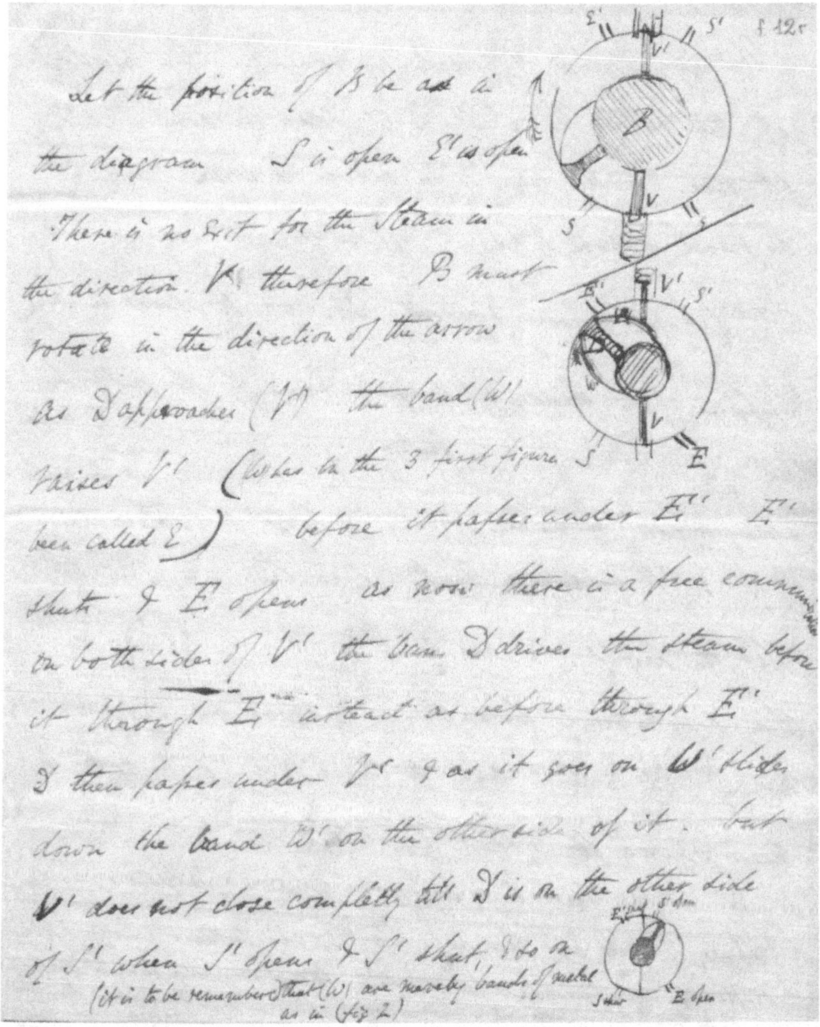

Let the position of B be as in
the diagram S is open E' is open
There is no exit for the steam in
the direction V therefore B must
rotate in the direction of the arrow
as D approaches (V) the band (W)
raises V' (What in the 3 first figure has
been called E) before it passes under E' E'
shuts & E opens as now there is a free communication
on both sides of V' the vane D drives the steam before
it through E instead as before through E'
D then passes under V' & as it goes on W' slides
down the band W' on the other side of it. but
V' does not close completely till D is on the other side
of S' when S' opens & S' shuts, & so on
(It is to be remembered that (W) are merely bands of metal
as in (fig 2)

Figure 2.32: 1841-02-03. FG to STG page 3.

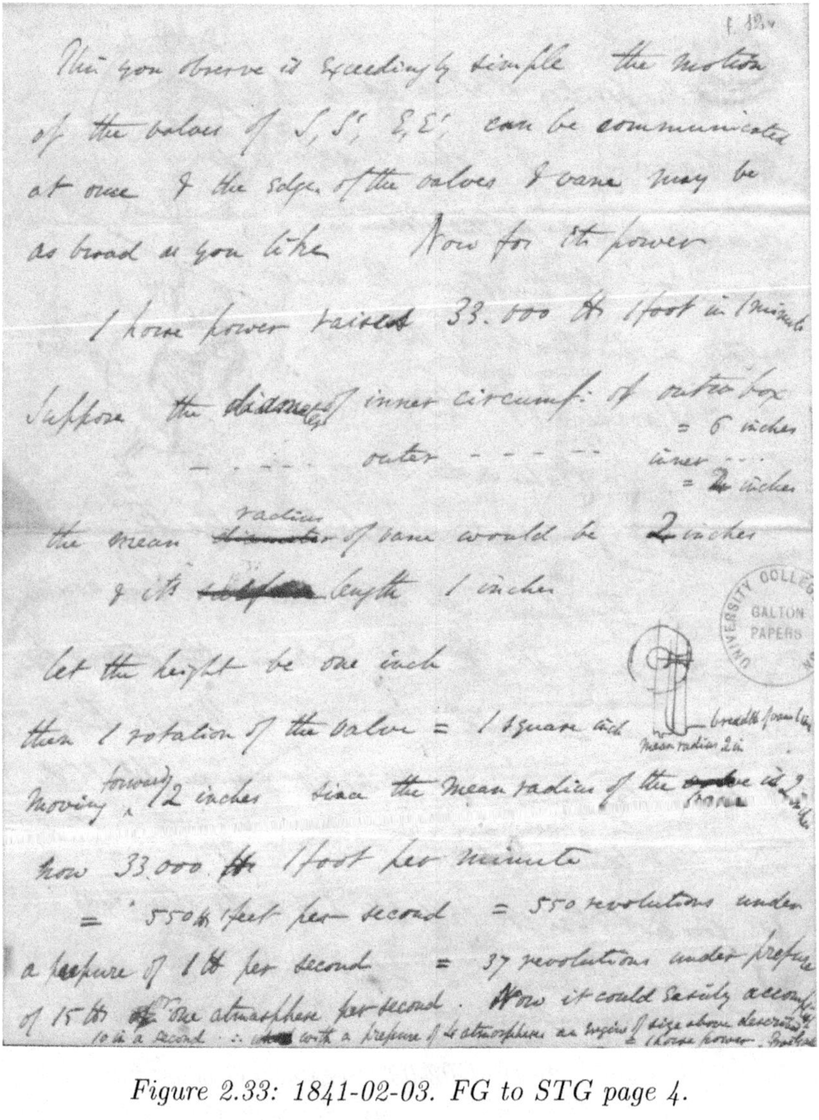

This you observe is exceedingly simple the motion of the valves of S, S', E, E' can be communicated at once & the edges of the valves & vane may be as broad as you like Now for the power

1 horse power raises 33.000 lb 1 foot in 1 minute

Suppose the ~~distance~~ radius of inner circumf: of outer box = 6 inches
—— outer —— inner —— = 2 inches

the mean ~~diameter~~ radius of vane would be 2 inches
& its ~~[illegible]~~ length 1 inches

let the height be one inch

then 1 rotation of the valve = 1 square inch

mean radius 2 in
breadth from 1 in

moving forward 12 inches since the mean radius of the ~~[illegible]~~ frame is 2 inches

now 33.000 lb 1 foot per minute
= 550 lb 1 feet per second = 550 revolutions under a pressure of 1 lb per second = 37 revolutions under pressure of 15 lbs or one atmosphere per second. Now it could easily accomplish 10 in a second ∴ with a pressure of 4 atmospheres an engine of size above described = 1 horse power. Brother

Figure 2.33: 1841-02-03. FG to STG page 4.

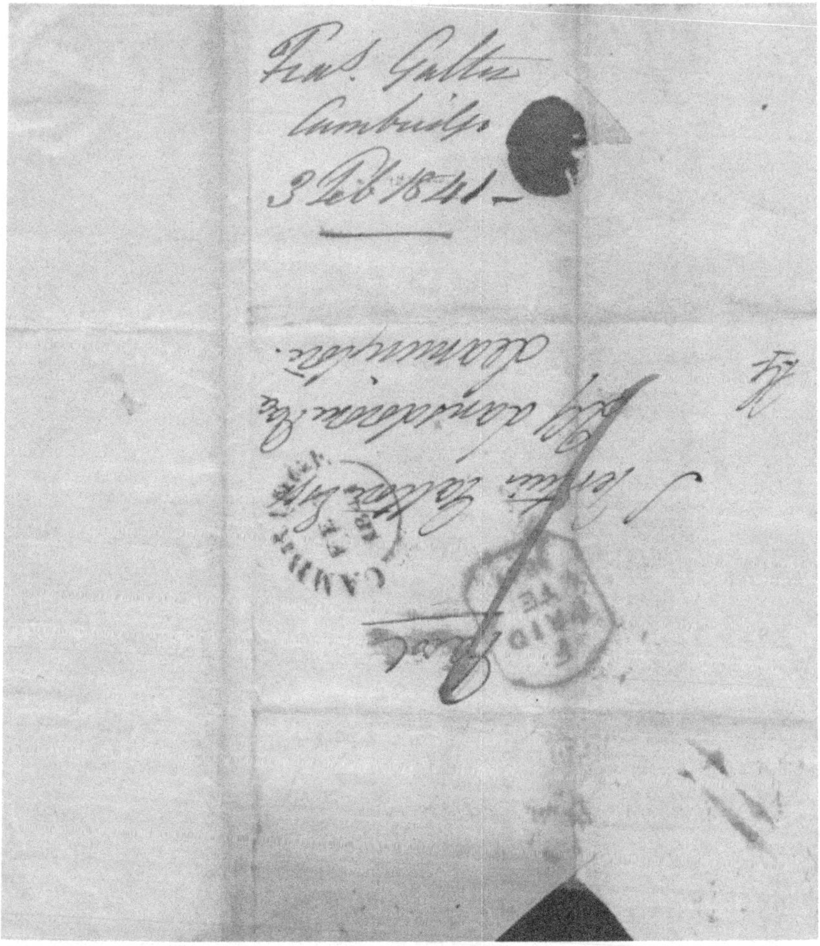

Figure 2.34: 1841-02-03. FG to STG page 5.

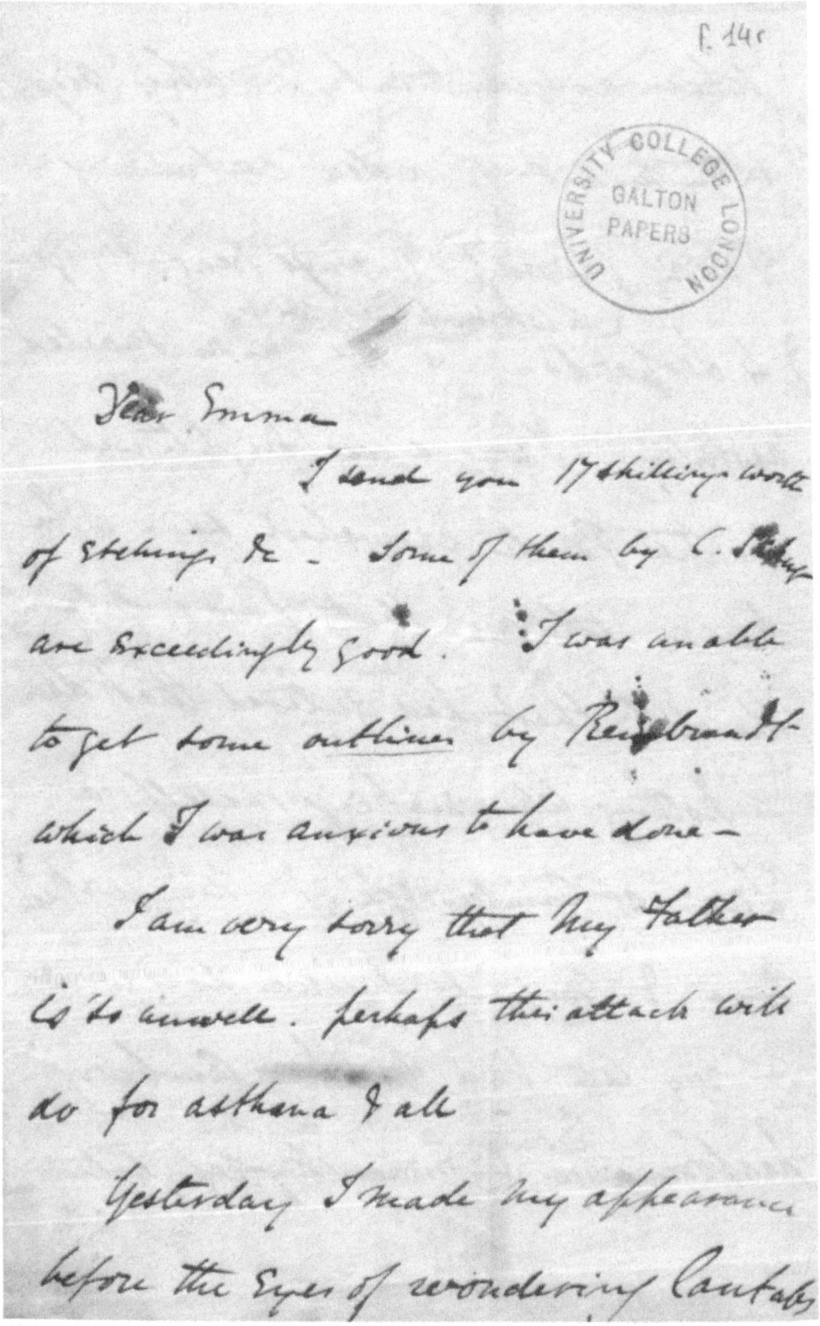
f. 14r

Dear Emma

I send you 17 shillings worth of etchings &c – some of them by C. [illegible] are exceedingly good. I was unable to get some outlines by Rembrandt which I was anxious to have done –

I am very sorry that my Father is so unwell. perhaps this attack will do for asthma & all

Yesterday I made my appearance before the Eyes of wondering Cantabs

Figure 2.35: 1841-03. FG to ESG page 1.

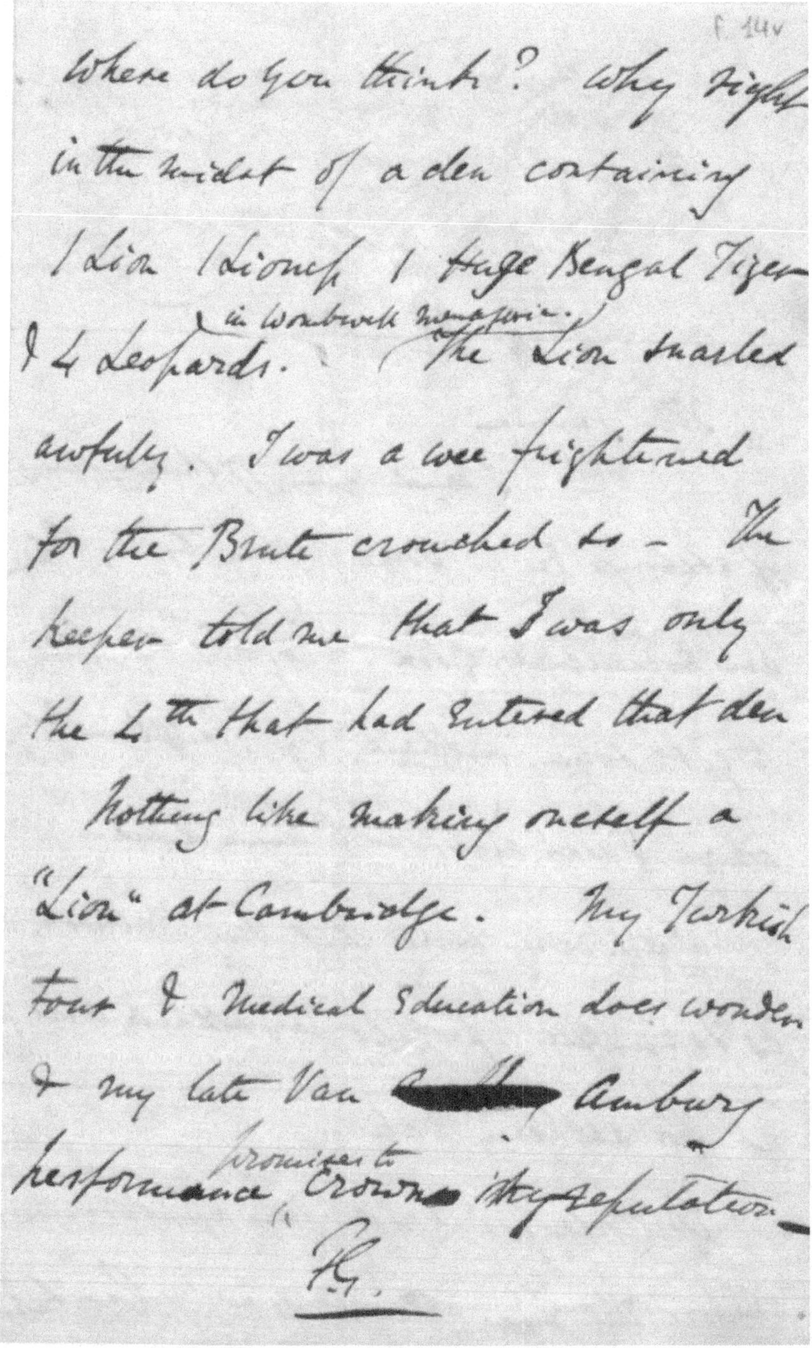
f. 14v

Where do you think? Why right
in the midst of a den containing
1 Lion 1 Lioness 1 Huge Bengal Tiger
& 4 Leopards. (in Wombwell menagerie) The Lion snarled
awfully. I was a wee frightened
for the Brute crouched so – The
Keeper told me that I was only
the 4th that had entered that den
Nothing like making oneself a
"Lion" at Cambridge. My Turkish
tour & Medical Education does wonders
& my late Van [redacted] Amburg
performance promises to crown its reputation –
FG.

Figure 2.36: 1841-03. FG to ESG page 2.

f. 15r

P.S. I send a view I had of a Street in Smyrna – thinking that it would make a very good picture à la Prout –

Figure 2.37: 1841-03. FG to ESG page 3.

Figure 2.38: 1841-03. FG to ESG page 4.

was ill yesterday Saturday morning, and going to see him after hall at 5 p.m. found him dying, with what I took to be typhus; called again at 9, he was much worse and evidently could not hold out 6 hours. An eruption of scarlet fever had broken out. He died at 12. He was one of the most kind-hearted fellows I ever met with; he had just taken his degree and was going into orders and had begun hospital attendance that he might be of use in his parish. He thus caught the scarlet fever and is dead. He was a very great friend of Hughes.

It is curious that I have as yet lost only 7 schoolfellows or fellow-collegiates that I really care for and every one of these have died of scarlet fever or typhus, and all except one within three days warning. How fearfully death intrudes in the midst of enjoyment like the skeleton at the Egyptian's feast. It is remarked by D'Israeli (I think) that the shock from the sudden loss of a friend is the only feeling which the mind cannot become callous to. The frequent sight of death seems in no way to diminish it. Patients in a hospital one looks upon as doomed men and their death takes place as in the natural order of things. A friend appears part of oneself, and when he dies, one contemplates the grave where he is laid in as ready to receive oneself; we then know that we are mortal. However this kind of language is out of place to you at home in the midst of marriage festivities, etc. I received no cake with your letter!

Good bye, Yours truly,
Fras. Galton.

1841/03/31–04/05. Diary of Samuel Tertius Galton

March 31st. Wednesday. Claverdon.
.. Francis arrived in the evening.

April 5th. Sunday. Claverdon.
I took Francis in the Sociable to meet the Birmingham Coach at Hatton at a quarter before nine.

1841/04/08. Francis Galton to Samuel Tertius Galton

My dear Father

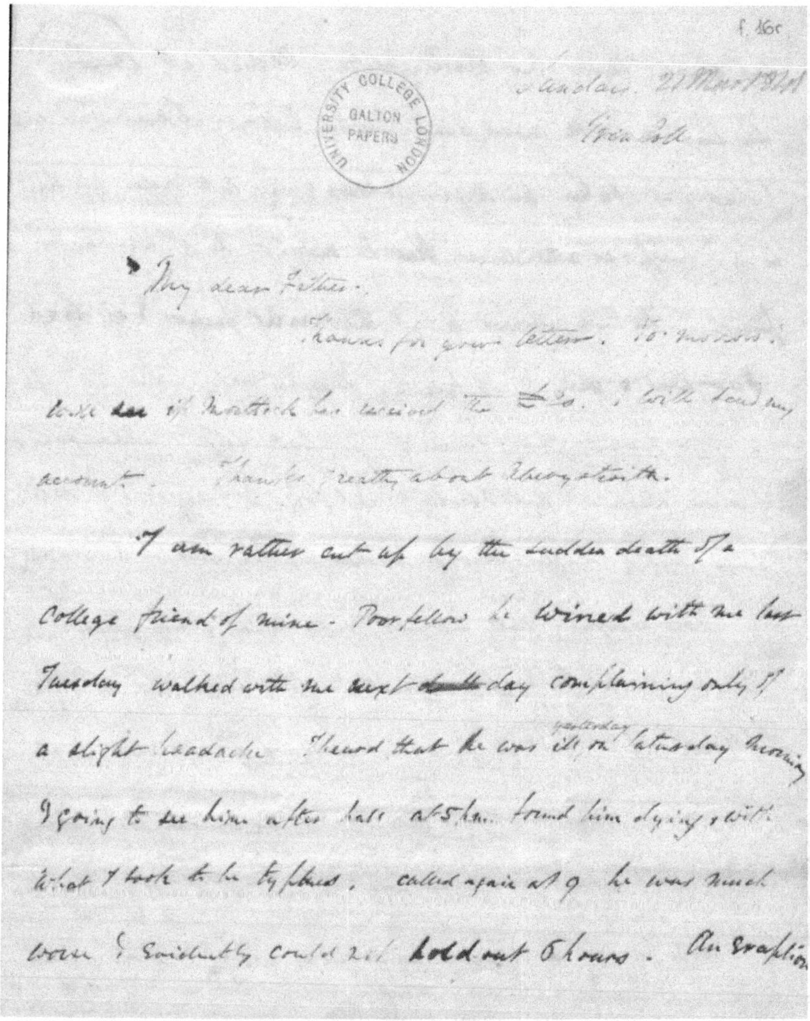
f. 16r

Sunday. 21 Mar 1841
Trin Coll

My dear Father.

Thanks for your letter. To-morrow I will see if Mortlock has received the £20. I will send my account. Thanks greatly about Aberystwith.

I am rather cut up by the sudden death of a College friend of mine. Poor fellow he wined with me last Tuesday walked with me next ~~…~~day complaining only of a slight headache. I heard that he was ill on yesterday Saturday morning & going to see him after hall at 5 p.m. found him dying & with what I took to be typhus. called again at 9 he was much worse & evidently could not hold out 6 hours. An eruption

Figure 2.39: 1841-03-21. FG to STG page 1.

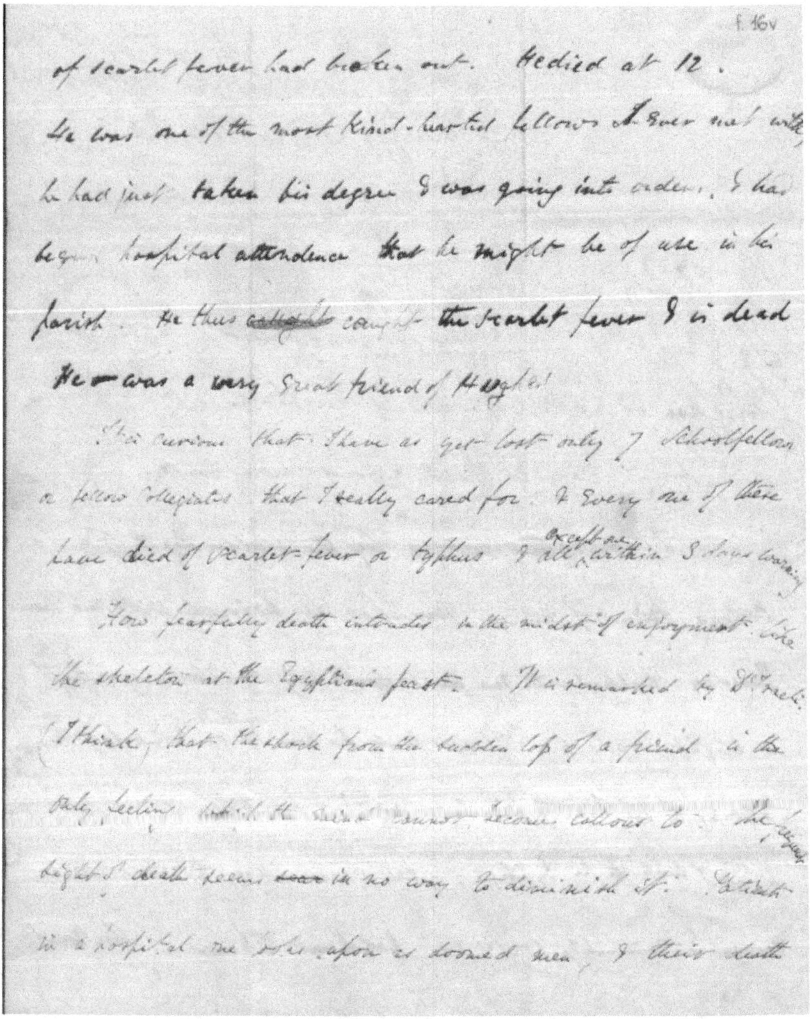

f. 16v

of scarlet fever had broken out. He died at 12.
He was one of the most kind-hearted fellows I ever met with,
he had just taken his degree & was going into orders, & had
begun hospital attendance that he might be of use in his
parish. He thus ~~caught~~ caught the scarlet fever & is dead
He was a very great friend of Hughes!
It is curious that I have as yet lost only 7 schoolfellows
& fellow collegiates that I really cared for. & every one of these
have died of scarlet fever or typhus & all except one within 3 days warning
How fearfully death intrudes in the midst of enjoyment like
the skeleton at the Egyptian's feast. It is remarked by Dr [illegible]
(I think) that the shock from the sudden loss of a friend is the
only feeling which the mind cannot become callous to. The frequent
sight of death seems ~~[illegible]~~ in no way to diminish it. Patients
in a hospital one looks upon as doomed men, & their death

Figure 2.40: 1841-03-21. FG to STG page 2.

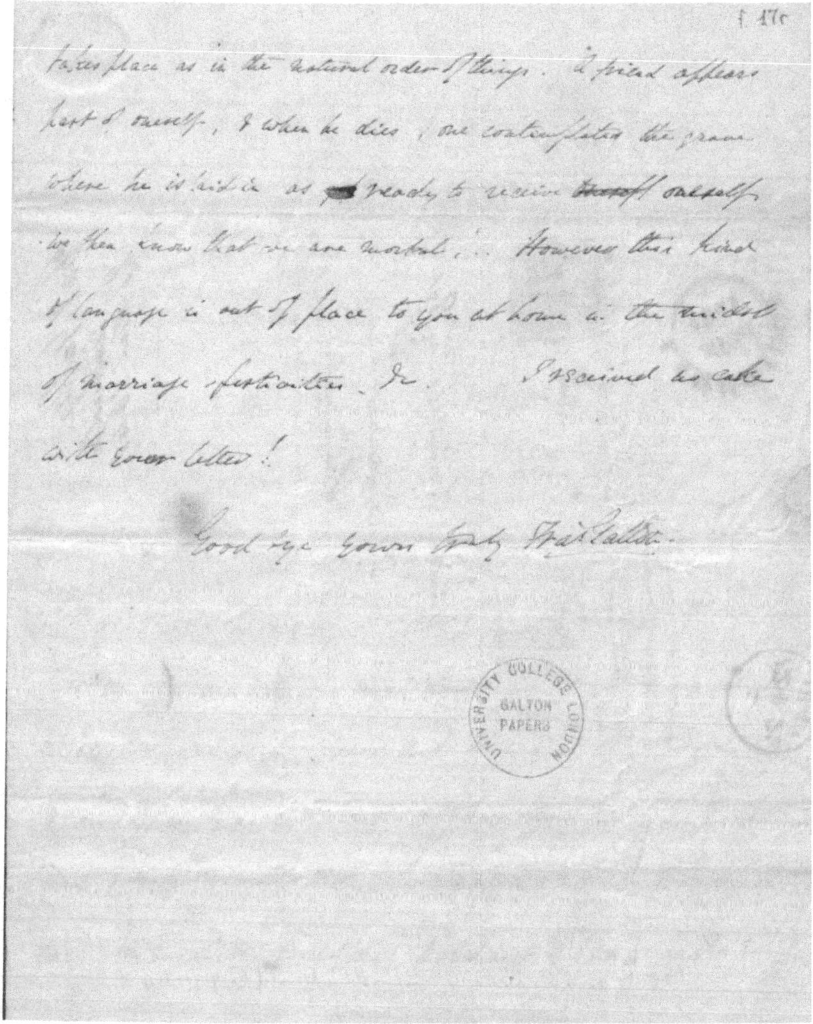
f. 47v

takes place as in the natural order of things. A friend appears part of oneself, & when he dies, one contemplates the grave where he is laid in as ready to receive oneself, we then know that we are mortal... However this kind of language is out of place to you at home in the midst of marriage festivities &c. I received no cake with your letter!

Good bye yours truly Frank Galton

Figure 2.41: 1841-03-21. FG to STG page 3.

Figure 2.42: 1841-03-21. FG to STG page 4.

I saw Hodgson[15] in Birmingham who approved altogether of my plans. Evans was not at home. Dr. Booth on my mentioning it to him lifted up both of his hands ejaculated Hawgh! & said is that at the instigation of Hodgson—hearing that Hodgson approved he launched out diametrically opposite & got so sulky that I beat a speedy retreat.

I have on reconsidering reconsulting &c. &c. determined to ~~not~~ abide the Trinity Examination, as I should be sure not to get the Caius prize as they have two very superior men there & as they take up slightly different subjects—but to do my best for a Trinity first class & to migrate afterwards.

On arriving at Cambridge I found Uncle Hubert's card in my rooms & himself with my Aunt Ma & Ju. at the Bull.[16] They set off early the next morning to Norfolk.

I did not send the wire because I could get none but will send some from Cambridge on the 1st opportunity. It is called Metallic wire.

Hodgson does not think Bessy well, he talked much about her, but thinks it desirable to wait till her return from London before making any alteration whatever in her plan of living &c. hoping that change will do her good.

Good bye & believe me ever yours truly
Fras. Galton.

Trin Coll April 1841

1841/04. Francis Galton to Samuel Tertius Galton

Trin Coll.
April 1841

My dear Father

No letter of yours, whether received at school announcing that I might come home a week before my time, or whether containing a cheque on Barclay, Bevan, Tritton & Co. ever came more gratefully than the one I have just received to say that my Mathematical MSS were at Claverdon. I have been latterly in despair

15 Hodgson, Joseph. See page 483.

16 His father's brother Hubert John Barclay Galton (1789-1864), & his wife Mary (1792-1859). "Ju." may refer to their daughter Julia Theresa Galton (1826-1851).

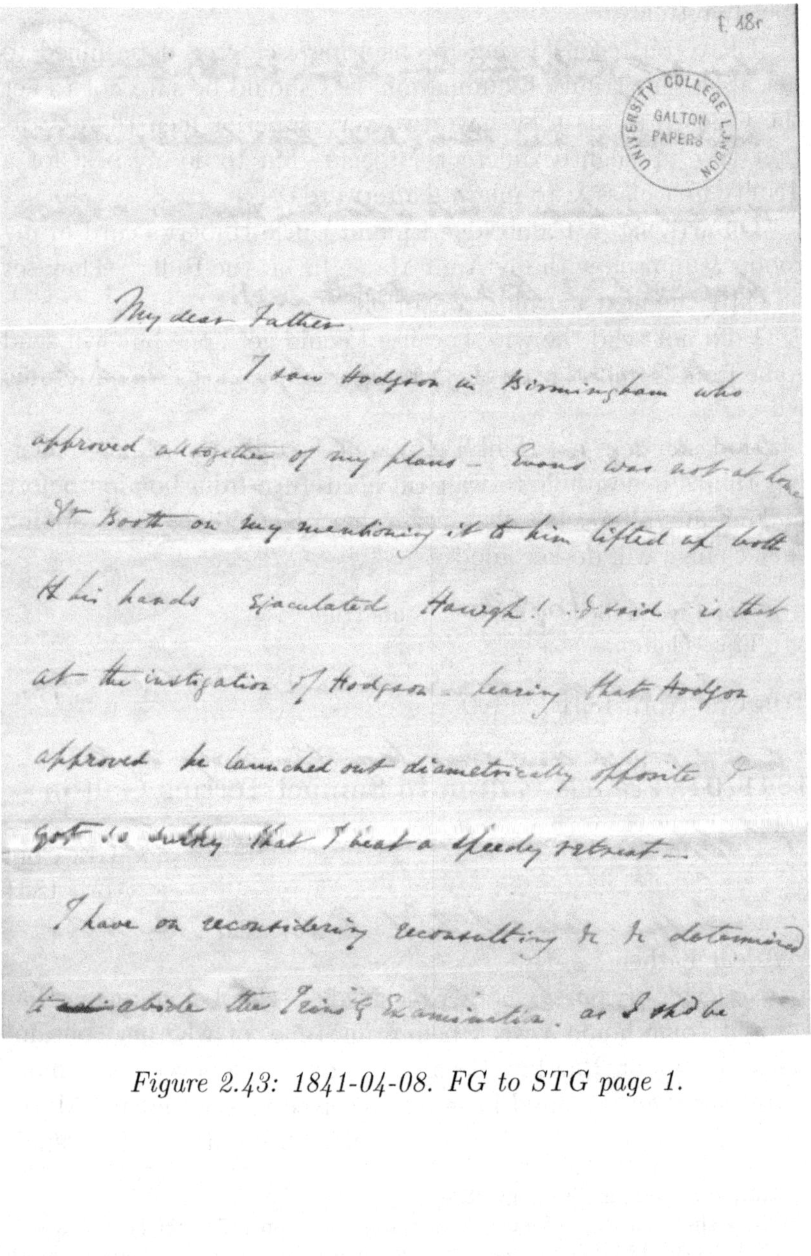

f. 18r

My dear Father

I saw Hodgson in Birmingham who approved altogether of my plans – Evans was not at home. Dr Booth on my mentioning it to him lifted up both of his hands & ejaculated Haugh! I said it was at the instigation of Hodgson – hearing that Hodgson approved he launched out diametrically opposite & got so sulky that I beat a speedy retreat –

I have on reconsidering & reconsulting &c &c determined to abide the Prize Examination, as I shd be

Figure 2.43: 1841-04-08. FG to STG page 1.

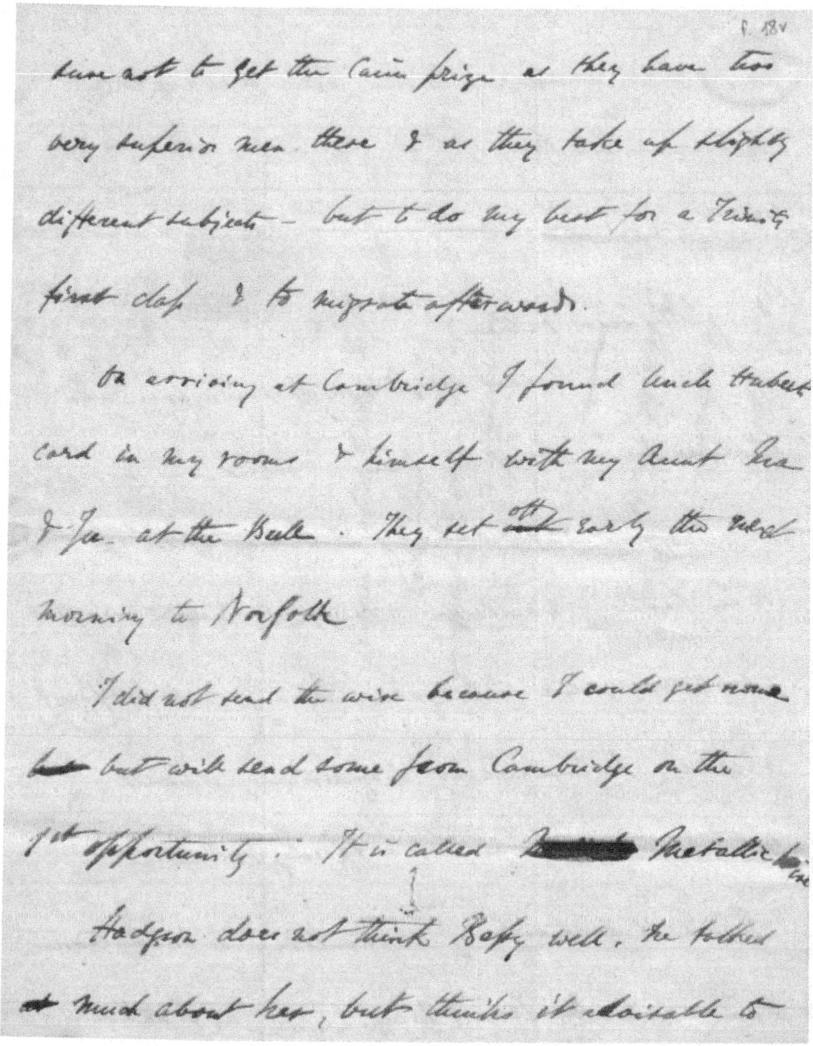

f. 18v

sure not to get the Caius prize as they have two very superior men there & as they take up slightly different subjects – but to do my best for a Trinity first class & to migrate afterwards.

On arriving at Cambridge I found Uncle Herbert's card in my rooms & himself with my Aunt Isa & Isa at the Bull. They set ~~out~~ off early the next morning to Norfolk

I did not send the wine because I could get none ~~here~~ but will send some from Cambridge on the 1st opportunity. It is called ~~[illegible]~~ Metallic wine

Hodgson does not think Baby well. he talked ~~at~~ much about her, but thinks it advisable to

Figure 2.44: 1841-04-08. FG to STG page 2.

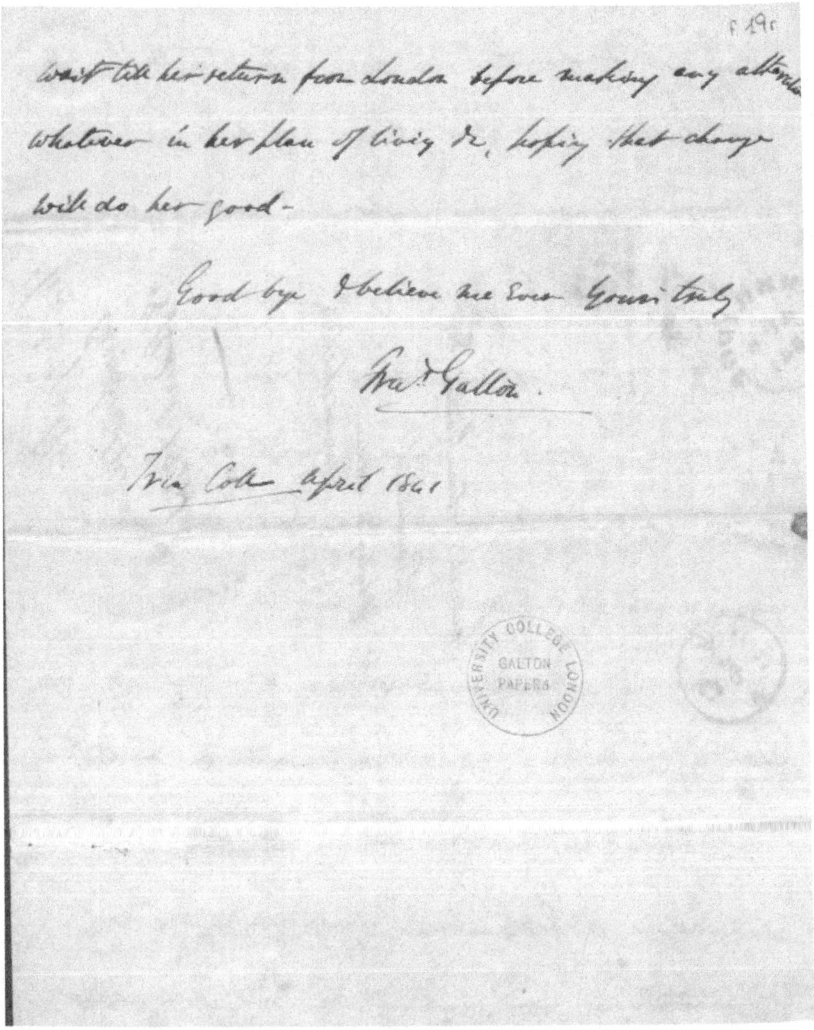

f. 19r

wait till her return from London before making any alteration whatever in her plan of living &c, hoping that change will do her good –

Good bye & believe me Ever Yours truly

Fran^s Galton.

Trin Coll April 1841

Figure 2.45: 1841-04-08. FG to STG page 3.

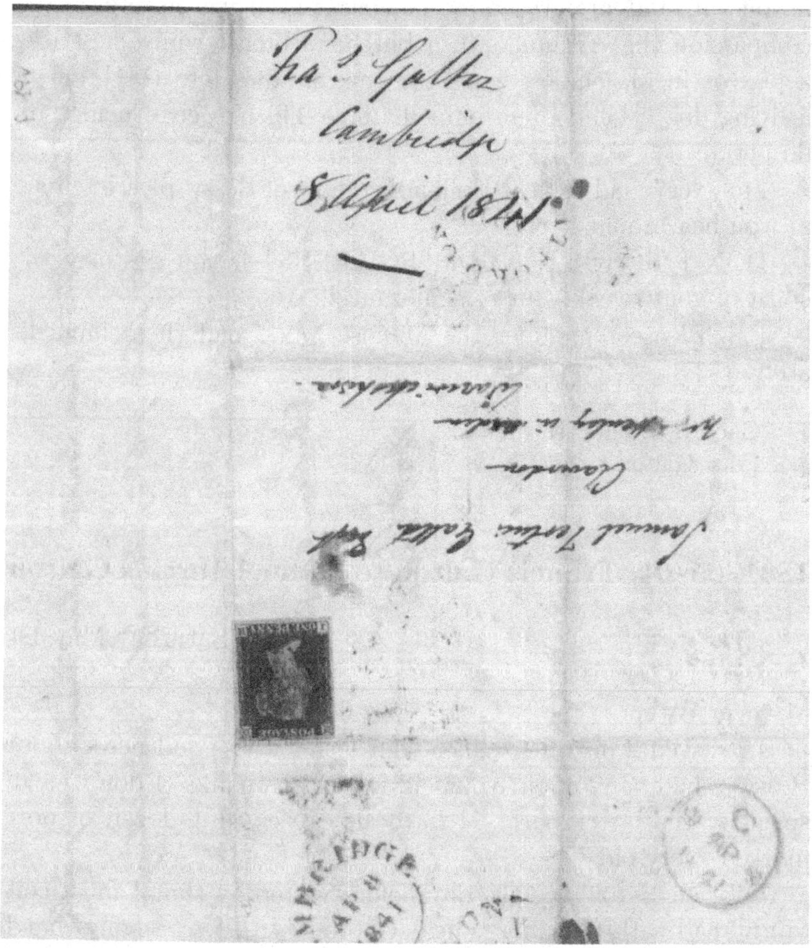

Figure 2.46: 1841-04-08. FG to STG page 4.

owing to their loss. I had hunted for them before leaving Claverdon, but found them not. I looked every where in Cambridge & was [...] disappointed.

They are invaluable, all the talent of Perry,[17] O'Brien[18] & Mathison[19] are condensed into those papers. Therefore please take the greatest care of them, burn the Duddeston title-deeds if you will, but preserve these manuscripts, if you have even the compassion that glimmers in a butcher's breast whilst he sticks a pig, or in Majendie's whilst he runs needles into the brain of a living dog,[20] send them immediately. Till I receive them I am desperate.

I am very glad to hear Holland's report of Bessy, please tell me all you hear about her.

DON'T FORGET THE MSS PAPERS—if you do, may the spirit of gout tweak your remembrance!!!

P.S. Please remember the Manuscripts—send them immediately.

Good bye yours truly
Fras. Galton.

1841/05/01. Francis Galton to Samuel Tertius Galton

Saturday May 1st

My dear Father

I received 2 or 3 days since your letter with good news about Bessy & bad news about yourself, for which thanks (I don't mean specially the latter part). I have also received 1 dozen of port marked "very old."

O'Brien[21] told me the day before yesterday that I must certainly read with Hopkins[22] next October, & on my saying that I would rather remain with him he strongly recommended me not. I own this has made me very bumptious, it does great credit to O'Brien for his openness, as of course tutors prefer to keep the

[17] Perry, Charles. See page 490.
[18] O'Brien, Matthew. See page 487.
[19] Mathison, William Collings. See page 486.
[20] François Magendie (1783-1855), the French physiologist.
[21] O'Brien, Matthew. See page 487.
[22] Hopkins, William. See page 484.

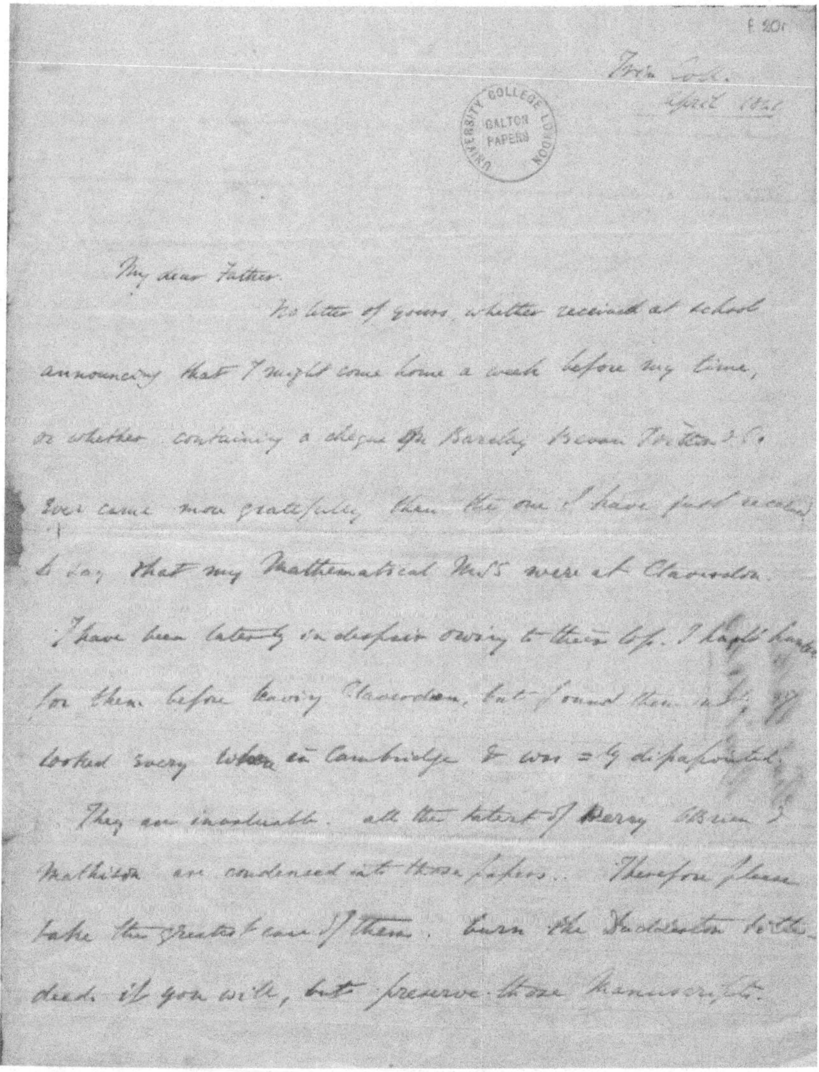

f 20r

Trin Coll.
April 1841

My dear Father.

No letter of yours, whether received at school announcing that I might come home a week before my time, or whether containing a cheque on Barclay Bevan Tritton & Co ever came more gratefully than the one I have just received to say that my Mathematical MSS were at Claverdon. I have been lately in despair owing to their loss. I hunted for them before leaving Claverdon, but found them not, & looked every where in Cambridge & was =ly disappointed. They are invaluable. all the talent of Berry, O'Brien & Mathison are condensed into those papers. Therefore please take the greatest care of them. burn the Declaration to the dead if you will, but preserve those manuscripts.

Figure 2.47: 1841-04. FG to STG page 1.

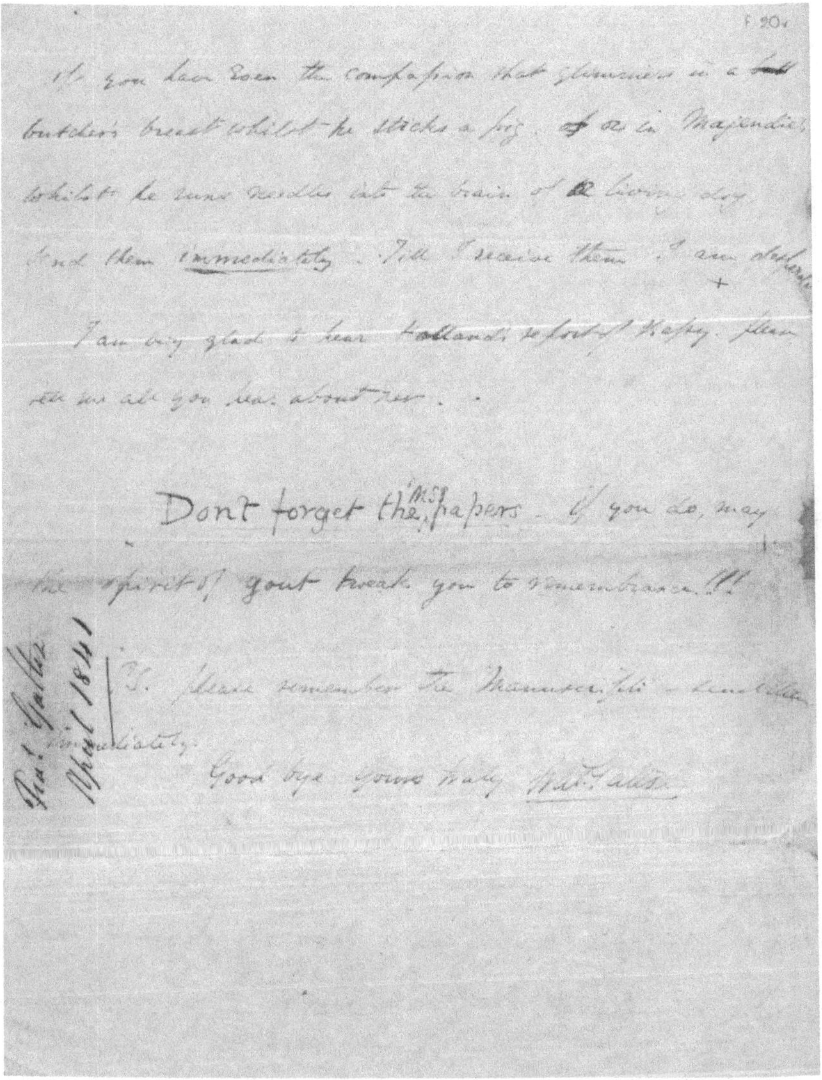

If you have seen the compassion that glimmers in a butcher's breast whilst he sticks a pig, or in Majendie's whilst he runs needles into the brain of a living dog, send them immediately. Till I receive them, I am desperate +

I am very glad to hear Holland's report of Happy. Please tell me all you hear about her. -

Don't forget the MSS papers - If you do, may the spirit of gout tweak you to remembrance!!!

PS. Please remember the Manuscripts - send them immediately.

Good bye yours truly Frs. Galton

Recd. Galton April 1841

Figure 2.48: 1841-04. FG to STG page 2.

better men. As he stays in Cambridge during the long Vacation (poor man, he is married), which is very dull & hot during Summer, I go with Mathison[23] our Mathematical Lecturer to Keswick in Cumberland—with a party to read. The terms are £30 for about 3 months & the life we lead a very pleasant & inexpensive one, certainly much cheaper than in College. By the bye we are turned out of our rooms during the Long Vacation.

I have been obliged to take a half classical coach for the approaching College examination (in about $4\frac{1}{2}$ weeks).

Now you must not expect me to be first Mathematic in Trinity. I do not expect it myself, as amongst other very good men, there are some who have already read exceedingly high, and who know their early subjects very well. I hope to do better in each succeeding Examination, but ill health, for I severely overstrained myself my first term, & I feel convinced that to have read during the X^{tmas} Vacation w^{d} have been madness has necessarily kept me back. But with no more excuses, as there is much in what Dr. Jeune used to say.

Good bye Your affectionate Son
Fras. Galton.

1841/05. Francis Galton to Elizabeth Anne Galton

Dear Bessy[24]

Many thanks for your letter received this morning. It wasn't bad—I am rather headachy now, will write soon.

Please ask Pater if he can~~not~~ find in my account at the end of the October term money paid for Medicine about £3. I am nearly sure that I paid it but have not the receipt. And the man has sent in his bill.

Your very affectionate & most highly grateful Brother
Fras Galton.

---

23 Mathison, William Collings. See page 486.

24 Annotated by EAG: Answered 26 May 1841.

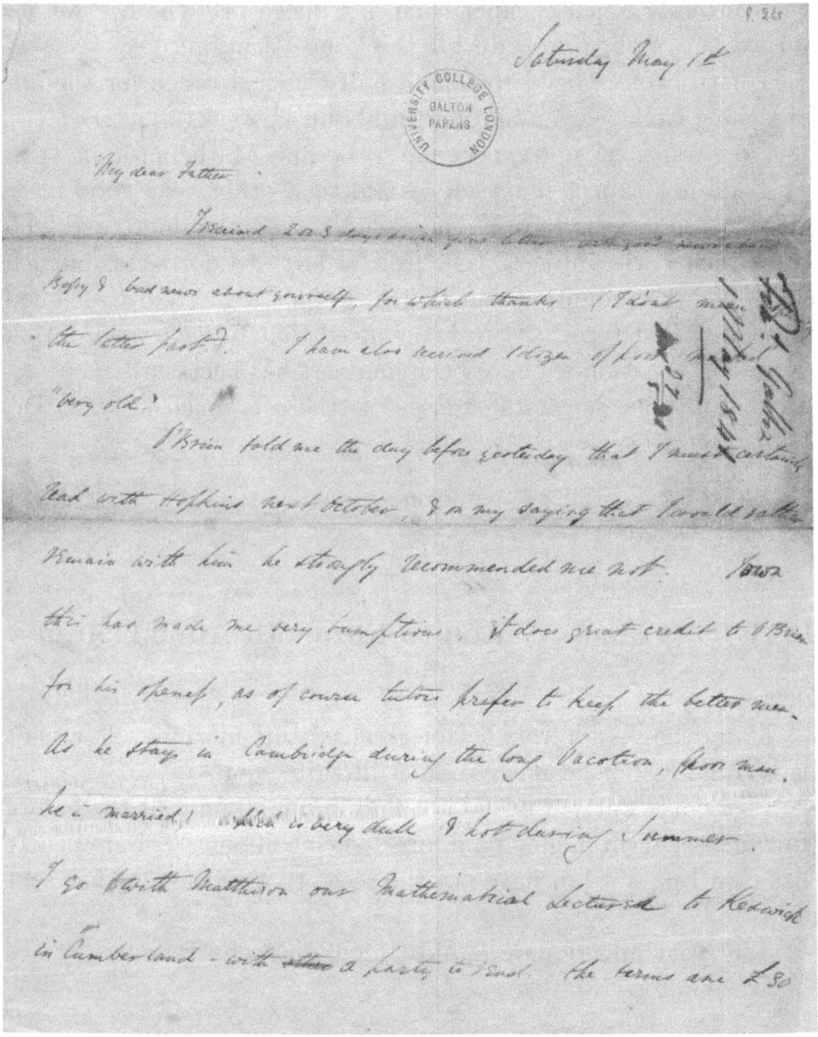

Saturday May 1st

My dear Father.

I received 2 or 3 days since your letter with your [illegible] happy & bad news about yourself, for which thanks (I don't mean the letter part). I have also received 1 dozen of port marked "very old."

O'Brien told me the day before yesterday that I must certainly read with Hopkins next October, & on my saying that I would rather remain with him he strongly recommended me not. Entre nous this has made me very bumptious. It does great credit to O'Brien for his openness, as of course tutors prefer to keep the better men. As he stays in Cambridge during the long Vacation, (poor man, he is married!) which is very dull & hot during Summer. I go with Mattheson our Mathematical Lecturer to Keswick in Cumberland - with a party to read. The terms are £30

Figure 2.49: 1841-05-01. FG to STG page 1.

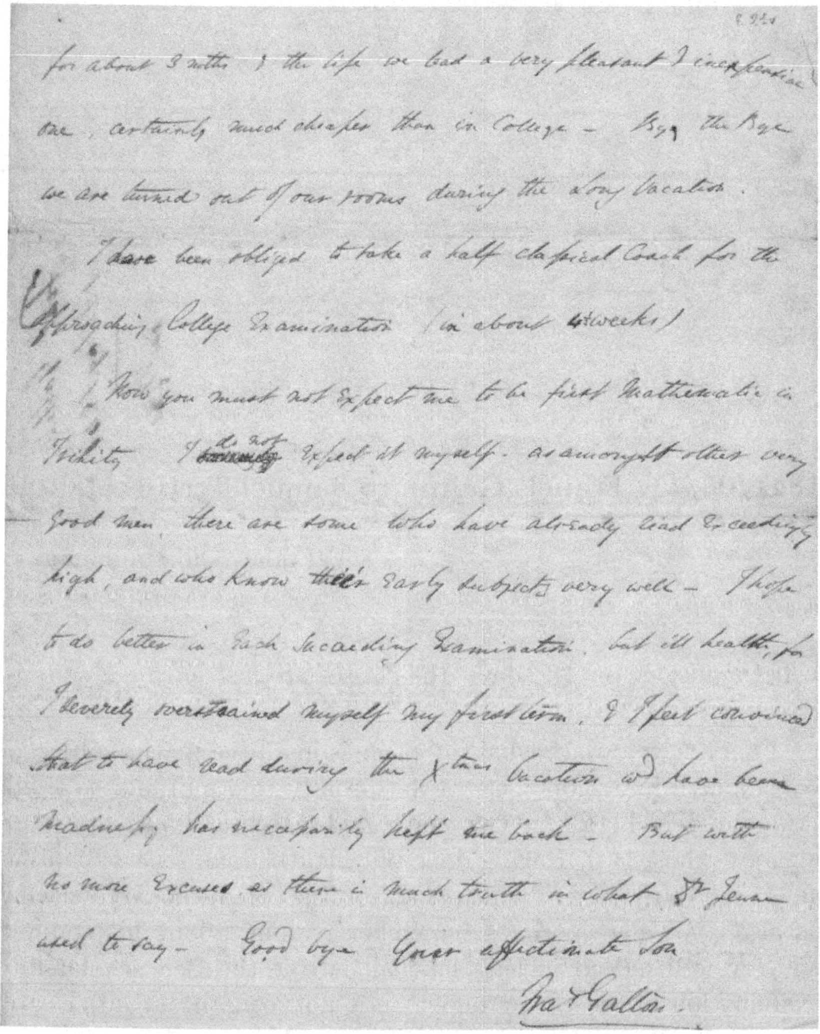
for about 3 mths of the life we had a very pleasant & inexpensive one, certainly much cheaper than in College - By the Bye we are turned out of our rooms during the Long Vacation.

I have been obliged to take a half Classical Coach for the approaching College Examination (in about 4 weeks)

Now you must not expect me to be first Mathematic in Trinity I ~~certainly~~ do not expect it myself - as amongst other very good men, there are some who have already read exceedingly high, and who know their early subjects very well - I hope to do better in each succeeding Examination. but ill health, for I severely overstrained myself my first term, & I feel convinced that to have read during the Xmas Vacation wd have been madness, has necessarily kept me back - But with no more Excuses as there is much truth in what Dr Jenan used to say - Good bye Your affectionate Son

Fras Galton.

Figure 2.50: 1841-05-01. FG to STG page 2.

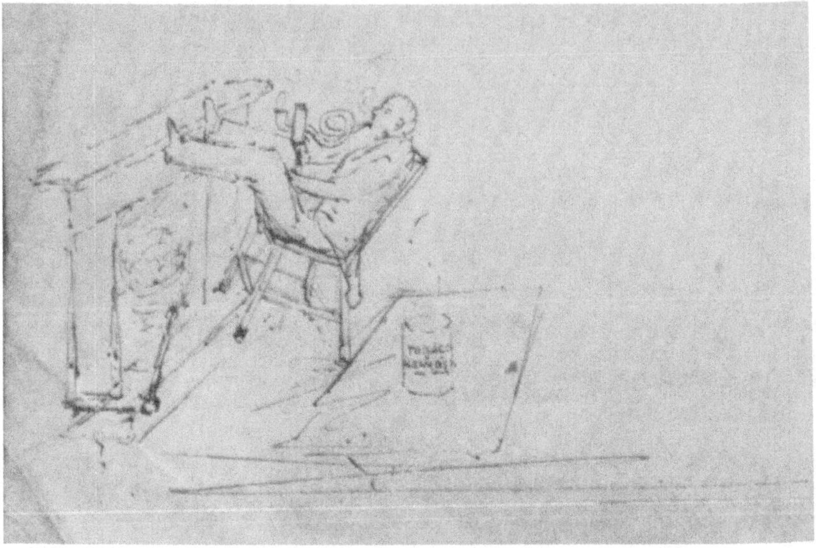

Figure 2.51: 1841-05 FG to EG: Pipe and beer.

1841/06/11. Francis Galton to Samuel Tertius Galton

Friday, Old Hummums[25]
Covent Garden

Left Cambridge on Tuesday—the classes are just out.
My dear Father

I am not yet aware what my place is in all the math: subjects. I was fourth in Trigonometry (Mathison[26] told me) and as I did comparatively better in Geometry and Algebra, I probably am higher in those two. Having done but little classics & that badly I am in the third class in which Hodgson the son of the Archdeacon [is also] (you may perhaps remember reading about his coming off very well indeed in the classical part of the Bell scholarship examination in the Birm: paper he is not a math. at all) I care scarcely at all about being where I am as I am as high in maths. as I expected. You must not forget that, as regards Classics more especially I have to compete with men who have spent that time on them which I have employed in medicine and it is therefore improbable that I should take a good place amongst them.

[25] A well-known hotel.
[26] Mathison, William Collings. See page 486.

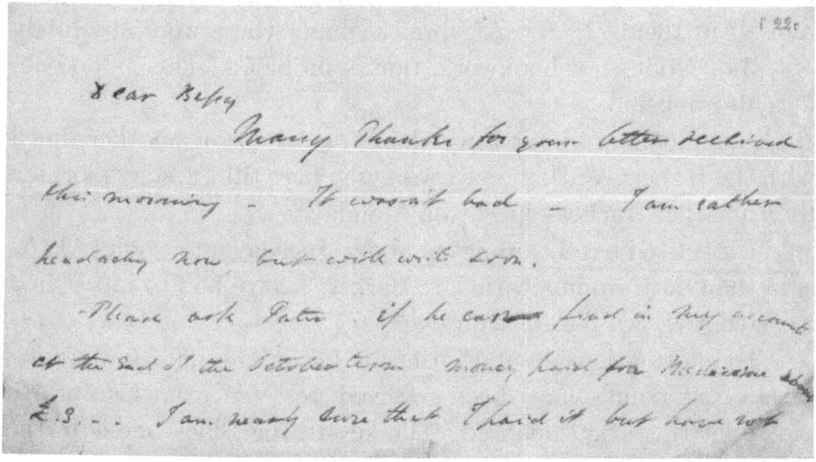

f 22r

Dear Bessy

Many thanks for your letter received this morning – It wasn't bad – I am rather headachy now, but will write soon.

Please ask Gates if he cannot find in my account at the end of the October term money paid for Medicine about £3.. I am nearly sure that I paid it but have not

Figure 2.52: 1841-05. FG to EAG page 1.

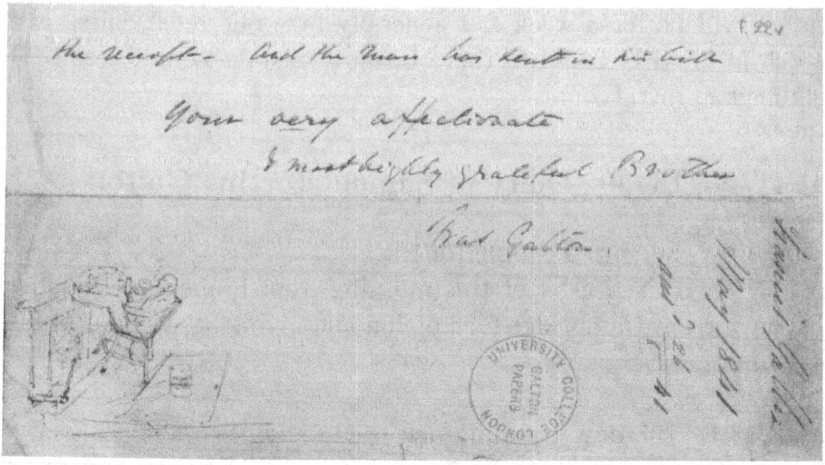

f 22v

the receipt – and the man has sent in his bill

Your very affectionate
& most highly grateful Brother

Frank Galton.

Francis Galton
May 1841
ans 2 26 / 41

Figure 2.53: 1841-05. FG to EAG page 2.

The Math. papers were excessively easy this year so that everybody who knows anything about them must of necessity do three fourths hence there was little room for a man to distinguish himself in them. In the Algebraical paper there were absolutely only 3 questions not bookwork, that is problems. This is too bad, it is also unusual.

I am mooning about town doing one thing or another dined with the Huberts & Homers. I ~~wait~~ stay here till Tuesday morning to hear Mlle. Rachel's [perf.] on Monday.[27]

I expect to be in Leamington about Tuesday afternoon. Alston is in the Old Hummus with Mr. Barker. I have had to invest in a frock coat & two pair of trousers

My information about Einishes is becoming nightly more extensive. If trunks have been received per Van from Cambridge please let them be unpacked. The key to the Black one is in the Deal one.

Your affectionate Son
Fras Galton

Hence as you observe I have not paid my Classical Tutor £7 who had left Cambridge without an address. I have not paid for my Frock Coat which will be about £5.

I should ∴ be obliged for £5-£10 as my bill at the Old Hummums will be for a week & I ~~generally~~ take one meal daily. My stock in hand is £6. 19. there being a mistake somewhere of 3 shillings in my account.

1841/06/13–29. Diary of Samuel Tertius Galton

June 13th. Sunday. Leamington.

... I received a letter this morning from Francis mentioning his having left Cambridge for London and giving an account of his examination.

June 15th. Tuesday. Leamington.

... On our return to Leamington we found Francis returned from London looking remarkably well. ...

[27] Elisa Félix, the French actress who performed as "Rachel" and made her London debut in June 1841 at Her Majesty's Theatre.

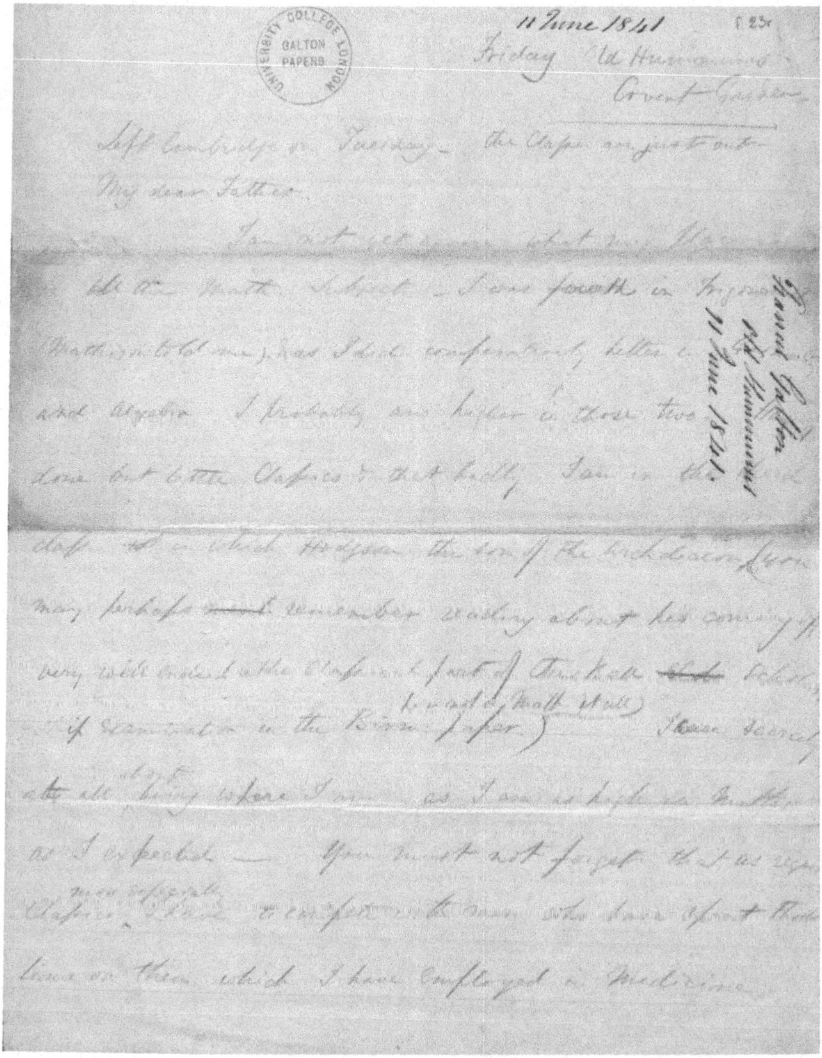

11 June 1841

Friday Old Hummums
Covent Garden

Left Cambridge on Tuesday – the Classes are just out

My dear Father,

I am not yet aware what my Place [illegible]
[illegible] the Math. Subject – I was fourth in [illegible]
(Mathematical [illegible]) [illegible] I did comparatively better [illegible]
and Algebra I probably am higher in those two [illegible]
done but little Classics & that badly. I am in the [illegible]
class [illegible] in which Hodgson the son of the Archdeacon (you
may perhaps ~~[illegible]~~ remember reading about his coming [illegible]
very well indeed in the Classical part of the [illegible] ~~[illegible]~~ [illegible]
[illegible] examination in the [illegible] paper.) [illegible] I have scarcely
[illegible] at all about being where I am – as I am as high in Math. [illegible]
as I expected – You must not forget that as regards
Classics especially I have to compete with men who have spent three
times on them which I have employed in Medicine.

Samuel Galton
Old Hummums
11 June 1841.

Figure 2.54: 1841-06-11. FG to STG page 1.

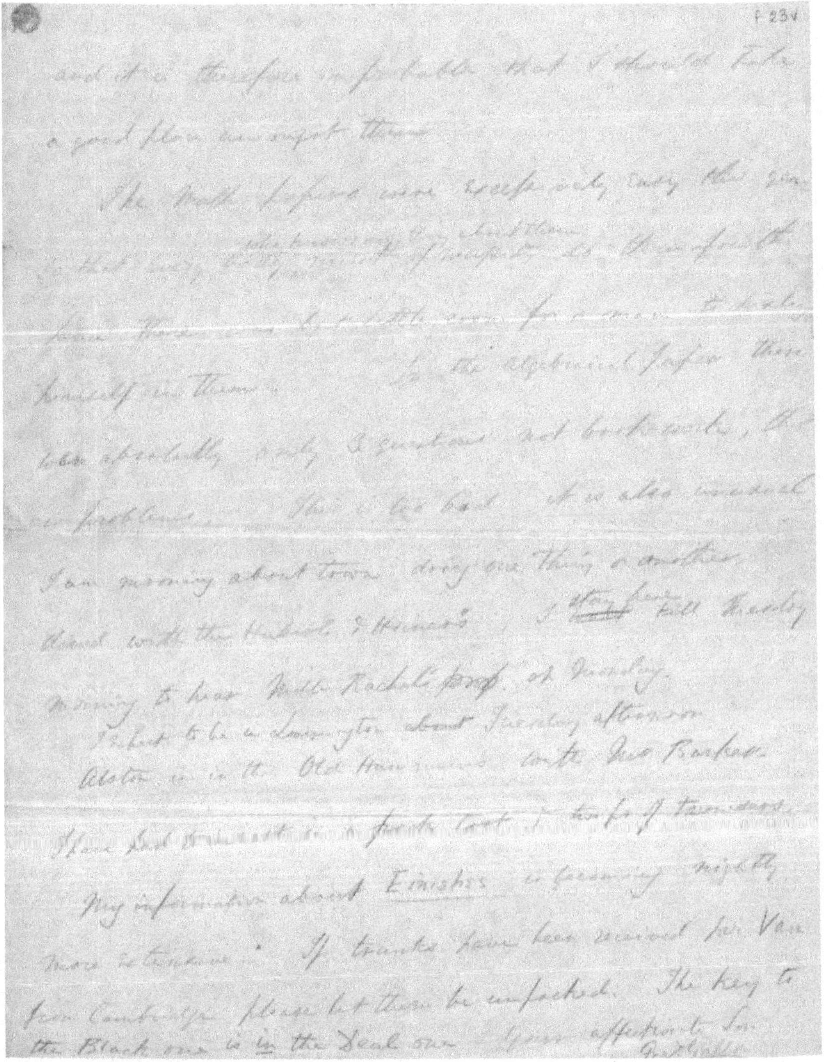

F 23v

and it is therefore improbable that I should take
a good place amongst them.
The Math papers were exceedingly easy the [illegible]
[illegible]
[illegible]
himself in them. [illegible] the Algebraical paper there
were absolutely only 3 questions not bookwork, the
[illegible] problems. This is too bad. It is also unusual
I am moving about town doing one thing or another
[illegible] I stay here till Monday
morning to hear Mlle Rachel's [illegible] on Monday
I expect to be in Leamington about Tuesday afternoon
Alston is in the Old [illegible] with Mr Barker
[illegible]
My information about Ermishes is becoming mightily
more extensive. If trunks have been received for [illegible]
from Cambridge please let them be unpacked. The key to
the Black one is in the Desk one. Your affectionate Son
[illegible]

Figure 2.55: 1841-06-11. FG to STG page 2.

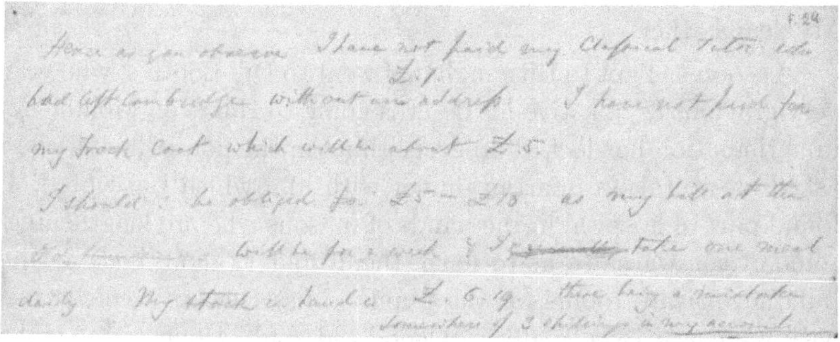

Here as you observe I have not paid my Clerical Tutor who £1 had left Cambridge without any address. I have not paid for my frock coat which will be about £5.
I should be obliged for £5 – £10 as my bill at the Ed. [illegible] will be for a week & I ~~generally~~ take one meal daily. My stock in hand is £. 6. 19. there being a mistake somewhere of 3 shillings in my account.

Figure 2.56: 1841-06-11. FG to STG page 3.

June 18th. Friday. Leamington.

... Francis went early this morning to Edstone and remained all night. ...

June 22nd. Tuesday. Leamington.

... Darwin brought Francis from Edstone and dined with us. ...

June 23rd. Wednesday. Leamington.

... Violetta, Emma, and Francis went to Claverdon and reported that most of the hay in that part of the country is cut and mostly got in ... I agreed this day to allow Francis £280 a year from the last day July next and also the sum of £6.13.4 each journey backwards and forwards between Cambridge and Leamington for each of the vacations. I also agreed to pay Matheson's charge £30.

June 25th. Friday. Leamington.

... Bessy and Francis drank tea at the Sherrards. ...

June 29th. Tuesday. Leamington.

Some thunder and rain. Francis went to Birmingham at 8 a.m. on his way to Keswick. ...

1841/07/01. Francis Galton to Samuel Tertius Galton

Thursday July 1st 1841
Browtop, Keswick.

My dear Father

As soon as I got to Birmingham I went to Dr. Booth's, who was very gracious & talkative his de-ar is going to Havre on Monday & in anticipation has locked up every thing in the house, he however asked me to take a Spartan dinner with him which I accepted. I found part of hospital in the hands of masons who are lengthening some of the wards so as to make them hold 12 more beds & are building private rooms for the pupils. Miss Hawkes received me with open arms she really is a good old soul. Hodgson[28] came to see his patients & I walked back to Temple Row with him, it seems that Mrs. Ledsam's death was by no means unexpected, however Mrs. Hodgson is much pulled down by it. They talk of making an expedition very shortly to the lakes & Mr. H. asked me to send accounts of Lodgings to him. After dining with the Booths the Dr. said that my Aunt was going in her phaeton to Selly Hall & offered to resign his place to me which I accepted. Lucy is ~~fatter &~~ flourishing & in deep mourning. James is about to begin cutting down 70 acres of hay, but I fear that the weather will not be as propitious as he might wish. On returning we met a large mob with flying colours music &c. some say that this approaching election will be anything but a quiet one. My Aunt of course wanted to be off to Havre before it began, but the Dr. thinks such flight ignominious, so they wait. I passed the evening, as my Aunt goes to bed in the afternoon, in listening to the stupidest play imaginable at the theatre, 7 persons were in the lower boxes which are calculated for about 300. I set off by the $\frac{1}{4}$2 o'clock train in the night & slept without once awaking until we were near Preston, we got to Lancaster at $7\frac{1}{2}$ the castle looks extremely well, as it is extensive & composed of square castellated towers, joined by walls covering the top & hill sides of a fine green hill—a florid gothic church has been built close to & in front of it. I set off at 8 by mail to Kendal. The town was in a bustle owing to the nomination, flags, trumpets & so forth. I had a very entertaining fellow-traveller he had a hooked nose gold spectacles was a member of the reform club and a ne-plus-ultra radical—he had travelled & had also been a rowing Cantab: we had a red hot argument on politics, which I firmly believe neither of us knew any-thing about, but he would talk about them & as I must answer yes or no even Bessy will excuse my not assenting to a radical's ideas, he knew

[28] Hodgson, Joseph. See page 483.

the lakes very well and told me many legends &c. about them. Windermere is said to be a beautiful lake. Wordsworth asserts that it is superior to anything abroad, but I humbly conceive that he thereby shews his patriotism rather than his taste. It is certainly pretty but very narrow in comparison with its length, & the high mountains on each side apparently reduce this width still more, so that it looks like a river & not a particularly broad one. We passed many other lakes of various sizes having all one common character, which is that the mountains descend very steeply to within 100 yards of the water & then there is a shelving grassy shore thus even more gently sloping than in the sketch so the lakes are never deep. It seems to me that the water must have stood a great deal higher once & that the whole country has had a lift-up. [See Figure 2.57, p. 116.]

We arrived at length at Keswick a small stone-built & scattered town on a plain hardly at all elevated above Derwent water which therefore swamps the land adjacent to it, some distance on in the same plain is Braithwaite water. Skiddaw is one side of this plain & a mountain chain on the other then. [See Figure 2.58, p. 116.]

All the mountains look very pretty from being green, in nearly every instance, up to the very top. To-day is horribly misty. Skiddaw is covered with clouds that reach low down as a Quakeress's dress but those on the chain opposite are more a la Taglioni.[29]

I had intended walking up Skiddaw last night to see the sun rise but it began to rain. Soon after my arrival yesterday I called on the St. Quintins who were much surprised at my appearance & very agreeable, they had taken Browtop but our party had turned them out so they left it the day before yesterday. Browtop is a very nice house, the habitable part is quite separate from the kitchen & outhouses, it is one story high & a passage down the middle with 4 bedrooms on one side & 4 sitting-rooms on the other side, then the whole house is £5 per week, so I pay 25 shillings. [See Figure 2.59, p. 116.] It has a beautiful view. I should have sketched it for you but there is nothing but clouds this morning. I am the first of the party that has come, the rest are expected to-morrow.

Good bye your affectionate Son,
Fras Galton.

It thus took me 13 hours from Birm. to Keswick. Direct Post Office as they do not deliver to Browtop.

[29] Marie Taglioni, the opera singer.

Figure 2.57: 1841-07-01 FG to STG: Lifted up.

Figure 2.58: 1841-07-01 FG to STG: Derwent Water to Skiddaw.

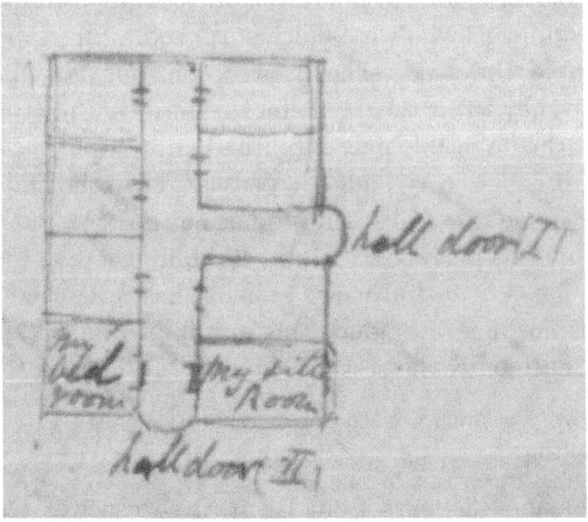

Figure 2.59: 1841-07-01 FG to STG: Rooms in Browtop.

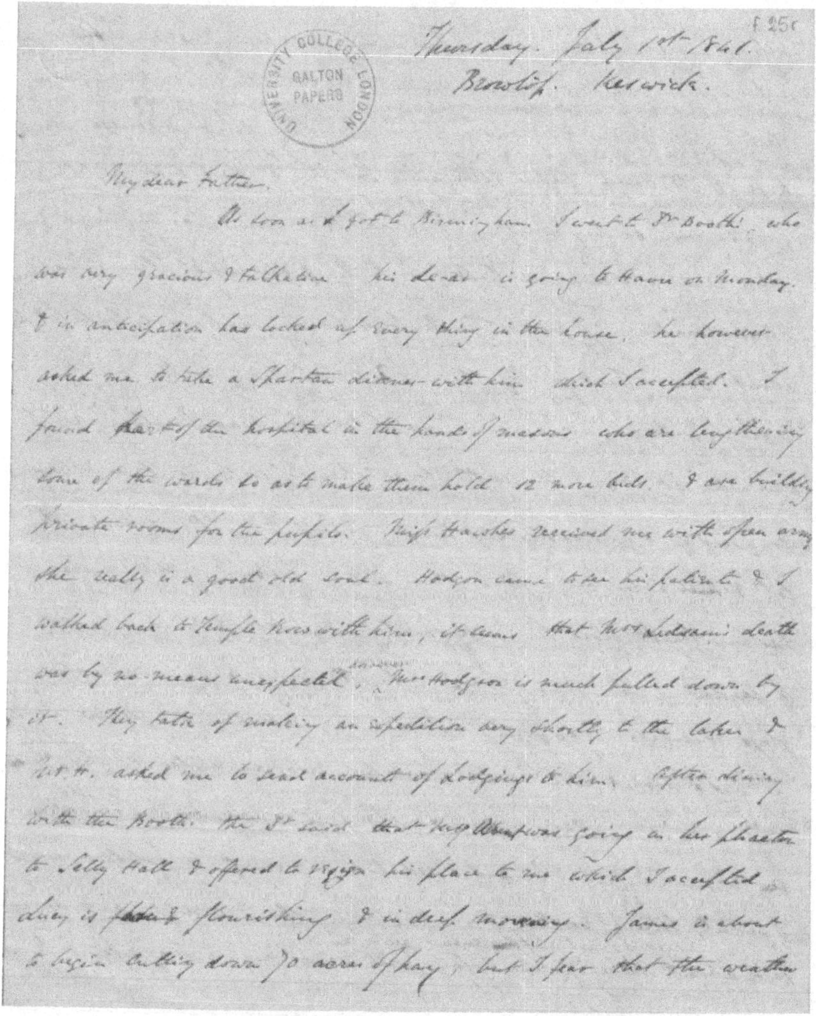

Thursday. July 1st 1841.
Bowtiff. Keswick.

My dear Father,

As soon as I got to Birmingham I went to Dr Booth's who was very gracious & talkative his dear is going to Havre on Monday & in anticipation has locked up every thing in the house, he however asked me to take a Spartan dinner with him which I accepted. I found part of the hospital in the hands of masons who are lengthening some of the wards so as to make them hold 12 more beds, & are building private rooms for the pupils. Miss Hawkes received me with open arms she really is a good old soul. Hodgson came to see his patients & I walked back to Temple Row with him, it seems that Mrs Lidsam's death was by no means unexpected, Mrs Hodgson is much pulled down by it. They talk of making an expedition very shortly to the lakes & Mr H. asked me to send account of lodgings to him. After dining with the Booths the Dr said that Miss Wimpson was going in her phaeton to Selly Hall & offered to resign his place to me which I accepted. Lucy is ~~[illegible]~~ flourishing & in deep mourning. James is about to begin cutting down 70 acres of hay, but I fear that the weather

Figure 2.60: 1841-07-01. FG to STG page 1.

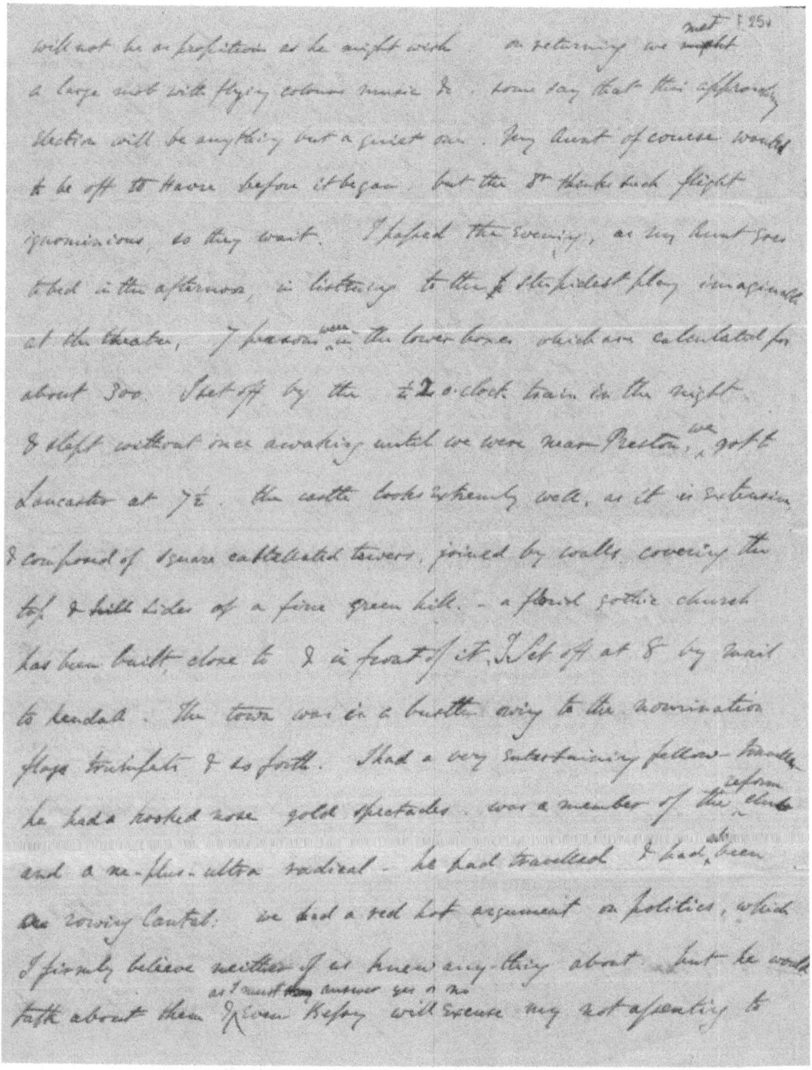

will not be so propitious as he might wish. on returning we met
a large mob with flying colours music &c. some say that this approaching
election will be anything but a quiet one. My Aunt of course wanted
to be off to Havre before it began. but the 8r thinks such flight
ignominious, so they wait. I passed the evening, as my Aunt goes
to bed in the afternoon, in listening to the stupidest play imaginable
at the theater, 7 persons were in the lower boxes which are calculated for
about 300. Set off by the ½ 2 o.clock train in the night
& slept without once awaking until we were near Preston, we got to
Lancaster at 7½. the castle looks extremely well, as it is extensive
& composed of square castellated towers, joined by walls covering the
top & hill sides of a fine green hill. - a florid gothic church
has been built close to & in front of it. I set off at 8 by mail
to Kendal. The town was in a bustle owing to the nomination
flags trumpets & so forth. I had a very entertaining fellow-traveller
he had a hooked nose gold spectacles. was a member of the reform club
and a ne-plus-ultra radical - he had travelled & had also been
a rowing Cantab: we had a red hot argument on politics, which
I firmly believe neither of us knew any-thing about. but he would
talk about them & as I must answer yes or no Bessy will excuse my not assenting to

Figure 2.61: 1841-07-01. FG to STG page 2.

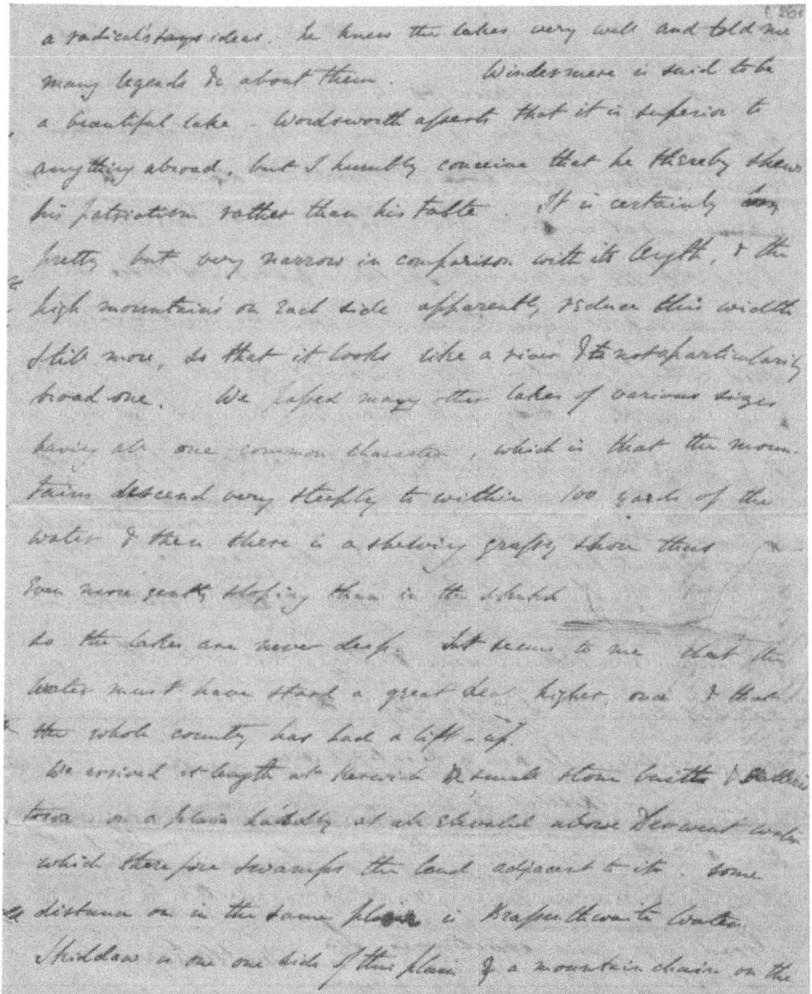

a radical's tory's ideas. he knew the lakes very well and told me
many legends &c about them. Windermere is said to be
a beautiful lake. Wordsworth asserts that it is superior to
anything abroad, but I humbly conceive that he thereby shews
his patriotism rather than his taste. It is certainly very
pretty, but very narrow in comparison with its length, & the
high mountains on each side apparently reduce this width
still more, so that it looks like a river & that not particularly
broad one. We passed many other lakes of various sizes
having all one common character, which is that the moun-
tains descend very steeply to within 100 yards of the
water & then there is a shelving grassy shore thus
even more gently sloping than in the sketch
so the lakes are never deep. It seems to me that the
water must have stood a great deal higher once & that
the whole country has had a lift-up.
We arrived at length at Keswick a small stone built & slated
town on a plain but slightly elevated above Derwent water
which therefore swamps the land adjacent to it. Some
distance on in the same plain is Bassenthwaite water.
Skiddaw is one side of this plain & a mountain chain on the

Figure 2.62: 1841-07-01. FG to STG page 3.

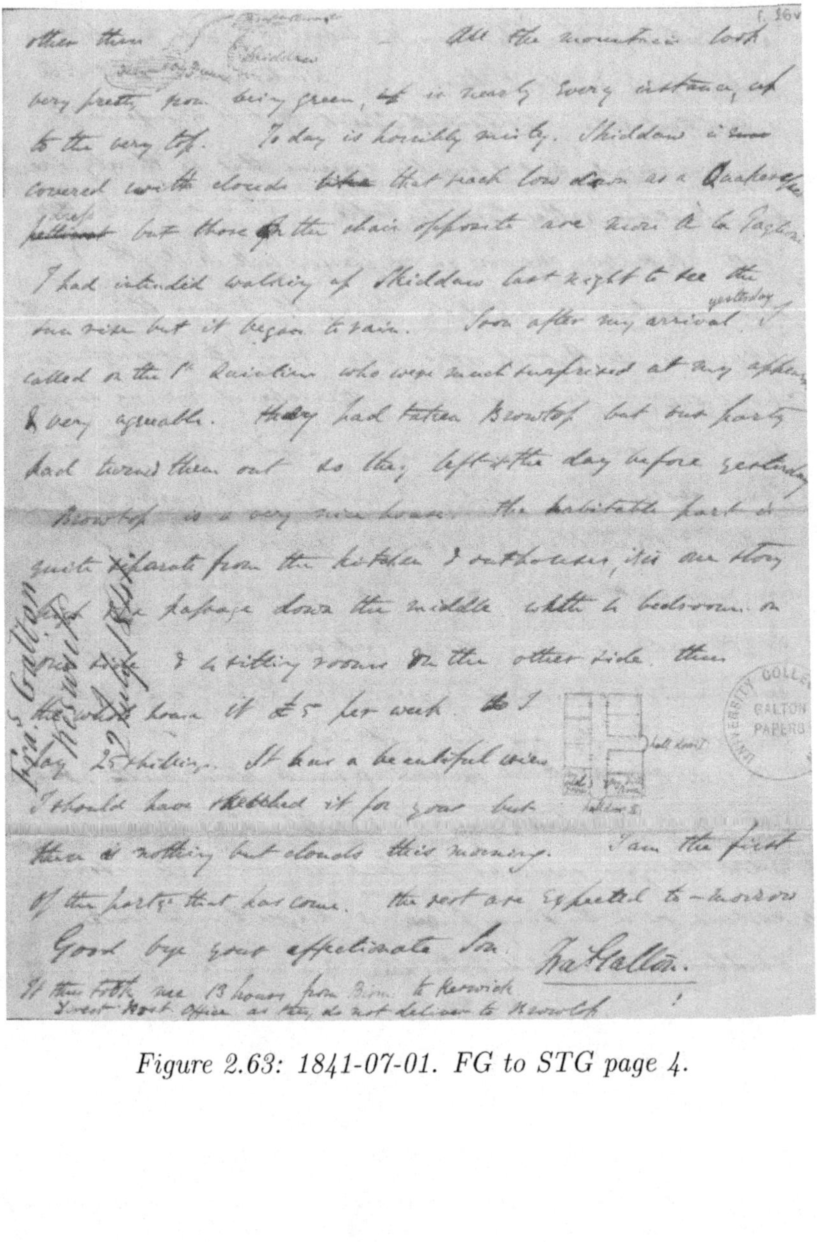
f. 16v

other than [sketch: Skiddaw] — All the mountains look
very pretty from being green, up in nearly every instance up
to the very top. To day is horribly misty. Skiddaw is
covered with clouds that reach low down as a Quaker's
~~petticoat~~ brim but those on the chain opposite are more à la Taglioni
I had intended walking up Skiddaw last night to see the
sun rise but it began to rain. Soon after my arrival yesterday I
called on the 1st Division who were much surprised at my appearance
& very agreeable. They had taken Browtop but our party
had turned them out so they left the day before yesterday
Browtop is a very nice house. the habitable part is
quite separate from the kitchen & outhouses, it is one story
high a passage down the middle with the bedrooms on
one side & the sitting rooms on the other side. then
the whole house is £5 per week & I
pay 25 shillings. It has a beautiful view
I should have sketched it for you but
there is nothing but clouds this morning. I am the first
of the party that has come. the rest are expected to-morrow
Good bye your affectionate son. Fra. Galton.
It thus took me 13 hours from Birm. to Keswick
Direct Post Office as they do not deliver to Browtop

Fra^s Galton Keswick 1 July 1841

Figure 2.63: 1841-07-01. FG to STG page 4.

Figure 2.64: Keswick Vale from Browtop, by W. Green.

1841/07/08. Francis Galton to Samuel Tertius Galton

Thursday, July 8th
Browtop.

My dear Father

Being upon my own allowance you must excuse letter paper. In answer to your question. The Apothecaries' bill ~~is correct~~ I owe, though differing I believe some shillings from the one he sent to Blakesley.[30] The Hosier's Bill is not mine probably's Theodore's. I do not want the boots at Keswick, certainly not at present as I always walk in thick shoes. I am very glad to hear about your asthma's want of punctuality this year, should you have an opportunity I wish you would enquire whether the other asthmatic Leamingtonians I think 3 in number have got off equally well, as it would be interesting to find out whether some years are more unfavourable than others as is apparently the case in typhus & influenza.

I never enjoyed myself so thoroughly as at present. Mathison[31]

---

30 Blakesley, Joseph Williams. See page 480.
31 Mathison, William Collings. See page 486.

and Eddis[32] are thoroughly good-natured. When it is fine we walk out in a body from 2 to 5 when wet play at battledore and shuttlecock or at fives in a most un-studious manner. Eddis you know is a senior medallist. Blomfield is the other undergraduate staying at Browtop.[33] The St Quintins are very kind & their son (Charles) very agreeable—he rattles off about the Himalaya Passes and the scenery of Thibet & totally condemns Howqua's Mixture.[34] They introduced me to the Parson, a Mr Myers[35] who again introduced me to a Count Osolinkski[36] (I purposely write the word illegibly & that for an obvious reason, it consists of "three sneezes and a ski"), who is the lion of these parts being Chemist Botanist Zoologist &c. &c. and last not least a top-sawyer Animal Magnetiser. I need scarcely say that we got the greatest allies immediately he is a very gentlemanly man, he shewed me through laboratories hot-houses & monkey cages. Made me look through a telescope of "hextra million magnifying power" at a lady in the drawing-room of a distant house who was innocently lifting up her eyes to a brass chandelier in poetic raptures &c. &c. He has moreover got me patients to magnetise lent me books & in short we are the greatest possible allies. He married a rich heiress £4000 a year, ~~and~~ who is a very pretty girl 18 years old, he is about 40.

I went to a tea-party with Mathison & Eddis yesterday at the parsonage (Mr. Myers') where Eddis and myself flirted with and blarneyed an innocent Keswickian who would recite poetry out of the Christian (year/era), to such an extent that she was quite overcome & evidently expected at least one proposal. I ought to say that as I walked with the Count through his laboratory he introduced me to a Dr. Schmidt who was staying with him & working there. I thought I knew both the face and name & found out that he was one of Liebig's class at Giessen & we had great amusement in talking over the wine-parties &c. after the table d'hôte. He was

[32] Eddis, Arthur Shelly. See page 483.

[33] Thomas Eardley Wilmot Blomefield (1820-1878) of Trinity.

[34] A popular brand of China tea, in a mixture supposedly obtained from a Cantonese merchant by that name.

[35] Rev. Frederic Myers (1811-1851).

[36] The Polish Count Vladimir Boris Ossolinski (?-1859), Chestnut Hill, Keswick. Sometimes given as Ossalinsky. He married the wealthy heiress Mary Jackson in 1839. He seems to have deserted his wife sometime in the 1840s. After his death in 1859 his widow was well-known in the Lake District as "Countess Ossalinsky". For Ossolinski, Vladimir Boris, see page 487.

a great phrenologist & I got him to paw my head, he gave me I think a very true character (self-esteem was remarkably full). I have not now the bump of constructiveness very large though he says it is large. Mary's bump is firmness. I described her character and he immediately said that he had observed equally well developed cases. I have just descended from Skiddaw it is a very seedy mountain to go up there being no difficulty whatever, the view is very extensive including the Isle of Man & Ben Lomond. It was very hot & we pitched into much whisky, and on the strength of it cheered 3 times 3 for God save the Queen, Trinity, &c. Whatever Father Mathew[37] may say there is nothing like vast quantities of whisky on a mountain top, it would be a splendid way to subject a convert to temptation. Please address in future Browtop as an oblique-eyed inmate of our skullion having nothing else to do brings up the letters. Love to all & each. May the critical spirit of Bessy smile on this epistle.

Good Bye. Your affectionate Son Fras Galton.

1841/07/18. Francis Galton to Samuel Tertius Galton and Elizabeth Anne Galton

Sunday
Browtop Keswick.
July 18th 1841

My dear Father

Thank you for your letter. I am very glad that you talk of coming towards the lakes, as they are well worth seeing & as a long course of fine weather almost invariably sets in about this time. I do not know of any house in Keswick which you could take (bonâ fide) as they are all small & mostly dark & dirty. The inn is very good, good bedrooms &c. & has been honoured by Queen Adelaide's sleeping in it, if that be any inducement to you, but Ambleside has many good lodging houses. I received the enclosed letter from Christian the day before yesterday. We have had beautiful weather the last 4 days & have already been making several excursions, to Buttermere, where the celebrated beauty whom Hatfield the robber eloped with some 20 years since, lived, & where several uncommonly pretty girls keep up the character of the place

[37] The Irish Capuchin Priest Father Theobald Mathew (1790-1856) preached abstinence, founding the "Cork Total Abstinence Society".

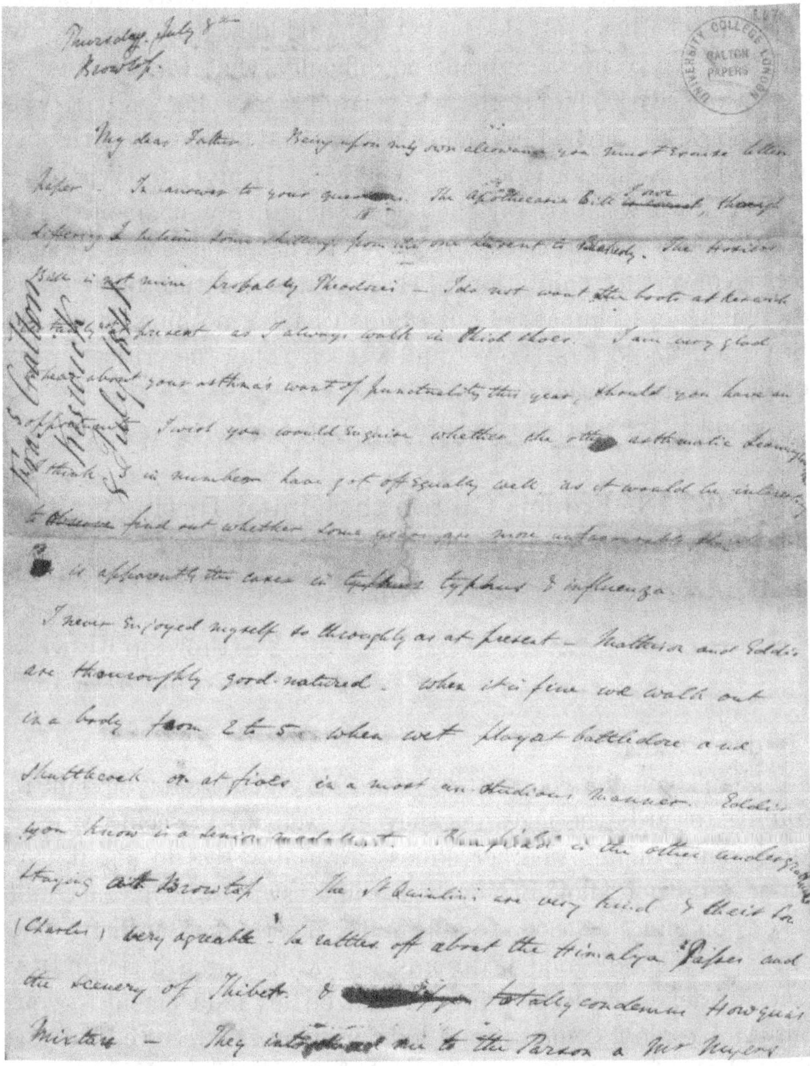
Thursday, July 8th
Browtop.

My dear Father. Being upon my own allowance you must excuse letter paper. In answer to your questions. The Apothecaries bill [illegible], though differing I believe some shillings from the one I sent is correct. The Hosier's bill is not mine probably Theodore's — I do not want the boots at Keswick at present as I always walk in thick shoes. I am very glad to hear about your asthma's want of punctuality this year, should you have an opportunity I wish you would enquire whether the other asthmatic [illegible] I think [illegible] in number have got off equally well, as it would be interesting to find out whether some years are more unfavourable [illegible] as is apparently the cases in typhus & influenza.

I never enjoyed myself so thoroughly as at present — Matheson and Eddis are thoroughly good-natured. When it is fine we walk out in a body from 2 to 5. When wet play at battledore and shuttlecock or at fives in a most un-studious manner. Eddis you know is a senior medallist. [illegible] is the other undergraduate staying at Browtop. The St Quintins are very kind & their son (Charles) very agreeable. he rattles off about the Himalaya [illegible] and the scenery of Thibet. & [illegible] totally condemns Howqua's Mixture — They introduced me to the Parson a Mr Myers

Fras. S. Galton
Browtop
8 July 1841

Figure 2.65: 1841-07-08. FG to STG page 1.

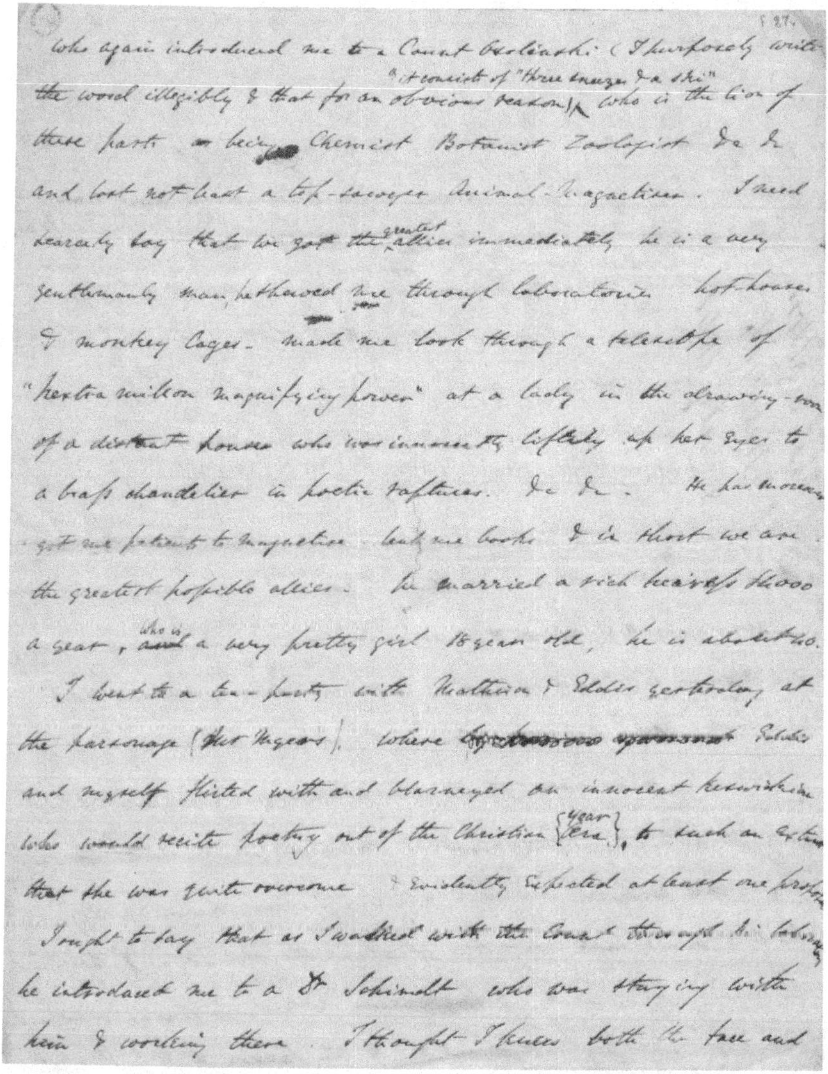
who again introduced me to a Count [illegible] (I purposely write the word illegibly & that for an obvious reason) "it consists of "three sneezes & a shi"" who is the lion of these parts as being Chemist Botanist Zoologist &c &c and last not least a top-sawyer Animal-Magnetiser. I need scarcely say that we got the greatest allies immediately he is a very gentlemanly man, he showed me through laboratories hot-houses & monkey Cages- made me look through a telescope "of "hextra million magnifying power" at a lady in the drawing-room of a distant house who was innocently lifting up her eyes to a brass chandelier in poetic raptures. &c &c. He has moreover got me patients to magnetise lent me books & in short we are the greatest possible allies. He married a rich heiress £1000 a year, who is a very pretty girl 18 years old, he is about 40.

I went to a tea-party with [illegible] & Elder yesterday at the parsonage ([illegible]), where ~~[illegible]~~ Elder and myself flirted with and blarneyed an innocent [illegible] who would recite poetry out of the Christian {Year / Era}, to such an extent that she was quite overcome & evidently expected at least one proposal.

I ought to say that as I walked with the Count through his laboratory he introduced me to a Dr Schmidt who was staying with him & working there. I thought I knew both the face and

Figure 2.66: 1841-07-08. FG to STG page 2.

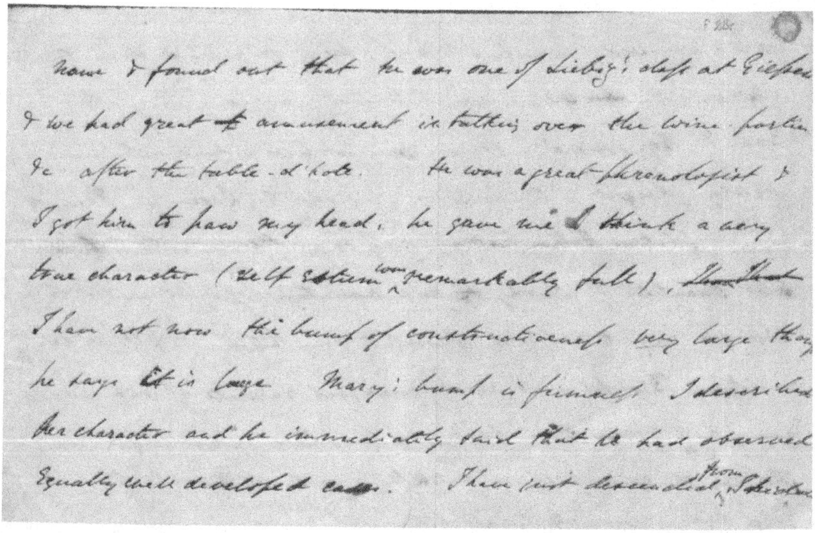

name & found out that he was one of Liebig's class at Giessen
& we had great amusement in talking over the wine parties
&c after the table-d'hote. He was a great phrenologist &
I got him to paw my head, he gave me I think a very
true character (self esteem was remarkably full),
I have not now the bump of constructiveness very large though
he says it is large. Mary's bump is firmness I described
her character and he immediately said that he had observed
Equally well developed case. I have just descended from Schiehallion

Figure 2.67: 1841-07-08. FG to STG page 3.

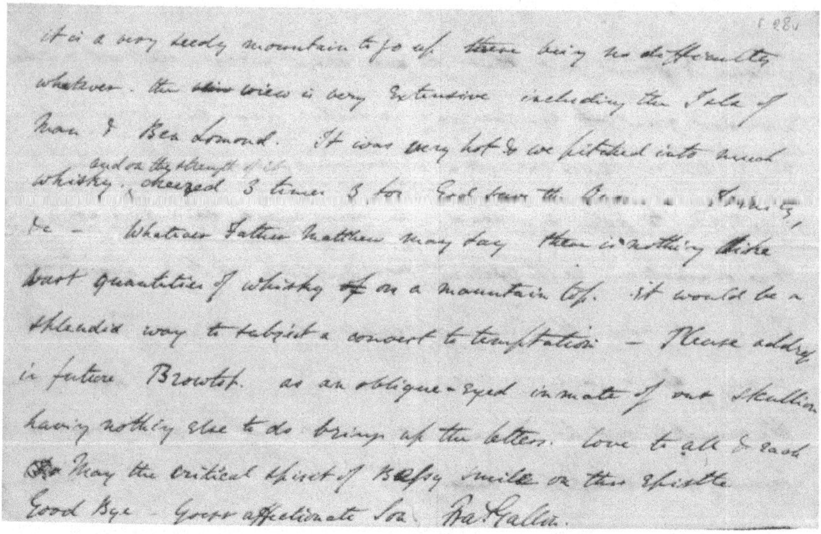

it is a very seedy mountain to go up there being no difficulty
whatever. the view is very Extensive including the Isle of
Man & Ben Lomond. It was very hot & we pitched into much
whisky, and on the strength of it cheered 3 times 3 for God save the Queen ... Toasts,
&c — Whatever Father Matthew may say there is nothing like
vast quantities of whisky on a mountain top. it would be a
splendid way to subject a convert to temptation — Please address
in future Browtop. as an oblique-eyed inmate of our Skullion
having nothing else to do brings up the letters. love to all & each
May the critical spirit of Bessy smile on this Epistle
Good Bye — Your affectionate Son Frat Galton.

Figure 2.68: 1841-07-08. FG to STG page 4.

still. We have also been up Helvellyn. They said it would take us 3 hours to ~~get to~~ reach the top, but ~~which took us~~ it occupied only $2\frac{3}{4}$ to get up stay 25 minutes at the top & be at the bottom on the other side. The country people are exceptionally hospitable, they give us no end of milk, oat cakes, home-made cheese &c. & it is difficult to make them take anything in return, so our plan is to ask if they have children & pick out one of the curly-headed young scrubs & visibly slip a shilling or two, according to the probable extent of our united appetites, into his hand. Yesterday after taking a stroll up Skiddaw we went to the perfection of a farm house. A very pretty Creole niece of the farmer chatted, smiled, gave us milk &c. & set off a musical snuff-box playing, then brought a nosegay & lastly sat down with us to grub. She asked if we were any of the Cambridge gentlemen, & on our saying that we were she told us that 2 or 3 years ago there was a large party at Keswick who were a "sad set of scamps." So you see what a reputation Cambridge has got. Whewell, wonderful to say, has fallen in love with an Ullswater beauty, Miss Marshall, & is going to marry her. She is very rich & of good family hence our continual conversation at Browtop is in surmises as to how Whewell would set to work to make love, he is nearly 50, she a little more than 20.

The St. Quintins have gone to Blackpool & by some mishap their pet terrier is left behind. I have had her chained up & have written for further orders.

So Miss Collinson & Mr. Aylmer are going to be spliced, which one is it? as I only know Catherine & Louisa by name. I suppose Kitty Sutton will start an Encyclopaedia of rejected addresses like Miss Ryland, & in folio one huge [I and line I]. Darwin's name & good qualities will be duly registered. How is Mary, & ~~when will~~ how soon will the baby linen be sorted & aired?

We like Browtop excessively, the only fault in the situation of Keswick being that it is in a wide valley, so that there are but few ~~near~~ walks within a short distance. The hills are all quite green so that we can run down them at a capital [pace]. Mathison tried sliding down one of them, but he reached the bottom a complete cherub having scarcely ~~anywhere~~ <u>whereon</u> to sit, owing to the friction.

I have come from church where there is a parson (Mr. [Hunter]) who cuts his hair like a fancy yew-tree, he is blessed with a wife half German who talks away intolerably. She is a Roman Catholic & is always arguing & snubbing every body, whether she knows them or not, who is talking near her, accordingly when her loving spouse got to that part of the evening lessons relating to wives, he dropped his hen-pecked, subdued tone, lifted himself on tip-toe, leant over the reading desk & shouted it all out into his wondering wife's ears—what a curtain lecture he'll get tonight!

We hear that a party of Cantabs at Ambleside think of migrating to Keswick, it will be a great bore if they come as we are enough already. Armitage the to-be senior wrangler of the February after next is one of them. (vide 1st Page)[38]

Dear Bessy

I know that this letter is a stupid one but I really have nothing to say for myself. Write soon & tell me about Scarborough.

Good-bye
& believe me ever
Your affectionate
{Son Brother} Fras. Galton.

1841/07/23. Samuel Tertius Galton to Francis Galton

N^r^9 Cliff Scarborough
23 July 1841

Francis,

We all arrived here, in ... well, tho' the cold wet weather [we] have had since our arrival [has] not suited her—Bessy seems better for the breezes. I am [an] hour from ...

1841/08/01. Francis Galton to Samuel Tertius Galton and Elizabeth Anne Galton

Browtop
Sunday Aug 1st 1841

My dear Father,

I received this morning £15 all correct enclosed in an envelope stamped "too late," for which thanks. The letter before this one was dated from "N.g. cliff".[39] Do they ticket the houses as

[38] For the cross-written note to Bessy transcribed below.

[39] A misreading of "N^r^ 9 Cliff".

Figure 2.69: 1841-07-18 FG to STG: A country road in Skiddaw.

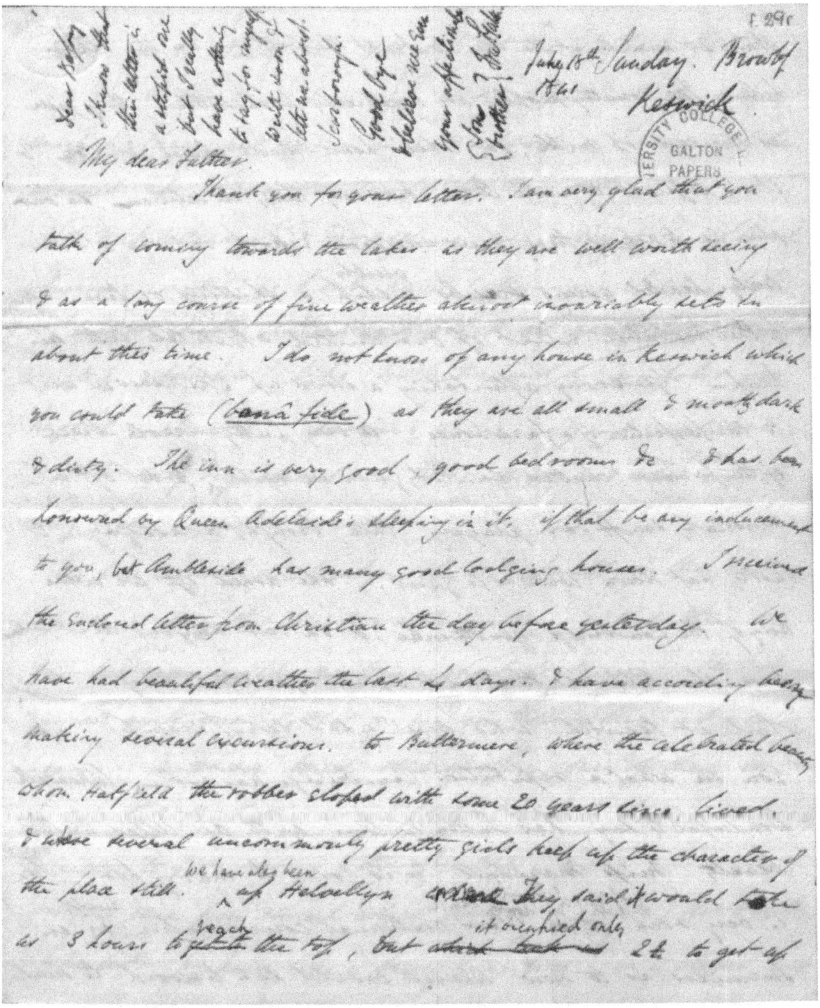

Figure 2.70: 1841-07-18. FG to STG page 1.

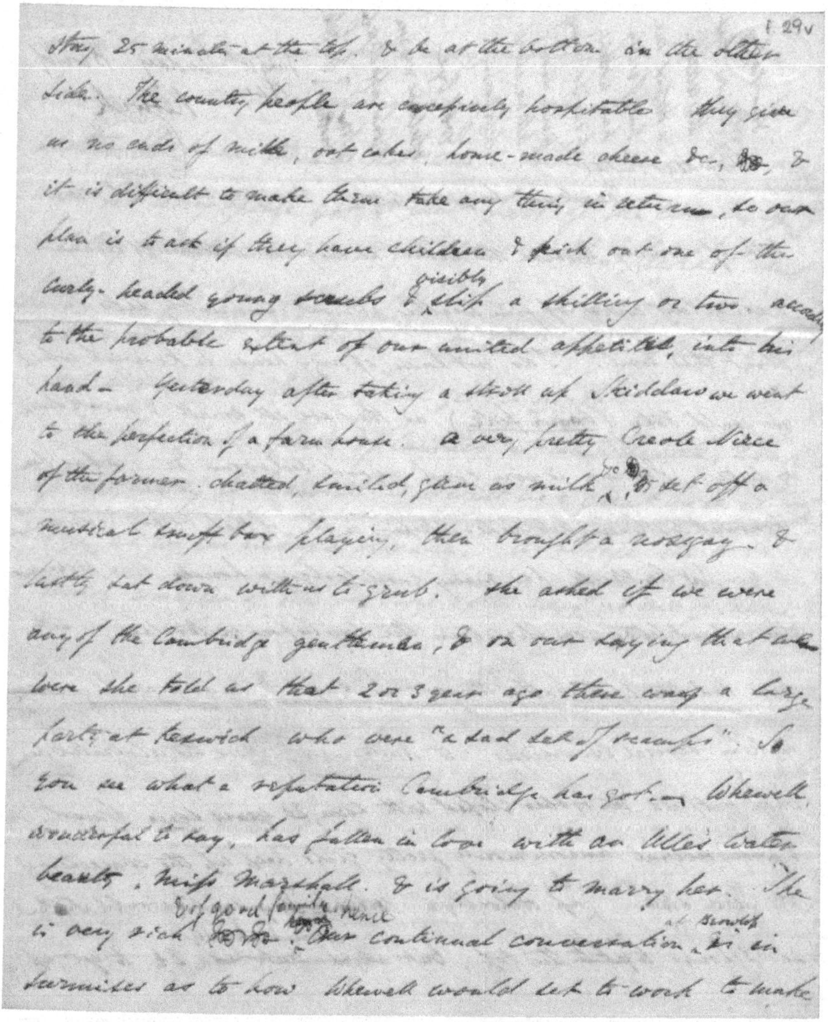

f.29v

stay 25 minutes at the top. & be at the bottom in the other side. The country people are exceedingly hospitable they give us no ends of milk, oat cakes, home-made cheese &c., &c, & it is difficult to make them take any thing in return, so our plan is to ask if they have children & pick out one of the curly-headed young rascals & visibly slip a shilling or two, according to the probable extent of our united appetites, into his hand— Yesterday after taking a stroll up Skiddaw we went to the perfection of a farm house. A very pretty Creole Niece of the farmer chatted, smiled, gave us milk &c &c, set off a musical snuff box playing, then brought a nosegay & lastly sat down with us to grub. She asked if we were any of the Cambridge gentlemen, & on our saying that we were she told us that 2 or 3 years ago there was a large party at Keswick who were "a sad set of scamps". So you see what a reputation Cambridge has got— Whewell, wonderful to say, has fallen in love with an Ulles Water beauty, Miss Marshall. & is going to marry her. She is very rich &c &c of good family & very nice. Our continual conversation at Bowness is in surmises as to how Whewell would set to work to make

Figure 2.71: 1841-07-18. FG to STG page 2.

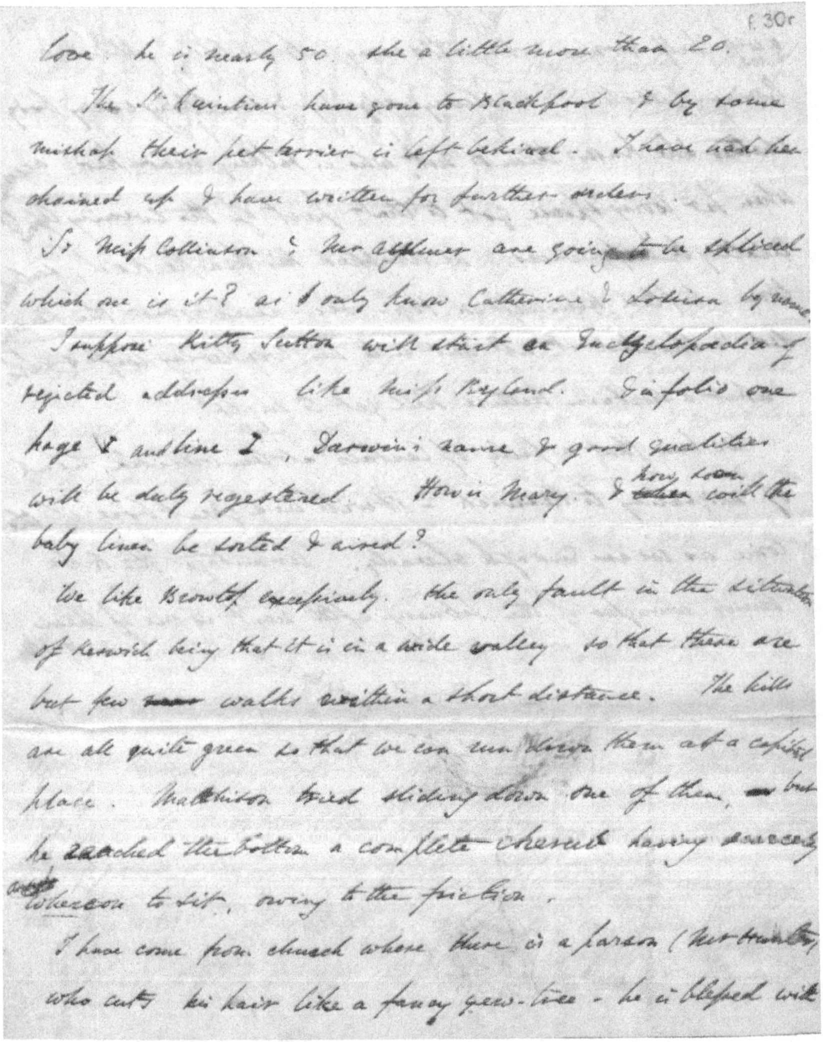

f. 30r

love. he is nearly 50. she a little more than 20.

The St Quintins have gone to Blackpool & by some mishap their pet terrier is left behind. I have had her chained up & have written for further orders.

So Miss Collinson & Mr Aylmer are going to be spliced which one is it? as I only know Catherine & Lothian by name.

I suppose Kitty Sutton will start an Encyclopaedia of rejected addresses like Miss Ryland. In a folio one page & outline I, Darwin's name & good qualities will be duly registered. How is Mary & how soon will the baby linen be sorted & aired?

We like Keswick exceedingly. the only fault in the situation of Keswick being that it is in a wide valley so that there are but few walks within a short distance. The hills are all quite green so that we can run down them at a capital place. Matthison tried sliding down one of them, but he reached the bottom a complete cherub having scarcely anything whereon to sit, owing to the friction.

I have come from church where there is a parson ([illegible]) who cuts his hair like a fancy yew-tree - he is blessed with

Figure 2.72: 1841-07-18. FG to STG page 3.

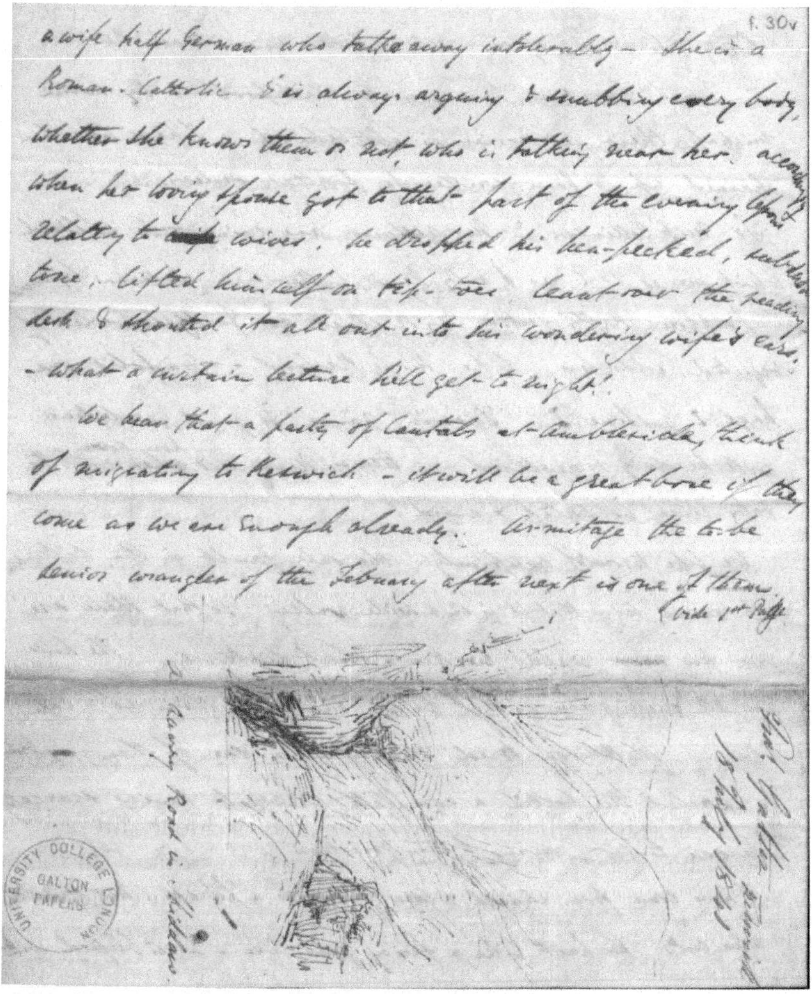

f. 30v

a wife half German who talks away intolerably – She is a Roman Catholic & is always arguing & snubbing every body, whether she knows them or not, who is talking near her. accordingly when her loving spouse got to that part of the evening [service] relating to wives, he dropped his hen-pecked, subdued tone, lifted himself on tip-toes, leant over the reading desk & shouted it all out into his wondering wife's ears. – What a curtain lecture he'll get to night!

We hear that a party of Cantabs at Ambleside, think of migrating to Keswick – it will be a great bore if they come as we are enough already. Armitage the to-be senior wrangler of the February after next is one of them.

(vide 1st page)

18 July 1841

UNIVERSITY COLLEGE LONDON GALTON PAPERS

Figure 2.73: 1841-07-18. FG to STG page 4.

N 9 - Cliff - Scarborough
23 July 1841
Francis,
We all ... ived here, in-
...

Figure 2.74: 1841-07-23. STG to FG page 1.

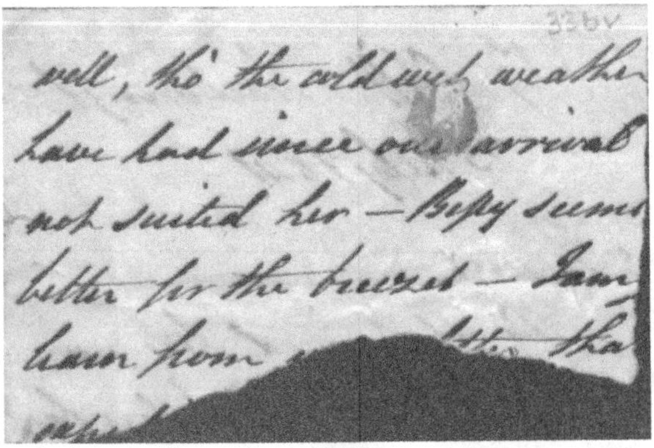
well, tho' the cold wet weather
have had since our arrival
not suited her — Bessy seems
better for the breezes — I am
... from ... tha
...

Figure 2.75: 1841-07-23. STG to FG page 2.

Cabinet curiosities? It is certainly novel and decidedly literary. I suppose that Lucy & James have already arrived, what with them & my two uncles The Galton family will inundate the place & if the ordinary appetite that accompanies the several branches of the family be ~~still~~ present at Scarboro' it will give a very satisfactory interpretation of the N.g. ticketing. Whewell is undoubtedly under the guardianship of Hymen, though an oyster may be affected by love, a Whewell can't, for he has (I understand) been so involved in the metaphysical line that he looks on the approaching event with the most philosophical indifference. In 3 weeks Keswick is to be turned topsy-turvy with amusements. Inprimis a 4-oared race in which your humble servant is to pull, as we get up a boat for the honour of Cambridge. (The names of the crew are Atkinson Strickland[40] Young Galton & Cooper steersman.) £12 prize. We really shall have a very fair chance, for though the Keswick boatmen are trained to pulling from their sash & petticoat age yet they are more in the cart horse line, whilst the description that I heard given of our crew was that they are "intensely plucky." We have great amusement here in scrambling about. Mathison[41] is a capital walker but not a dab at climbing consequently he occasionally sticks amongst the rocks like a saint in a niche & immovable without a miracle. We get on capitally at Browtop. The order of the day is—Breakfast finished as the clock strikes 9. Reading till 1 or $1\frac{1}{2}$. Lunch walk till 5 dine and chat till 8, Read till 10 tea to $10\frac{1}{2}$ Read to 11 or 12.

When a long walk is taken we eschew dinner put our leathern whisky flasks in our pockets, which I am convinced is the true interpretation of "seven leagued boots," and walk from 1 to 8 or 9. We certainly do great things in the walking line instead of "managing our forces" after the Swiss régime we scamper up the hills & somehow or other don't get tired, today I ascended more than $\frac{2}{3}^{\text{ds}}$ of the height of Skiddaw (driven back by mist) in 35 minutes about 2200 feet, this was in Sunday costume & without puggyfying to any extent.

Dear Bessy

How's the bathing? & how are Emma's freckles? The amount of sunshine here is by no means dangerous to beauty as the sun has generally a mass of mist rather thicker than Skiddaw is high to shine through. It has been miserably cold so that I read with a pea-coat on & with my feet on the fire-hob. In your letter in a quotation from Aunt Hubert a word occurred "oddm...ts." I have looked in Johnson's Dictionary but can find nothing corresponding to it, so I presume that it must have been coined since his time. Again, there was a passage in your letter ending with 4 notes of admiration combined. This is an excess. Thirdly, I should recommend a more refined choice of phrases than such as the "weather taking up" and others of a similar nature. You state "The air is delightful, and a beautiful walk along the cliff." I am not aetherial enough to enter into your delights (I must bully you). How has Lucy's bazaar gone off? Please write oftener. Tell me what you had for dinner if you

---

[40] Strickland, Frederick. See page 490.

[41] Mathison, William Collings. See page 486.

can think of nothing else but do write.

Good-bye Your affect Brother, Fras Galton.

1841/08/13. Francis Galton to Samuel Tertius Galton

Keswick Browtop
August 1841

My dear Father

Thank you for your letter. The title of the Guide book that you lent me is "Leigh's Guide to the Lakes." I should not advise you to invest in ~~another~~ a second one as it is the worst published. The best is "Black's guide" 5 shillings 6_d. It contains one large & 3 or 4 small maps very accurate. You talk of "fear of annoying me with a formal visit" &c. I can assure you that I should enjoy nothing so much as Atwood-izing you over the country. We can give you dinner occasionally at Browtop. You will find Eddis[42] & Mathison[43] very agreeable & I really think that the very best thing that you can do is to settle in Keswick for a fortnight ~~of~~ or 3 weeks. If you will give me a commission for lodgings, I will make every enquiry. Our boat-racing scheme has been given up for on enquiry we find that the competitors must pull on pins not in rollocks to pull in that manner we unanimously decreed was below the dignity of a Cambridge "oar" as all the beauty of & skill of rowing consists in correct feathering which of course is impracticable with pins. It is altogether a ridiculous piece of business. There actually is no practising on the lake & consequently the pulling at the race must be wretched. [See Figure 2.80, p. 141.]

I enclose you the ~~addr~~ date I was speaking of and appeal to your unprejudicial eyes at to its legibility.

Poor Chance my old schoolfellow & chum is dead. He was my chief friend at Dr. Jeune's & also at King's College where he read classics. He was one of the best fellows that I have met with, he was expected to have distinguished himself. Poor fellow—he died of consumption.

We have had a visit here from a late fellow commoner of Trinity named Hope. He is son to Lady Beresford & left Cambridge in his 2nd year (last term) to get made MP which he now is being about

[42] Eddis, Arthur Shelly. See page 483.
[43] Mathison, William Collings. See page 486.

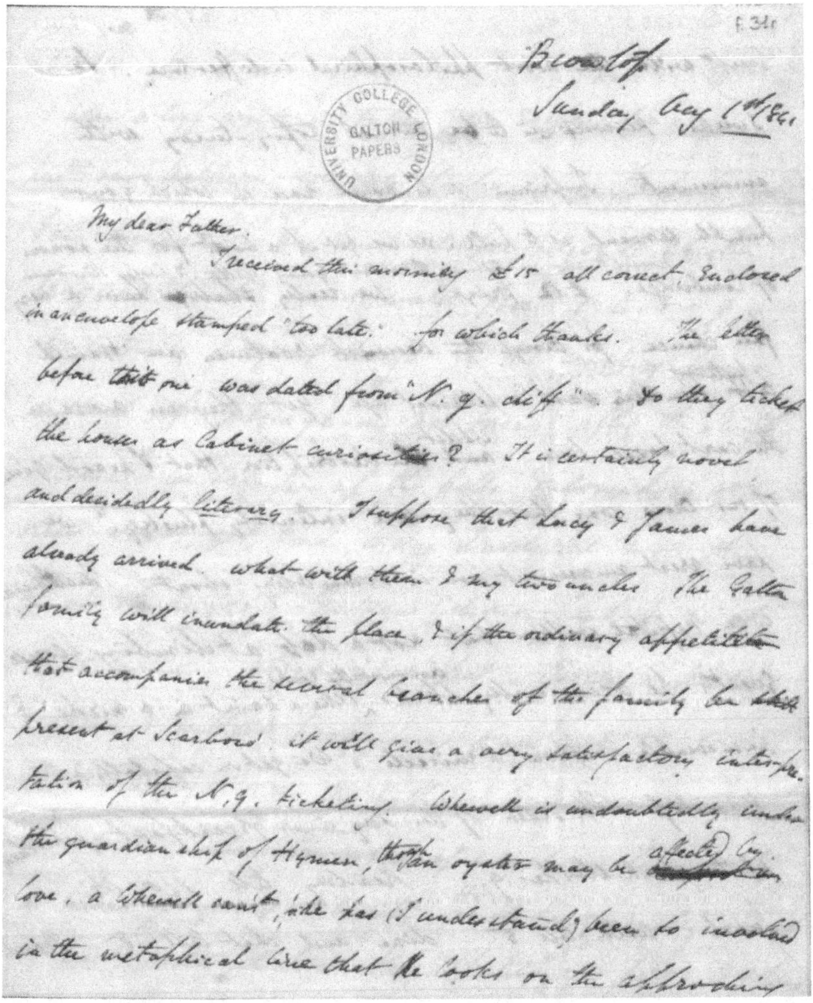

f. 34r

Bcowtop
Sunday Aug 1st/1841

My dear Father,

I received this morning £15 all correct, enclosed in an envelope stamped "too late" for which thanks. The letter before this one was dated from "N. G. cliff". So they ticket the houses as Cabinet curiosities? It is certainly novel and decidedly literary. I suppose that Lucy & James have already arrived, what with them & my two uncles the Galton family will inundate the place, & if the ordinary appetites that accompanies the several branches of the family be present at Scarboro' it will give a very satisfactory interpretation of the N. G. ticketing. Whewell is undoubtedly under the guardianship of Hymen, though an oyster may be affected by love, a Whewell can't, he has (I understand) been so involved in the metaphysical line that he looks on the approaching

Figure 2.76: 1841-08-01. FG to STG and EAG page 1.

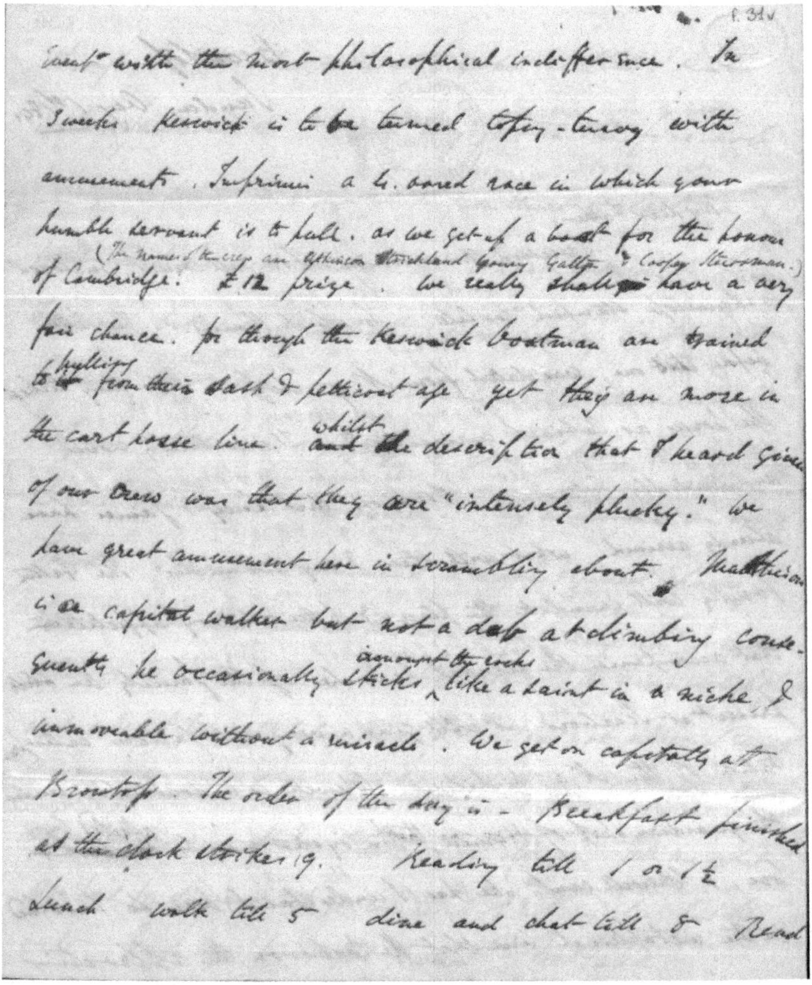
event with the most philosophical indifference. In
3 weeks Keswick is to be turned topsy-turvy with
amusements. Imprimis a 4-oared race in which your
humble servant is to pull. as we get up a boat for the honour
(The names of the crew are Atkinson Strickland Young Galton & Cooper Steersman.)
of Cambridge: £12 prize. We really shall have a very
fair chance. for though the Keswick boatmen are trained
to sculling from their sash & petticoat age yet they are more in
the cart horse line. whilst the description that I heard given
of our Crew was that they are "intensely plucky." We
have great amusement here in scrambling about. Matthieson
is a capital walker but not a dab at climbing conse-
quently he occasionally sticks amongst the rocks like a saint in a niche, &
immoveable without a miracle. We get on capitally at
Brontop. The order of the day is - Breakfast finished
at the clock strikes 9. Reading till 1 or 1½
Lunch walk till 5 dine and chat till 8 Read

Figure 2.77: 1841-08-01. FG to STG and EAG page 2.

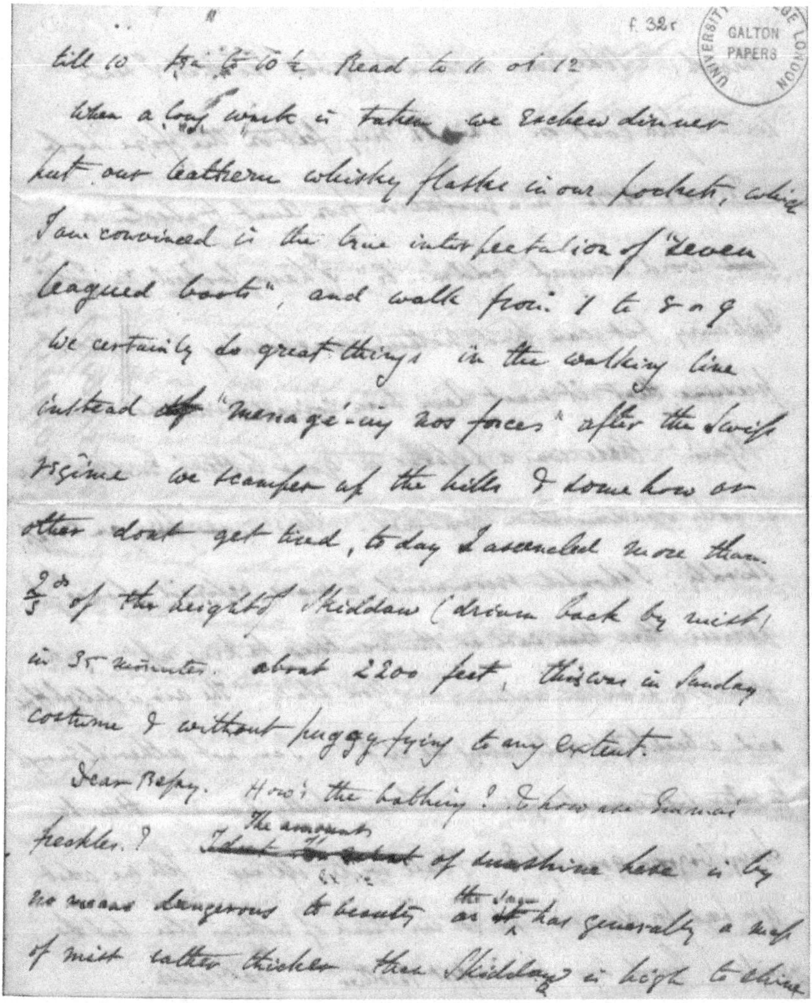

f. 32r

till 10 Bkst to 10½. Read to 11 or 12
When a long walk is taken we eschew dinner
put our leathern whisky flasks in our pockets, which
I am convinced is the true interpretation of "seven
leagued boots", and walk from 1 to 8 or 9
We certainly do great things in the walking line
instead of "menage-ing nos forces" after the Swiss
régime we scamper up the hills & somehow or
other don't get tired, to day I ascended more than
2/3rds of the height of Skiddaw (driven back by mist)
in 35 minutes about 2200 feet, this was in Sunday
costume & without puggyfying to any extent.
Dear Bessy. How's the bathing? & how are Emma's
freckles? The amount of sunshine here is by
no means dangerous to beauty as the Sun has generally a map
of mist rather thicker than Skiddaw is high to shine

Figure 2.78: 1841-08-01. FG to STG and EAG page 3.

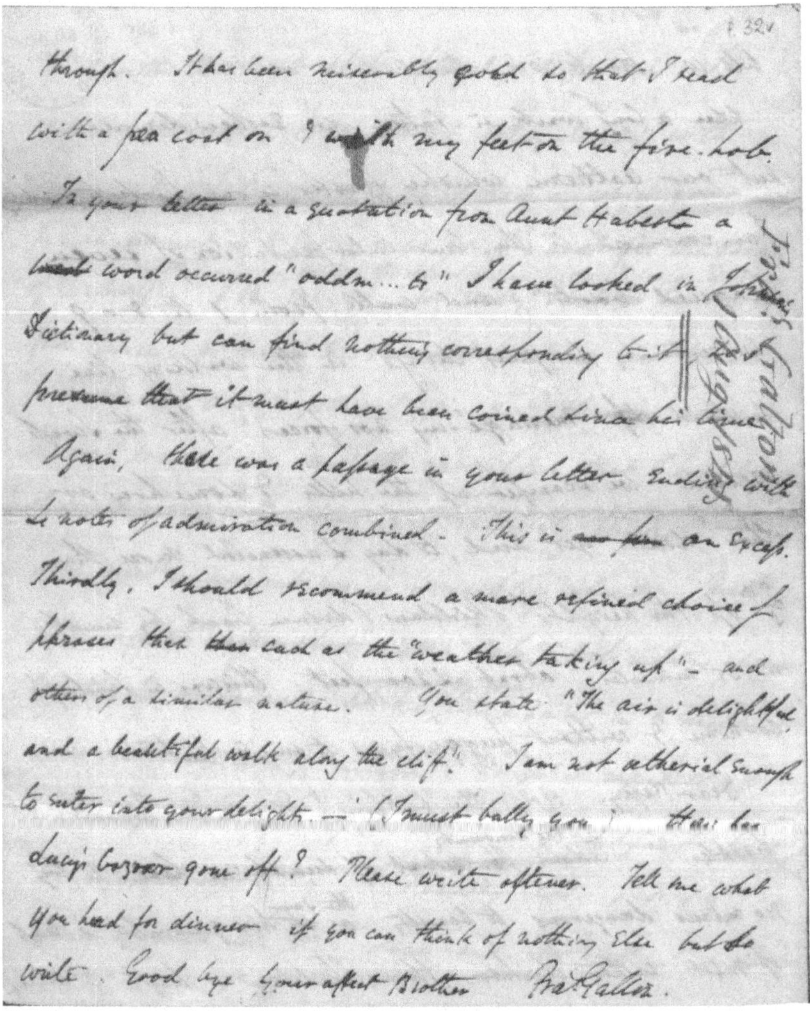

f. 32v

through. It has been miserably cold so that I read
with a pea coat on & with my feet on the fire-hob.
In your letter is a quotation from Aunt Haberts a
word occurred "oddm...ts" I have looked in Johnson's
Dictionary but can find nothing corresponding to it & so
presume that it must have been coined since his time.
Again, there was a passage in your letter ending with
2 notes of admiration combined. This is an excep.
Thirdly, I should recommend a more refined choice of
phrases than such as the "weather taking up" - and
others of a similar nature. You state "The air is delightful
and a beautiful walk along the cliff". I am not ætherial enough
to enter into your delight — (I must bully you) — Has Mrs
Lucy's bazaar gone off? Please write oftener. Tell me what
you had for dinner if you can think of nothing else but do
write. Good bye Your affect Brother Fra[?] Galton.

Figure 2.79: 1841-08-01. FG to STG and EAG page 4.

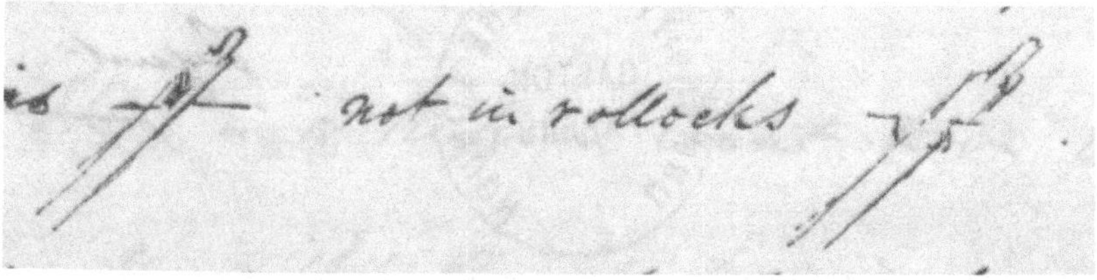

Figure 2.80: 1841-08-13 FG to STG: Pins versus Rollocks.

22 years old.[44] I think Theodore knew him. Yesterday morning I walked up Skiddaw to see the sun rise. I got to the top of the eastern peak which is not 150 feet lower than the highest one in 40 minutes. Of course saw nothing but mist. I shall however try it again tonight. We have got some sails to our boat at Keswick, it is curious how frightened all the boatmen here are of them, they never use them, even the attendant "cad" upon the party, a man ready to poach, knock down, do any thing on an emergency, refused to go into the boat on the ground of having "a wife & 5 small children."

Dear Bessy. I want to ask you about a point of etiquette. There is a Russian Count here with whom I am a great ally. The Count has a very coxy ~~very intensely~~ very insipid, exceedingly rich, but quite uneducated wife. I have been introduced to her by the Count but she merely tosses her head & says "no I do not think so" to every thing that I say, this is her ordinary behaviour to every body. Now I and other strangers (all previously introduced to her) dined there the other day. She did not receive us in the drawing room, but sat with her feet on the fender — & with her back to the door leaving the Count to do the honours. Now to the point. The count (a very punctillious man) has left the town leaving the Countess behind. I dined there the other day & have not since called. Ought I to call upon the coxy countess?

Good by your affectionate
Brother Fras. Galton.

[44] Sir Alexander James Beresford Hope (1820–1887), M.P. for Maidstone. However he appears to have been a pensioner and not a fellow commoner at Trinity.

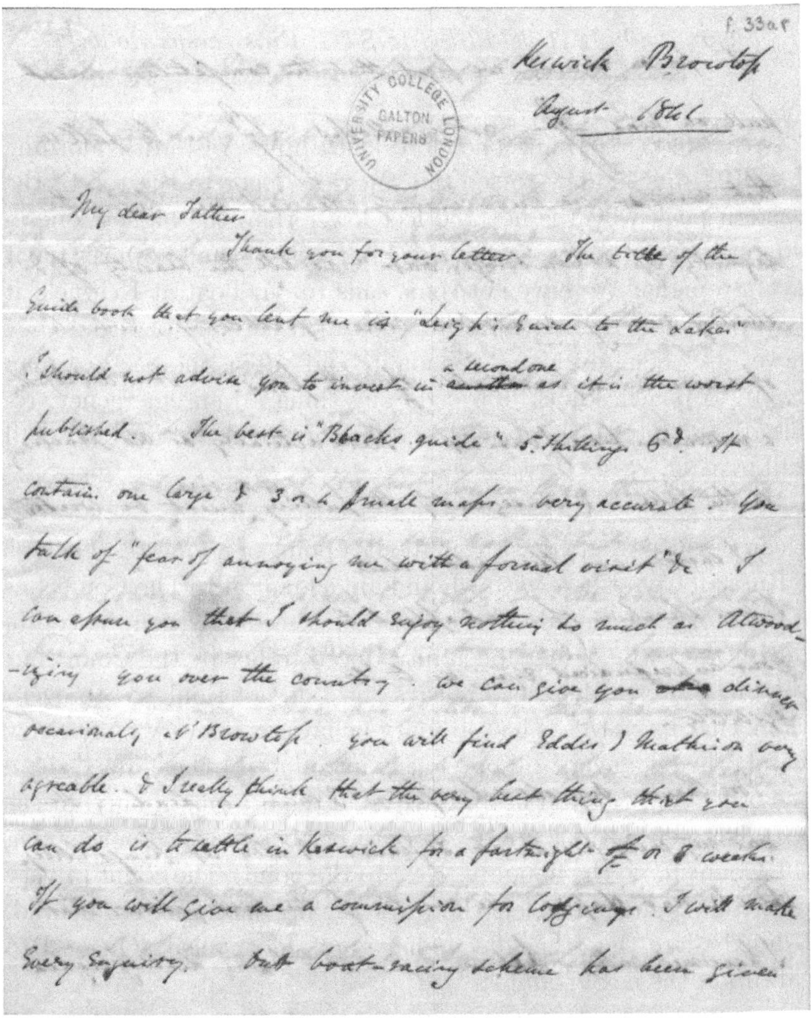

Keswick Browtop
August 1841

My dear Father

Thank you for your letter. The title of the Guide book that you lent me is "Leigh's Guide to the Lakes" I should not advise you to invest in ~~another~~ a second one as it is the worst published. The best is "Blacks guide" 5 Shillings 6d. It contains one large & 3 or 4 small maps very accurate. You talk of "fear of annoying me with a formal visit" &c. I can assure you that I should enjoy nothing so much as Atwood-izing you over the country. We can give you ~~…~~ dinners occasionally at Browtop. You will find Elder & Mathison very agreable. & I really think that the very best thing that you can do is to settle in Keswick for a fortnight [illegible] or 8 weeks. If you will give me a commission for lodgings I will make every enquiry. Our boat-racing scheme has been given

Figure 2.81: 1841-08-13. FG to STG page 1.

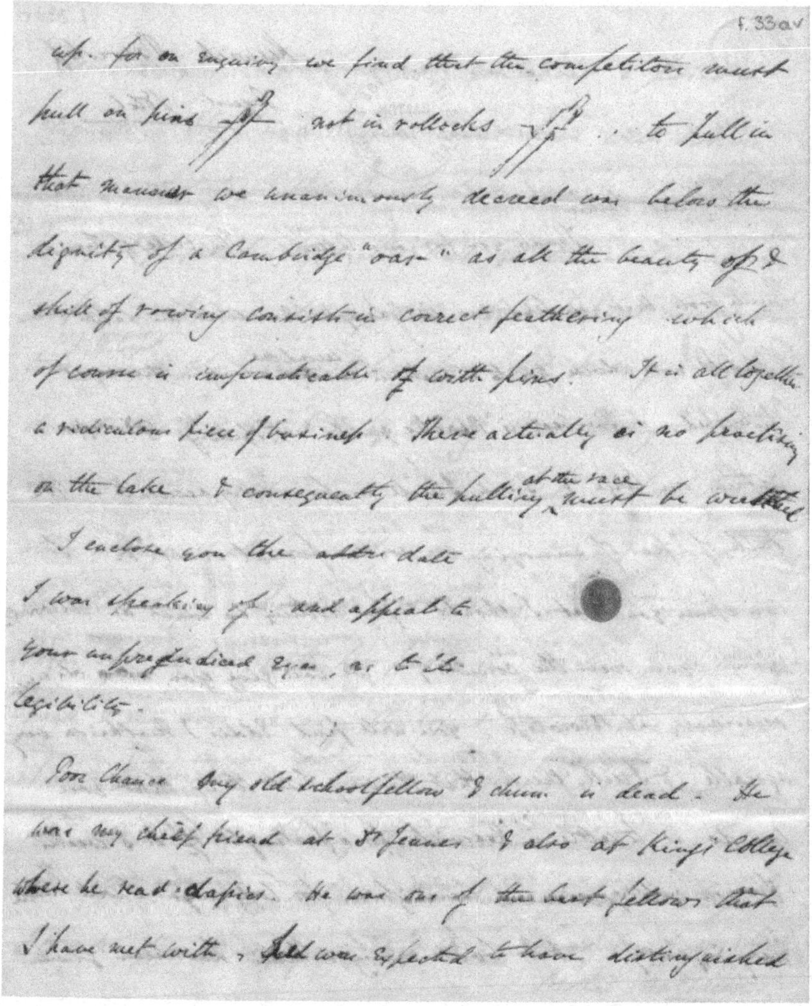

up. for on enquiry we find that the competitors must
pull on pins not in rollocks. to pull in
that manner we unanimously decreed was below the
dignity of a Cambridge "oar" as all the beauty of &
skill of rowing consists in correct feathering which
of course is impracticable with pins. It is all together
a ridiculous piece of business. There actually is no practicing
on the lake & consequently the pulling at the race must be wretched
I enclose you the [illegible] date
I was thinking of read appeal to
your unprejudiced eyes, as to its
legibility.

Poor Chance my old schoolfellow & chum is dead. He
was my chief friend at St James & also at King's College
where he read classics. He was one of the best fellows that
I have met with, & was expected to have distinguished

Figure 2.82: 1841-08-13. FG to STG page 2.

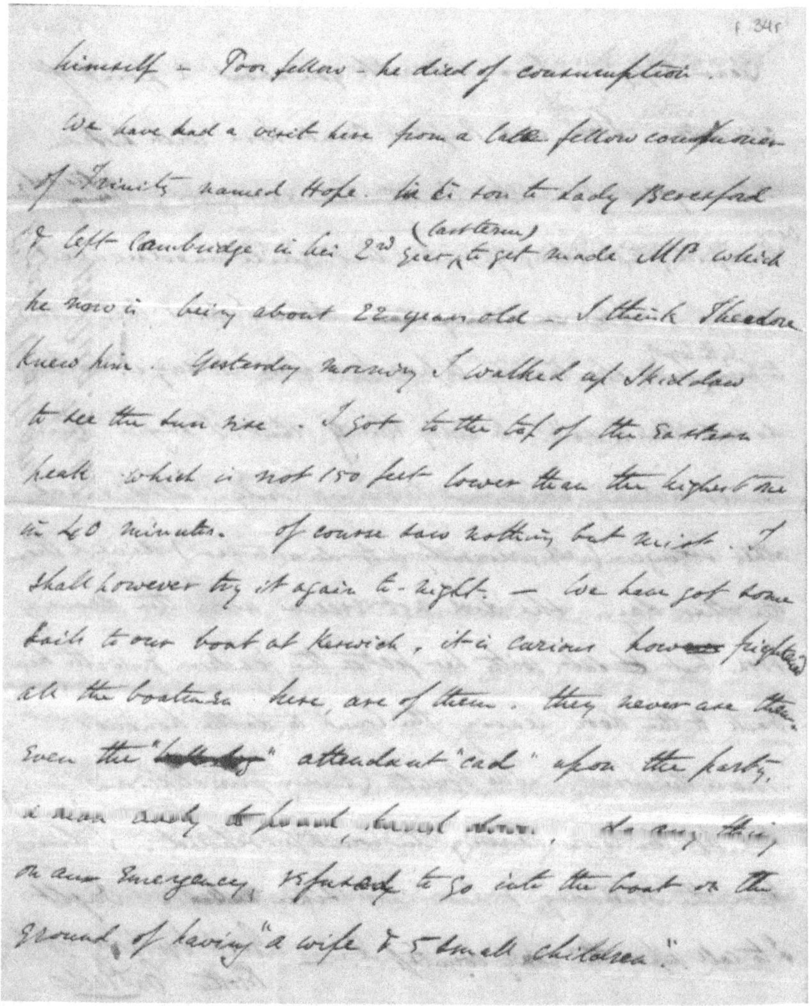

f. 34r

himself – Poor fellow – he died of consumption.

We have had a visit here from a late fellow commoner of Trinity named Hope. he is son to Lady Beresford & left Cambridge in his 2d (last term) year, to get made MP which he now is being about 22 years old – I think Theodore knew him. Yesterday morning I walked up Skiddaw to see the sun rise. I got to the top of the eastern peak which is not 150 feet lower than the highest one in 60 minutes. Of course saw nothing but mist. I shall however try it again to-night. – We have got some sails to our boat at Keswick, it is curious how frightened all the boatmen here, are of them. they never use them. Even the "[illegible]" attendant "cad" upon the party, a man ready at hand, [illegible], to do any thing on an emergency, refused to go into the boat on the ground of having "a wife & 5 small children".

Figure 2.83: 1841-08-13. FG to STG page 3.

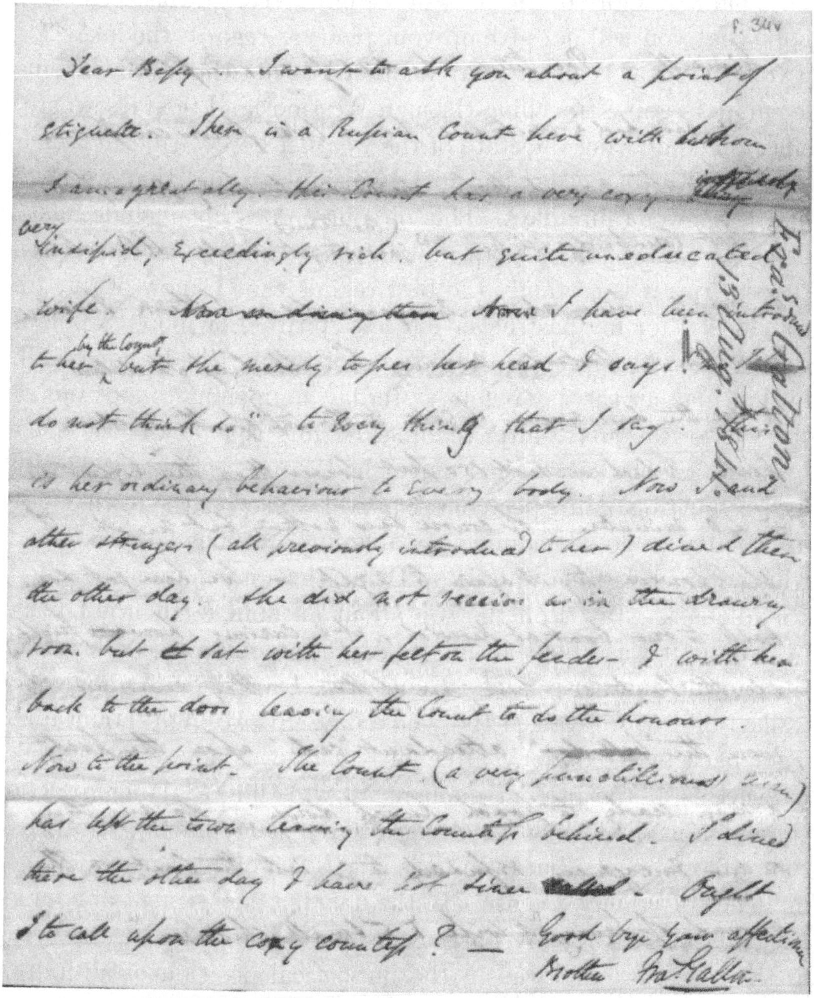

f. 34v

Dear Bessy. I want to ask you about a point of etiquette. There is a Russian Count here with whom I am a great ally. This Count has a very cozy [illegible] very insipid, exceedingly rich but quite uneducated wife. ~~Now on dining there~~ Now I have been introduced to her by the Count, but she merely tosses her head & says "I do not think so" to every thing that I say. This is her ordinary behaviour to every body. Now I and other strangers (all previously introduced to her) dined there the other day. She did not receive us in the drawing room but sat with her feet on the fender & with her back to the door leaving the Count to do the honours. Now to the point. The Count (a very gentlemanlike man) has left the town leaving the Countess behind. I dined there the other day & have not since called. Ought I to call upon the cozy countess? — Good bye your affectionate Brother Frs. Galton.

Fra. 5 Galton. 13 Aug. 1841

Figure 2.84: 1841-08-13. FG to STG page 4.

1841/08/19. Francis Galton to Samuel Tertius Galton

Browtop. August 19th 1841.

My dear Father

Thank you for your letter which I have received this evening. I hope that you will not give up your plans as regards the lakes, for if your only fear is about rainy weather I do not think you will suffer more in ~~going~~ travelling through Westmorland than elsewhere, since on comparing the state of the weather here with that which a Cambridge party has had in Devonshire we find that we have had the fewer rainy days. This of course does not include misty days. I have been deluded enough lately to climb mountains to see the sun-rise it is certainly the best regime that I know of to cure romance. I for my part never felt less spiritual or more corporeal than I did when I got to the bottom of them. I had a long walk in that manner the day before yesterday, happening to look out of the window about 12 after reading, I found that it was the most beautiful night that we had as yet had, so pocketing my whisky flask & putting on my pea-coat & plaid I walked to the town & got up a party to go, slept under a table for 35 minutes drank some whisky punch & then walked up Blencathra, ignominiously called "Saddleback" stayed on the top about an hour & then got back by 7 a.m. it was about 16 miles. As the morning was splendid I then got up another party for Ennerdale. Then slept 25 minutes & walked off—& we walked the whole day, up two high mountains. I got back by $8\frac{1}{2}$ p.m. & after all I really was not so very tired. Keswick is at the present moment all wrestling & dancing. The champion in the former has been declared today having "thrown" some 10 opponents, but even he is now taken off his legs under the influence of brandy & water. In the dancing department of course I assisted & had for my partner a damsel whom I had observed in the morning employed in the unpoetical position of all-fours, scrubbing stone steps with great diligence. I have today committed a most fearful offence in the eye of the law. I happened to be walking in a field when I saw a bull looking intensely ferocious, so I picked up a stone ~~that~~ of a size corresponding to my fears which was therefore very large, thus armed I ran to the nearest gate for escape when up jumped a hare. All thoughts of the Game laws vanished, as also of the bull. I threw the stone with a most lucky aim, & knocked the gentleman over, & then I soon got over the gate & gave him the coup-de-grace with my shillelagh. I shall eat him to-morrow or the next day. We are getting very dull. We read the Times through, advertisements & all everyday and often ask for the catalogue of the circulating library. Under these circumstances, Good bye and believe me ever

Your affectionate Son Fras Galton.

P.S. I forgot whether I have ever told you that the meaning of my initials (F.G.) in the corner of the letter ~~means~~ is that they are family letters & may be opened without the presence of the person to whom they are addressed being necessary.

Remember me to Theodore.

1841/09/15. Diary of Samuel Tertius Galton

September 15th Wednesday. Pattersdale.

... We arrived at the Royal Oak, kept by Hudson, and had comfortable apartments. Francis came to us before 12 o'clock and after lunching we drove along Borrowdale, passing along the side of Derwentwater. ... [we] stopped at the Bowder stone, 62 feet long 89 round and 36 high, which Francis clambered up in a few minutes ... we were afterwards much struck by the celerity with which he ran by the carriage side and also up and down the hills.

1841/10/14–19. Diary of Samuel Tertius Galton

October 14th. Thursday. Leamington.

Francis is gone to Birmingham. ...

October 16 Saturday. Leamington.

... Francis returned in the evening. ...

October 19th. Tuesday. Leamington.

.. Francis started by coach for Cambridge at 9 a.m. ...

1841/10/20. Francis Galton to Samuel Tertius Galton

Wednesday 20th 1841
Trin Coll.

My dear Father

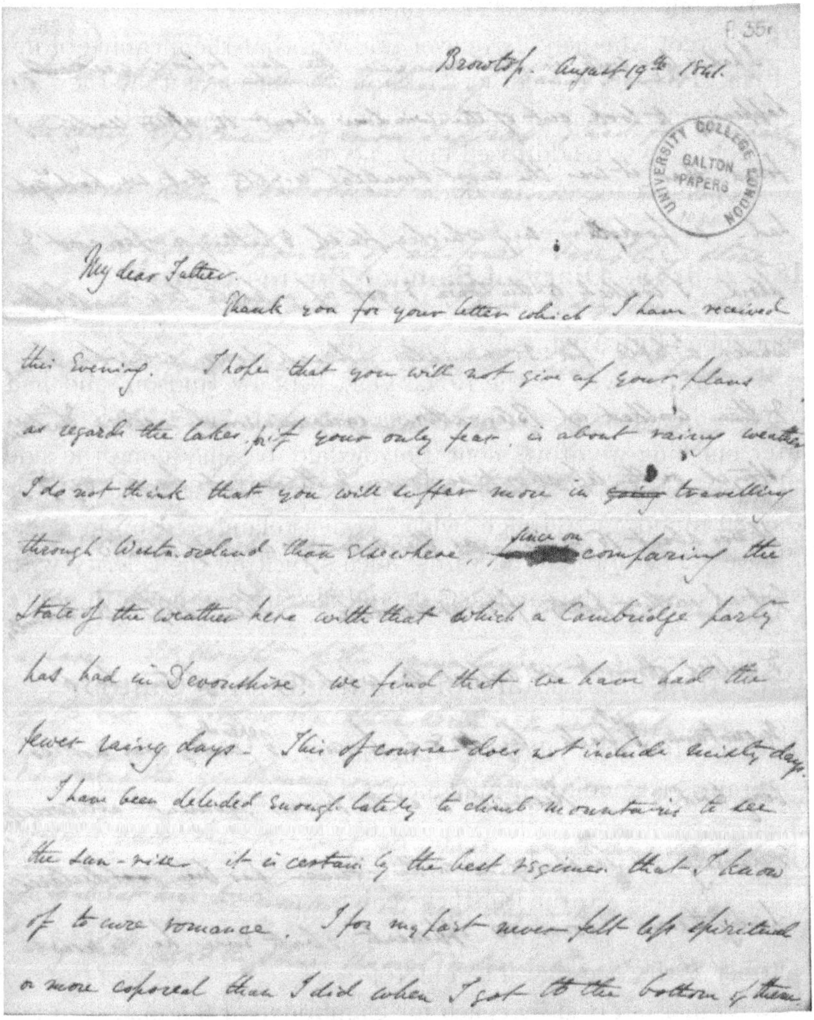

Browtop. August 19th 1841.

My dear Father

Thank you for your letter which I have received this Evening. I hope that you will not give up your plans as regards the lakes, if your only fear is about rainy weather I do not think that you will suffer more in ~~going~~ travelling through Westmoreland than elsewhere, since on comparing the State of the weather here with that which a Cambridge party has had in Devonshire, we find that we have had the fewer rainy days. This of course does not include misty days.

I have been deluded enough lately to climb mountains to see the sun-rise, it is certainly the best regimen that I know of to cure romance. I for my part never felt less spiritual or more corporeal than I did when I got to the bottom of them

Figure 2.85: 1841-08-19. FG to STG page 1.

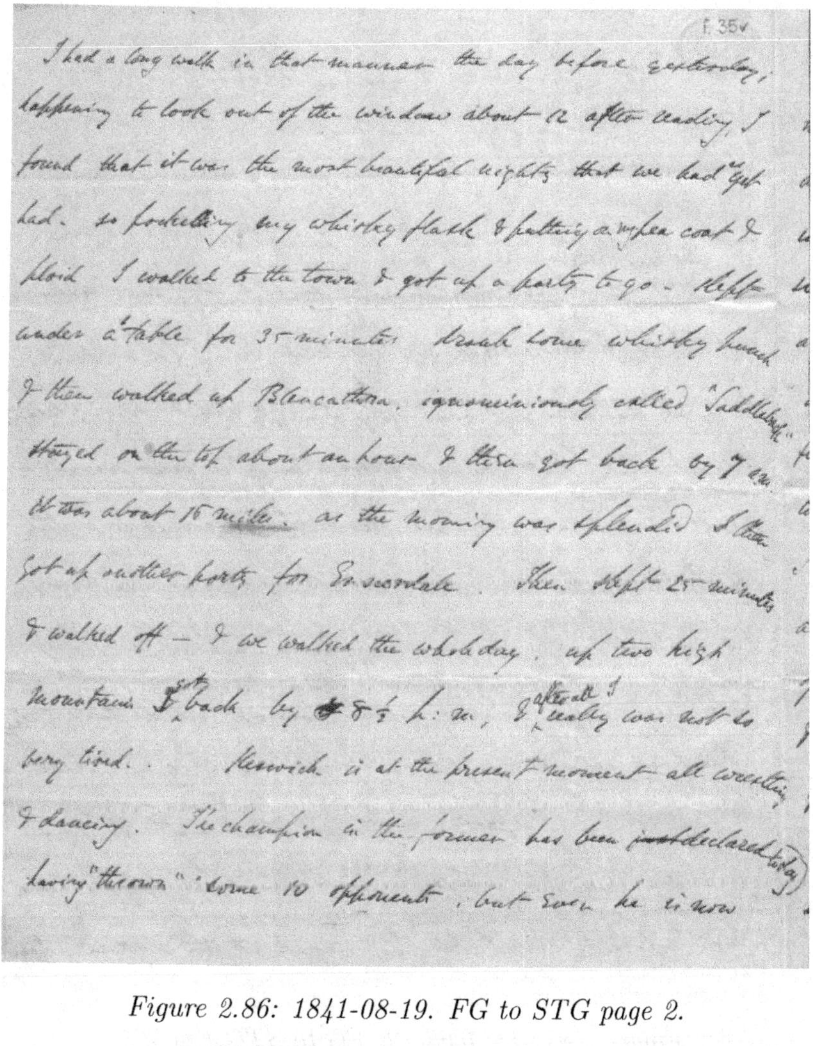

I had a long walk in that manner the day before yesterday,
happening to look out of the window about 12 after reading, I
found that it was the most beautiful nights that we had yet
had. so pocketting my whisky flask & putting on my pea coat &
plaid I walked to the town & got up a party to go. Slept
under a table for 35 minutes drank some whisky punch
& then walked up Blencathra, ignominiously called "Saddleback"
stayed on the top about an hour & then got back by 7 a.m.
It was about 15 miles. as the morning was splendid I then
got up another party for Ennerdale. Then slept 25 minutes
& walked off — & we walked the whole day, up two high
mountains & got back by ~~7~~ 8½ p.m., & after all I really was not so
very tired. Keswick is at the present moment all wrestling
& dancing. The champion in the former has been ~~just~~ declared today
having "thrown" some 10 opponents, but even he is now

Figure 2.86: 1841-08-19. FG to STG page 2.

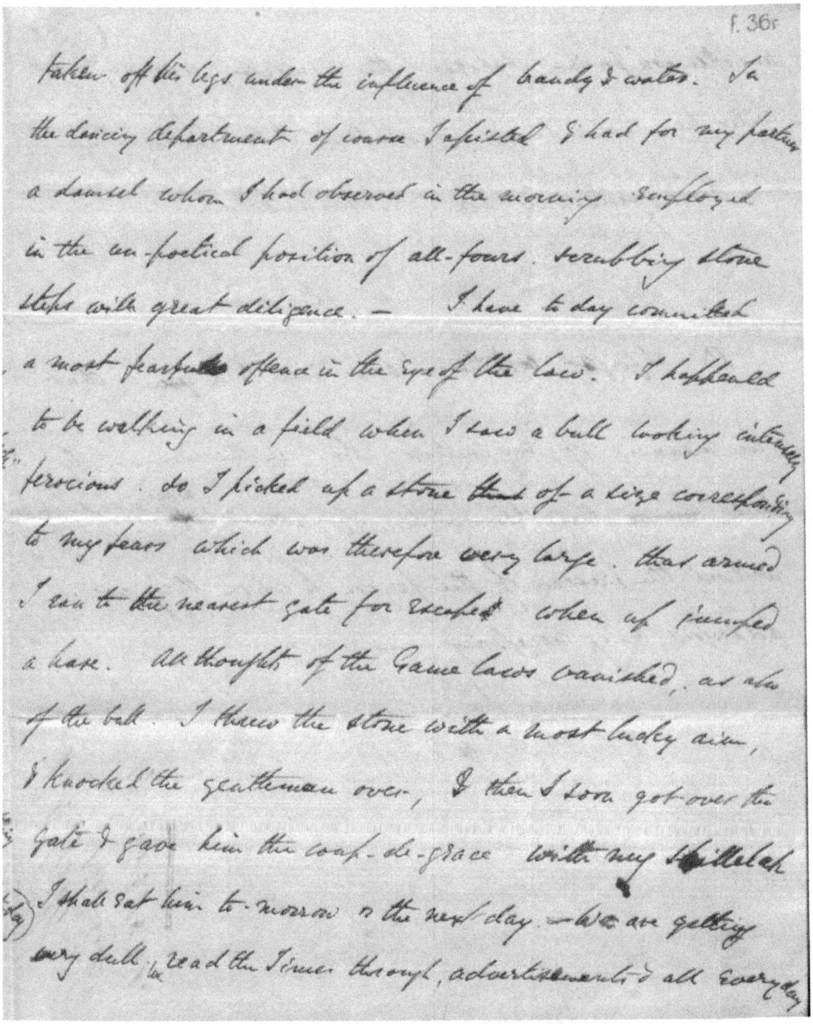
f. 36r

taken off his legs under the influence of brandy & water. In
the dancing department of course I assisted & had for my partner
a damsel whom I had observed in the morning employed
in the un-poetical position of all-fours. scrubbing stone
steps with great diligence. — I have to day committed
a most fearful offence in the eye of the law. I happened
to be walking in a field when I saw a bull looking intensely
ferocious. so I picked up a stone of a size corresponding
to my fears which was therefore very large. thus armed
I ran to the nearest gate for escape when up jumped
a hare. All thoughts of the Game laws vanished, as also
of the bull. I threw the stone with a most lucky aim,
& knocked the gentleman over, & then I soon got over the
gate & gave him the coup-de-grace with my shillelah
I shall eat him to-morrow or the next day. — We are getting
very dull, we read the Times through, advertisements & all everyday

Figure 2.87: 1841-08-19. FG to STG page 3.

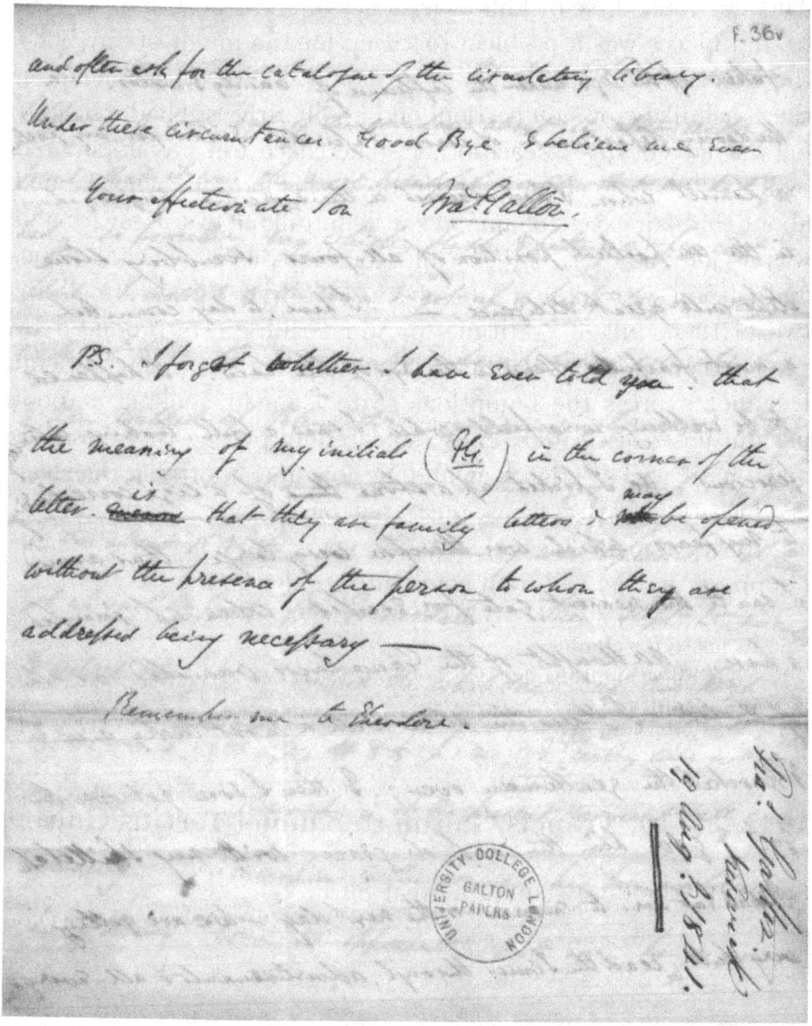
F. 36v

and often ask for the catalogue of the circulating library.
Under these circumstances Good Bye & believe me ever

Your affectionate Son Fras Galton.

PS. I forget whether I have ever told you that
the meaning of my initials (FG) in the corner of the
letter. ~~means~~ is that they are family letters & may be opened
without the presence of the person to whom they are
addressed being necessary —

Remember me to Theodore.

Dr Galton
Warwick
19 Aug 1841.

Figure 2.88: 1841-08-19. FG to STG page 4.

I left Leamington the only Cantab on the coach by the side of a jolly old Coachman who had been a horse dealer at Northampton & had sold horses to Uncle Hubert he made sundry enquiries after you. On arriving at ~~Keswick~~ Weedon a complete shoal of Cambridge men poured out from one of the trains amongst whom was Theodore and three or four other allies of mine, how they all were to find places was a problem too deep for the minds of any body there except the coachman's to surmount. However they hung on the coach like crows on carrion, and a jolly drive we had, recounting our adventures to each other. The coach top was unpolluted by a freshman. I called on Mathison[45] this morning who skipped about through excess of animal spirits in talking over Keswick and was as jolly as ever. I then called on Hopkins[46] who takes me, and I begin with him on Monday. My Keswick friends are all up, two of them full of gratitude for wonderful effects produced by prescriptions which I had left them & I have got a new patient. I cannot express the bumptious state I am in, looking at poor bashful unsophisticated "cubs" so carefully pulling their gown to make it sit well & fidgety at finding how uncomfortable their cap is which they have unconsciously put on the wrong way, all over as "fresh" as paint bless their innocencies. So Whewell is Master I suppose that he will soon come into residence.

In haste for chapel,
Your affectionate son,
Fras Galton

1841/11/10. Francis Galton to Samuel Tertius Galton

Excuse my blots as I am in a great hurry.

<u>Tuesday</u>

My dear Father

I am very sorry that from having been ~~half the~~ either too lazy or too much occupied ~~that~~ I have not written sooner, though I consider you too little of an invalid to be ~~any~~ further anxious about your health. Emma has probably given you a full account of her

45 Mathison, William Collings. See page 486.
46 Hopkins, William. See page 484.

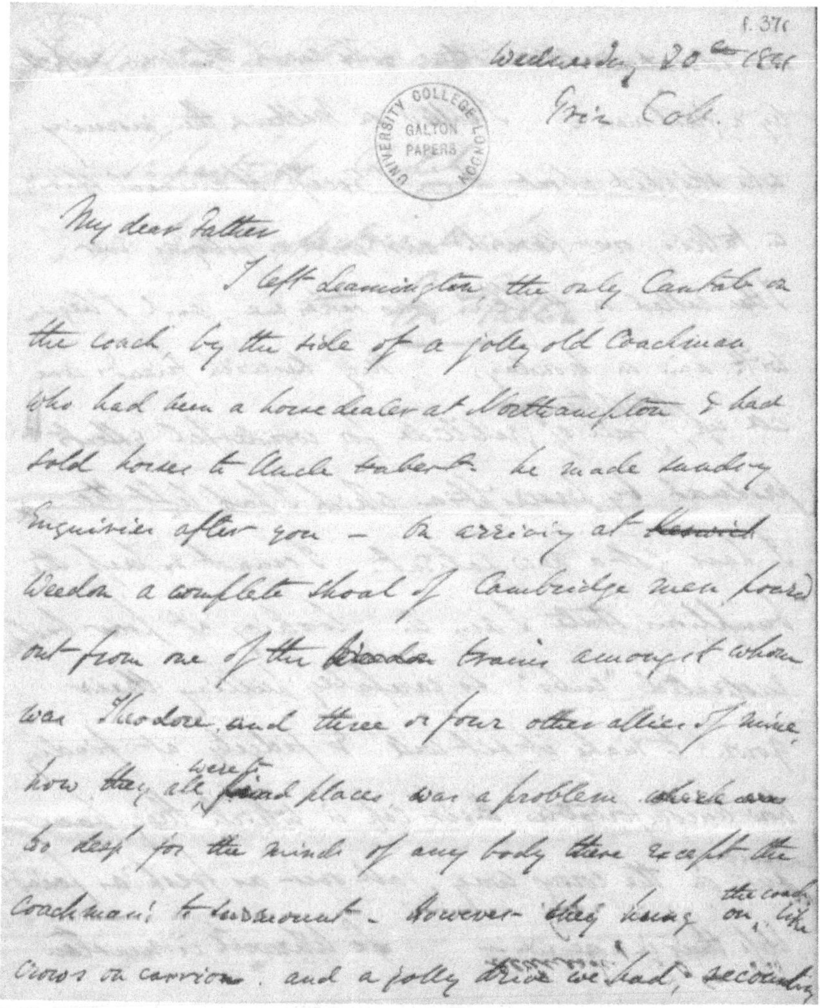
f. 37

Wednesday 20th 1841
Trin. Coll.

My dear Father
I left Leamington, the only Cantab on the coach, by the side of a jolly old Coachman who had been a horse dealer at Northampton & had sold horses to Uncle Hubert. he made leading enquiries after you — On arriving at ~~Norwich~~ Weedon a complete shoal of Cambridge men poured out from one of the ~~London~~ trains amongst whom was Theodore, and three or four other allies of mine, how they all were ~~found~~ fit places, was a problem ~~which was~~ so deep for the minds of any body there except the Coachman's to understand. however they hung on the coach, like crows on carrion. and a jolly drive we had, according

Figure 2.89: 1841-10-20. FG to STG page 1.

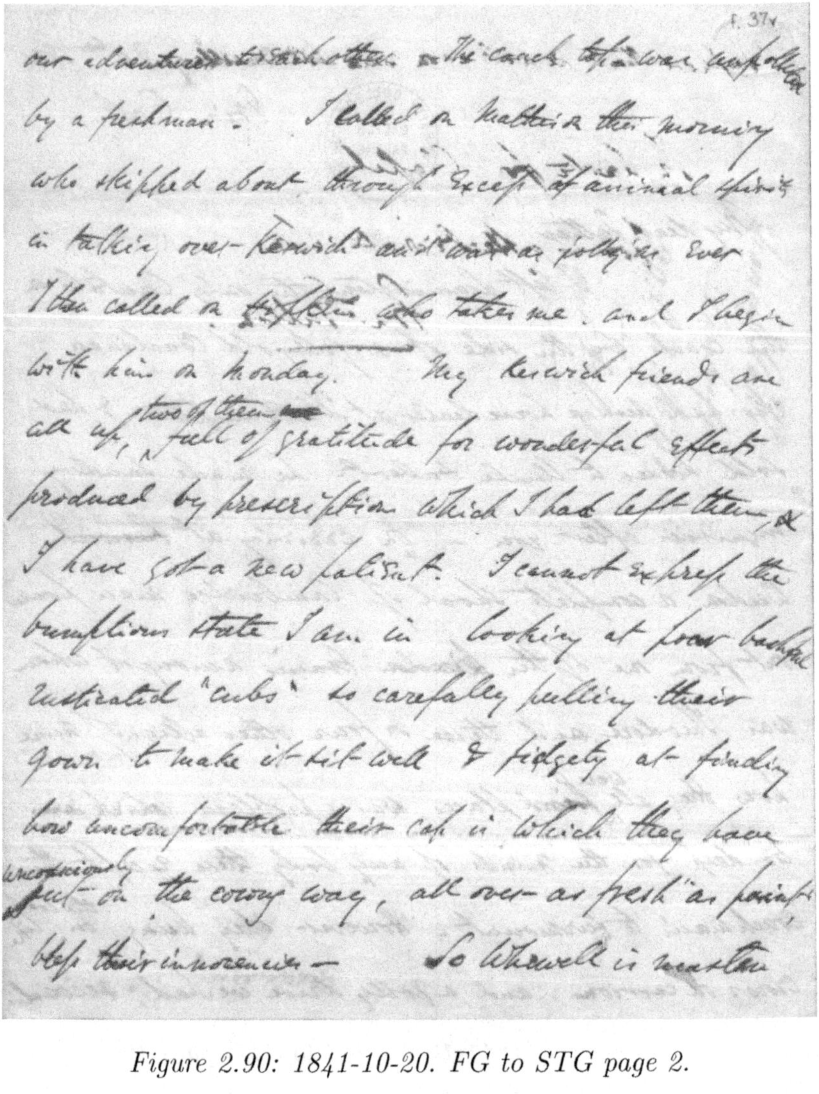
f. 37v

our adventures to each other. The coach top was unpolluted by a freshman. I called on Mathison this morning who skipped about through excess of animal spirits in talking over Keswick and was as jolly as ever. I then called on Hopkins who takes me, and I begin with him on Monday. My Keswick friends are all up, two of them full of gratitude for wonderful effects produced by prescription which I had left them, & I have got a new patient. I cannot express the bumptious state I am in looking at poor bashful rusticated "cubs" so carefully pulling their gown to make it sit well & fidgety at finding how uncomfortable their cap is which they have unconsciously put on the wrong way, all over as fresh as paint, bless their innocence — So Whewell is master

Figure 2.90: 1841-10-20. FG to STG page 2.

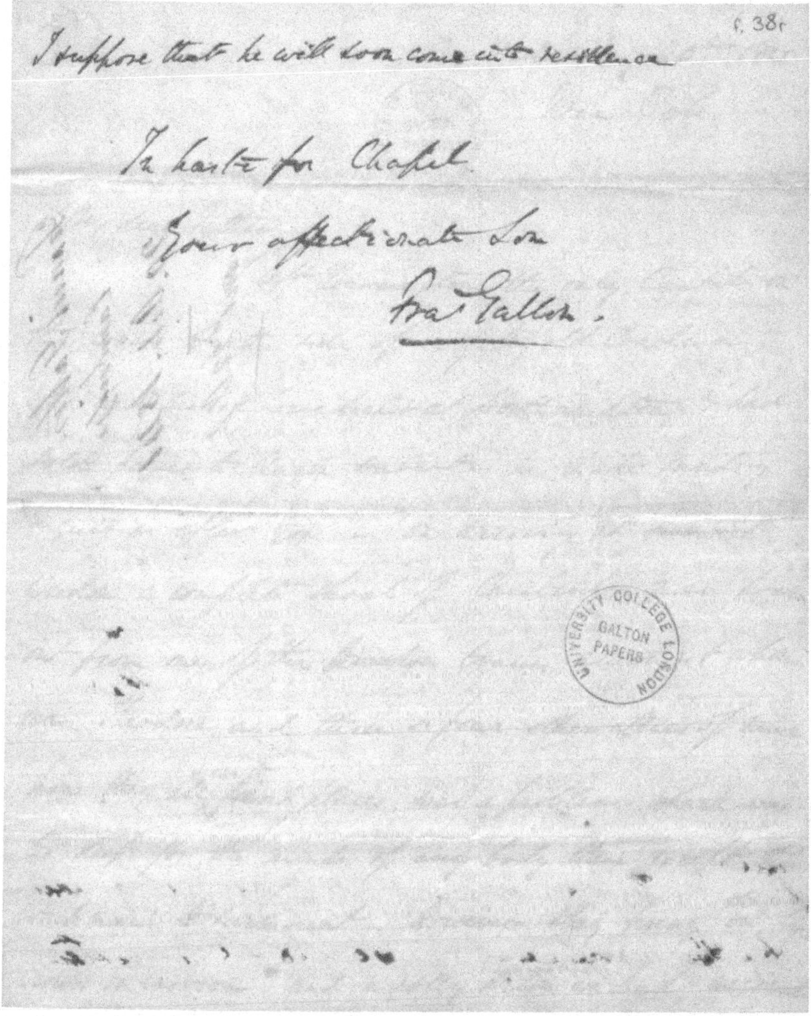

Figure 2.91: 1841-10-20. FG to STG page 3.

Francis Galton
Cambridge
20. Oct. 1841
ansd 22 Oct.

Figure 2.92: 1841-10-20. FG to STG page 4.

proceedings in Cambridge & I trust rescued me from Mr. Hodgson's malicious charge.[47] She was most active & tired down both Theodore & myself. Thank you very much for your statement of my accounts they were certainly most beautifully written out and quite a model; in one point however they were not quite as useful as otherwise they might have been for after reading it through carefully 5 times I abandoned all hope of making out the meaning of any one single line in utter despair. The words appear all bewitched for I can't make out which is substantive & which is nominative case &c. &c.

As a specimen "Debtor to balance agreed £.." Should you have spare time, would you be so kind as to write two or three lines in account book style with their interpretation in popular English & then I have no doubt but that I shall make out the rest of the paper wh. you so kindly sent me.

My box arrived safely and the books inside in good preservation, with the exception of one book right through which like Sisera's temples a hob nail was driven—however it was only some temperance tracts bound up & therefore it is quite immaterial.

Emma was not sure whether my D'Israeli's Curiosities of Literature was packed up—it is.

I like Hopkins[48] more and more every day, and I never enjoyed Cambridge so thoroughly. Loves to all.

Goodbye, Your affectionate Son
Fras. Galton

1841/11/11. Francis Galton to Samuel Tertius Galton

Trinity.
Thursday Evening
Nov. 1841.

My dear Father

Thanks for the 2nd edition of my accounts which I received this morning, and still greater thanks for the explanatory notes by Bessy thereunto attached.

Hopkins progresses capitally. I had forgotten to tell you that I find that his charges are only £72 per annum instead of £100 as

47 Hodgson, Joseph. See page 483.
48 Hopkins, William. See page 484.

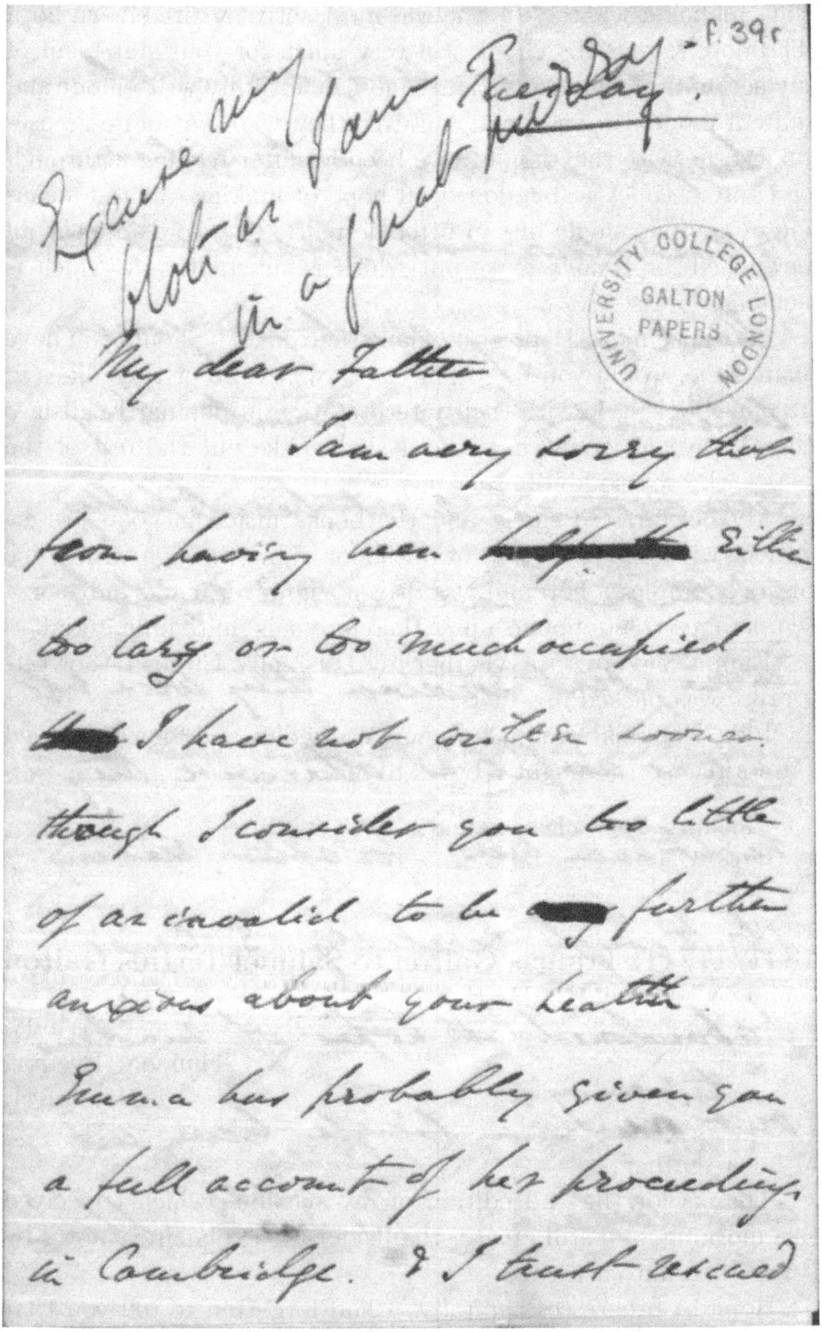

f. 39r

Excuse my blots I am in a great hurry Tuesday.

My dear Father

I am very sorry that
from having been [redacted] either
too lazy or too much occupied
[redacted] I have not written sooner.
though I consider you too little
of an invalid to be [redacted] further
anxious about your health.
Emma has probably given you
a full account of her proceedings
in Cambridge. & I trust accused

Figure 2.93: 1841-11-10. FG to STG page 1.

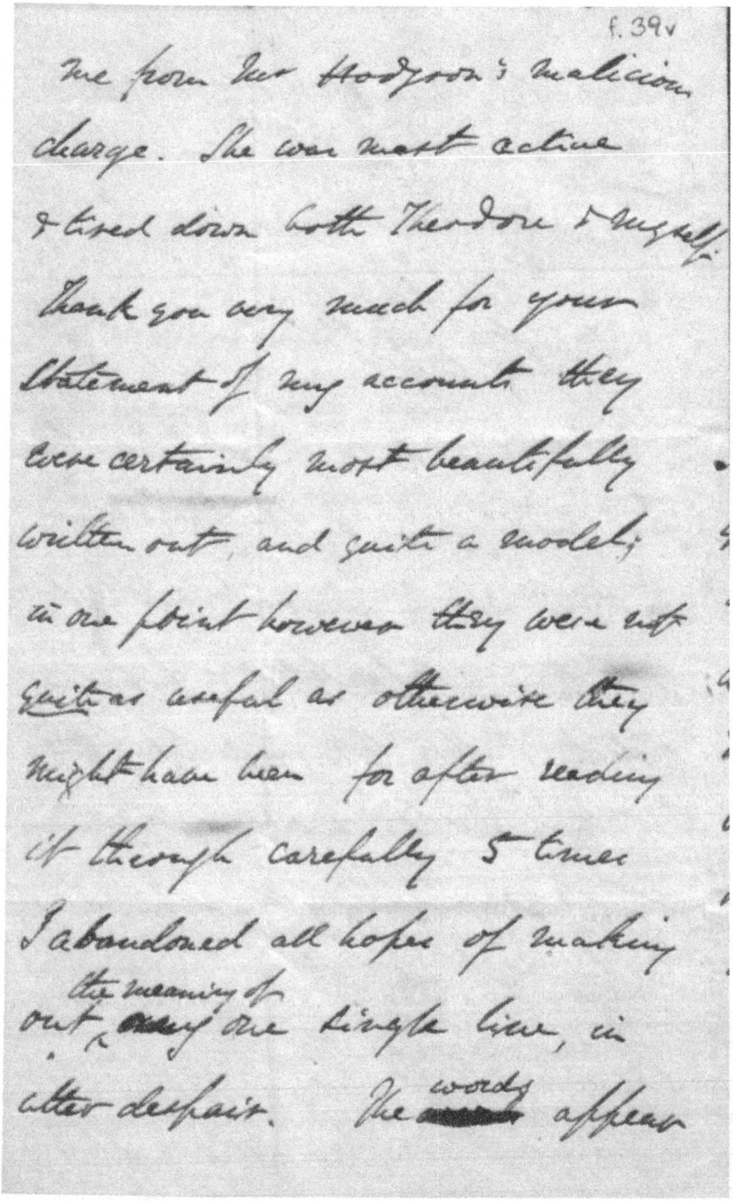
f. 39v

me from Mr Hodgson's malicious
charge. She was most active
& tired down both Theodore & myself.
Thank you very much for your
Statement of my accounts they
were certainly most beautifully
written out, and quite a model;
in one point however they were not
quite as useful as otherwise they
might have been for after reading
it through carefully 5 times
I abandoned all hopes of making
out the meaning of one single line, in
utter despair. The words appear

Figure 2.94: 1841-11-10. FG to STG page 2.

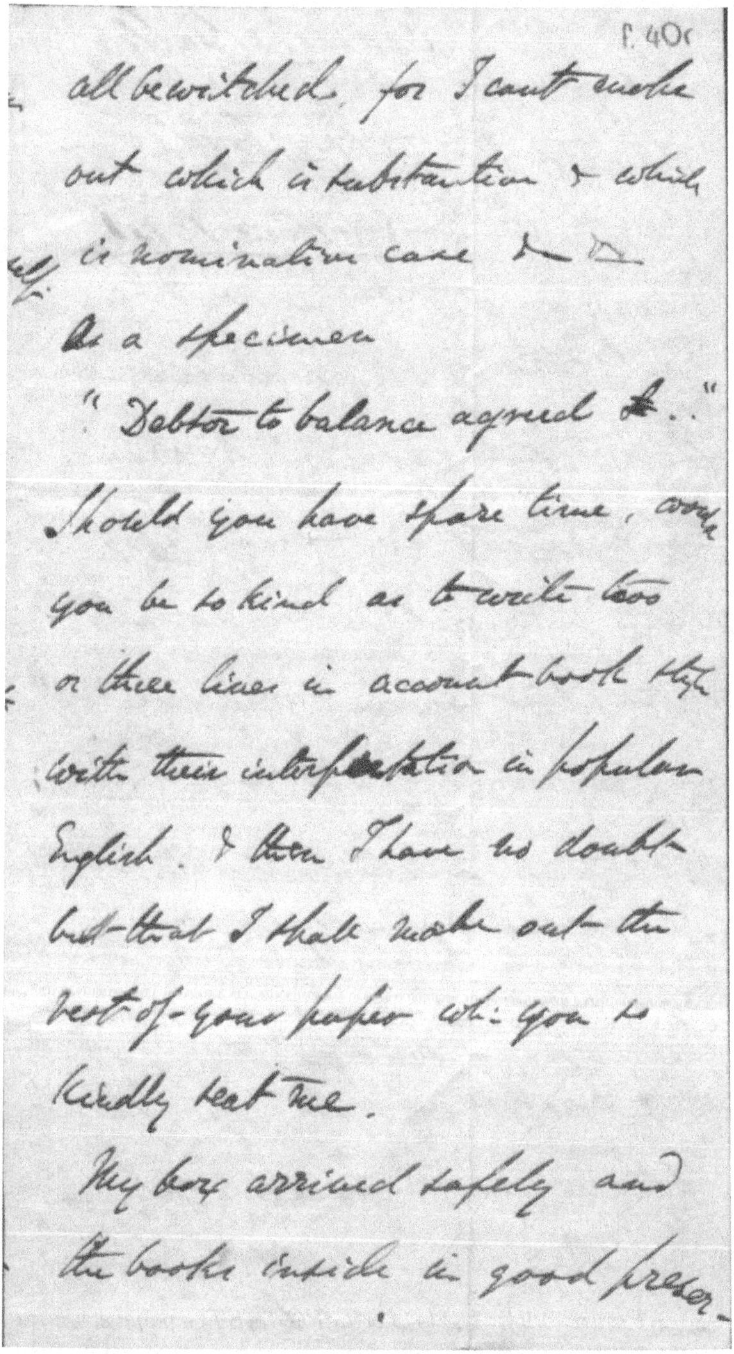
f. 40

all bewitched, for I cant make
out which is substantive & which
is nominative case &c &c
As a specimen
"Debtor to balance agreed &c.."
Should you have spare time, would
you be so kind as to write two
or three lines in account book style
with their interpretation in popular
English. & then I have no doubt
but that I shall make out the
rest of your paper wh: you so
kindly sent me.
My box arrived safely and
the books inside in good preser-

Figure 2.95: 1841-11-10. FG to STG page 3.

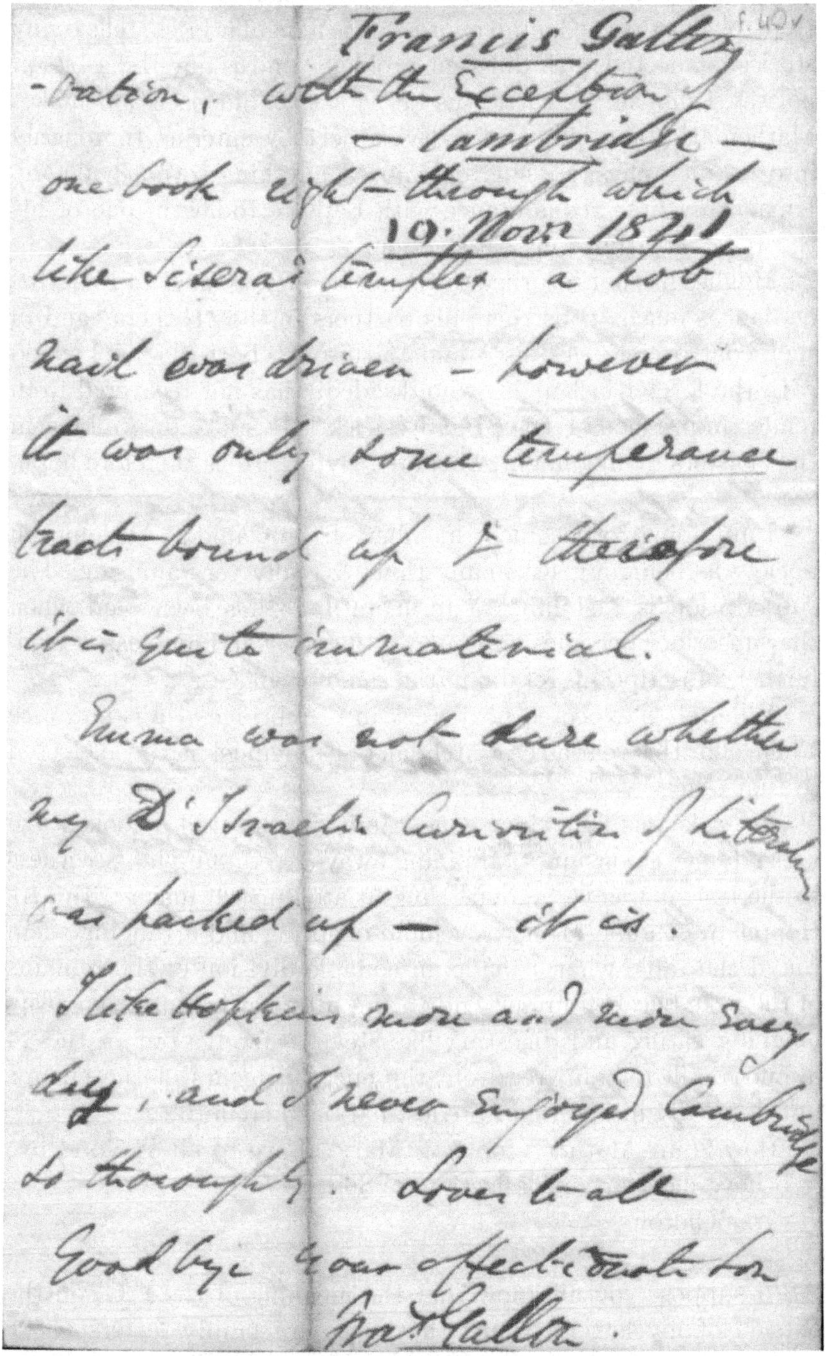
Francis Galton f. 40v

-vation, with the exception of
— Cambridge —
one book right-through which
19. Nov 1841
like Sisera's temples a hob
nail was driven - however
it was only some temperance
tracts bound up & therefore
it is quite immaterial.

Emma was not sure whether
my D'Israeli's Curiosities of Literature
was packed up — it is.

I like Hopkins more and more every
day, and I never enjoyed Cambridge
so thoroughly. Love to all
Good bye your affectionate son
Frans Galton.

Figure 2.96: 1841-11-10. FG to STG page 4.

currently reported, this will make a jolly difference to my finances. Hopkins[49] to use a Cantab expression is a regular brick, tells funny stories connected with different problems and is ~~any thing but~~ in no way Donnish, he rattles us on at a splendid pace and makes Mathematics anything but a dry subject by entering thoroughly into its metaphysics. I never enjoyed anything so much before. I made my first acquaintance with Laplace today, in one of his theorems, greatly to my satisfaction.

Hughes has not returned to Cambridge. He is an utter Puseyite, he dances much & instructs his partners in the "Fathers" and in their controversies. Eddis[50] and Mathison[51] both bloom. I wined with the former last night, who decidedly has not recovered from tender impressions received at Keswick. He spoke on walking in the cloisters by moonlight, & quoted Byron. He is therefore hopeless.

I am going to become a member of the Camden Antiquarian Society as being a gentlemanly thing & really very amusing. The subscription is 7 shillings a term until £3 has been paid when the subscriber becomes an honorary member and is released from further subscription, retaining the same privileges.

Whewell is expected next week in Cambridge. He is not Vice Chancellor this year as Dr. Archdall of Emmanuel has been elected to that office.

My cake has long since gone. As it disappeared it looked like a girl dying of consumption pining away & retaining its sweetness to the last; it was very good. I ingratiated myself much with Mrs Hoppit by sending for her sevenfold offspring and arranging them round the table when I made an equable division of the remains of the cake between myself and them, afterwards sent them away laughing eating and digesting like steam. Twelve out of the 14 jampots still remain free from the encroachments of the enemy, but decidedly in a state of predestination thereunto.

How is my Mother's foot? & Mary? Love to all & Good bye & believe me ever your affectionate Son

Fras Galton.

P.S. I suppose you all know that the meaning of the F.G. on the outside of my letters means that they are family letters to be

---

[49] Hopkins, William. See page 484.
[50] Eddis, Arthur Shelly. See page 483.
[51] Mathison, William Collings. See page 486.

opened by anybody and merely addressed to some particular person for the satisfaction of the postmaster.

1841/12/16–27. Diary of Samuel Tertius Galton

December 16th. Thursday. Leamington.
... Francis arrived from Cambridge, at 6 p.m.

December 18th. Saturday. Leamington.
... Darwin took Francis to Edstone. ...

December 23rd. Thursday. Leamington.
... I attended Lafontaine's lecture upon Mesmerism with Francis. He succeeded well with two of his patients but had little success with volunteer patients. He and Francis mutually failed in endeavoring to mesmerize each other.

December 26th. Sunday. Leamington.
... I went to Chapel in the morning with Violetta, Bessy and Francis. ...

December 27th. Monday. Leamington.
I rode to Warwick before breakfast. I attended Lafontaine's lecture on mesmerism with Francis when a Mr. Frant addressed the lecturer in rather rude terms upon the partial failure of his experiments upon his own subjects.

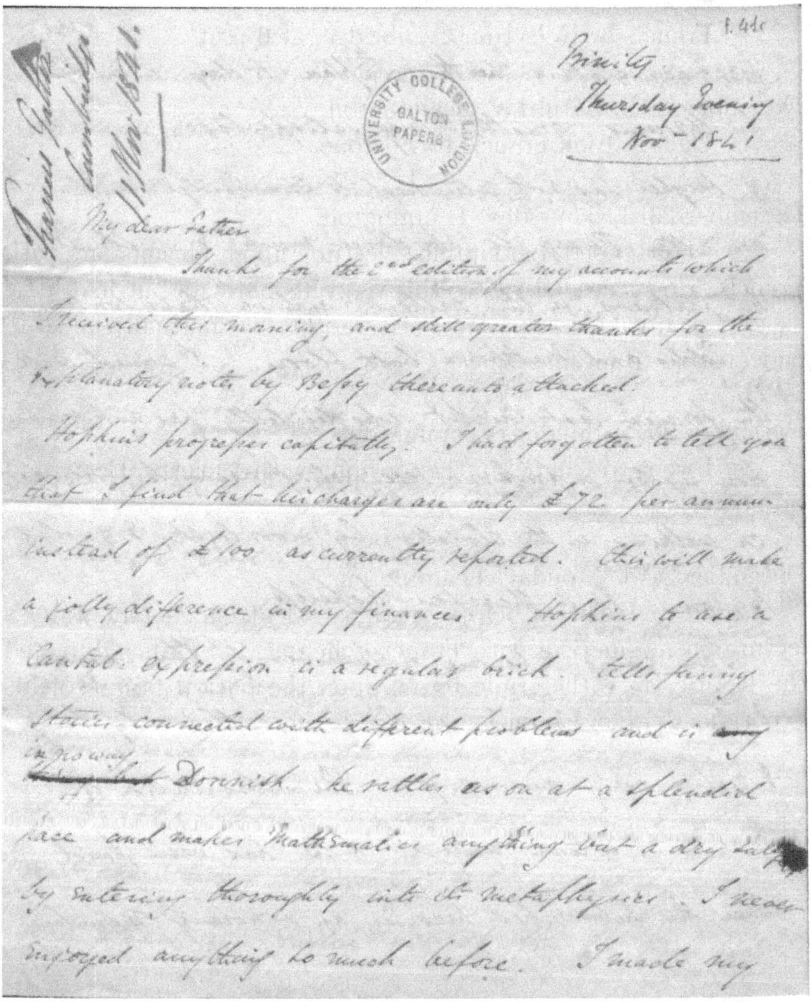
Francis Galton
Cambridge
11 Nov 1841.

f. 41c

Trinity
Thursday Evening
Nov 1841

My dear Father

Thanks for the 2nd edition of my accounts which I received this morning, and still greater thanks for the explanatory note by Bessy thereunto attached.

Hopkins progresses capitally. I had forgotten to tell you that I find that his charges are only £72 per annum instead of £100 as currently reported. This will make a jolly difference in my finances. Hopkins to use a Cantab expression is a regular brick, tells funny stories connected with different problems and is in no way Donnish he rattles us on at a splendid pace and makes Mathematics anything but a dry study by entering thoroughly into its metaphysics. I never enjoyed anything so much before. I made my

Figure 2.97: 1841-11-11. FG to STG page 1.

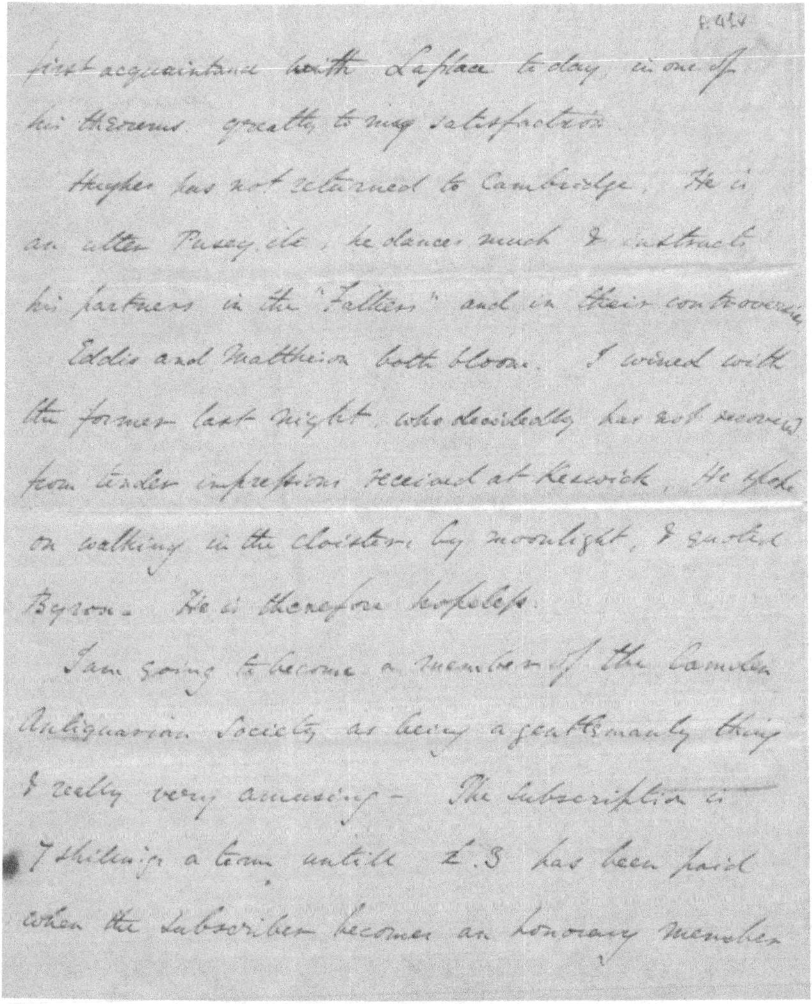

first acquaintance with Laplace to day, in one of his theorems greatly to my satisfaction.

Hughes has not returned to Cambridge. He is an utter Pusey-ite, he dances much & instructs his partners in the "Fathers" and in their controversy.

Eddis and Mathison both bloom. I wined with the former last night who decidedly has not recovered from tender impressions received at Keswick. He spoke on walking in the cloisters by moonlight, & quoted Byron. He is therefore hopeless.

I am going to become a member of the Camden Antiquarian Society as being a gentlemanly thing & really very amusing – The subscription is 7 shillings a term until £.3 has been paid when the subscriber becomes an honorary member

Figure 2.98: 1841-11-11. FG to STG page 2.

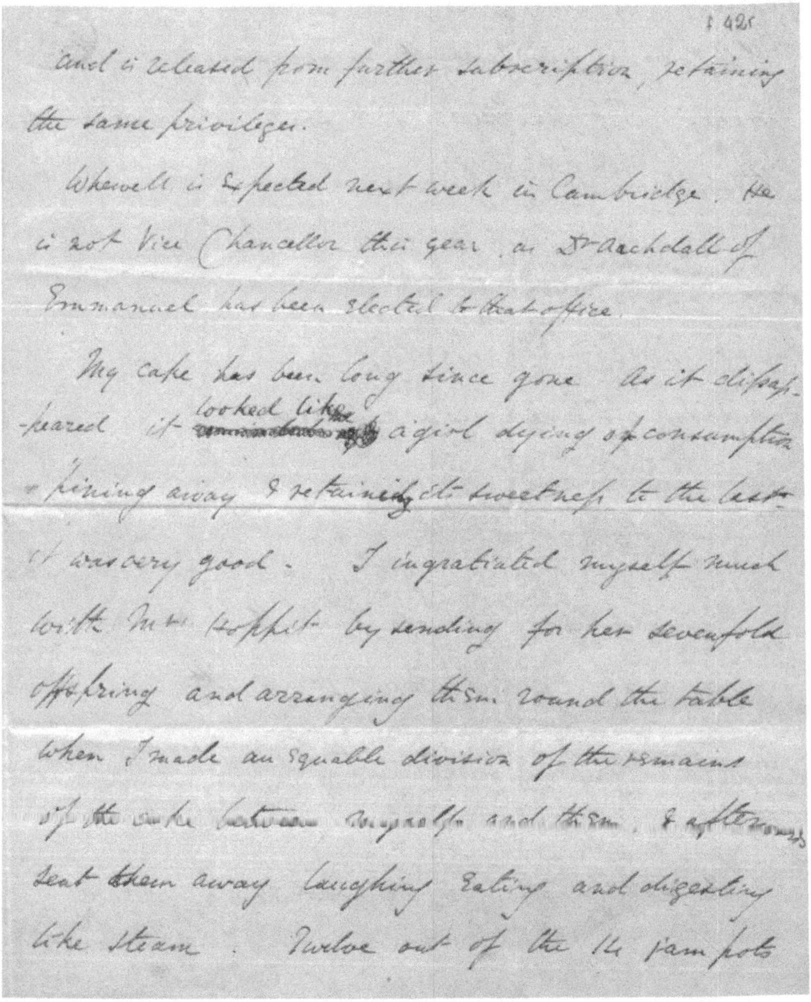

f 42r

and is released from further subscription, retaining the same privileges.

Whewell is expected next week in Cambridge. He is not Vice Chancellor this year, as Dr Archdall of Emmanuel has been elected to that office.

My cake has been long since gone. As it disappeared it ~~[illegible]~~ looked like a girl dying of consumption – pining away & retaining its sweetness to the last – it was very good. I ingratiated myself much with Mrs Hoppit by sending for her sevenfold offspring and arranging them round the table when I made an equable division of the remains of the cake between myself and them & afterwards sent them away laughing eating and digesting like steam. Twelve out of the 14 jam pots

Figure 2.99: 1841-11-11. FG to STG page 3.

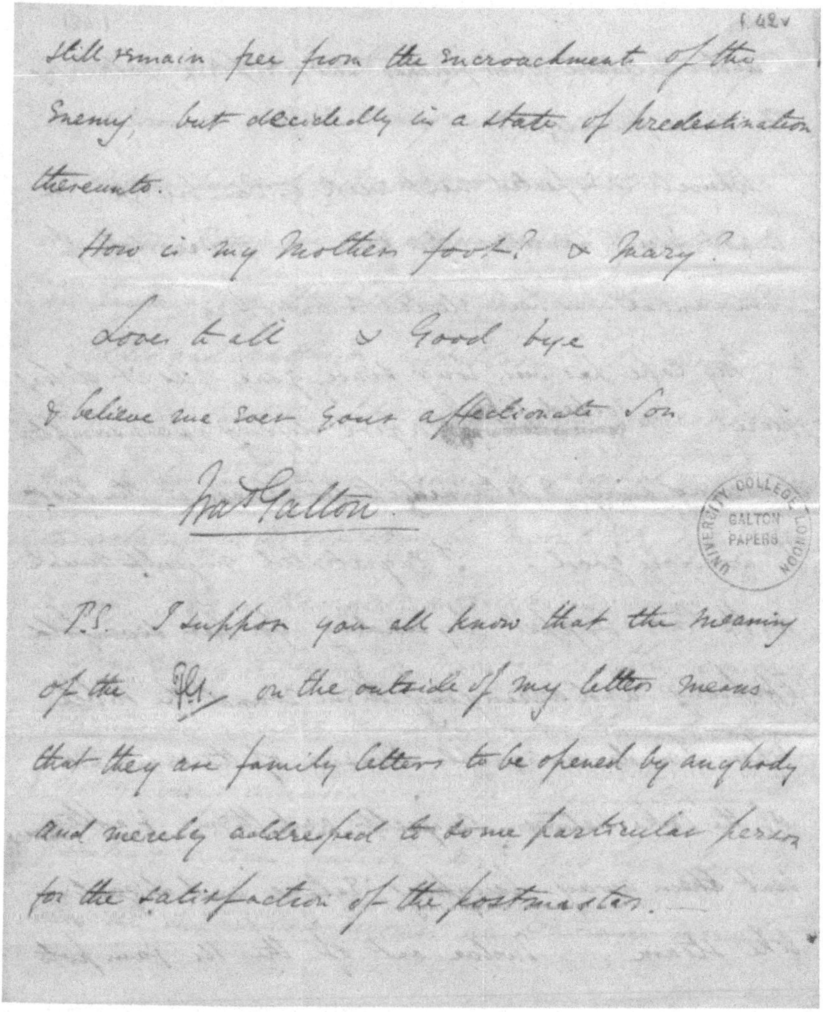
f. 42v

still remain free from the encroachments of the Enemy, but decidedly in a state of predestination thereunto.

How is my Mother's foot? & Mary?

Love to all & Good bye

& believe me ever your affectionate Son

Frank Galton.

P.S. I suppose you all know that the meaning of the F.L. on the outside of my letters means that they are family letters to be opened by anybody and merely addressed to some particular person for the satisfaction of the postmaster.

Figure 2.100: 1841-11-11. FG to STG page 4.

Figure 2.101: Arthur Cayley is installed as Senior Wrangler, 1842.

Chapter 3

1842

1842/01/03–08. Diary of Samuel Tertius Galton

January 3rd. Monday. Leamington.
... Francis returned from Birmingham. ...

January 6th. Thursday. Leamington.
... We all dined with Mrs. Phillips where we met the Rev. Mr. Goldney whom Francis attempted to mesmerise.

January 8th. Saturday. Leamington.
Francis started at 9 a.m. by the Eagle coach for Cambridge. ...

1842/01/21. Francis Galton to Samuel Tertius Galton

Jany 21 1842
Trin: Coll:

My dear Father

Thank you for your letter which I received this morning. The Math: Examination for degrees is just over & as you will see by the accompanying paper Cayley is 1st & Simpson[1] 2nd wrangler; these were very far superior to the rest. Hopkins[2] told me today that Simpson was 1000 marks ahead of the 3d wrangler & the getting of 500 marks only entitles a man to be a wrangler. The 1st & 2nd wranglers are considered to be the most superior men,

[1] Charles Turner Simpson (1819-1902), who went on to practice at the Bar.
[2] Hopkins, William. See page 484.

for at least many years that Cambridge has produced. The rest of the Tripos as usual.

The examination papers this year are much easier than usual.

The Glaciarium is composed of a mixture of carbonate & sulphate of potash which deliquesces in their water of crystallisation & afterwards hardens. This Miller told me, who himself had heard it from Faraday. I have not seen the specification.

I shall be quite dissipated next week. Monday I dine with Dr Fisher[3] and go to an evening party at Hopkins Tuesday Bachelors' Ball Wednesday I go to Keswick &c.

Good bye, your affectionate Son
Francis Galton

1842/02/28. Diary of Samuel Tertius Galton

February 28th. Monday Edgbaston.

... I sat up late with Hodgson[4] speaking of Hubert and Francis.

1842/03/08. Diary of Samuel Tertius Galton

March 8th. Tuesday. Leamington.

... I received a letter from Francis announcing his wish to go to Heidelberg to read.

1842/03/09. Francis Galton to Samuel Tertius Galton

Hurrah. I'm through.[5]

Fras Galton.
Wed^y 1842

1842/03/11. Diary of Samuel Tertius Galton

March 11th. Friday. Leamington.

I received a letter last night from Francis announcing his having passed the Little Go at Cambridge on Wednesday last. ...

3 William Webster Fisher.

4 Hodgson, Joseph. See page 483.

5 The "Little Go" examinations.

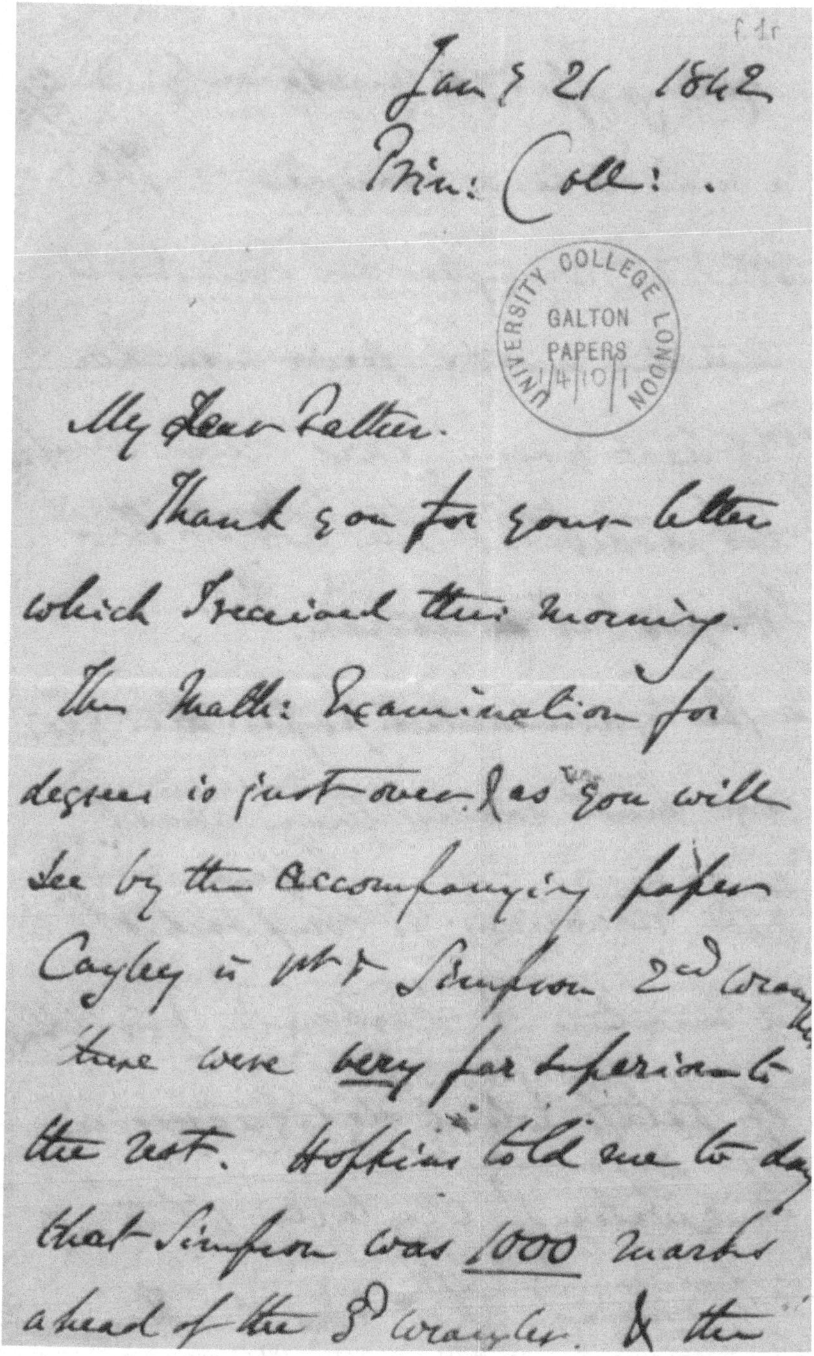

f. 1r

Jany 21 1842
Trin: Coll:.

UNIVERSITY COLLEGE LONDON GALTON PAPERS 1/4/10/1

My dear Father.

Thank you for your letter which I received this morning. The Math: Examination for degree is just over & as you will see by the accompanying paper Cayley is 1st & Simpson 2nd Wrangler; these were very far superior to the rest. Hopkins told me to day that Simpson was 1000 marks ahead of the 3rd Wrangler. & the

Figure 3.1: 1842-01-21. FG to STG page 1.

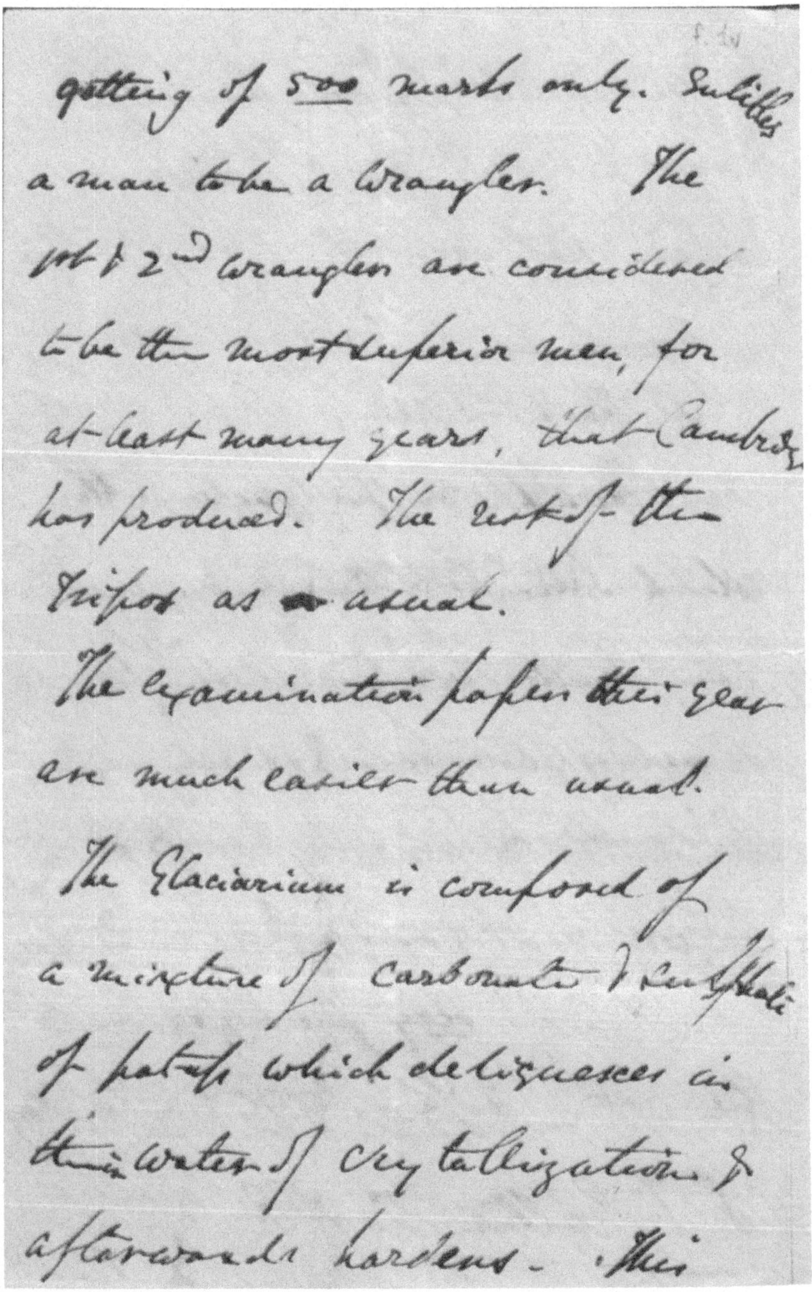

f. 1v

getting of 500 marks only. Entitles
a man to be a Wrangler. The
1st & 2nd Wranglers are considered
to be the most superior men, for
at least many years, that Cambridge
has produced. The rest of the
Tripos as usual.
The examination papers this year
are much easier than usual.
The Glaciarium is composed of
a mixture of carbonate & sulphate
of potash which deliquesces in
their water of crystallization &
afterwards hardens. This

Figure 3.2: 1842-01-21. FG to STG page 2.

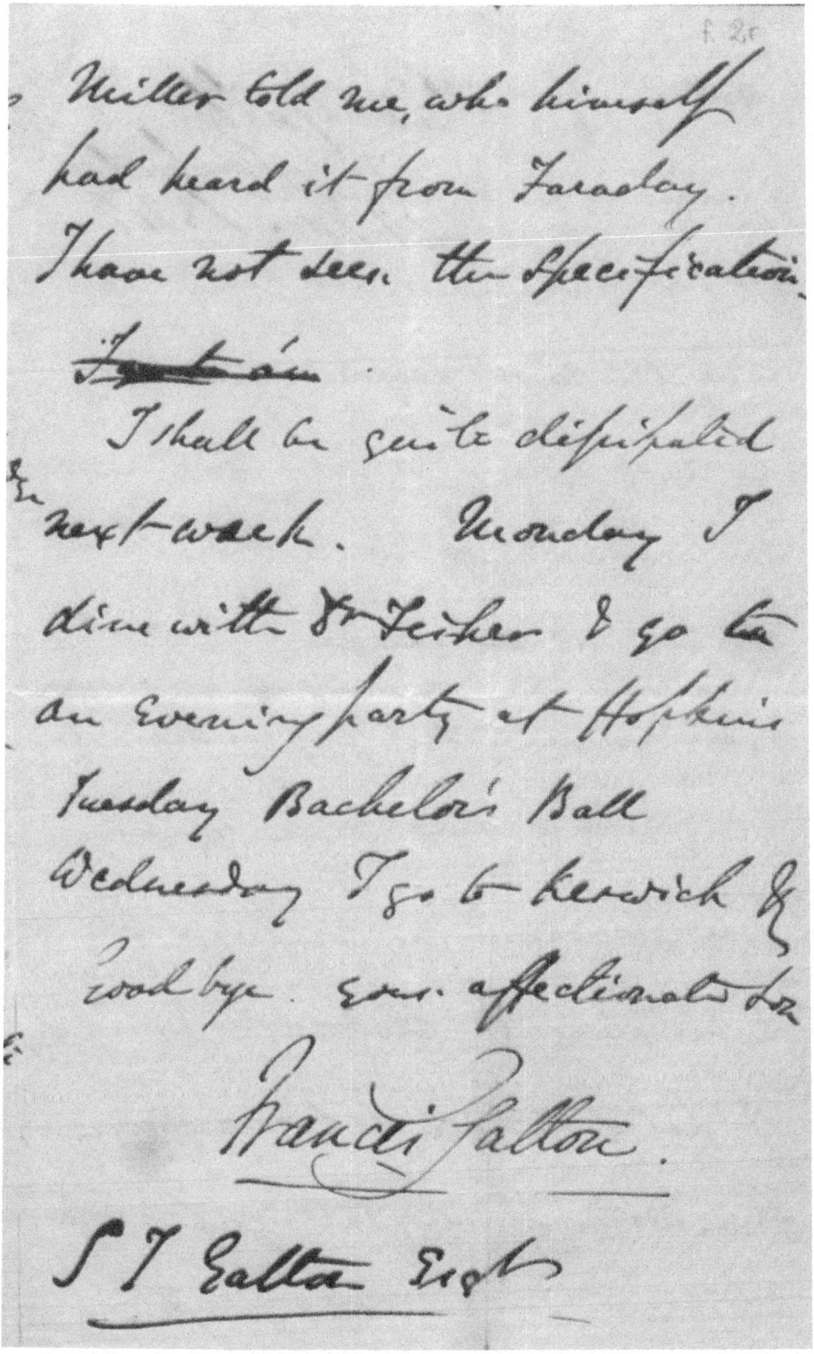

f. 2r

Miller told me, who himself
had heard it from Faraday.
I have not seen the specification.

I shall be quite dissipated
next week. Monday I
dine with Dr Fisher & go to
an Evening party at Hopkins
Tuesday Bachelor's Ball
Wednesday I go to Keswick &
Good bye. Your affectionate Son

Francis Galton.

S T Galton Esqr

Figure 3.3: 1842-01-21. FG to STG page 3.

f. 2v

Fras Galton
Cambridge
21 Jan 1842

UNIVERSITY COLLEGE LONDON
GALTON PAPERS

Figure 3.4: 1842-01-21. FG to STG page 4.

Figure 3.5: 1842-03-09. FG to STG page 1.

1842/03/19. Francis Galton to Samuel Tertius Galton

Trinity Cambridge
March 19th

My dear Father,

I enclose you the Little Go list. I ought to have sent it yesterday but was not in my rooms till after post-time. I have, I consider, had 3 grand escapes in my lifetime in walking into a Lion's den & coming out undigested, 2ndly bathing in a frosty stream at moonlight & not remaining at its bottom in an apoplexy, 3rdly going into the Little-go ~~having~~ when I had not read over half my subjects & coming out unplucked, not however that the pluck w^{d} be of any consequence. I have ~~however~~ nevertheless been the gainer by this examination, for calculating the advantage of being in the first class to be estimated at sixpence I went to a man yesterday & bet a shilling I w^{d} be in the 2nd thus leaving a sixpenny pieces to luck, which as you see I have won. I am much obliged to Bessy for the melancholy news about Miss H: I was, however, aware of my misfortune the same week that the engagement took place, but had not been informed who Mr Edwards was, there is one thing that acts as a poultice to my wounded feelings which is that that small chimpanzee Mr Savile is not the happy man. Time has done wonders for me in soothing &c. but, oh Bessy! ! Miss D: has done much more, she is without exception the most beautiful &c.

&c. &c. I have ever seen. I was at a hop at her Ma's house the other night (I know most of the families in Cambridge now), I was dancing with her (the daughter not the Mama) in a quadrille with one of my Little-Go Examiners for a vis-a-vis. Today, as the sun was shining beautifully I decked myself out in resplendent summer apparel, light trowsers light waistcoat (those that I had last year) to make a call upon this fair creature, but as I was fast finishing my toilette & was "throwing a perfume over the violet" in the way of arranging my cravat ties the wind blew & the rain fell horribly & the streets were one mass of mud, I was in despair, but reflecting that if Leander swam the Hellespont for Hero, I was in duty bound to wade as far as the Fitz William for Miss D:, off I set. When however I arrived at their door, I wisely reflected on the splashed state of my trousers before I knocked & then retreated crest-fallen.

Tonight I hang out, (otherwise give a spread), in oysters; ~~yesterday~~ I have been all anxiety to get a dish of frogs as an adjunct & yesterday I made tremendous efforts to catch some with the annexed apparatus [see Figure 3.6, p. 177] but quite unsuccessfully. I could not see one although with another man I patrolled every imaginable ditch within reach. I firmly believe St Patrick had got the start of ~~him~~ me, for Cambridge without frogs is quite an anomaly. Hopkins[6] gave up for the term on Wednesday; before he left he called me to him and complimented me no end on my mechanics wh: has made me quite jolly, I wish though that I were a better analyst. Buxton & Kay are going to leave his class as their health won't stand it. I shall certainly stay with him during the long Vacation, & if the Dons won't let me stay in Cambridge I propose quietly taking lodgings in Granchester ($1\frac{1}{4}$ mile off) & coming over to him every morning. For want of room

Good bye & believe me ever your affectionate Son
(It's vacation time now!)
Fras Galton

1842/03/22. Francis Galton to Samuel Tertius Galton

Trinity Tuesday

My dear Father

6 Hopkins, William. See page 484.

Figure 3.6: 1842-03-19 FG to STG: Catching frogs.

On thinking over about the approaching Scholarship Examination, I so plainly see that I have no chance whatever of getting one this year that I really think it quite useless to compete. The mere going into an examination for a few days is a thing of no great labour, but I am of course anxious that if examined I should do the mathematical part as well as I am qualified to & not ~~find~~ place myself below my level for want of preparation as w^{d} be the case with me now. I have at present read 14 different mathematical subjects now to get up all these sufficiently well to undergo a good examination in them, w^{d} necessarily require very considerable application which w^{d} be ~~much~~ far better bestowed on the subject (Mechanics) that I have now in hand, inasmuch as the fact of having them now well prepared would in no way assist me in any future examination, after a year's interval for instance. I spoke to Hopkins[7] about it a fortnight ago, he strongly recommends me not to put myself out of the way for the examination sh^{d} I be inclined to go in for it. I find that it is impossible to get up my subjects without doing so & therefore think it preferable not to go in at all. I w^{d} however have possibly tried my best, but my head is already rather bad from having overworked myself in attempting to get up these subjects, as well as doing what Hopkins has set us for this Vacation, so that in the short time that is left I could do but very little. There is one other reason remaining namely that I should not know my standing in Maths. in the College from the result of this examination any better than I do now as those with whom I have to compete read with Hopkins (viz. Walker, Hotham

[7] Hopkins, William. See page 484.

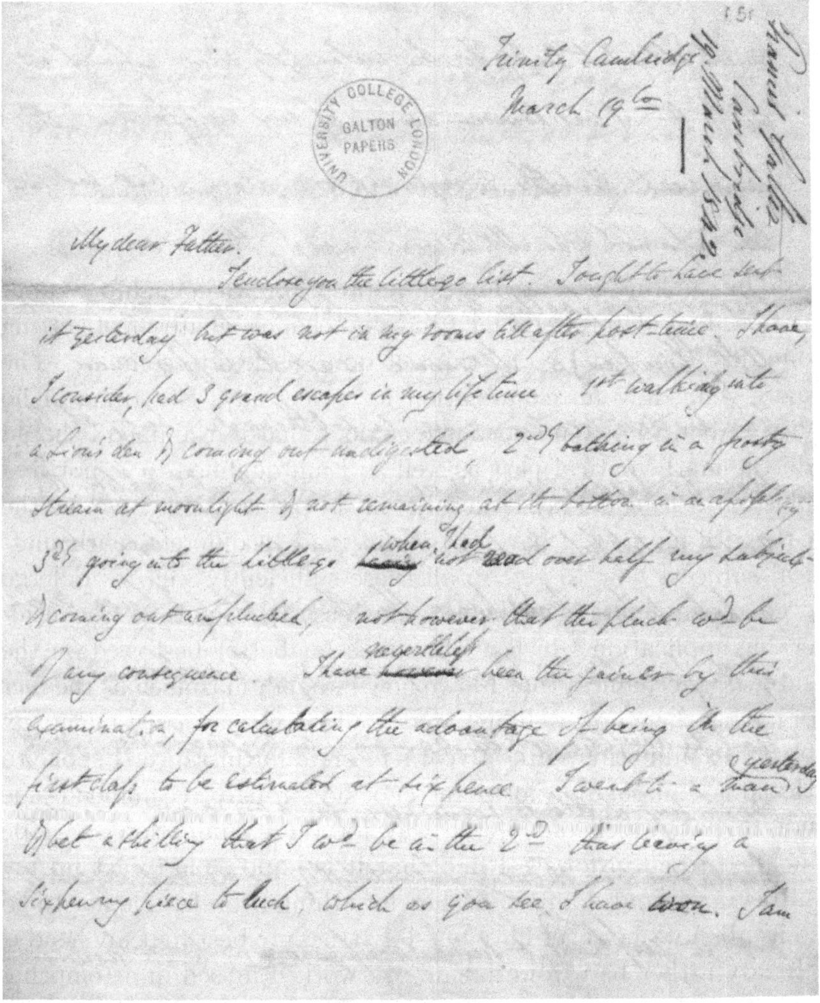

Trinity Cambridge
March 19th

Francis Galton
Cambridge
19 March 1842.

UNIVERSITY COLLEGE LONDON GALTON PAPERS

My dear Father.

I enclose you the littlego list. I ought to have sent it yesterday but was not in my rooms till after post-time. I have, I consider, had 3 grand escapes in my lifetime 1st walking into a Lion's den & coming out undigested, 2nd bathing in a frosty stream at moonlight & not remaining at the bottom in an apoplexy 3rd going into the littlego when I had not read over half my subjects & coming out unplucked, not however that the pluck wd be of any consequence. I have nevertheless been the gainer by this examination, for calculating the advantage of being in the first class to be estimated at sixpence I went to a man yesterday & bet a shilling that I wd be in the 2nd thus leaving a sixpenny piece to luck, which as you see I have won. I am

Figure 3.7: 1842-03-19. FG to STG page 1.

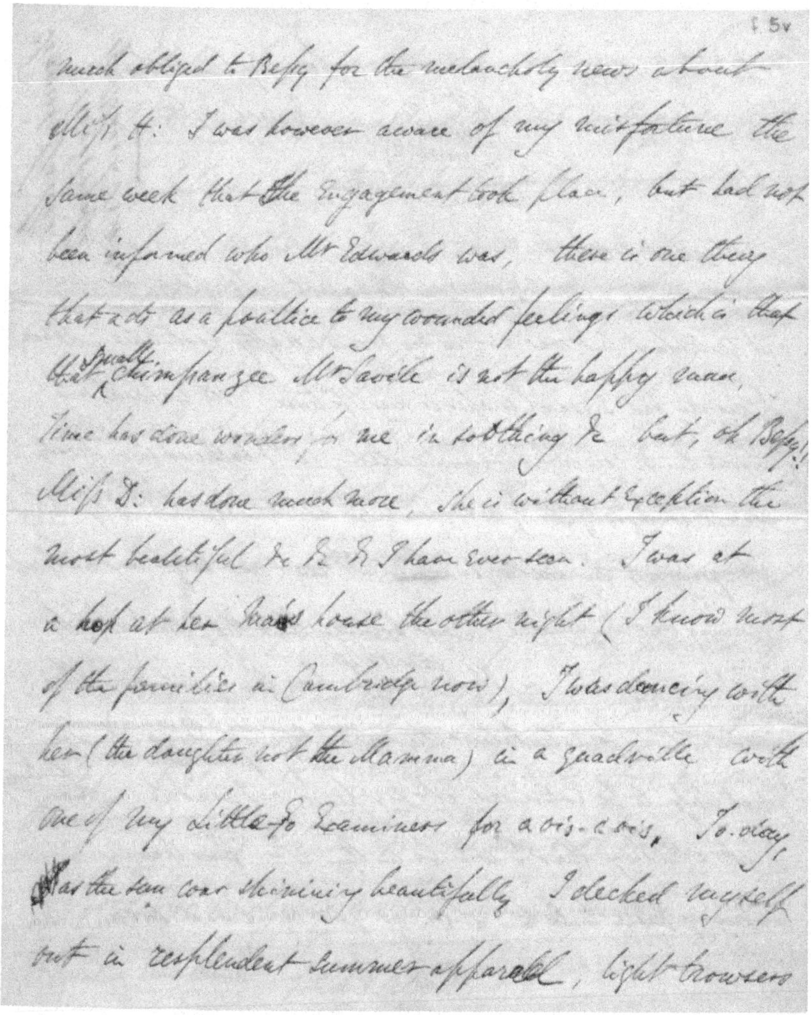

f. 5v

much obliged to Bessy for the melancholy news about
Miss H: I was however aware of my misfortune the
same week that the engagement took place, but had not
been informed who Mr Edwards was, there is one thing
that acts as a poultice to my wounded feelings which is that
that small chimpanzee Mr Savile is not the happy man,
Time has done wonders for me in soothing &c but, oh Bessy!
Miss D: has done much more, she is without exception the
most beautiful &c &c &c I have ever seen. I was at
a hop at her Ma's house the other night (I know most
of the families in Cambridge now), I was dancing with
her (the daughter not the Mamma) in a quadrille with
one of my Little-go Examiners for a vis-a-vis. To-day
as the sun was shining beautifully I decked myself
out in resplendent summer apparel, light trowsers

Figure 3.8: 1842-03-19. FG to STG page 2.

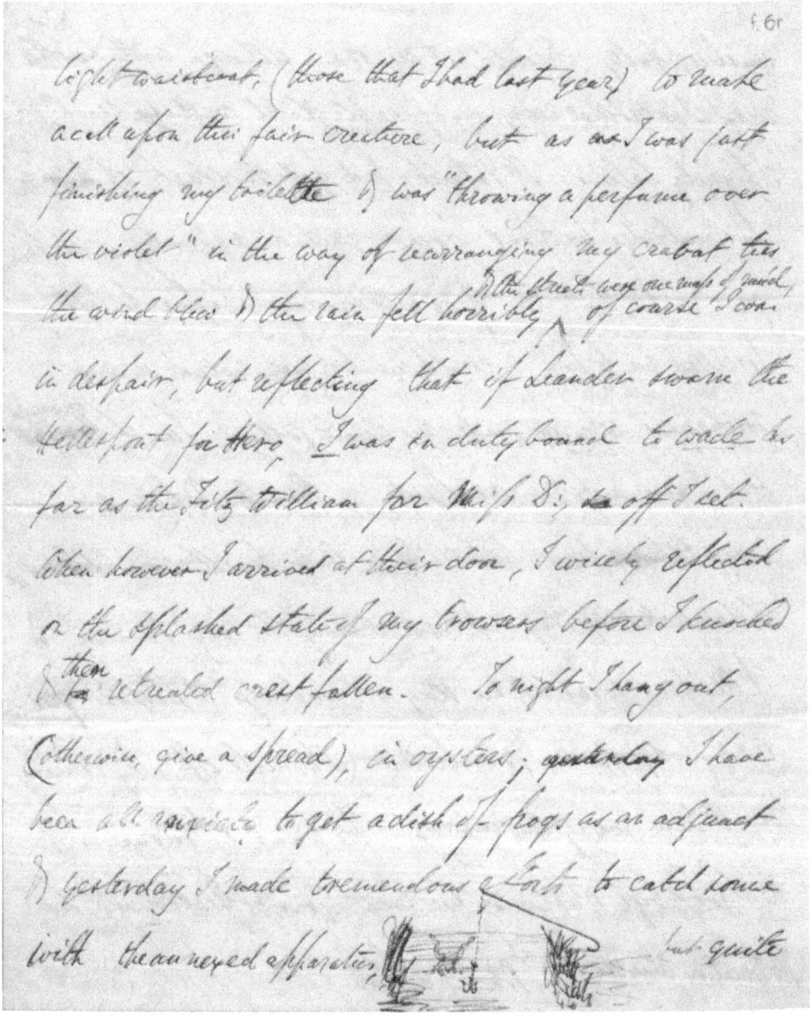
f. 6r

light waistcoat, (those that I had last year) to make
a call upon this fair creature, but as ~~as~~ I was just
finishing my toilette & was "throwing a perfume over
the violet" in the way of rearranging my cravat ties
the wind blew & the rain fell horribly ^ the streets were one mass of mud, of course I was
in despair, but reflecting that if Leander swam the
Hellespont for Hero, I was in duty bound to wade as
far as the Fitz William for Miss D:, ~~so~~ off I set.
When however I arrived at their door, I wisely reflected
on the splashed state of my trowsers before I knocked
& then retreated crest fallen. To night I hang out,
(otherwise give a spread), in oysters, ~~yesterday~~ I have
been all impatience to get a dish of frogs as an adjunct
& yesterday I made tremendous efforts to catch some
with the annexed apparatus, but quite

Figure 3.9: 1842-03-19. FG to STG page 3.

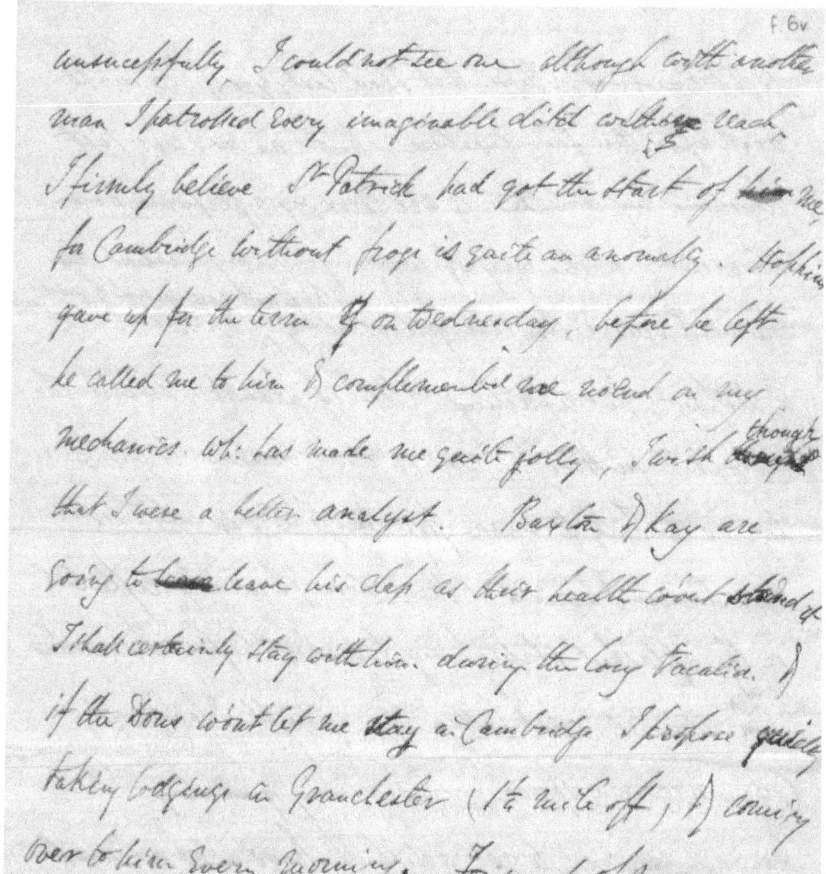
f. 6v
unsuccessfully. I could not see one although with another
man I patrolled every imaginable ditch with a [?] each.
I firmly believe St Patrick had got the start of me,
for Cambridge without frogs is quite an anomaly. Hopkins
gave up for the term [?] on Wednesday, before he left
he called me to him & complimented me on my
mechanics wh: has made me quite jolly, I wish though
that I were a better analyst. Baxter & Kay are
going to leave his class as their health won't stand it
I shall certainly stay with him during the long vacation. &
if the Dons won't let me stay in Cambridge I propose
taking lodgings in Grantchester (1½ mile off) & coming
over to him every morning. for want of room
Good bye & believe me ever your affectionate son
(It's vacation time now) Frank Galton

Figure 3.10: 1842-03-19. FG to STG page 4.

PREVIOUS EXAMINATION.

LENT TERM, 1842.

EXAMINERS:

CHARLES HENRY MATURIN, M.A. *King's College.*
JAMES EDWARD DALTON, B.D. *Queens' College.*
CHARLES STYLE DRAKE, M.A. *Jesus College.*
JAMES HEMERY, M.A. *Trinity College*

The plucked men are about 1/4 th of those that passed.

FIRST CLASS.

Aldersey	Corpus.	Crowther	Cath.	Hedley	Trin.	Magnay	Clare.	Smith	Christ's.
Andrews	Joh.	Curtis	Joh.	Hemming	Joh.	Maine	Pemb.	Snowball	Joh.
Badger	Queens'.	Dalyell	Trin.	Henery	Caius.	Marshall	Queens'.	Somerville	Queens'.
Baker	Trin.	Darby	Emman.	Heslop	Trin.	Martineau	Caius.	Staley	Queens'.
Barker	Joh.	Davie	Joh.	Heygate	Trin.	Mason	Joh.	Stallard	Joh.
Barker	Christ's.	Davies	Trin.	Hicks	Down.	Mason	Pemb.	Stedman	Christ's.
Barker	Caius.	Davies, A. G.	Queens'.	Hiley	Joh.	Massey	Cath.	Stephen	Joh.
Barker	Clare.	Day	Trin. H.	Hill	Caius.	Maxwell	Corpus.	Steventon	Corpus.
Barrett	Joh.	Dean, W. S.	Trin.	Hill	Corpus.	Miles	Magd.	Steuart	Trin.
Barrow	Joh.	Delacour	Pet.	Hillman	Clare.	Monkhouse	Trin.	Stewart	Joh.
Bateson	Trin.	Dikes	Clare.	Hoare	Joh.	Murton	Joh.	Stokes	Trin.
Baumgartner	Caius.	Dixon, T.	Joh.	Hodgson	Pet.	Nash	Trin.	Sutcliffe	Cath.
Bell	Jesus.	Dove	Caius.	Hodson	Trin.	Nicholson	Emman.	Sutcliffe	Magd.
Bennett	Joh.	Dykes	Caius.	Holmes	Trin.	Norman	Corpus.	Sweet	Joh.
Beresford	Joh.	Edwards, L. C.	Trin.	Homfray	Cath.	Nunn	Jesus.	Swindells	Joh.
Best	Pet.	Ellis	Magd.	Hooper	Trin.	Oldham	Trin.	Tarr	Sid.
Blaker	Joh.	Evans	Queens'.	Hopkins	Caius.	Oliver	Queens'.	Taylor	Magd.
Blathwayt	Corpus.	Evans	Caius.	X Hotham	Trin.	O'Reilly	Joh.	Thomas	Trin.
Bodley	Queens'.	Faulkner	Trin.	Hotham	Jesus.	Parr	Joh.	Thomas	Queens'.
Booker	Magd.	Field, J. W.	Joh.	Houghton	Caius.	Paterson	Trin.	Thompson	Cath.
Bowring	Trin.	Field, T.	Joh.	Housman	Joh.	Patey	Cath.	Thompson	Queens'.
Bray	Queens'.	Fitzinger	Pemb.	Hughes	Trin.	Pearocke	Trin.	Trevelyan	Caius.
Broadmead	Trin.	Fisher	Emman.	Ingleby	Joh.	Pinder	Trin.	Tritton	Joh.
Bromby	Sid.	Forbes	Queens'.	Inman	Queens'.	Platt	Trin.	Tryon	Clare.
Bromhead	Caius.	Ford	Trin.	Jermy	Trin.	Procter	Queens'.	Underwood	Joh.
Brooke	Caius.	Formby	Cath.	Jervis	Pet.	Purchas	Christ's.	Waddingham	Joh.
Brunnell	Joh.	Fowler	Joh.	Jones	Trin.	Purches	Corpus.	X Walker	Trin.
Buck	Christ's.	Frampton	Clare.	Jones	Jesus.	Pyper	Cath.	Wall	Joh.
Budd	Pemb.	French	Emman.	Keary	Trin.	Randolph	Jesus.	Wallas	Trin.
Burchell	Corpus.	Frost	Pemb.	Kewley	Magd.	Raven	Caius.	Warren	Trin.
Burton	Joh.	Gascoigne	Cath.	Killpack	Corpus.	Reade	Cath.	Watson, J.	Caius.
X Buxton	Trin.	Geldart	Clare.	King	Trin.	Richards	Trin.	Watson, T.	Caius.
Byers	Christ's.	Girdlestone	Christ's.	King	Christ's.	Richardson	Clare.	X Watson, W. G.	Caius.
Byers	Pet.	Gisborne	Pet.	Kingdon	Trin.	Rigg	Pemb.	Westmore	Queens'.
Chamberlain	Corpus.	Goodman	Christ's.	Kirby	Joh.	Robertson	Caius.	Weston	Christ's.
Chauncy	Pet.	Gorham	Joh.	X Kite	Pet.	Rodgers	Trin. H.	Westropp	Caius.
Chandless	Trin.	Graham	Trin.	Knapp	Sid.	Rogers	Trin.	Whish	Queens'.
Charlesworth	Trin.	Gray	Joh.	Knott	Christ's.	Rowley	Corpus.	White	Pet.
Chawner	Joh.	Green	Corpus.	Lathbury	Jesus.	Rushton	Jesus.	Whittaker	Joh.
Child	Magd.	Greensmith	Joh.	Leader	Corpus.	Russell	Trin.	Whyte	Down.
Chisholm	Joh.	Greenwood	Pet.	Leading	Joh.	Sabine	Jesus.	Wilbraham	Pet.
Chorley	Caius.	Gregory	Pet.	Levett	Cath.	Sabonadiere	Trin.	X Wilkinson	Joh.
Churton	Cath.	Gutch	Sid.	Lighton	Joh.	Sandford	Clare.	Williams	Trin.
Clark	Trin.	Hall, J. E.	Trin.	Linthwaite	Jesus.	Sells	Clare.	Wilson	Queens'.
Clark, J.	Corpus.	Hall	Joh.	Lister	Cath.	Seymour	Trin.	Wilson	Joh.
Clark, G. N.	Corpus.	Hamilton	Trin.	Lloyd	Christ's.	Shakespear	Trin.	Wing	Queens'.
Clarke, W. H.	Queens'.	Hardwick	Cath.	Loftus	Caius.	Sharp, W.	Trin.	X Woodhouse	Caius.
Cole	Joh.	Harris	Trin.	Lower	Pet.	Sharples	Queens'.	Woolley	Pet.
Cooper, G. H.	Trin.	Harrison	Corpus.	Lugg	Clare.	Simcockes	Trin.	Wratislaw	Trin.
Cooper, J.	Trin.	Hawkins	Magd.	Lynes	Christ's.	Smith	Trin.	Yates	Pemb.
Crawley	Joh.	Hayward	Trin.	Lyon	Trin.	Smith	Joh.		

SECOND CLASS.

Alexander	Trin.	Fearenside	Joh.	Idle	Clare.	Mould	Trin.	Swann	Trin.
Allestree	Joh.	Fellowes	Joh.	X Kay, J.	Trin.	Nicholson	Pet.	Tatham	Joh.
Andrews	Emman.	Fenn	Trin.	Kendall	Caius.	Nuttall	Queens'.	Thompson	Pemb.
Banks	Cath.	Fife	Trin.	Kent	Corpus.	Oliver	Trin.	Tompkins	Joh.
Bere	Emman.	Freeman	Joh.	Kingdon	Sid.	Parkinson	Corpus.	Tonkin	Corpus.
Beresford	Pet.	Frewer	Joh.	Law, G.	Joh.	Pickering	Pemb.	Treweeke	Jesus.
Blake	Pet.	X Galton	Trin.	Lawson	Joh.	Pollexfen	Queens'.	Turner	Sid.
Bridges	Down.	Graver	Trin. H.	Loyd	Trin.	Porter	Corpus.	Walker	Cath.
Brocklebank	Queens'.	Green	Trin.	Mann	Cath.	Prichard	Pet.	Walker	Joh.
Brown	Trin.	Harvey	Corpus.	Margitson	Joh.	Priestley	Trin.	Walton	Christ's.
Campbell	Trin.	Heale	Corpus.	Marsham	Magd.	Pritt	Trin.	X Walton	Pet.
Clarke	Trin.	Hendy	Joh.	Mellor	Joh.	Rudd	Trin.	Weildon	Queens'.
Clifford	Trin.	Hilton	Caius.	Milligan	Cath.	Skipwith	Trin.	Wheeler	Jesus.
Davies, W.	Queens'.	Hinde	Joh.	Morse	Trin.	Snow	Trin.	Willoughby	Trin.
De Crespigny	Magd.	Holmes	Corpus.	Moseley	Christ's.	Staines	Trin. H.	Wren	Joh.
Dennis	Pemb.	Hart	Joh.	Mould	Clare.	Suckling	Caius.	Wright	Joh.

Printed at the PITT PRESS, for J. TAYLOR, Senate-House and School- Keeper.

Those marked with X are Hopkins' pupils. — [illegible]

Figure 3.11: 1842-03-19. FG to STG page 5.

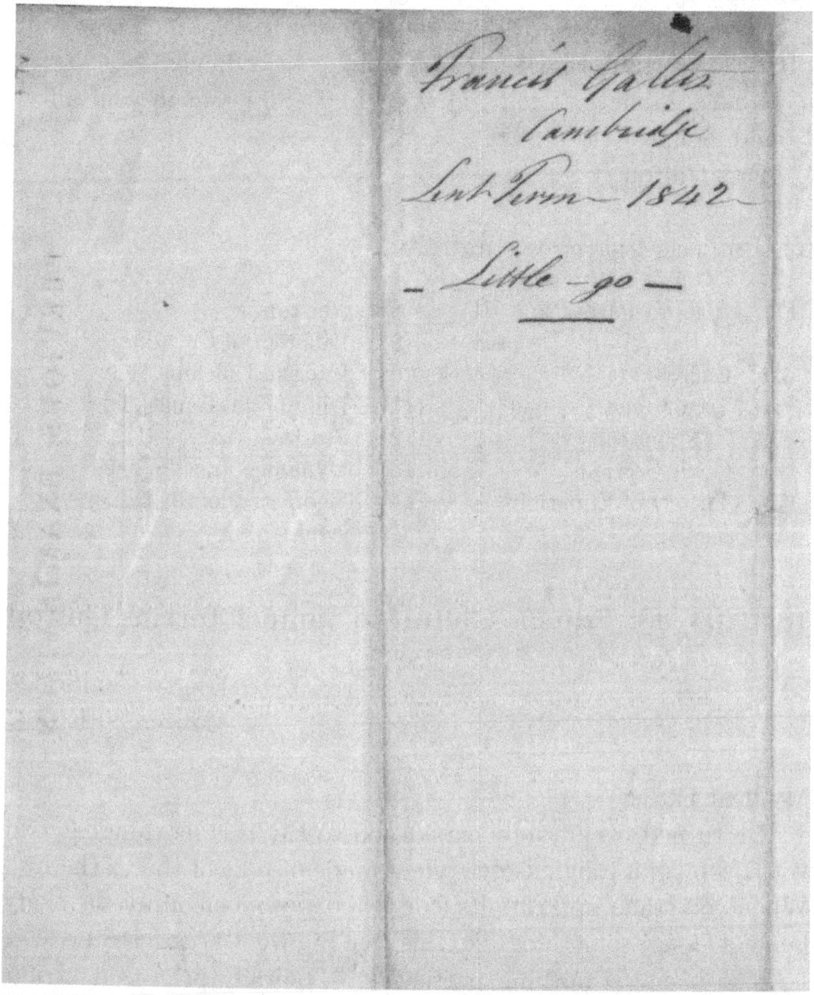

Figure 3.12: 1842-03-19. FG to STG page 6.

& Bowring[8]) {Kay & Buxton & Edwards have been obliged to leave off}. I should therefore much prefer not to go in at all, subject however of course to your wishes. I can easily get off on the plea of ill health which will be in a considerable degree a true one, & can leave Cambridge for 5 or 6 days during the time of the examination.

Should you agree with ~~my ideas~~ me will you let me know your plans, for certain, that I may make mine accordingly.

I have no news to tell you at present. So I remain your affectionate Son,

Fras Galton

The subjects I have read are

(1) Algebra Parts I and III
(2) ... Part II
(3) Euclid
(4) Trigonometry, Plane
(5) ... Spherical
(6) Conic Sections
(7) Theory of Equations
(8) Newton
(9) Differential Calculus
(10) Integral Calculus
(11) Differential Equations
(12) Statics
(13) Dynamics (not finished)
(14) Geometry of 3 Dimensions.

1842/03/30. Francis Galton to Samuel Tertius Galton

Trinity Cambridge
March 30th 1842

My dear Father

I returned from a jolly expedition to Ely this morning. I send you a copy of a rough sketch wh. I made of part of the Cathedral. When you stand opposite its side it looks very ungainly the body being so long, something like this [see Figure 3.17, p. 188.]

The inside is nothing particular. The only part that struck me as pretty is the subject of the inclosed sketch which is highly worked (Norman pillars)

I don't think that I shall leave Cambridge again this Vacation.

Your affectionate son
Fras. Galton

8 H.J. Hotham, Frederick Hermann Bowring.

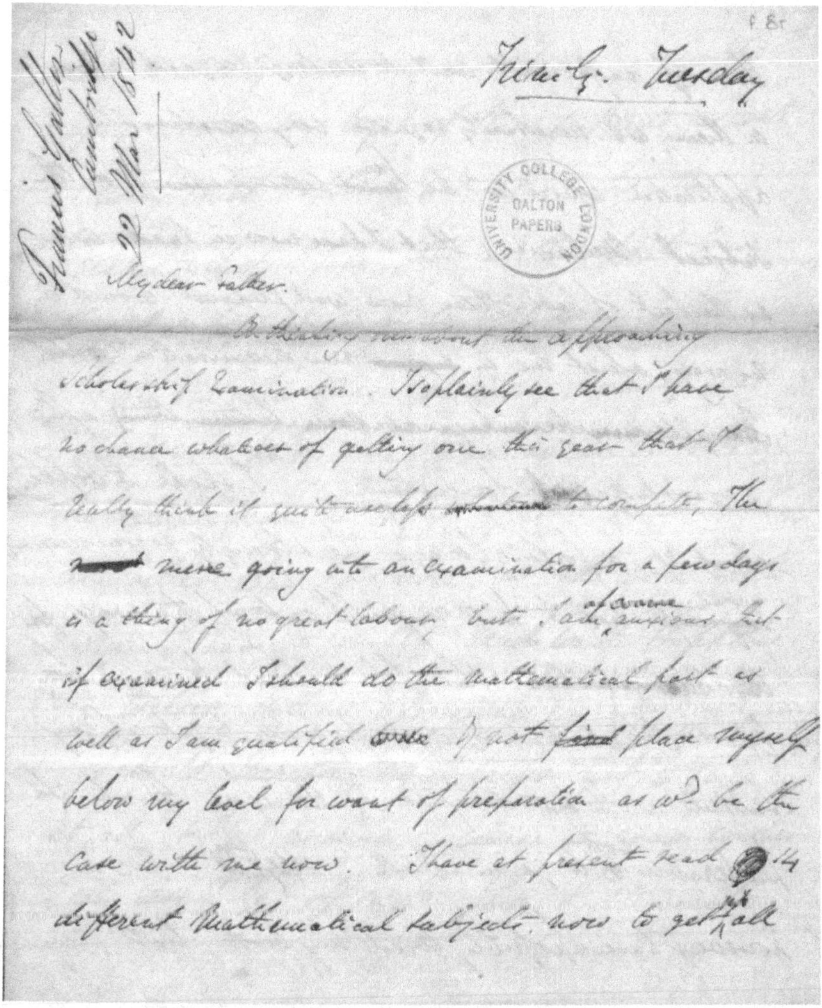
f. 8r

Francis Galton
Cambridge
22 Mar 1842

Trinity. Tuesday

My dear Father.

On thinking over about the approaching Scholarship Examination. I so plainly see that I have no chance whatever of getting one this year that I really think it quite useless ~~whatever~~ to compete. The ~~[illegible]~~ mere going into an Examination for a few days is a thing of no great labour but I am of course anxious that if examined I should do the mathematical part as well as I am qualified ~~[illegible]~~ & not ~~[illegible]~~ place myself below my level for want of preparation as wd be the case with me now. I have at present read [illegible] 14 different Mathematical subjects now to get up all

Figure 3.13: 1842-03-22. FG to STG page 1.

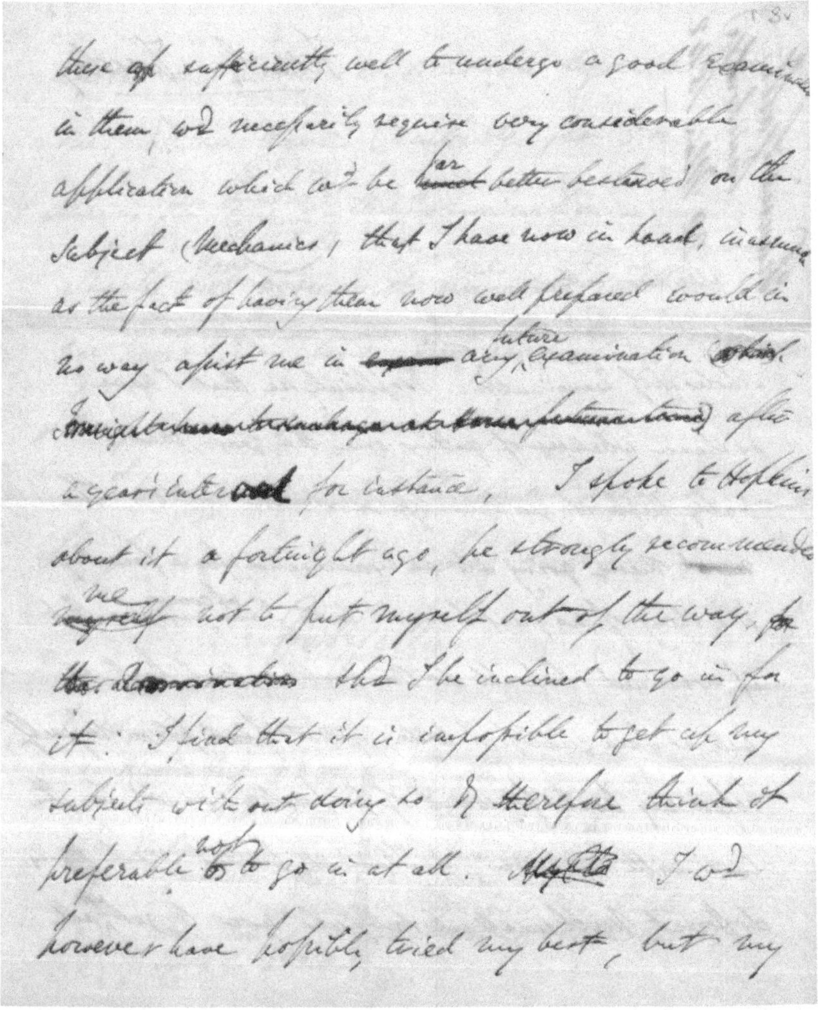

these up sufficiently well to undergo a good Examina-
in them, wd necessarily require very considerable
application which wd be far ~~most~~ better bestowed on the
Subject (Mechanics) that I have now in hand, inasmuch
as the fact of having them now well prepared would in
no way assist me in ~~exam~~ any future examination ~~(which~~
~~[illegible])~~ after
a years interval for instance, I spoke to Hopkins
about it a fortnight ago, he strongly recommended
~~myself~~ me not to put myself out of the way, ~~for~~
~~the examination~~ tho' I be inclined to go in for
it: I find that it is impossible to get up my
subjects without doing so & therefore think it
preferable not to go in at all. ~~[illegible]~~ I wd
however have possibly tried my best, but my

Figure 3.14: 1842-03-22. FG to STG page 2.

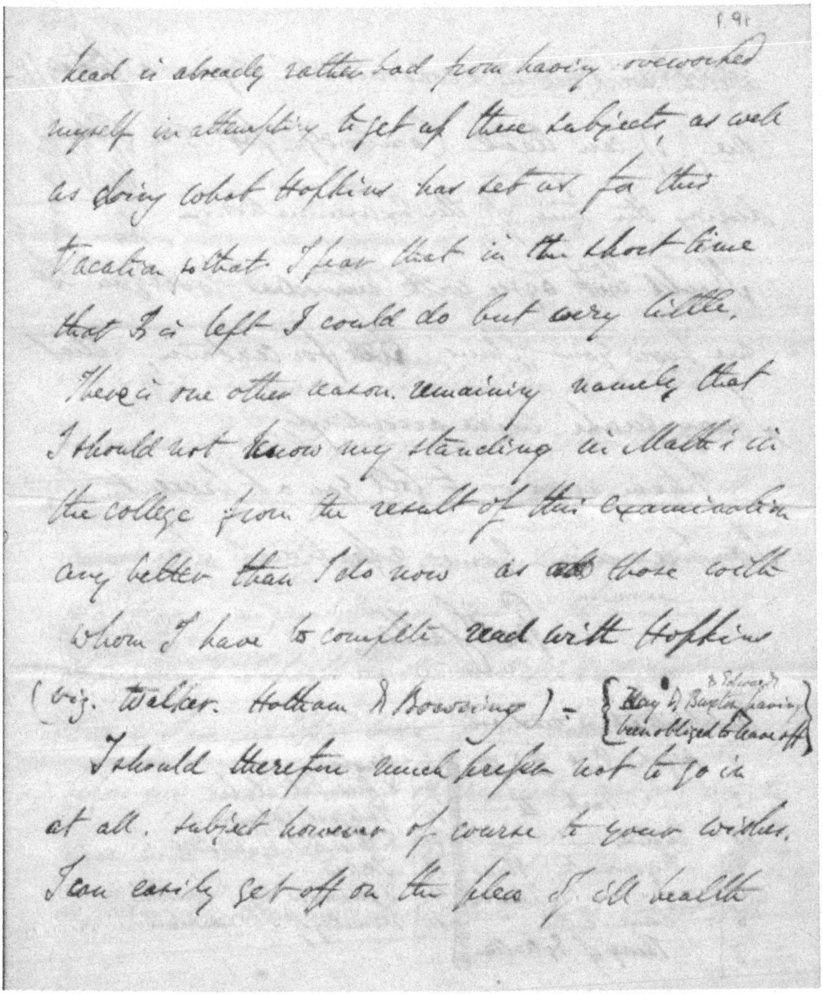
f. 9r

head is already rather bad from having overworked
myself in attempting to get up these subjects, as well
as doing what Hopkins has set us for this
Vacation, so that I fear that in the short time
that is left I could do but very little.
There is one other reason remaining namely that
I should not know my standing in Maths in
the college from the result of this examination
any better than I do now as all those with
whom I have to compete read with Hopkins
(viz. Walker. Hotham & Bowring) – {Blay & Barton & Edwards having been obliged to leave off}
I should therefore much prefer not to go in
at all, subject however of course to your wishes.
I can easily get off on the plea of ill health

Figure 3.15: 1842-03-22. FG to STG page 3.

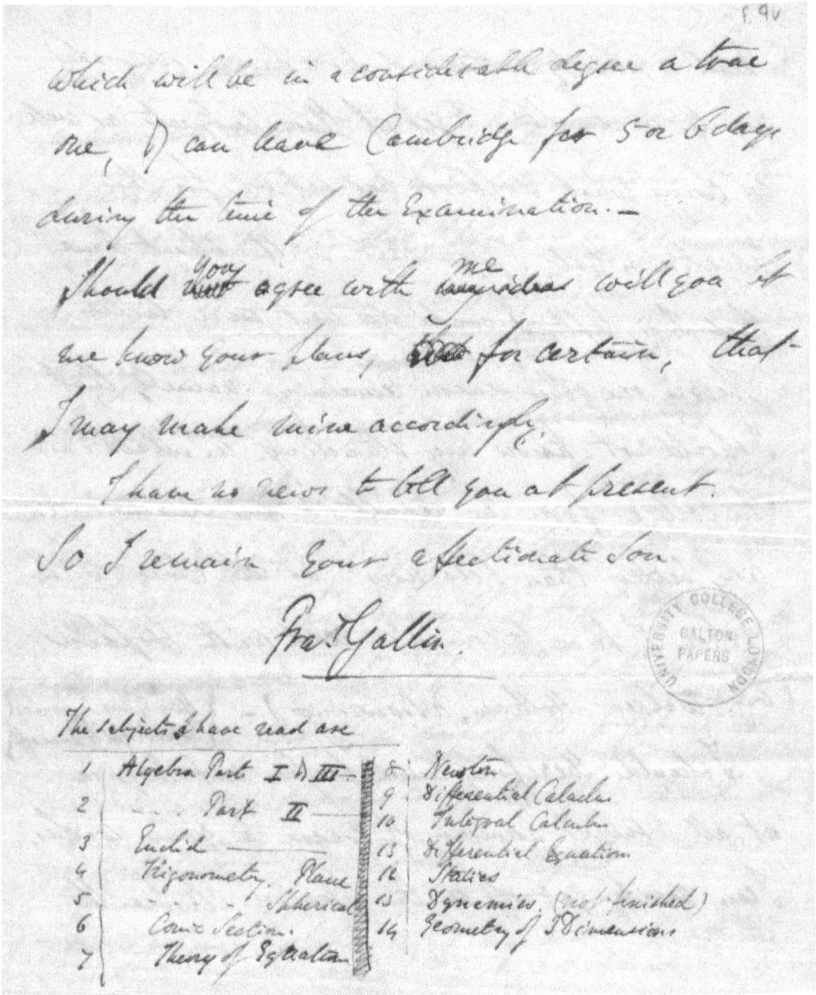

f. 9v

which will be in a considerable degree a true one, I can leave Cambridge for 5 or 6 days during the time of the Examination.—

Should you agree with me will you let me know your plans, for certain, that I may make mine accordingly.

I have no news to tell you at present.

So I remain Your affectionate Son

Fras. Galton.

UNIVERSITY COLLEGE LONDON GALTON PAPERS

The subjects I have read are

1 Algebra Part I & III
2 — Part II
3 Euclid
4 Trigonometry Plane
5 — Spherical
6 Conic Sections
7 Theory of Equations
8 Newton
9 Differential Calculus
10 Integral Calculus
11 Differential Equations
12 Statics
13 Dynamics (not finished)
14 Geometry of 3 Dimensions

Figure 3.16: 1842-03-22. FG to STG page 4.

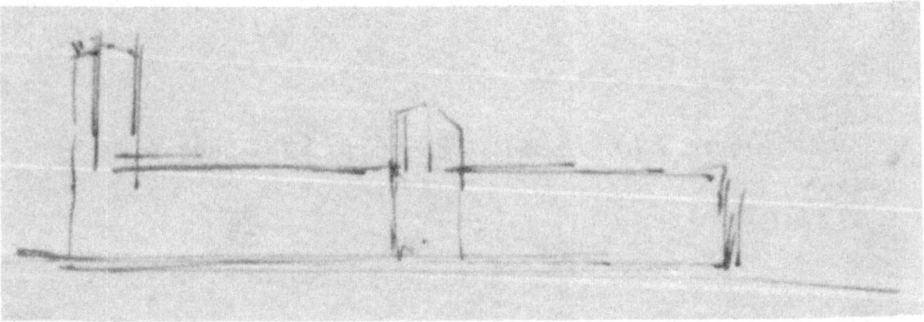

Figure 3.17: 1842-03-30 FG to STG: Side view of Ely Cathedral.

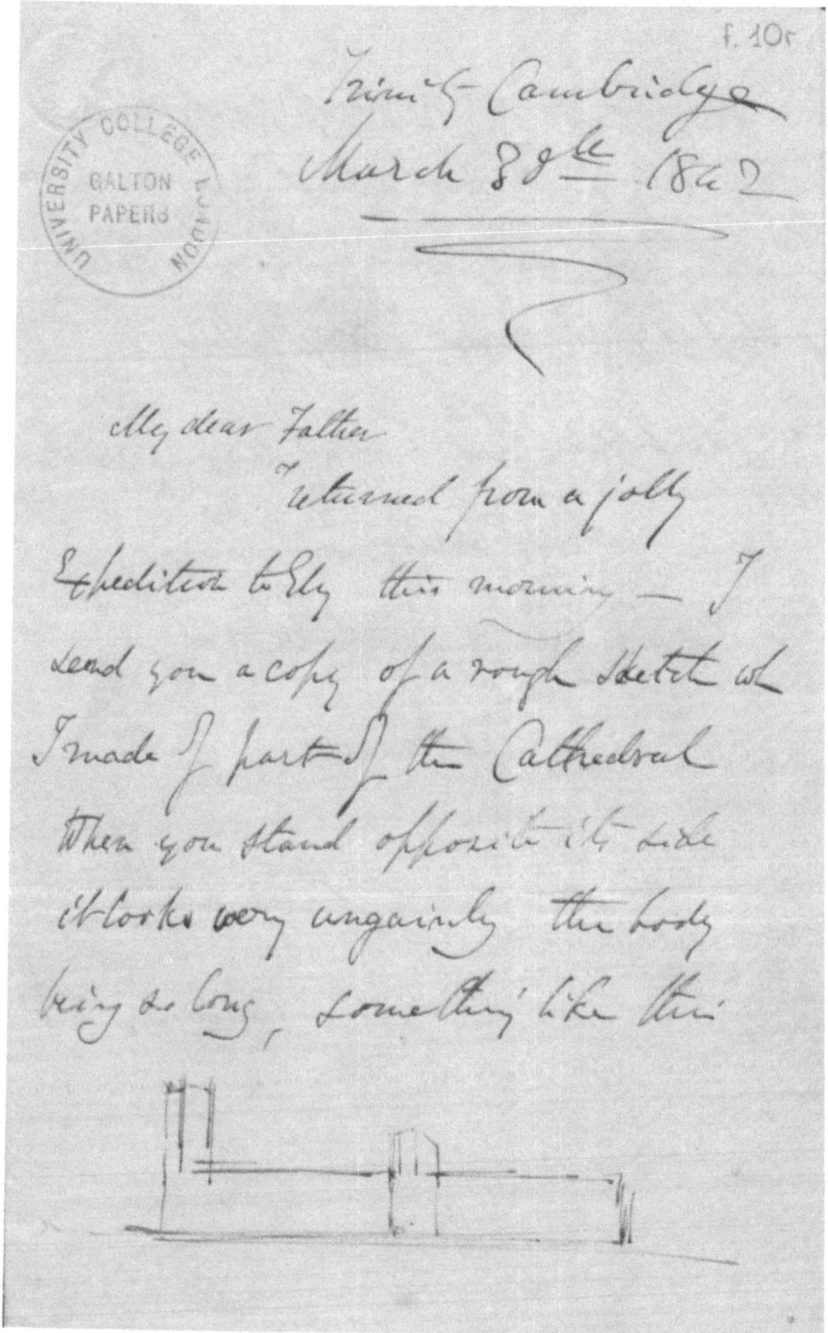

f. 10r

UNIVERSITY COLLEGE LONDON GALTON PAPERS

Trinity Cambridge
March 30th 1842

My dear Father

I returned from a jolly Expedition to Ely this morning — I send you a copy of a rough sketch wh I made of part of the Cathedral When you stand opposite its side it looks very ungainly the body being so long, something like this

Figure 3.18: 1842-03-30. FG to STG page 1.

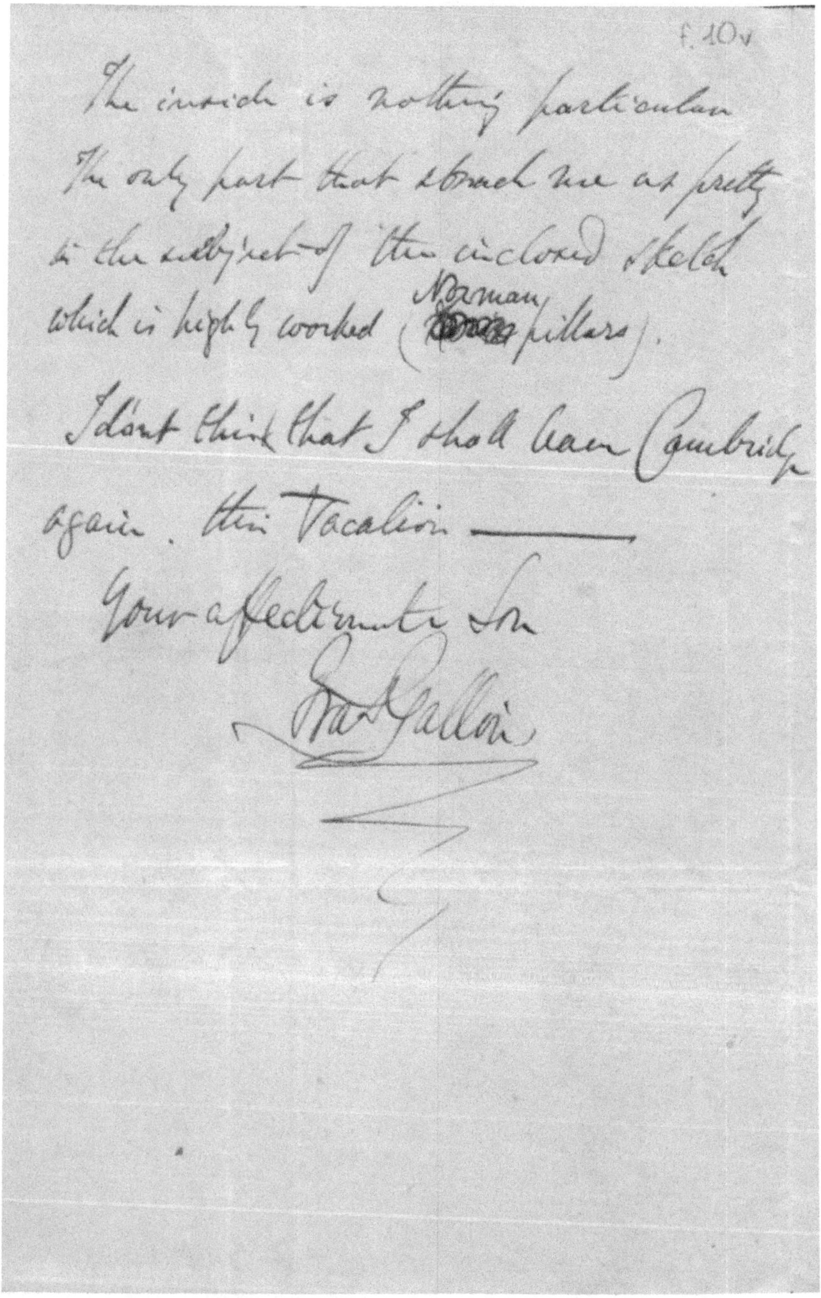
f. 10v

The inside is nothing particular
The only part that struck me as pretty
is the subject of the inclosed sketch
which is highly worked (Norman pillars).
I don't think that I shall leave Cambridge
again this Vacation ——
Your affectionate Son
Fran Galton

Figure 3.19: 1842-03-30. FG to STG page 2.

Figure 3.20: 1842-03-30. FG to STG page 3.

f. 11v

Francis Galton
30 Mar 1842

with Sketch of
Ely Cathedral

Figure 3.21: 1842-03-30. FG to STG page 4.

1842/04/12–29. Diary of Samuel Tertius Galton

April 12th. Tuesday. Hadzor.

... received our Britsker Barouche from Holmes and drove in it to Claverdon where we found Francis returned from Cambridge on account of ill health.

April 14th. Thursday. Claverdon.

Violetta, Francis and myself drove to Edstone where we saw Lucy and her children and Bessy. Weather very cold.

April 16th. Saturday. Claverdon.

I took Francis to Edstone to go on with Lucy to Birmingham that he might consult Dr. Hodgson.[9] ...

April 18th. Monday. Claverdon.

... Francis returned from Birmingham. ...

April 19th. Tuesday. Claverdon.

... Francis went out shooting rabbits in the afternoon. First warm day.

April 20th. Wednesday. Claverdon.

... Bessy, Francis and myself drove to Leamington and found Emma. ...

April 21st. Thursday. Claverdon.

... Francis much better.

April 22nd. Friday. Claverdon.

... Francis much better. Beautiful day.

April 29th. Friday. Claverdon.

Bessy, Emma, Francis and myself drove to Birmingham calling at the Oaklands on our way. ... Francis bought a rifle and appeared altogether much better.

---

9 Hodgson, Joseph. See page 483.

1842/05/02–17. Diary of Samuel Tertius Galton

May 2nd. Monday. Claverdon.

... Francis at Henley making a model for a lake. ...

May 7th. Saturday. Claverdon.

Cold damp morning. I took Francis to Hatton at 7:45 to meet the Birmingham coach. He told me that his college companions consisted of two Kays, two Buxtons, Nicholls, Baker and Spring Rice and Brodie. ... Francis retuned in the evening having seen Hodgson.[10] Stormy day.

May 11th. Wednesday. Claverdon.

Bessy, Emma, Francis and myself went to Leamington where some of the Edstone party joined us. ...

May 16th. Monday. Claverdon.

Bessy and Francis went to Birmingham and back to see Mr. Hodgson.[11] ...

May 17th. Tuesday. Claverdon.

... Hines cut her thumb and was doctored by Francis. ...

1842/06/01–15. Diary of Samuel Tertius Galton

June 1st. Wednesday. Leamington.

... Bessy and Francis came and we all returned together to Claverdon. ...

June 4th. Saturday. Claverdon.

I drove with Francis to Hatton on his way to Birmingham. ...

June 6th. Monday. Claverdon.

... Francis and Mr. Hisons had a post-mortem examination of Page [deceased servant]. ...

June 8th. Wednesday. Claverdon.

... Francis was sent for at night to attend Mr. Commeline. ...

---

[10] Hodgson, Joseph. See page 483.

[11] Hodgson, Joseph. See page 483.

June 11th. Saturday. Leamington.

Hot dry weather. Francis went to Birmingham and returned. ...

June 13th. Monday. Leamington.

... Francis started at 9 a.m. for London to visit his friend Kay at Battersea on his way to Aberfeldy in Perthshire. ...

June 15th. Wednesday. Leamington.

... Francis embarked this day for Scotland.

1842/06/15. Francis Galton to Samuel Tertius Galton

Terrace House
Battersea

My Dear Father

I have enjoyed myself no end, got through nearly all that I had to do. I start for [Dundee] in an hour & a half. On leaving Leamington nothing particular occurred except one of the horses tumbling down & a fracas with a gig. I however read in a newspaper which the coachman had in his pocket the unsettling intelligence that Dr. Buckland considered Box Hill Tunnel as well as all the other tunnels in the line to be in a highly dangerous state & that some fearful accident w^{d} speedily occur unless great repairs were made. Thus edified I left Weedon & thanked my stars each successive tunnel that I passed that Dr. Buckland was not infallible. By the way I acted as escort, for at Blissworth a girl about 16 very clever & naïve & evidently of good family was put in by the rector of Blissworth, she was very communicative & said she was going up to London for one night, to a ball I suspect & intended taking her friends by surprise, so I offered my services to look after her luggage & grub at Wolverton which were readily accepted & really, I think, made myself useful.

When in London I went & dined with Partridge.[12] Saw the Missourian, which struck me as falsely articulated in order to make it appear larger than it was.[13] The shoulder-blade is much out of

---

12 Partridge, Richard. See page 490.

13 The "Missourium", a mastodon assembled inexpertly by Albert Koch of St Louis, MO. It was sold to the British Museum in 1843 and eventually rebuilt correctly for display in the Natural History Museum, where it is

place [sketch] {A the shoulder-blades B the first pr. of ribs } and on mentioned it to Partridge he said that he thought so too & that Professor Owen had given it also as his opinion that the exhibition has increased the animal's height fictitiously by nearly 3 feet. In the Evening I went to the German Opera. Robert le Diable. I did not like the dresses or the acting though the singing was splendid. The amorous duke of Normandy was represented by a man of uncommon width & his face looks very much like Welch the farmer & the heroine was the perfect image of Boswell cut ~~down~~ off at the knee.

Well the amorous Duke, from his personal qualifications I suppose, fails to make any impression on the Lady so by the aid of a magician, the Devil in disguise he obtains a talismanic bough which he shakes before the Lady's eyes. She first looks disgusted to a degree, but the frown gradually relaxed and a smile appears which waxes sweeter & sweeter till at last quite overcome she rushes to him, full flop into the old porpoise's waistcoat with such force as to make his flesh quiver like blanc-mange.

I arrived at the Kay's last night. Lady Lovelace's two children called after I came, that girl is a beautiful little creature. Mrs Davenport & Miss Hurt also called they had just ~~left~~ seen Uncle Howard. Mr. & Miss Kay are very nice persons exceedingly kind & full of teaching schools. Owing to their brother's energy & their own in these matters they have got into very high circles & talk less about them than any persons similarly circumstanced that I have ever seen they are most good natured. I have no more room. Love to Mater & an anathema to her headache.

Your affect Son
Fras Galton.

1842/06/16. Diary of Samuel Tertius Galton

June 16th. Thursday. Leamington.

... I received a letter from Francis who is now in a steamer on his way to Scotland.

1842/06/19. Francis Galton to Samuel Tertius Galton

Aberfeldy June 19th 1842.

---

today.

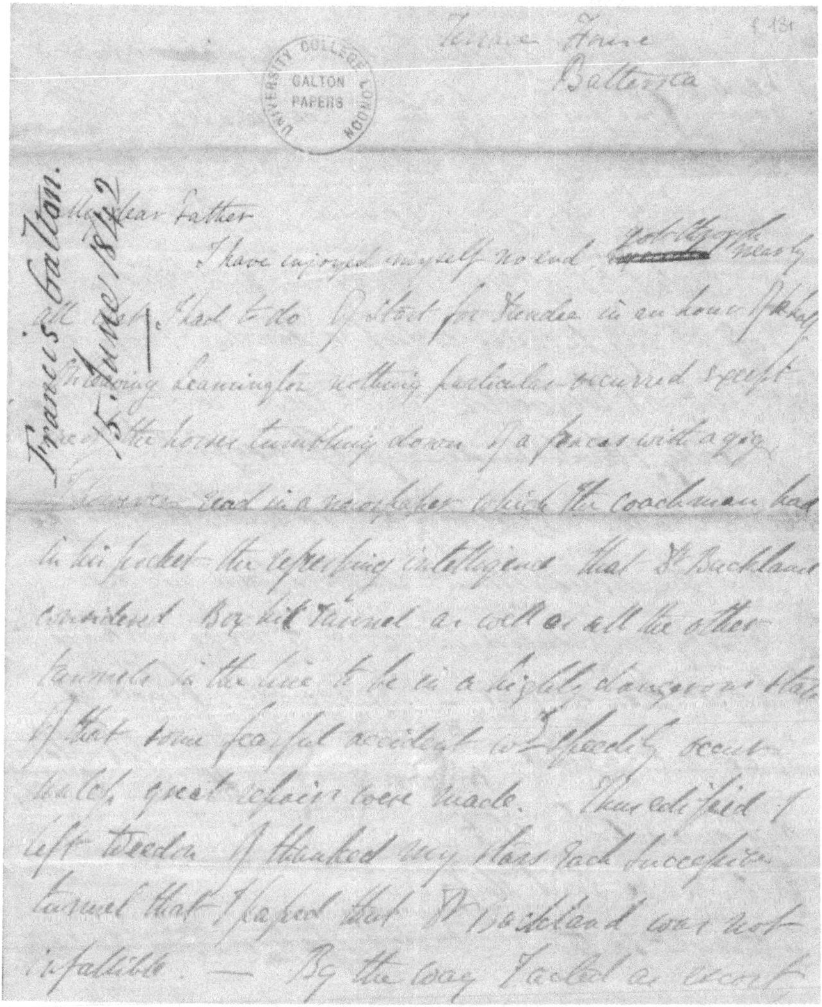

Terrace House
Battersea

Francis Galton. 15 June 1842

My dear Father
I have enjoyed myself no end [illegible] nearly all [illegible] I had to do [illegible] start for Dundee in an hour [illegible] leaving Leamington nothing particular occurred except the horses tumbling down & a fracas with a gig. I however read in a newspaper which the coachman had in his pocket the refreshing intelligence that Dr Buckland considered Boxhill tunnel as well as all the other tunnels in the line to be in a highly dangerous state & that some fearful accident wd speedily occur unless great repairs were made. Thus edified I left Weedon & thanked my stars each successive tunnel that I passed that Dr Buckland was not infallible. — By the way I acted as escort

Figure 3.22: 1842-06-15. FG to STG page 1.

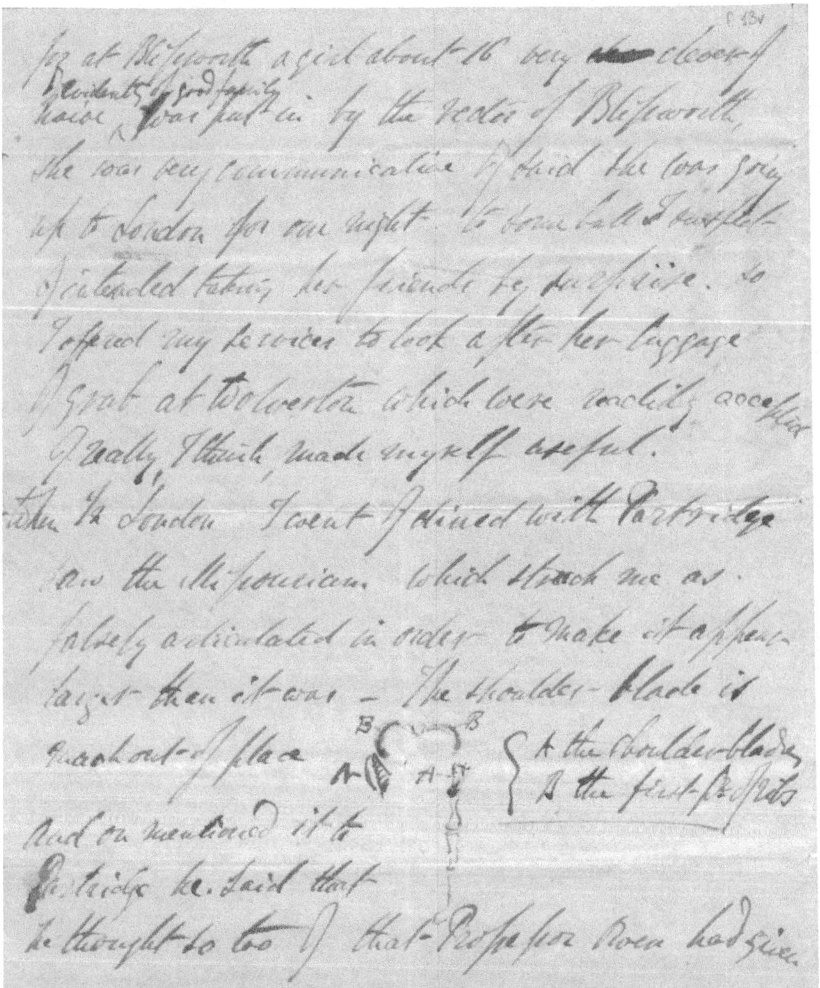

fee at Blisworth a girl about 16 very clever & naive (evidently of good family) was put in by the Rector of Blisworth, she was very communicative & said she was going up to London for one night to some ball I suspect & intended taking her friends by surprise. so I offered my services to look after her luggage &c at Wolverton which were readily accepted & really I think made myself useful.
When in London I went & dined with Partridge saw the Missourium which struck me as falsely articulated in order to make it appear larger than it was — The shoulder-blade is much out of place

And on mentioning it to Partridge he said that he thought so too & that Professor Owen had given

Figure 3.23: 1842-06-15. FG to STG page 2.

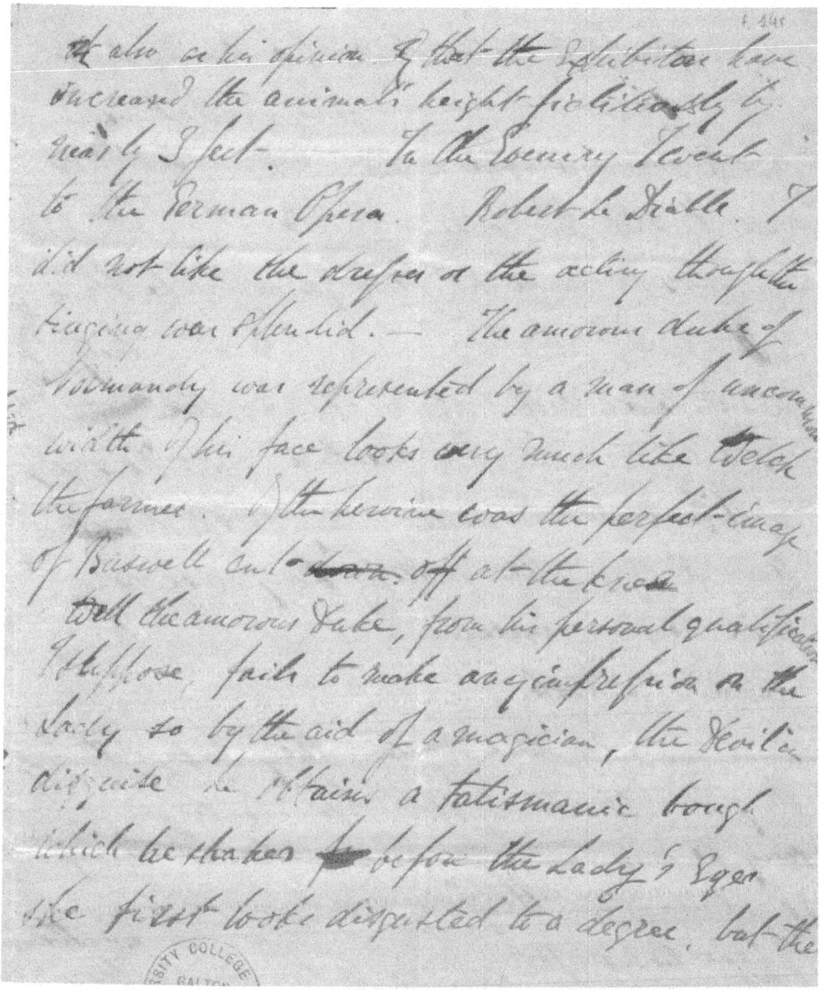

it also in his opinion & that the Exhibitors have
increased the animal's height fictitiously by
nearly 3 feet. In the Evening I went
to the German Opera. Robert le Diable. I
did not like the dresses or the acting though the
singing was splendid. — The amorous duke of
Normandy was represented by a man of uncommon
width of his face looks very much like Welch
the farmer. & the heroine was the perfect image
of Boswell cut ~~down~~ off at the knees
Well the amorous Duke, from his personal qualifications
I suppose, fails to make any impression on the
lady so by the aid of a magician, the Devil in
disguise he obtains a talismanic bough
which he shakes ~~be~~ before the Lady's Eyes
she first looks disgusted to a degree, but the

Figure 3.24: 1842-06-15. FG to STG page 3.

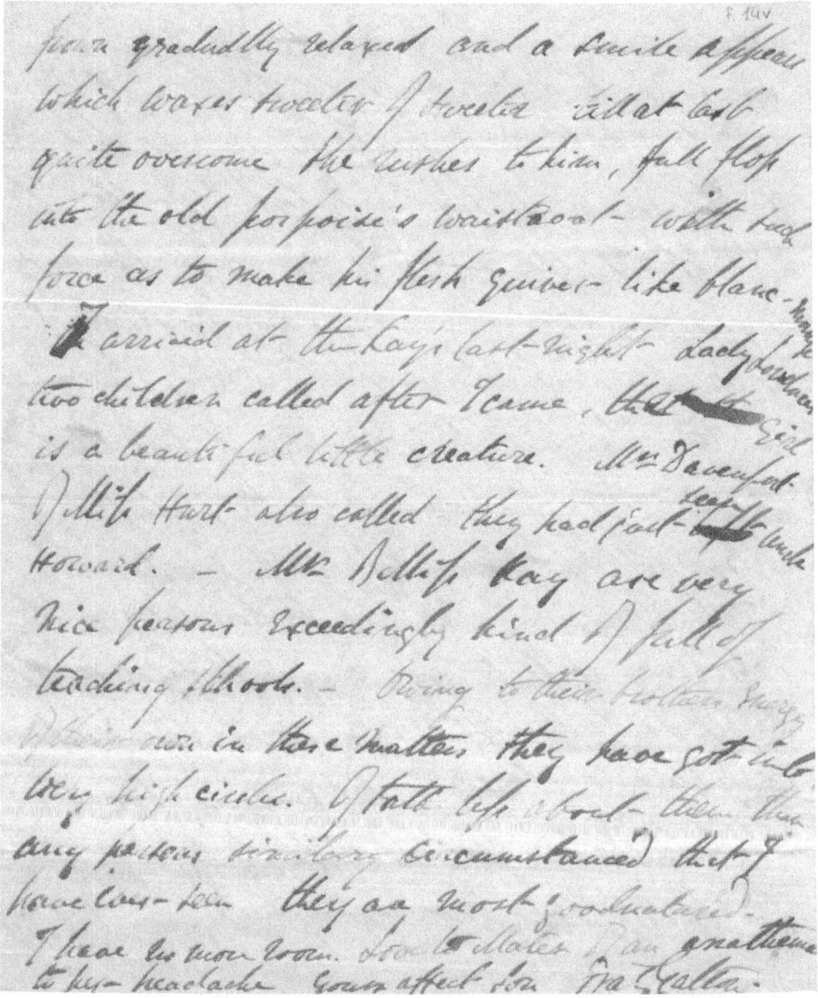

f. 14v

form gradually relaxed and a smile appear
which waxes sweeter & sweeter till at last
quite overcome she rushes to him, falls flop
into the old porpoise's waistcoat with such
force as to make his flesh quiver like blanc-mange
I arrived at the Kay's last night. Lady [illegible] &
two children called after I came, the [illegible] girl
is a beautiful little creature. Mrs Davenport
& Miss Hart also called they had just seen the uncle
Howard. — Mr & Mrs Kay are very
nice persons exceedingly kind & full of
teaching [illegible]. — Owing to their brother [illegible]
[illegible] in these matters they have got into
very high circles. & talk less about them than
any persons similarly circumstanced that I
have ever seen they are most goodnatured.
I have no more room. Love to Mater & an anathema
to her headache. Your affect son Fra Galton.

Figure 3.25: 1842-06-15. FG to STG page 4.

My dear Father

My proceedings have gone on splendidly but the voyage from London to Dundee was all that could be horrible; instead of taking 36 we were 50 hours, a swell of a most abominable description inclined slightly to our course so that the rolling was dreadful. Everybody ~~sick~~ (104 in all, 30 was the usual number) was wretchedly sick. I as usual dreadfully so. Otherwise we had a jolly voyage, most of our party on board & the two tutors. Cayley is unanimously voted a brick & a most gentlemanly-minded man. Some of the passengers too had seen much life. One was a traveller in the interior of Africa, shot Elephants Lions, &c.&c.

Dundee is nothing particular the houses are stone & there is a tower built by the Knights Templar in commemoration of a shipwreck of a ship full of their order, a pretty tale about it, which I must tell Emma some day. [See Figure 3.26, p. 202.]

Perth beautiful to a degree, ditto the lady inhabitants Dunkeld ugly town but beautiful environs. I will get Mrs Phillips' pine cones as soon as I go to Blair Athol, for which place I had not then time.[14] Thence we arrived at Aberfeldy. It is not nearly so pretty a place as we had expected, the walls not being various. The falls of Moness are beautiful & I hear Sir Neil Menzies is a most hospitable person, but I have not yet sent my letter of introduction. In hurry

Your affectionate Son
Fras. Galton.

What are my Mother's Genoa plans, but of course she will go—how is your asthma.
My address is

F. Galton
Aberfeldy
Perthshire
Scotland

1842/08/01. Francis Galton to Samuel Tertius Galton

Aberfeldie
Monday

[14] Mary Elizabeth Phillips married Darwin Galton in 1840. Francis must mean her mother Mary.

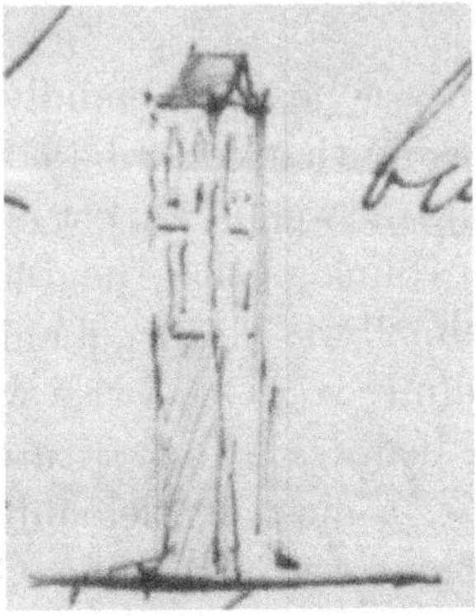

Figure 3.26: 1842-06-19 FG to STG: Dundee Knights Templar Tower.

My dear Father

We are enjoying ourselves very much at Aberfeldie, there is unfortunately much monotony in the walks, as the village is situated on the side of a broad strath through which runs the Tay, & itself formed by high barren moorlands. We have just witnessed a true Highland wedding & absolutely danced with only $1\frac{1}{2}$ hours altogether intermission from 3 in the afternoon till 4 the next morning. I myself was pretty considerably knocked up, but several of the villagers did not go to bed at all and really did not seem much the worse for it the next day. The Scotch reels are great fun, for after every one is ready & before the reel is played, a particular squeak is given on the fiddle & every one kisses his partner & if they are obstreperous there is a fine chase & scramble. We are really very much liked at Aberfeldie, and have been huzzaed more than once as we walked up the town. When we were invited to the wedding, we each subscribed 2 shillings & so bought the bride a very good looking tea-tray, 2 jolly brass candlesticks & snuffers, which overwhelmed the lady. The Scotch air has done wonders for my general health but my head scarcely improves. I have been able to do but little reading since I have been here & altogether am very low about myself.

I am quite glad about Emma's intended tour in this direction but I wish much that you w^{d} follow her. Would you ask my Mother to say to Mrs. Phillips that the fir cones will not do to gather yet, & also to say that there are older larches at Castle Menzies than at Blair Athol, but those at the latter place were brought from

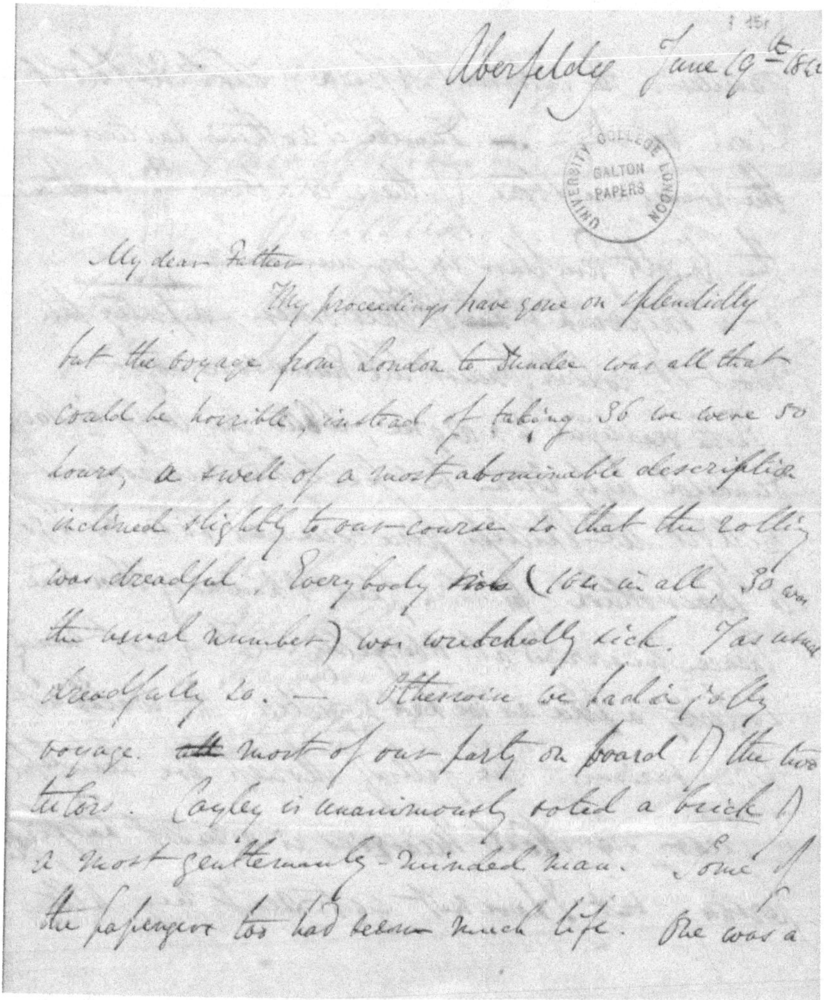

Aberfeldy June 19th 1842

My dear Father

My proceedings have gone on splendidly but the voyage from London to Dundee was all that could be horrible, instead of taking 36 we were 50 hours, a swell of a most abominable description inclined slightly to our course so that the rolling was dreadful. Everybody ~~sick~~ (164 in all, 30 was the usual number) was wretchedly sick. I was also dreadfully so. — Otherwise we had a jolly voyage. ~~all~~ most of our party on board of the two tutors. Cayley is unanimously voted a brick of a most gentlemanly-minded man. Some of the passengers too had seen much life. One was a

Figure 3.27: 1842-06-19. FG to STG page 1.

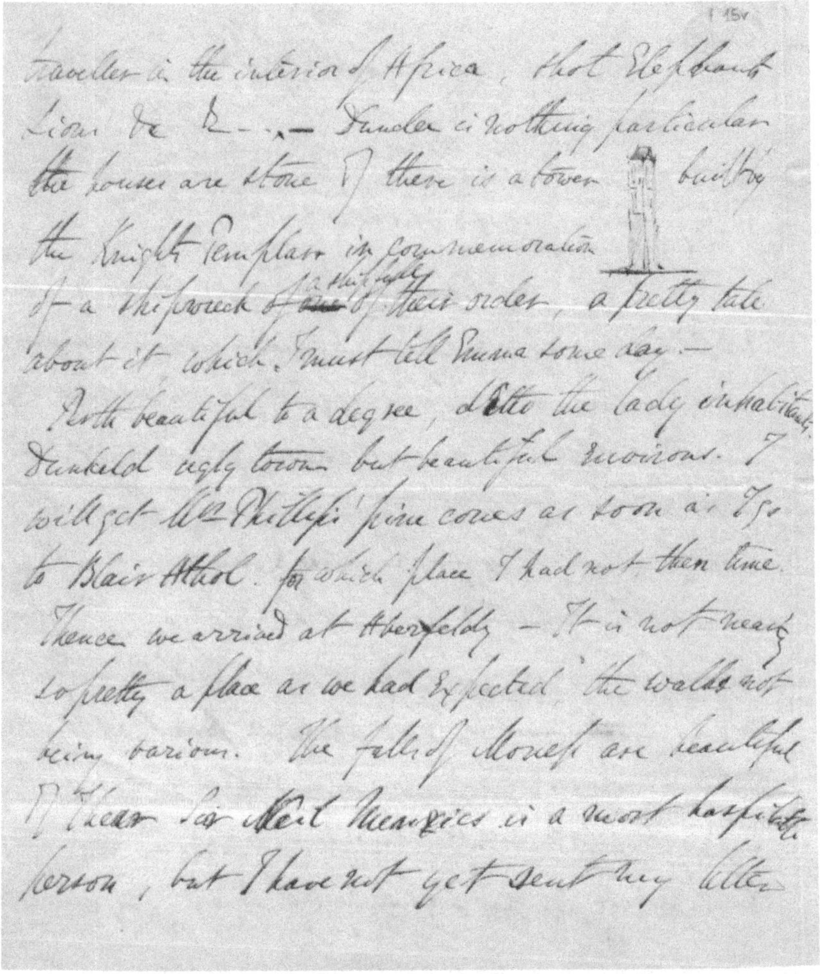

f 15v

traveller in the interior of Africa, shot Elephants
Lions &c &c — Dundee is nothing particular
the houses are stone & there is a tower built by
the Knights Templars in commemoration
of a shipwreck of ~~one~~ [illegible] of their order, a pretty tale
about it which I must tell Emma some day —
Perth beautiful to a degree, ditto the lady inhabitants.
Dunkeld ugly town but beautiful environs. I
will get Mr Phillips' pine cones as soon as I go
to Blair Athol. for which place I had not then time.
Thence we arrived at Aberfeldy — It is not nearly
so pretty a place as we had expected, the walks not
being various. The falls of Moness are beautiful
& I hear Sir Neil Menzies is a most hospitable
person, but I have not yet sent my letter

Figure 3.28: 1842-06-19. FG to STG page 2.

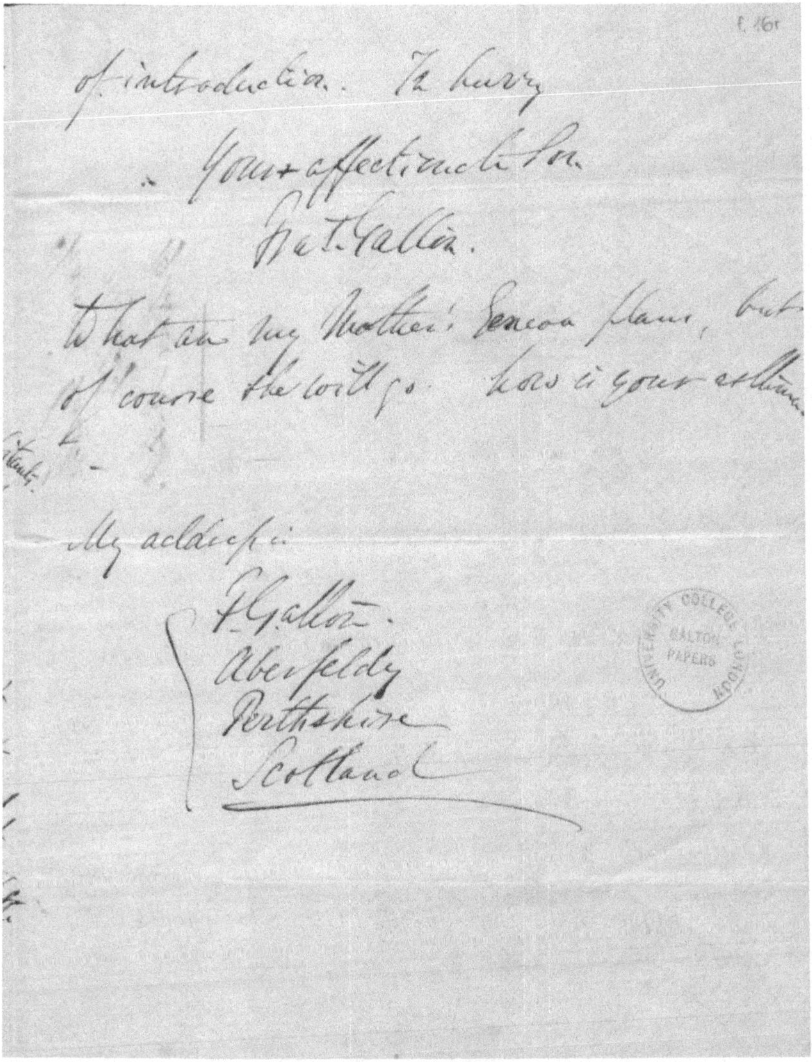

f. 16r

of introduction. In hurry

Your affectionate Son
Frat. Galton.

What are my Mother's [illegible] plans, but of course she will go. how is your [illegible]

My address:
F. Galton
Aberfeldy
Perthshire
Scotland

Figure 3.29: 1842-06-19. FG to STG page 3.

Figure 3.30: 1842-06-19. FG to STG page 4.

abroad, so I will send some cones from both places just before I leave when they will be in ~~their best~~ a better gathering condition.

By the way tell Pemmy[15] that if that Claverdon yellow rose is called "Menziana" it was first reared at Castle Menzies.

Lady Menzies has been most kind to me & other neighbouring residents have been exceedingly hospitable to the party. Will Bessy thank Mrs. Cameron for the note of introduction when next she sees her.

Goodbye your affectionate Son
Fras. Galton.

1842/08/20. Diary of Samuel Tertius Galton

August 20th. Saturday. Claverdon.

... At breakfast I received a letter from Darwin mentioning ... the increased indisposition of his baby, and afterwards received a letter from Adèle announcing the death of it at 1:30 after having been attended by Dr. and Mrs. Thomson. ... went to Edstone and saw the placid corpse of the baby. God's will be done. ...

1842/09/14. Francis Galton to Samuel Tertius Galton

Edinburgh
M'Queen's Hotel

My dear Father

I left Aberfeldy very early yesterday morning with Eben Kay[16] & went thro' Crief[17] and Stirling to Edinbro' really quite sorry to part with the Highlands. We left in high feather, knowing every ~~person~~ family well in the Strath some of them intimately and altogether have I really think left a very good name for Cambridge. Our ball went off superbly. I wrote a description of it to Emma who has possibly forwarded it on to you. We went to the Queen's reception at Taymouth. Major & Mrs. Menzies took me with them in their carriage. The Highlanders looked very well drawn up in files round a large quadrangle in 4 bodies dressed in the Campbell & Menzies dress & hunting tartans respectively. The Queen

[15] Emma.
[16] Kay, Edwin Ebenezer. See page 485.
[17] Crieff.

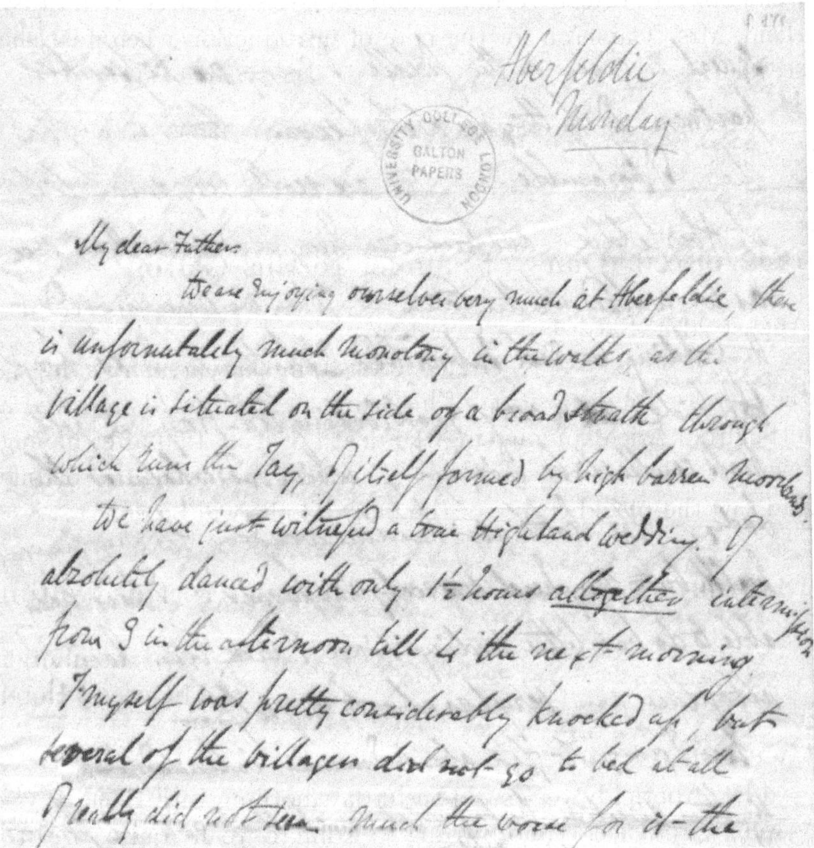

Aberfeldie
Monday

My dear Father

We are enjoying ourselves very much at Aberfeldie, there is unfortunately much monotony in the walks, as the village is situated on the side of a broad strath through which runs the Tay, & itself formed by high barren moorlands.

We have just witnessed a true Highland wedding. & absolutely danced with only 1½ hours altogether intermission from 3 in the afternoon till 4 the next morning.

I myself was pretty considerably knocked up, but several of the villagers did not go to bed at all & really did not seem much the worse for it the next day; The Scotch reels are great fun for after every one is ready & before the reel is played a particular

Figure 3.31: 1842-08-01. FG to STG page 1.

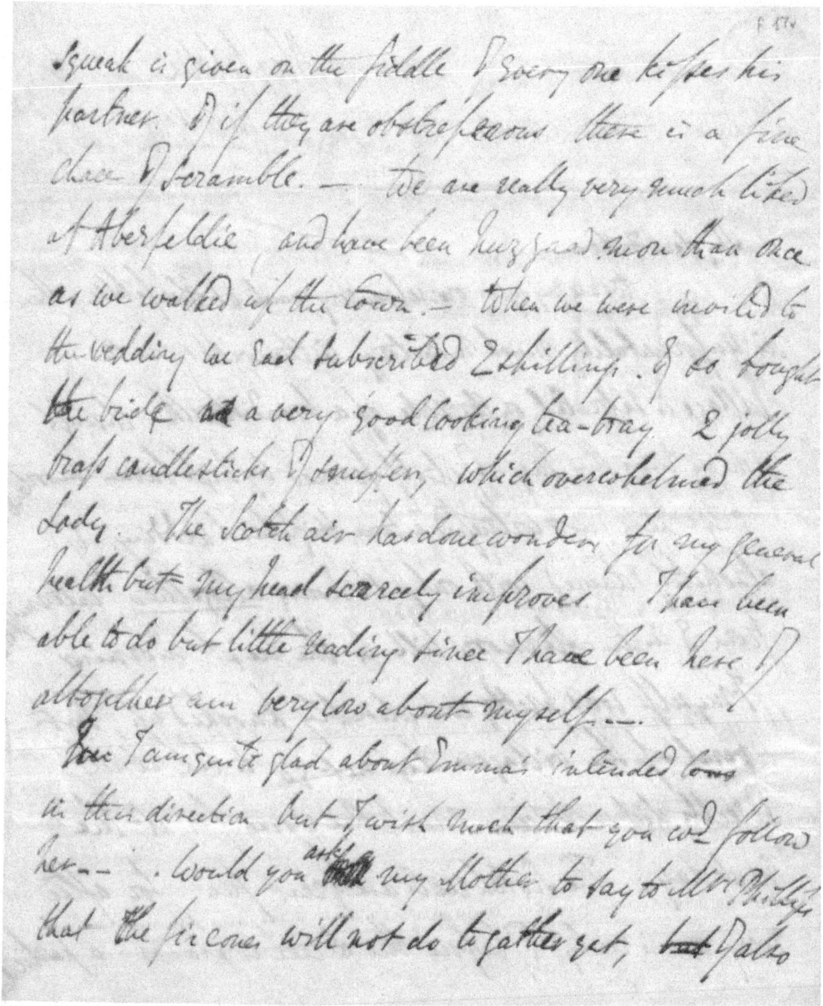
squeak is given on the fiddle & every one kisses his partner & if they are obstreperous there is a fine chace & scramble. — We are really very much liked at Aberfeldie and have been huzzaed more than once as we walked up the town. — When we were invited to the wedding we had subscribed 2 shillings. & so bought the bride ~~an~~ a very good looking tea-tray 2 jolly brass candlesticks & snuffers, which overwhelmed the lady. The Scotch air has done wonders for my general health but my head scarcely improves. I have been able to do but little reading since I have been here & altogether am very low about myself. —
~~But~~ I am quite glad about Emma's intended ~~tour~~ in this direction but I wish much that you wd follow her — . Would you ~~tell~~ ask my Mother to say to Mr Phillips that the pieces will not do together yet, ~~but~~ & also

Figure 3.32: 1842-08-01. FG to STG page 2.

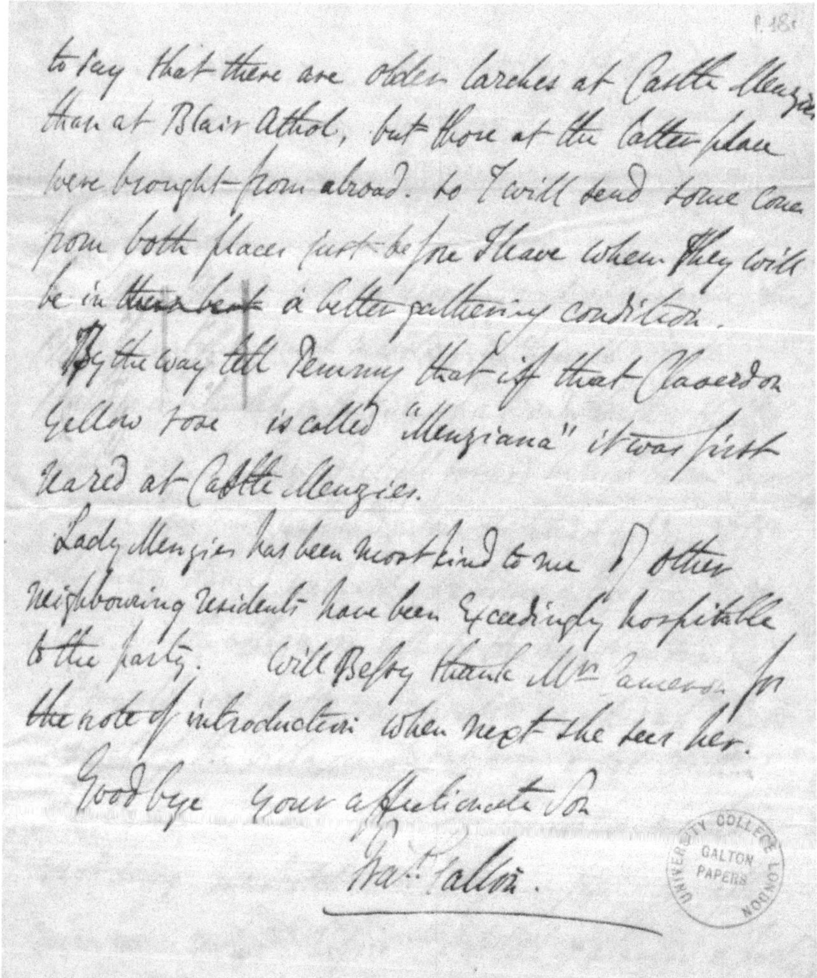
f. 18

to say that there are older larches at Castle Menzies than at Blair Athol, but those at the latter place were brought from abroad. So I will send some cones from both places just before I leave when they will be in ~~their best~~ a better gathering condition.

By the way tell Emmy that if that Clarendon yellow rose is called "Menziana" it was first reared at Castle Menzies.

Lady Menzies has been most kind to me & other neighbouring residents have been exceedingly hospitable to the party. Will Bessy thank Mrs Cameron for the note of introduction when next she sees her.

Good bye your affectionate Son

Frans Galton.

Figure 3.33: 1842-08-01. FG to STG page 3.

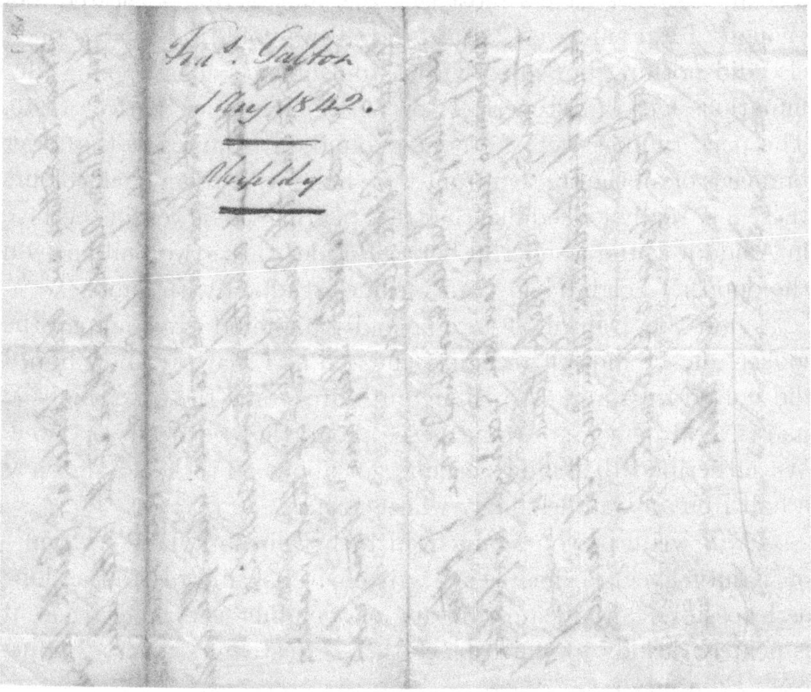
Fra^s. Galton
1 Aug 1842.

Figure 3.34: 1842-08-01. FG to STG page 4.

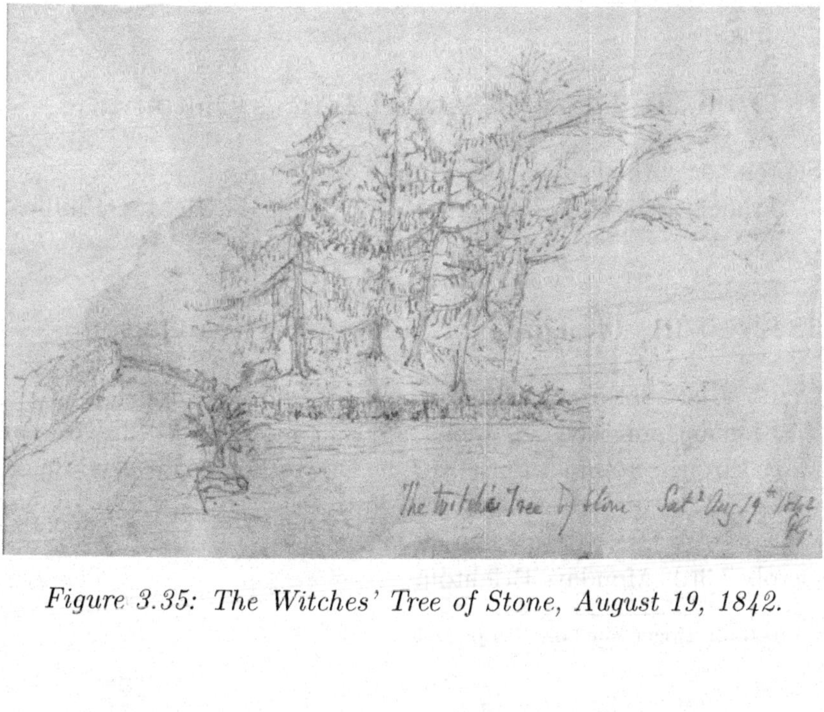

Figure 3.35: The Witches' Tree of Stone, August 19, 1842.

& Prince Albert looked most gracious, but were not cheered half enough. I am sure we Cantabs did all we could, but every body else did nothing but gape with astonishment. The evening illuminations were most perfect, every thing in such perfect taste. The park railings were festooned round with lamps & the lower branches of all the trees in front were hung with them of all colours these last really looked like the trees bearing the jewels in the cave in Aladdin & the wonderful lamp. Besides these two buildings on the opposite heights had their outlines studded with lamps.

I saw Mr. Dalrymple in a splendid highland dress among the lookers on, he did not recognise me and as I scarcely knew him, I did not address him. We "hung out" fire-works the other night & had several persons to come & see them, they went off very well. We subscribed 10 shillings each & got about 30 rockets a dozen & a half Roman Candles many wheels &c.&c.

I will write to you again from Inch Dairnie (where Aytoun[18] of Trinity lives) he has ~~asked~~ written to ask me to stay as long as I possibly can & I am thinking of spending a few days there it is near Kirkcaldy to which place there are steamers every 2 hours from Edinburgh.

Goodbye
Your affectionate Son
Francis Galton.

1842/09/30. Diary of Samuel Tertius Galton

September 30th. Friday. Brighton.

Francis arrived after breakfast. We all dined with Mrs. Phillips. ...

1842/10/01–14. Diary of Samuel Tertius Galton

October 1st. Saturday. Brighton.

Fine autumn day. ... Violetta, Francis and myself drove with Mary through Stanmer Park over the Downs home. I walked much with Violetta and afterwards with Francis.

October 3rd. Monday. Brighton.

[18] Aytoun, Roger Sinclair. See page 480.

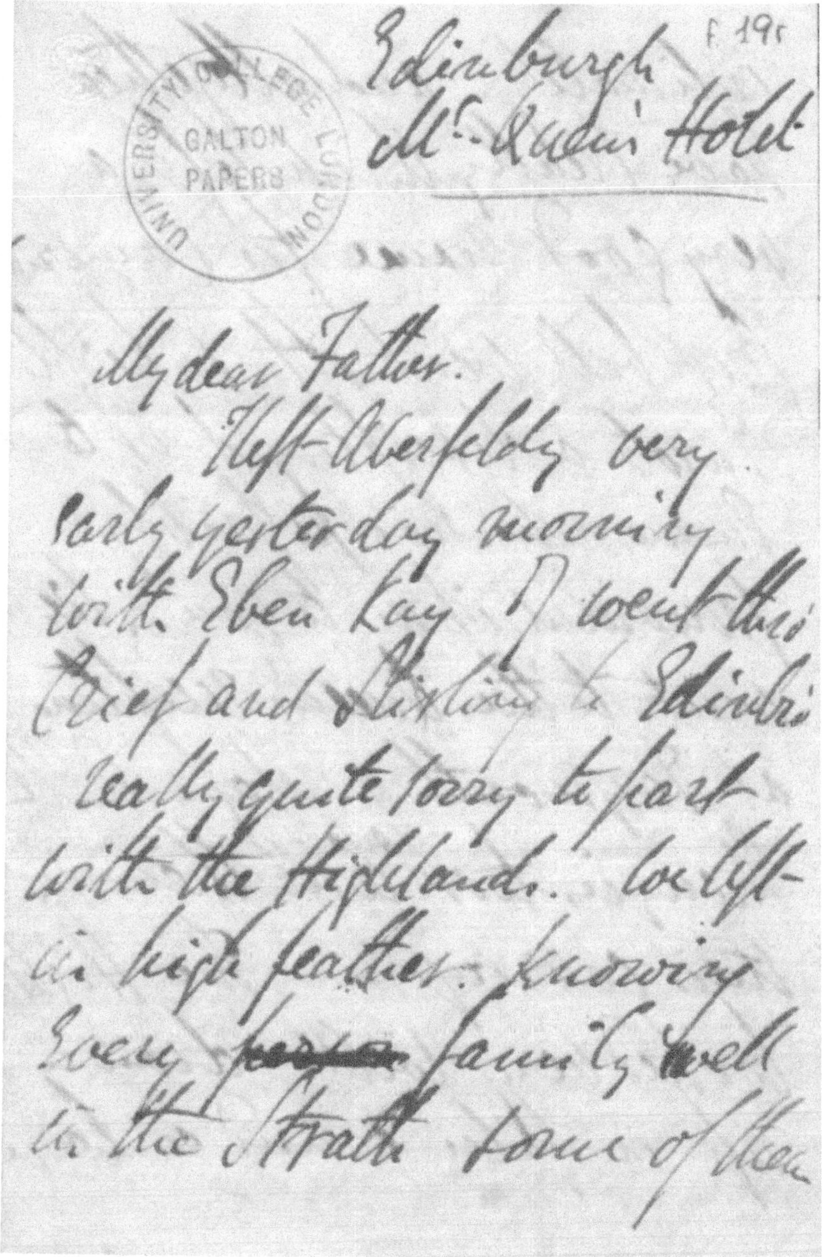

f 195

Edinburgh
Mc Queen's Hotel

My dear Father.
I left Aberfeldy very early yesterday morning with Eben Kay & went thro' Crieff and Stirling to Edinbro' really quite sorry to part with the Highlands. We left in high feather knowing every family well in the Strath some of them

Figure 3.36: 1842-09-14. FG to STG page 1.

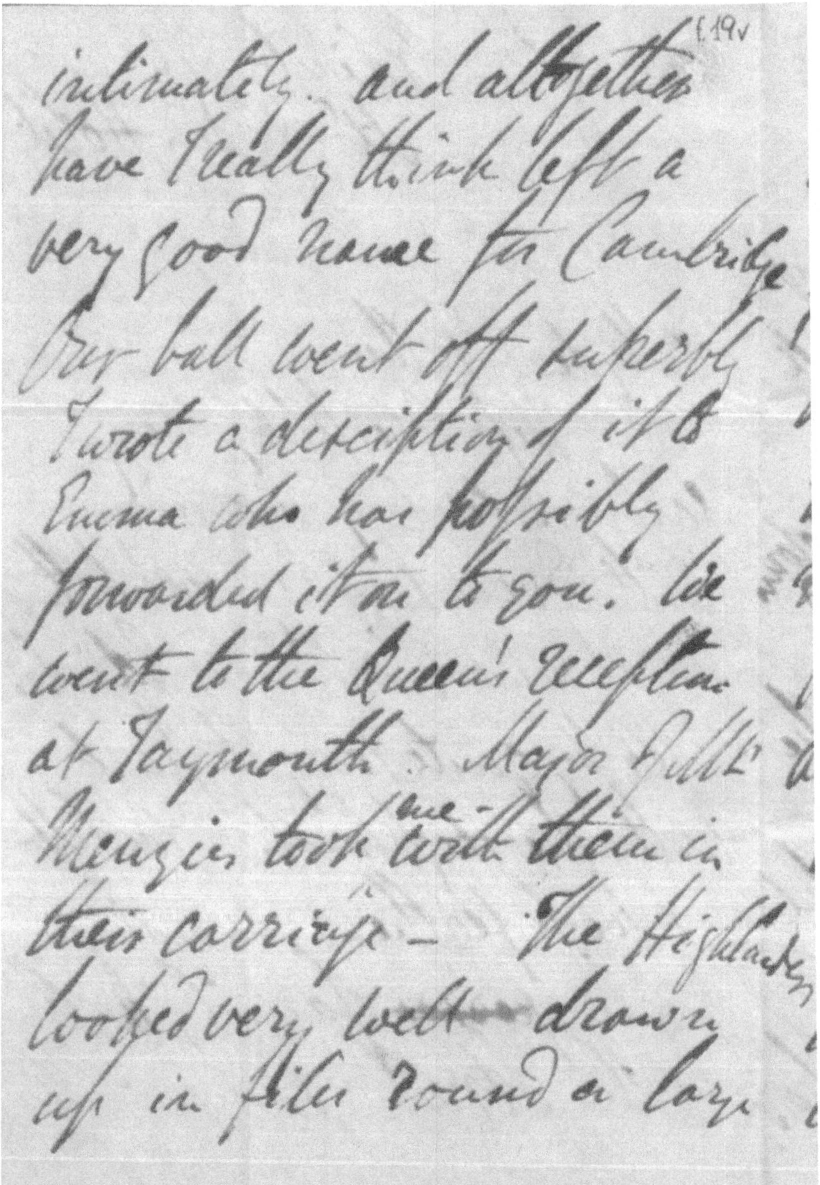
f.19v
intimately — and altogether
have I really think left a
very good name for Cambridge
Our ball went off superbly
I wrote a description of it to
Emma who has possibly
forwarded it on to you. We
went to the Queen's reception
at Taymouth — Major & Mrs
Menzies took [me] with them in
their carriage — The Highlanders
looked very well — drawn
up in files round a large

Figure 3.37: 1842-09-14. FG to STG page 2.

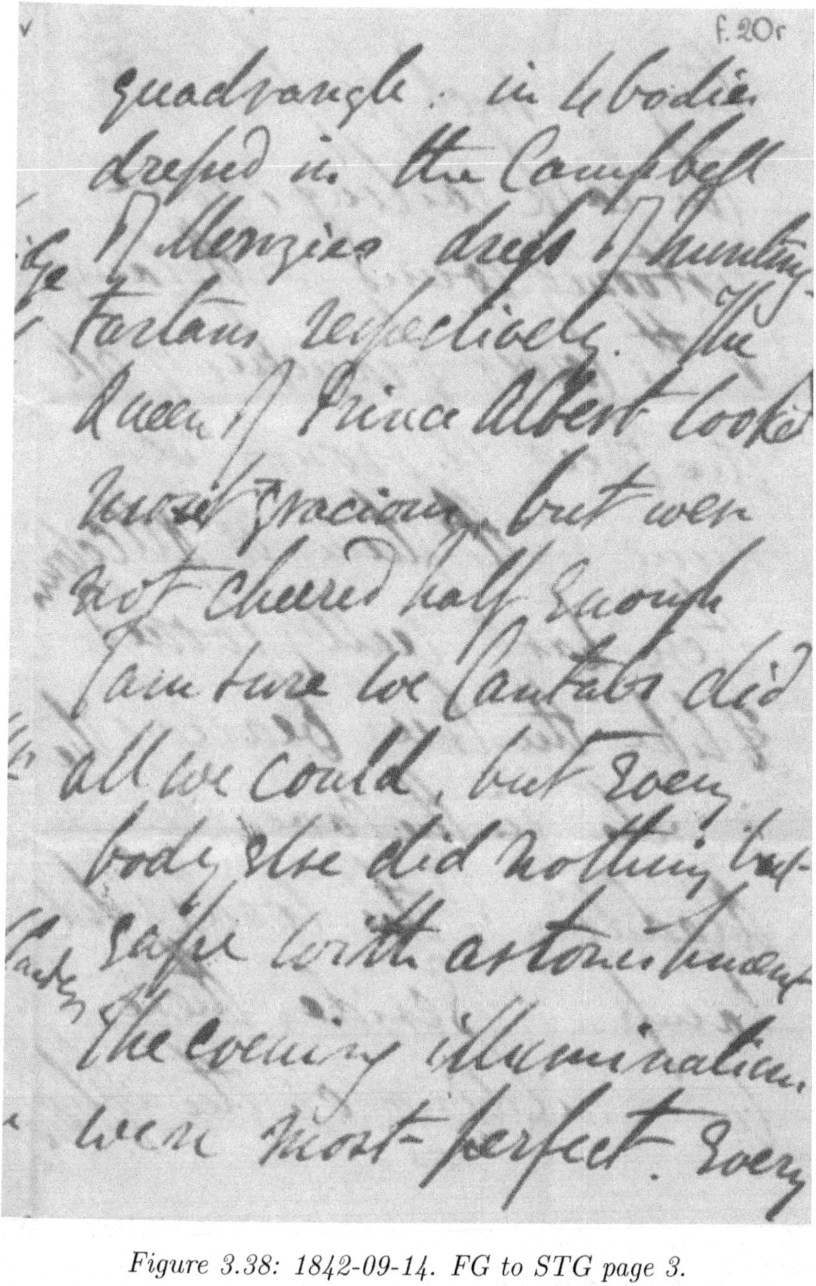

f. 20r

quadrangle: in le bodies
dressed in the Campbell
& Menzies dress & hunting
Tartans respectively. The
Queen & Prince Albert looked
most gracious but were
not cheered half enough
I am sure we Cantabs did
all we could, but every
body else did nothing but
gape with astonishment
The evening illuminations
were most perfect. Every

Figure 3.38: 1842-09-14. FG to STG page 3.

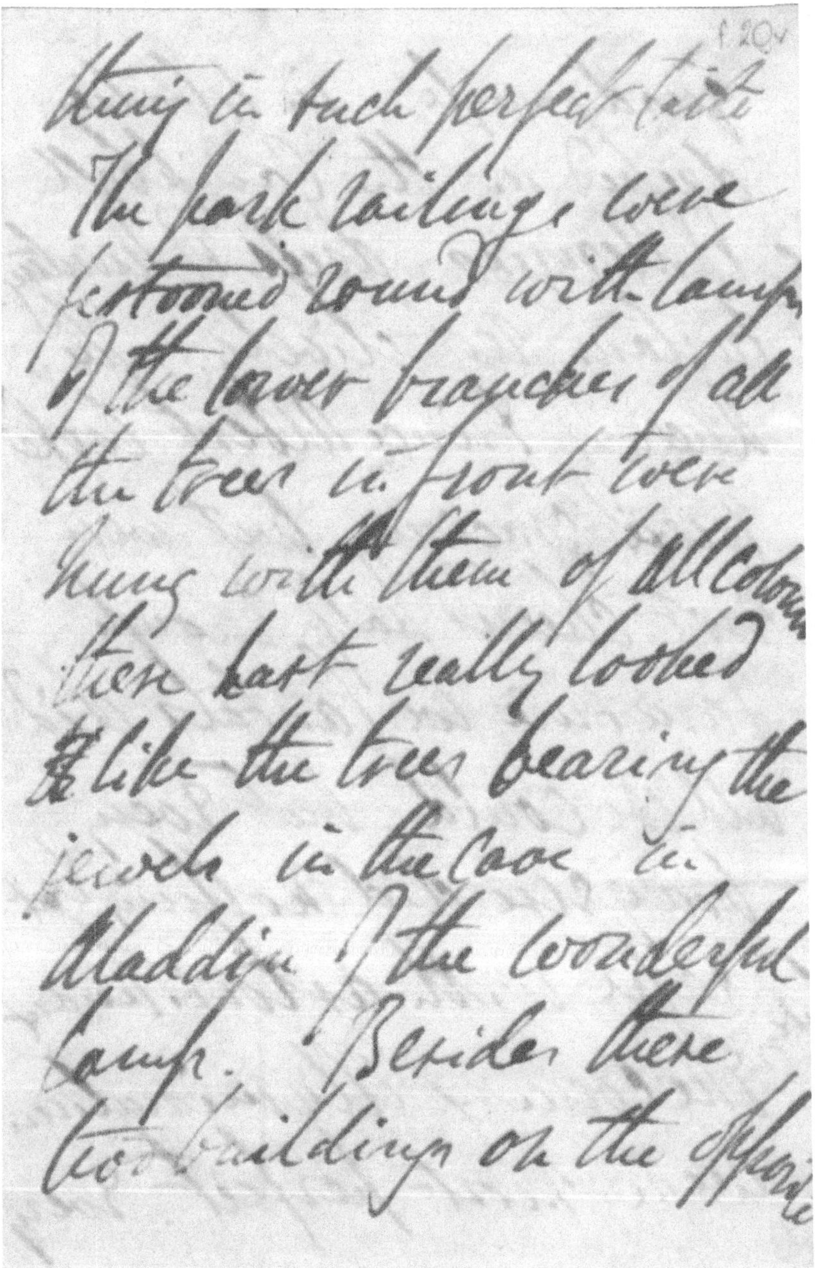

f. 204

thing in such perfect taste
The park railings were
festooned round with lamps
& the lower branches of all
the trees in front were
hung with them of all colours
the part really looked
like the trees bearing the
jewels in the cave in
Aladdin & the wonderful
lamp. Besides there
two buildings on the opposite

Figure 3.39: 1842-09-14. FG to STG page 4.

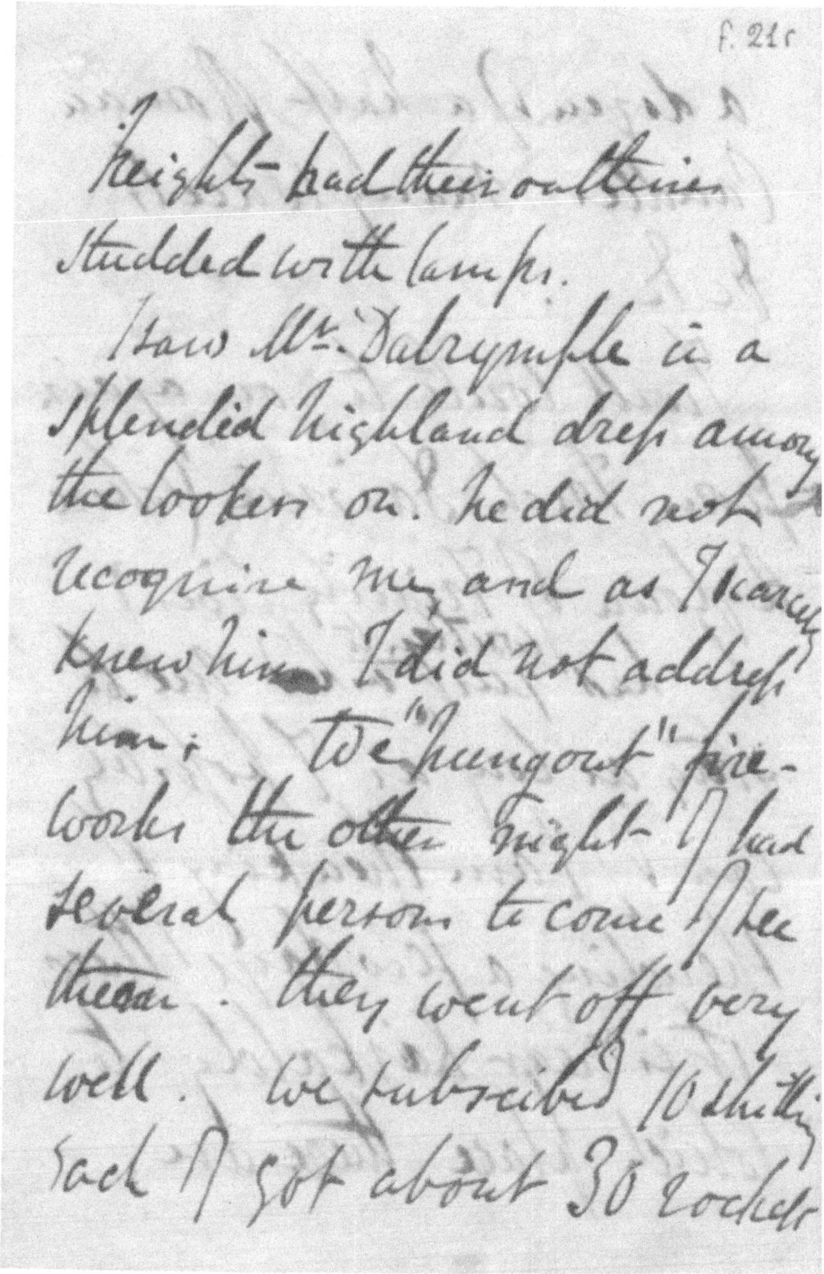

f. 21r

height - had their outlines
studded with lamps.
I saw Mr. Dalrymple in a
splendid highland dress among
the lookers on. He did not
recognize me and as I scarcely
knew him I did not address
him. We "hung out" fire-
works the other night & had
several persons to come & see
them. they went off very
well. We subscribed 10 shillings
each & got about 30 rockets

Figure 3.40: 1842-09-14. FG to STG page 5.

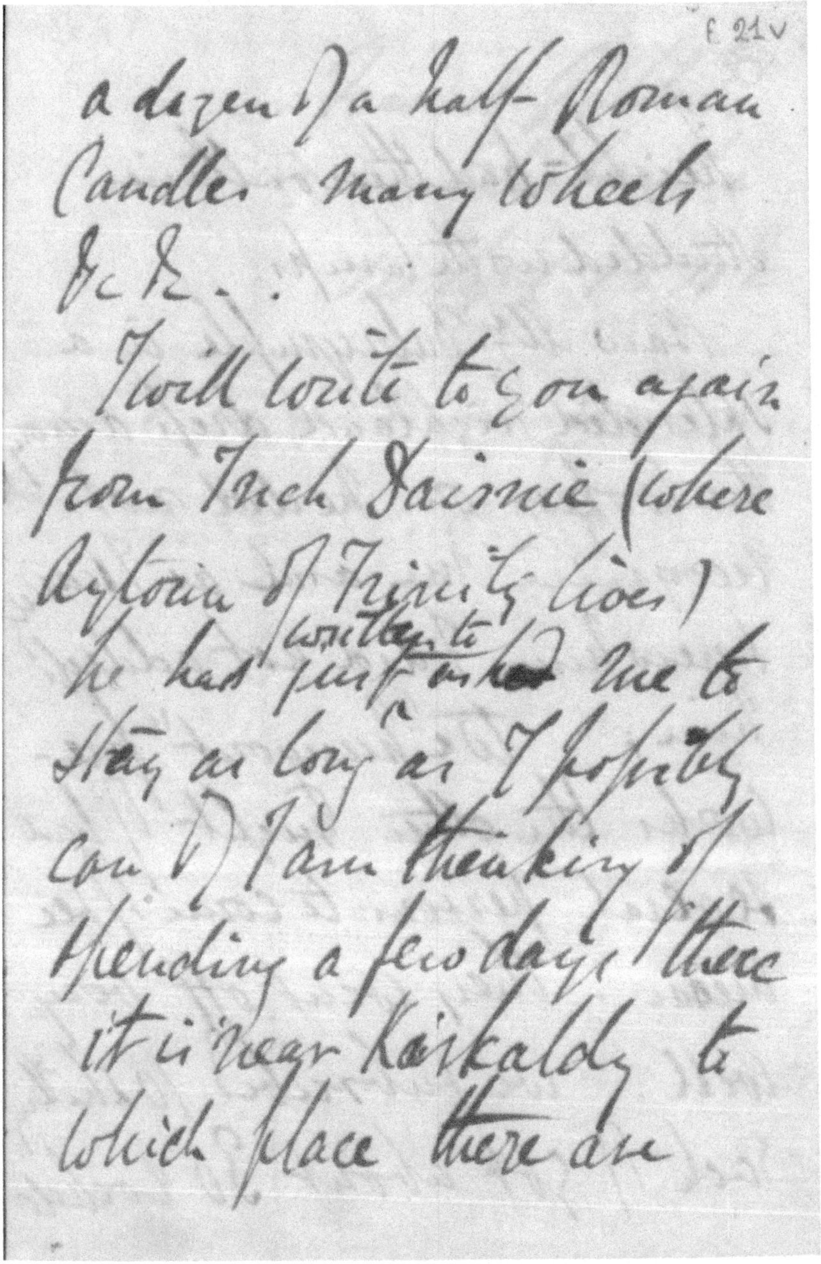

f. 21v

a dozen & a half Roman
Candles many wheels
&c &c.

I will write to you again
from Inch Dairnie (where
Aytoun of Trinity lives)
he has written to ~~just asked~~ me to
stay as long as I possibly
can & I am thinking of
spending a few days there
it is near Kirkaldy to
which place there are

Figure 3.41: 1842-09-14. FG to STG page 6.

f. 22r

steamer soon 2 hours
from Edinburgh.

Good bye.
Your affectionate son
Francis Galton.

Figure 3.42: 1842-09-14. FG to STG page 7.

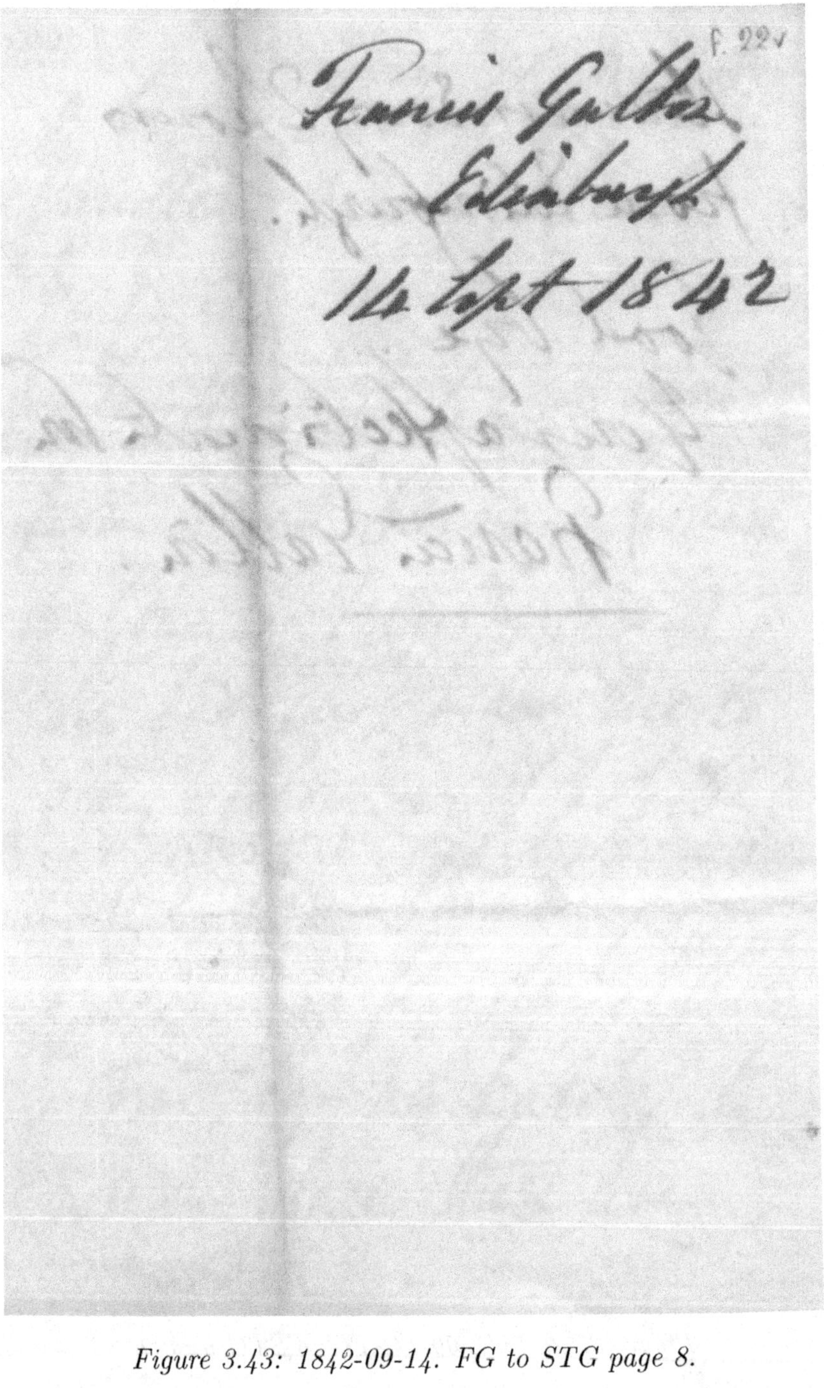

f. 22v

Francis Galton
Edinburgh
14 Sept 1842

Figure 3.43: 1842-09-14. FG to STG page 8.

Figure 3.44: The view from M'Queens' Hotel, Edinburgh, September 15, 1842. By F.G.

.. I played at chess with Francis. . . .

October 4th. Tuesday. Brighton.

. . . Violetta took Mrs. Phillips, Mary and Francis to the Dyke when the latter was taken rather ill with the exertion of running up and down the declivity. . . .

October 5th. Wednesday. Brighton.

I rode with Francis upon the Race Course Hill. . . . Mr. Lawrence called upon Francis.

October 6th. Thursday. Brighton.

I rode with Bessy and drove with Violetta and Francis.

October 7th. Friday. Brighton.

Continued fine weather. I rode to Rottingdean with Francis and we afterwards took Mary [for] a drive on the London Road. . . .

October 9th. Sunday. Brighton.

Fine weather. Bessy, Francis and myself attended St. Andrew's Chapel in Waterloo Street where a sermon was preached by the Rev. John Nelson in favour of the charity schools of the Parish of Hove. Bessy and I went again in the afternoon. Francis dined with the Edstone party.

October 11th. Tuesday. Brighton.
I walked with Violetta, Erasmus and Darwin and afterwards rode with Bessy and drove with Violetta, Erasmus and Francis. Bessy, Erasmus, and Francis dined with Mrs. Phillips.

October 12th. Wednesday. Brighton.
... Francis went to spend the day in London. ...

October 13th. Thursday. Brighton.
I walked with Violetta, Bessy and Erasmus to the chain pier where we were weighed. Samuel Tertius Galton weighed 13 st. 12 lbs., Violetta Galton 12 st. 13 lbs. ... Francis went to a concert at the Town Hall.

October 14th. Friday. Brighton.
Francis left us by the 9:30 train for London on his way to Cambridge. ...

1842/11/02. Francis Galton to Samuel Tertius Galton

Trin: Coll:
Novr 2nd 1842

My dear Father
I forgot in my last note to go into proper raptures about Stultz he really is a wonderful man. I had no idea that it was in human power to make such extraordinary improvements in my personal attire as the combined geniuses of Stultz & Gobby have effected,[19] consultations of course had to be frequent during the course of the Friday & Saturday that I was in Town but I at length emerged from my chrysalis crust of Cartwright and James' manufacture to the butterfly adornments of Messrs Stultz. My head is very

[19] Stultz and co., 10 Clifford Street, made coats (Stultz himself had died in 1832). Gobby made trousers.

uncertain so that I can scarcely read at all, however I find that I am not at all solitary in that respect. Of the year above me the first 3 men in their College examinations are all going out in the poll the first 2 from bad health & the third, Boulton,[20] from finding that he could not continue reading as he used to do without risking it. Fowell Buxton is quite knocked up & goes out in the poll so does Bristed[21] ~~the~~ one of the first classics in our year, in fact the whole of Trinity seems quite crank, two other men Hotham & Edwards who read with Hopkins[22] and at the same time were very superior classics (Hotham was Newcastle scholar at Eton, which is the highest classical honor they can get there) have both given up classics finding their two subjects are too much for them. It is quite melancholy too to see the men who stood high in their College, but did not get scholarships this year in May they seem most of them quite broken spirited. Our man Stokes who was considered sure of being Senior Classic of his year who used to be the merriest fellow going, lost his scholarship from not doing his Mathematics, scarcely ever perpetrates a laugh, & so also with the other men. Johnson also (Adèle knows Mrs. Johnson) is quite cut up. Joe Kay[23] has left from illness produced by reading & won't come back till next term. I feel more convinced every day that if there is any thing more to be repressed than another it is certainly the system of competition for the satisfaction enjoyed by the gainers is very far from counter balancing the pain it produces among the others.

I have not after all entered a boat club but patronise hockey and made my first debut yesterday at it.

Montague Boulton is a very nice fellow & uncommonly sharp, I do not know what his chance is considered to be in Honours.

Charley Buxton's & my debating club gets on famously, we have just enrolled Hallam[24] (the youngest & only surviving son of "Middle Ages") among our members & spout away most learnedly once a week on subjects in the Ethical line, there are 9 of us altogether & I shall be president next week & shall array myself in Stultz for the occasion. We keep it very secret and meet ~~alternately~~ in each others' rooms in rotation under the pretext of a wine party;

---

[20] Boulton, Matthew Piers Watt. See page 480.
[21] Bristed, Charles Astor. See page 482.
[22] Hopkins, William. See page 484.
[23] Kay, Joseph. See page 486.
[24] Hallam, Henry Fitzmaurice. See page 483.

then the man who gives the spread is president & at $\frac{1}{4}$ before 5 sports his door and the debate begins. The president of one meeting has to propose a subject & open the debate on it in the meeting a week after.

Mr. Hodgson[25] sent me my certificate for degrading the day before yesterday. I will let you know how the proceedings about it progress in my next letter. I have invested in such a jolly second hand arm chair. I really believe it is the most comfortable in Cambridge, it cost £3. 10. the seat is only 10 inches from the ground, so it is thoroughly luxurious. [Sketch]

How is Capt. Bowater, wedded or single. By the way one can imagine the following scene.

Scene. Interior of a church, marriage procession bridesmaids &c. rather an elderly bride & bridegroom. Priest. Wilt thou have this man to be thy wedded husband— Bride. Not a doubt about it.

How are the various family ailments? Does Mater find bannocks lie very lightly on the stomach? Thank you for taking care of the papers there were some books also that I left. Guizot's Civilisation &c. & there are some at Claverdon. Now if Mater had such a redundancy of preserves & jams from the garden that she was really obliged to manage with them as she used to with the cats of old, viz. give sixpence to whomsoever w^{d} take them, then the strength of my filial affection w^{d} rejoice at the opportunity of being useful and of taking 2 or 3 pots out of her way. And should all this be the case it would be worth while to put them with the books in a box and send them at once to Cambridge, where they would be severally eaten & read.

I am learning singing after the Hullah fashion, but alas notwithstanding maternal prophesies I find the Galton ear is as slightly developed inside my skull as it is largely on the outside and although I keep up the credit of the family failing yet I am afraid I shall not at the same time qualify for the professorship of Music.

I read with no tutor at all at present, as I question the advantages of doing so, but shall attend the University lectures on mechanics by Willis, which begin on the 10th.

Theodore is reading hard for his degree which comes off in about 10 weeks.

Goodbye & believe me your affectionate Son

[25] Hodgson, Joseph. See page 483.

Fras. Galton

1842/11/25 circa. Francis Galton to Samuel Tertius Galton

My dear Father

Thank you much for yours and Bessy's letter which last I received yesterday. I am quite ashamed at not having written oftener but my head generally is not as well as might be at the orthodox letter writing time namely about 7 p.m. I find myself quite unable to do anything in reading for by really ~~serious~~ deep attention to ~~them~~ Maths I can bring on my usual dizziness &c. almost immediately, though ~~ordinary~~ generally I feel much better than I used to do. Palpitations of the heart have lately come on when I read ~~any~~ more than I ought to do which I am rather glad of than otherwise, as it saves my head. What annoys me ~~so much~~ most is that my powers of reading vary so much on consecutive days, at one time being able to read some hours, at another not half one, and the dizziness &c. when it does come on comes on so rapidly, that I have no forewarning symptoms to tell me when to stop, except occasionally the palpitations. I have been rather diffuse a la Leamington about my health so by way of change do thank Hines particularly for the cake, & tell her that if the ~~rapidity~~ quickness of its disappearance be any sign of its excellence, that it decidedly must rank among the very best that culinary science has produced. It was capital & at least 20 individuals concurred in that remark.

Riddle—Why is the ball at the top of St Paul's like the being cook to her majesty?

Answer—Because it's a high {cooling airy / culinary} situation.

Did I read you the answer to the Archbishop's riddle? I had one but do not know if it was right, it was very ingenious & made it out to be a Tapir or taper. The Kings of the earth kneeling in the ground they had trod, is made out to be the Mexican & Peruvian &c. princes hunting the tapir following his trail and kneeling to take aim. The weight & length w^{d} apply to those tapers that are rolled up like string. ...[26]

[26] Closing page(s) missing.

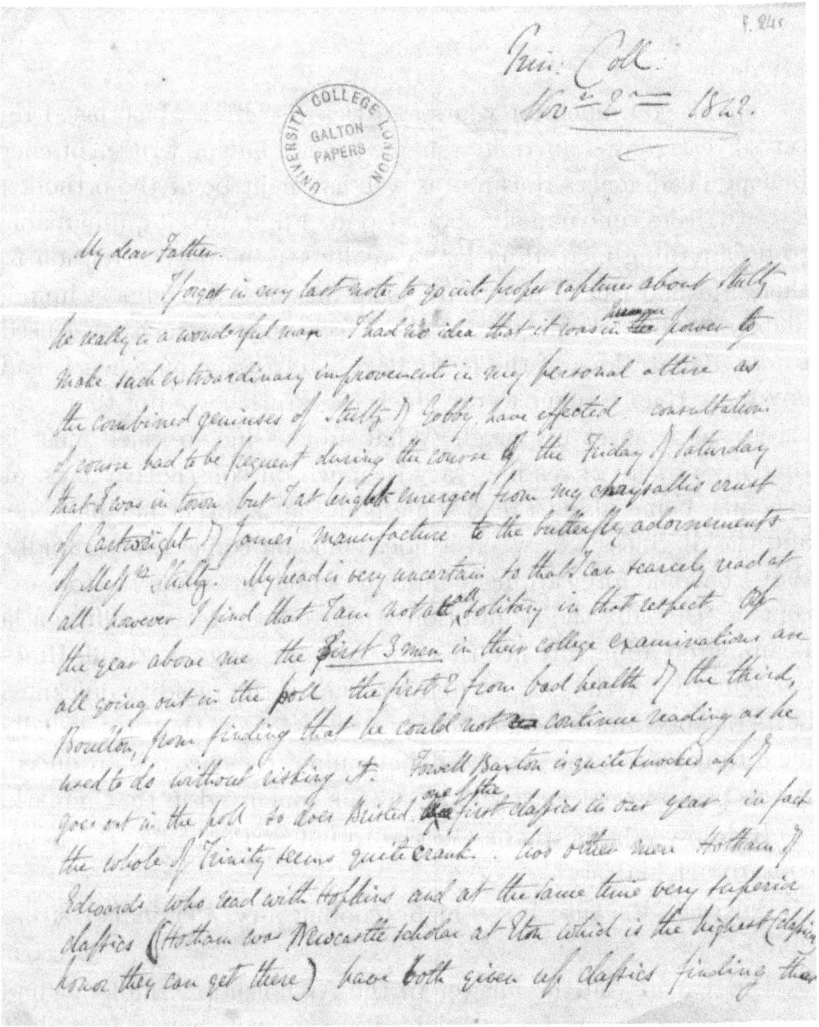

Trin: Coll:
Nov 2 1842.

My dear Father.

I forgot in my last note to go into proper raptures about Stultz he really is a wonderful man I had no idea that it was in human power to make such extraordinary improvements in my personal attire as the combined geniuses of Stultz & Gobby have effected consultations of course had to be frequent during the course of the Friday & Saturday that I was in town but I at length emerged from my chrysallis crust of Cartwright & James' manufacture to the butterfly adornements of Messrs Stultz. My head is very uncertain so that I can scarcely read at all however I find that I am not at all solitary in that respect. Of the year above me the first 3 men in their college examinations are all going out in the poll the first 2 from bad health & the third, Boulton, from finding that he could not continue reading as he used to do without risking it. Fowell Buxton is quite knocked up & goes out in the poll so does Birkett one of the first classics in our year, in fact the whole of Trinity seems quite crank. two other men Hotham & Edwards who read with Hopkins and at the same time very superior classics (Hotham was Newcastle scholar at Eton which is the highest classical honor they can get there) have both given up classics finding that

Figure 3.45: 1842-11-02. FG to STG page 1.

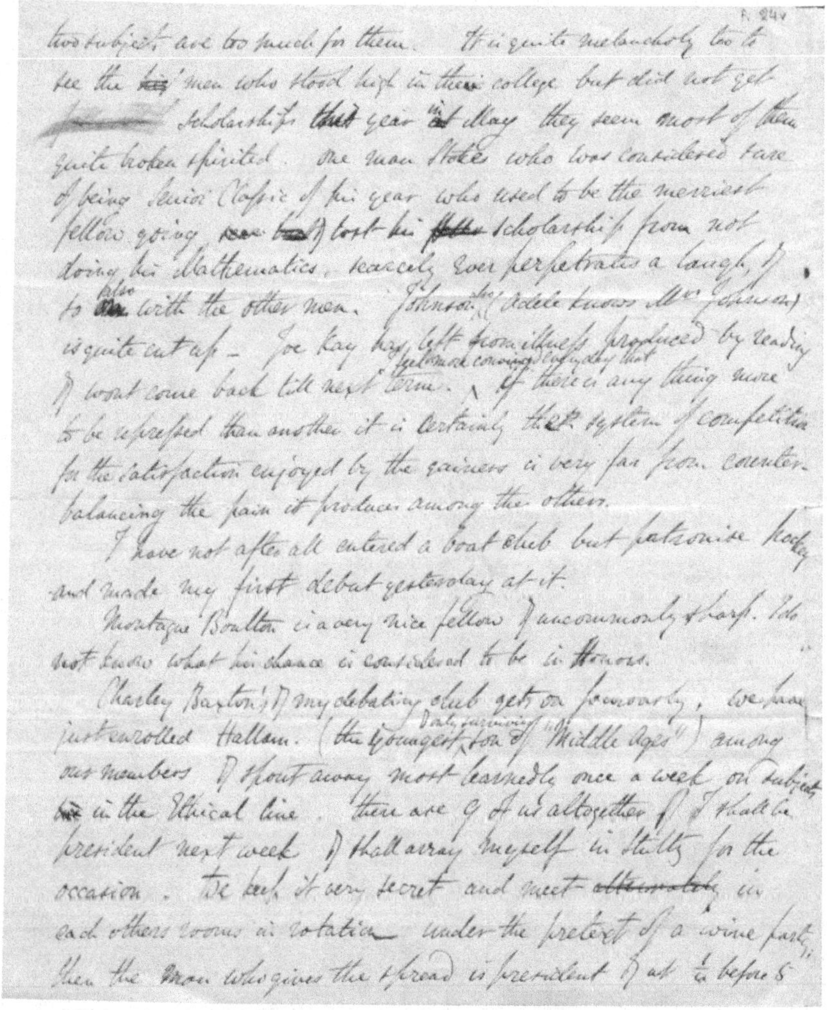

two subjects are too much for them. It is quite melancholy too to see the men who stood high in their college but did not get scholarships this year at May they seem most of them quite broken spirited. One man Stokes who was considered sure of being Senior Classic of his year who used to be the merriest fellow going & lost his scholarship from not doing his Mathematics, scarcely ever perpetrates a laugh, & so also with the other men. Johnson (Adele knows Mrs Johnson) is quite cut up — Joe Kay has left from illness produced by reading & wont come back till next term. I feel more convinced every day that if there is any thing more to be repressed than another it is certainly the system of competition for the satisfaction enjoyed by the gainers is very far from counterbalancing the pain it produces among the others.

I have not after all entered a boat club but patronise hockey and made my first debut yesterday at it.

Montague Boulton is a very nice fellow & uncommonly sharp. I do not know what his chance is considered to be in Honors.

Charley Buxton's & my debating club gets on famously, we have just enrolled Hallam (the youngest only surviving son of "Middle Ages") among our members & spout away most learnedly once a week on subjects in the Ethical line. there are 9 of us altogether & I shall be president next week & shall array myself in stilts for the occasion. We keep it very secret and meet ~~alternately~~ in each others rooms in rotation under the pretext of a wine party then the man who gives the spread is president & at ½ before 8

Figure 3.46: 1842-11-02. FG to STG page 2.

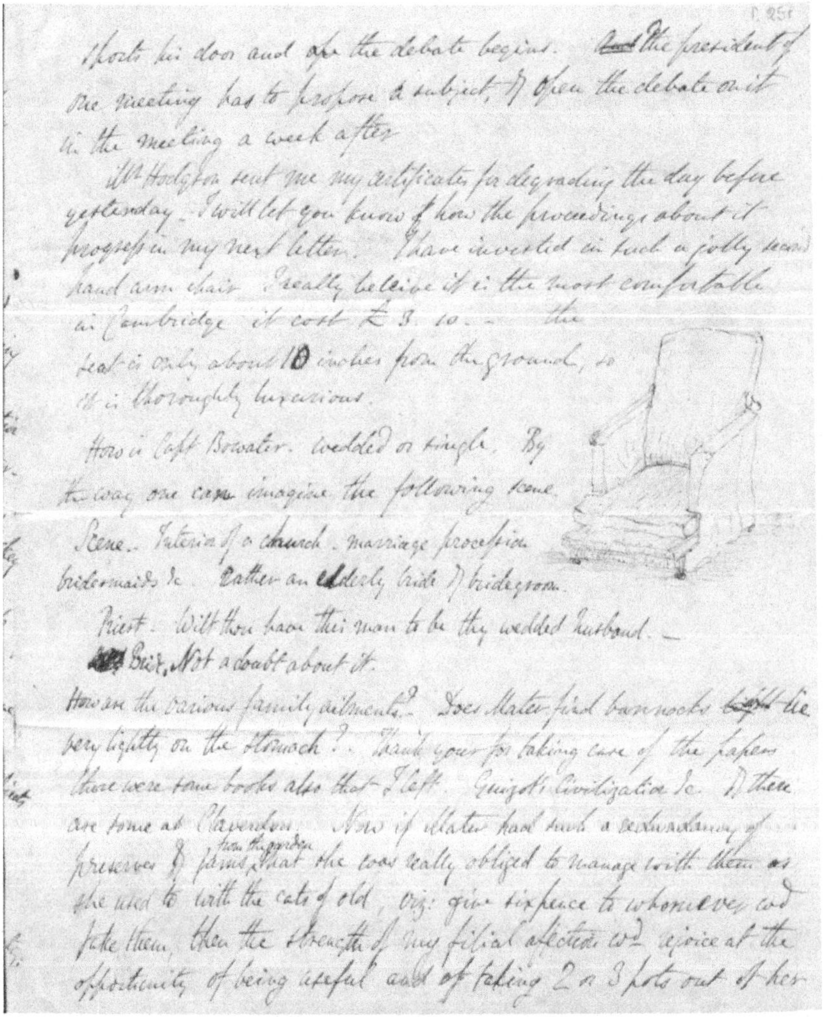
shuts his door and off the debate begins. ~~And~~ The president of
the meeting has to propose a subject, & open the debate on it
in the meeting a week after
Mr Hodgson sent me my certificate for degrading the day before
yesterday. I will let you know of how the proceedings about it
progress in my next letter. I have invested in such a jolly second
hand arm chair I really believe it is the most comfortable
in Cambridge it cost £3. 10. – the
seat is only about 10 inches from the ground, so
it is thoroughly luxurious.
How is Capt Bowater. wedded or single. By
the way one can imagine the following scene
Scene. Interior of a church. marriage procession
bridesmaids &c. rather an elderly bride & bridegroom.
Priest. Wilt thou have this man to be thy wedded husband. –
Bride. Not a doubt about it.
How are the various family ailments? Does Mater find bannocks ~~lift~~ lie
very lightly on the stomach? Thank you for taking care of the papers
there were some books also that I left. Guizot's Civilization &c. & there
are some at Claverdon. Now if Mater had such a redundancy of
preserves & jams from the garden that she was really obliged to manage with them as
she used to with the cats of old, viz: give sixpence to whomever wd
take them, then the strength of my filial affection wd rejoice at the
opportunity of being useful and of taking 2 or 3 pots out of her

Figure 3.47: 1842-11-02. FG to STG page 3.

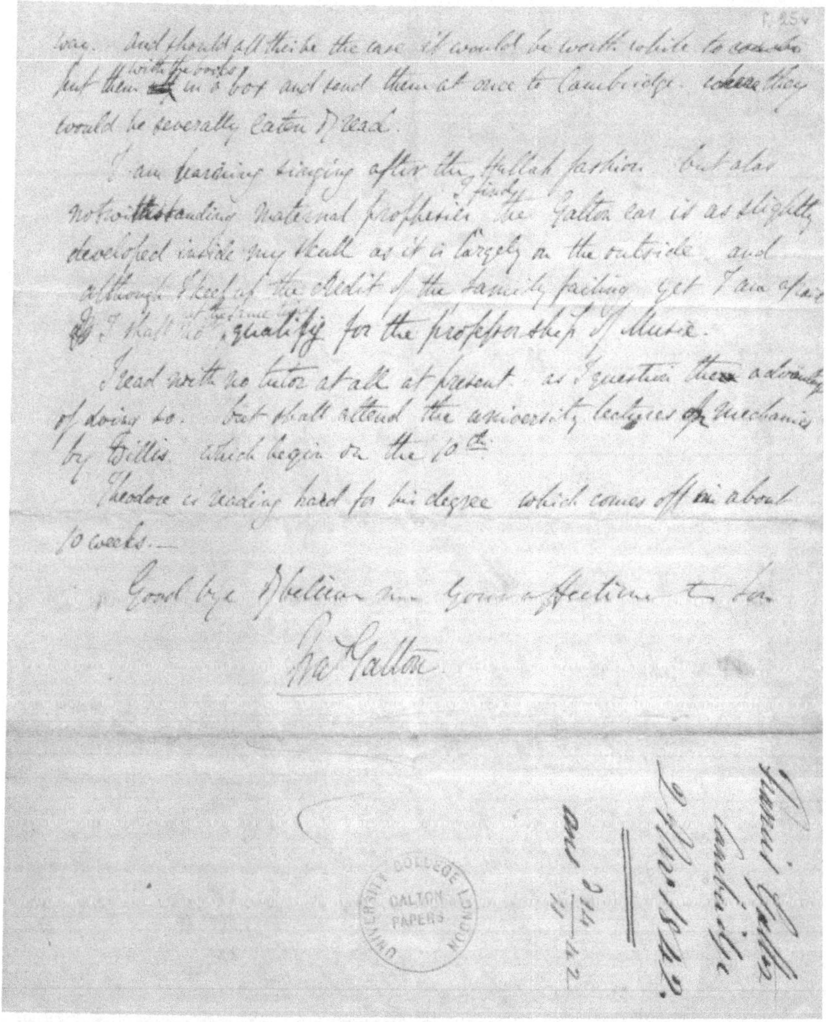
way. And should all this be the case it would be worth while to consider
put them with the books in a box and send them at once to Cambridge where they
would be severally eaten & read.
I am learning singing after the Fellah fashion but alas
notwithstanding maternal prophesies the Galton ear is as slightly
developed inside my skull as it is largely on the outside and
although I keep up the credit of the family failing yet I am afraid
I shall not at the time qualify for the professorship of Music.
I read with no tutor at all at present as I question the advantage
of doing so. but shall attend the university lectures of mechanics
by Willis which begin on the 10th.
Theodore is reading hard for his degree which comes off in about
10 weeks.—
Good bye Ibelieve me Your affectionate brother
Fra^s Galton

Figure 3.48: 1842-11-02. FG to STG page 4.

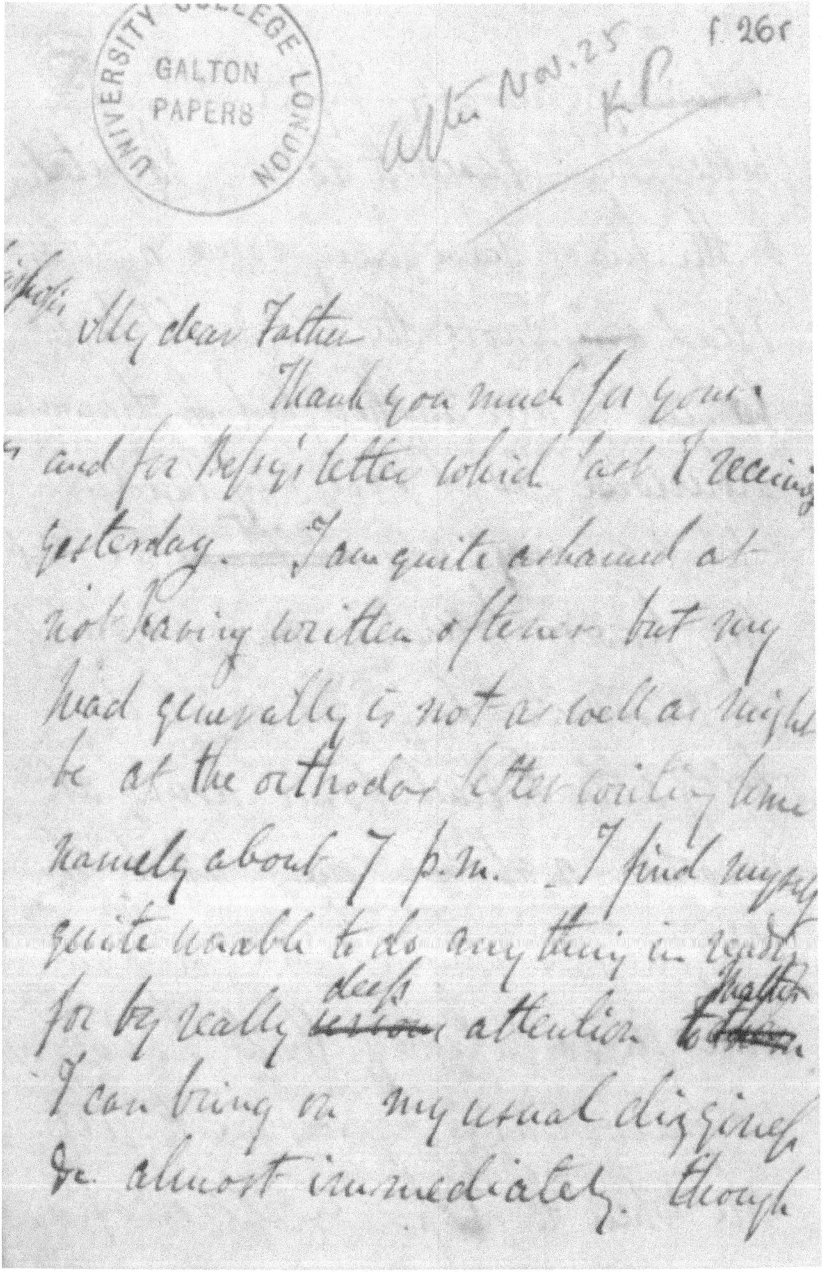

My dear Father

Thank you much for your and for Bessy's letter which last I received yesterday. I am quite ashamed at not having written oftener but my head generally is not as well as might be at the orthodox letter writing time namely about 7 p.m. I find myself quite unable to do anything in reading for by really deep attention matter I can bring on my usual dizziness &c. almost immediately though

Figure 3.49: 1842-11-25 circa. FG to STG page 1.

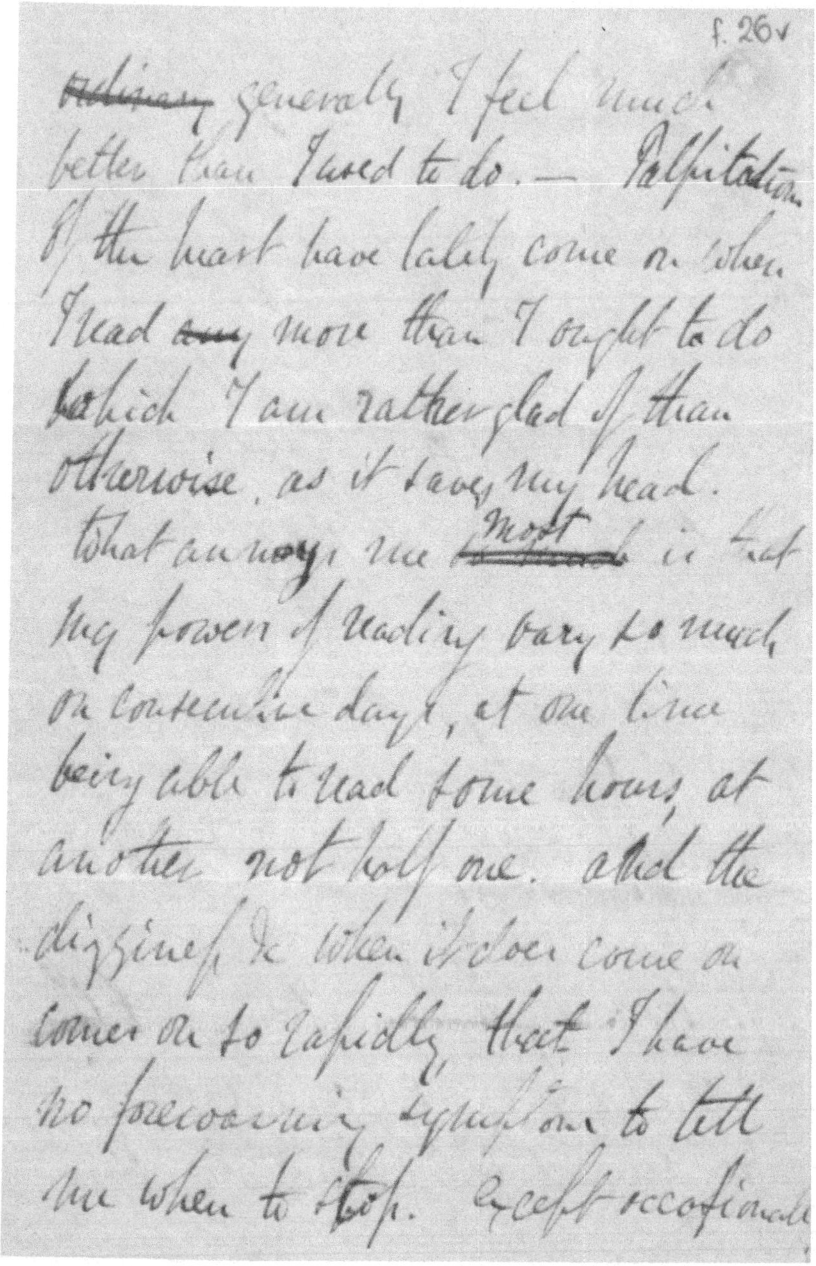

f. 26v

~~ordinary~~ generally I feel much
better than I used to do.— Palpitation
of the heart have lately come on when
I read ~~any~~ more than I ought to do
which I am rather glad of than
otherwise, as it saves my head.
What annoys me most is that
my powers of reading vary so much
on consecutive days, at one time
being able to read some hours, at
another not half one. and the
dizziness &c when it does come on
comes on so rapidly that I have
no forewarning symptoms to tell
me when to stop. Except occasionally

Figure 3.50: 1842-11-25 circa. FG to STG page 2.

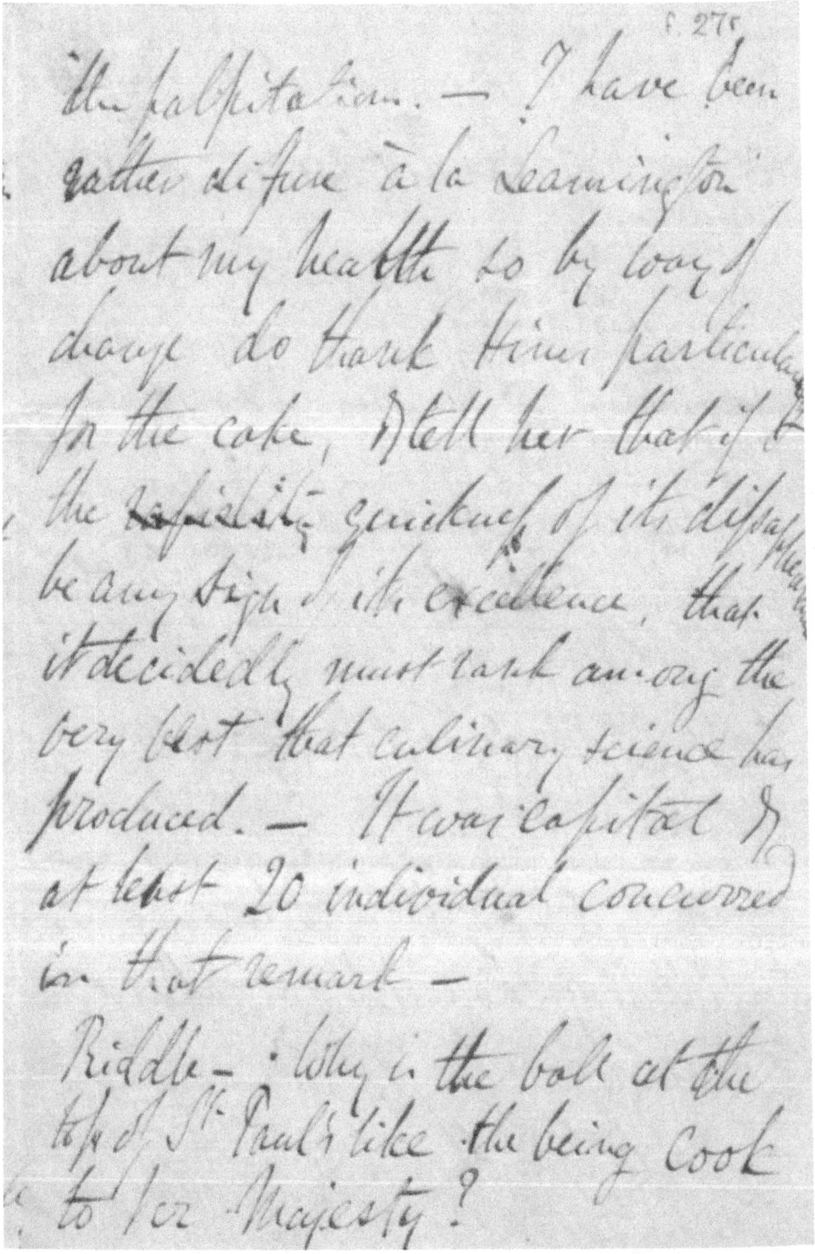
f. 275

the palpitation. — I have been
rather diffuse à la Leamington
about my health so by way of
change do thank Hime particularly
for the cake, & tell her that if
the ~~rapidity~~ quickness of its disappearance
be any sign of its excellence, that
it decidedly must rank among the
very best that culinary science has
produced. — It was capital &
at least 20 individuals concurred
in that remark —
Riddle — : Why is the ball at the
top of St Paul's like the being Cook
to Her Majesty?

Figure 3.51: 1842-11-25 circa. FG to STG page 3.

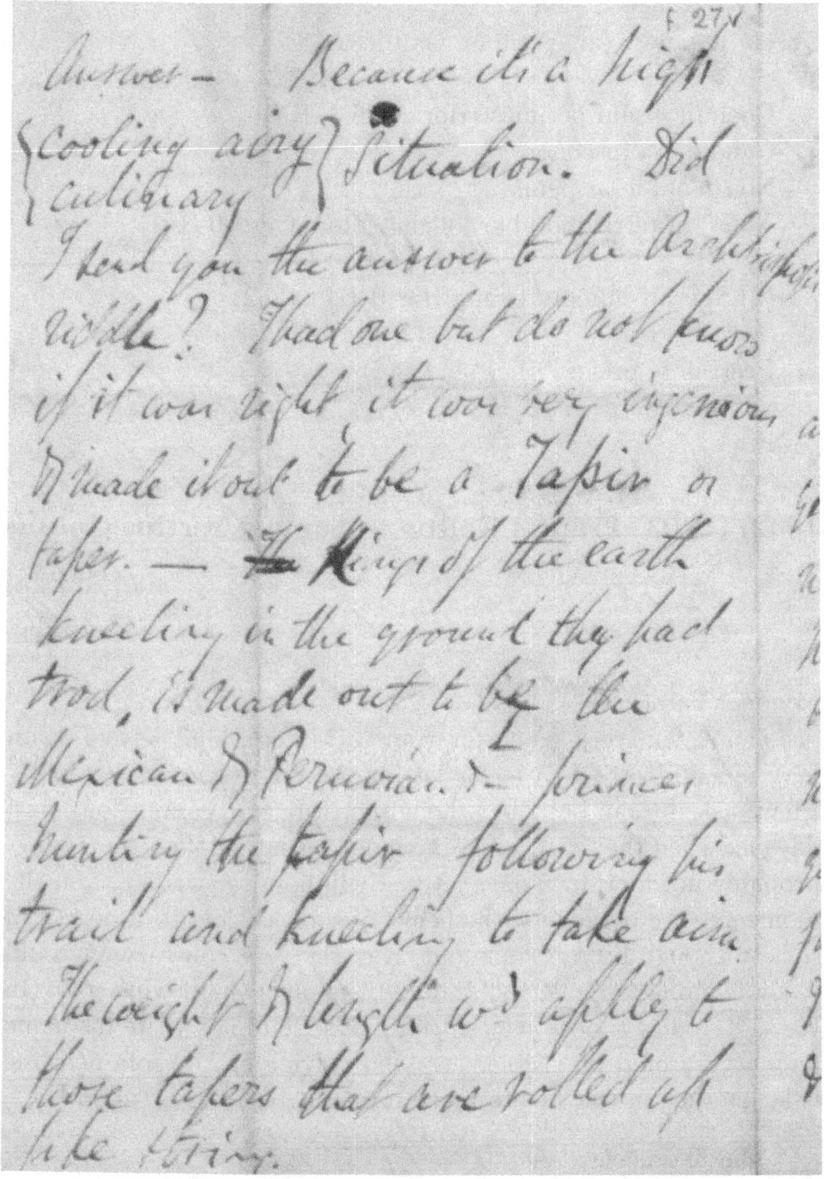
f 27v

Answer— Because it's a high
{cooling airy / culinary} Situation. Did
I send you the answer to the Archbishop's
riddle? I had one but do not know
if it was right, it was very ingenious
I made it out to be a Tapir or
taper. — The Kings of the earth
kneeling in the ground they had
trod, is made out to be the
Mexican & Peruvian princes
hunting the tapir following his
trail and kneeling to take aim
The weight & length w^d apply to
those tapers that are rolled up
like string.

Figure 3.52: 1842-11-25 circa. FG to STG page 4.

1842/11. Francis Galton to Samuel Tertius Galton

Mottoes for the Galton Crest

(1) Lag not. (anagram on Galton).

Coelum ipsum petimus (Hor: Ode 3, B I)[27]
Candidiora insequor
Non tenui ferar pennâ
Non[28] in terris morabor longius (Hor. Ode 30 B II)

(2) Spernit humum pennâ Hor B III Ode 2.

I certainly prefer either (1) or (2)

F. G.

1842/12/03. Francis Galton to Samuel Tertius Galton

Trin Coll.
Saturday

My dear Father

Thank you very much for your & Mater's kind letter. I am very much obliged to you for letting me invite Kay & Aytoun.[29] I have asked them but they, though most happy, cannot for 3 or 4 days fix when they can come, it cannot be earlier than the 14th & probably not for 4 or 5 days later. I will however write to you the moment I can arrange with them. Aytoun is slow to a degree in society though he can come out, Kay the very contrary & would enjoy Leamington gaieties. I am writing at a small wine party in Kay's[30] rooms that is very splendid & so dare not write his name large as it would be visible. [...] who is lying on the sofa near me is kicking me in the ribs to make myself agreeable. So good bye.

Your affectionate Son
Fras. Galton

---

27 Heaven itself we seek.
28 Neque.
29 Aytoun, Roger Sinclair. See page 480.
30 Kay, Edwin Ebenezer. See page 485.

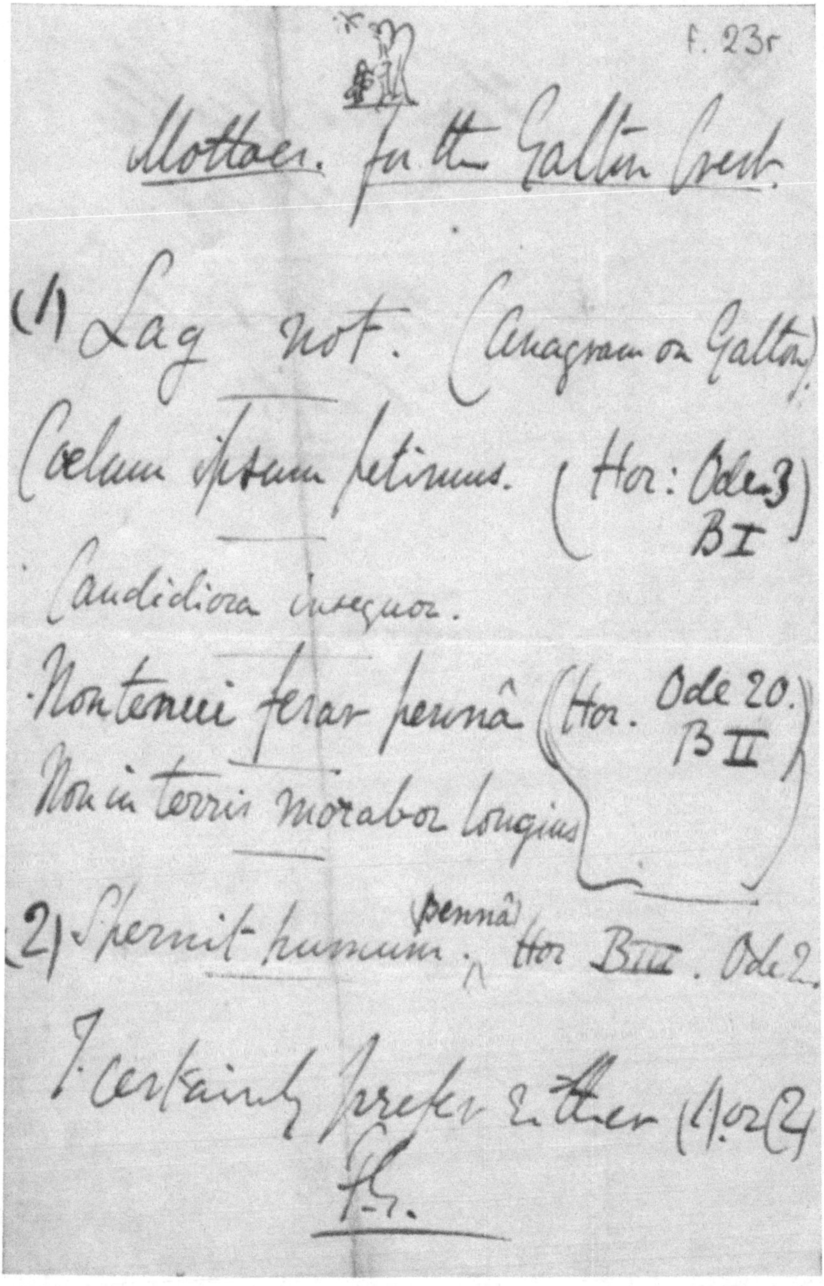
F. 23r

Mottoes. for the Galton Crest.

(1) Lag not. (Anagram on Galton)

Coelum ipsum petimus. (Hor: Ode 3 B I)

Candidiora insequor.

Non tenui ferar pennâ (Hor. Ode 20. B II)

Non in terris morabor longius

(2) Spernit humum pennâ Hor B III. Ode 2.

I certainly prefer either (1) or (2)

FG.

Figure 3.53: 1842-11. FG to STG page 1.

f. 23v

Francis Galton
Nov 1842

Matters for
Cash

Figure 3.54: 1842-11. FG to STG page 2.

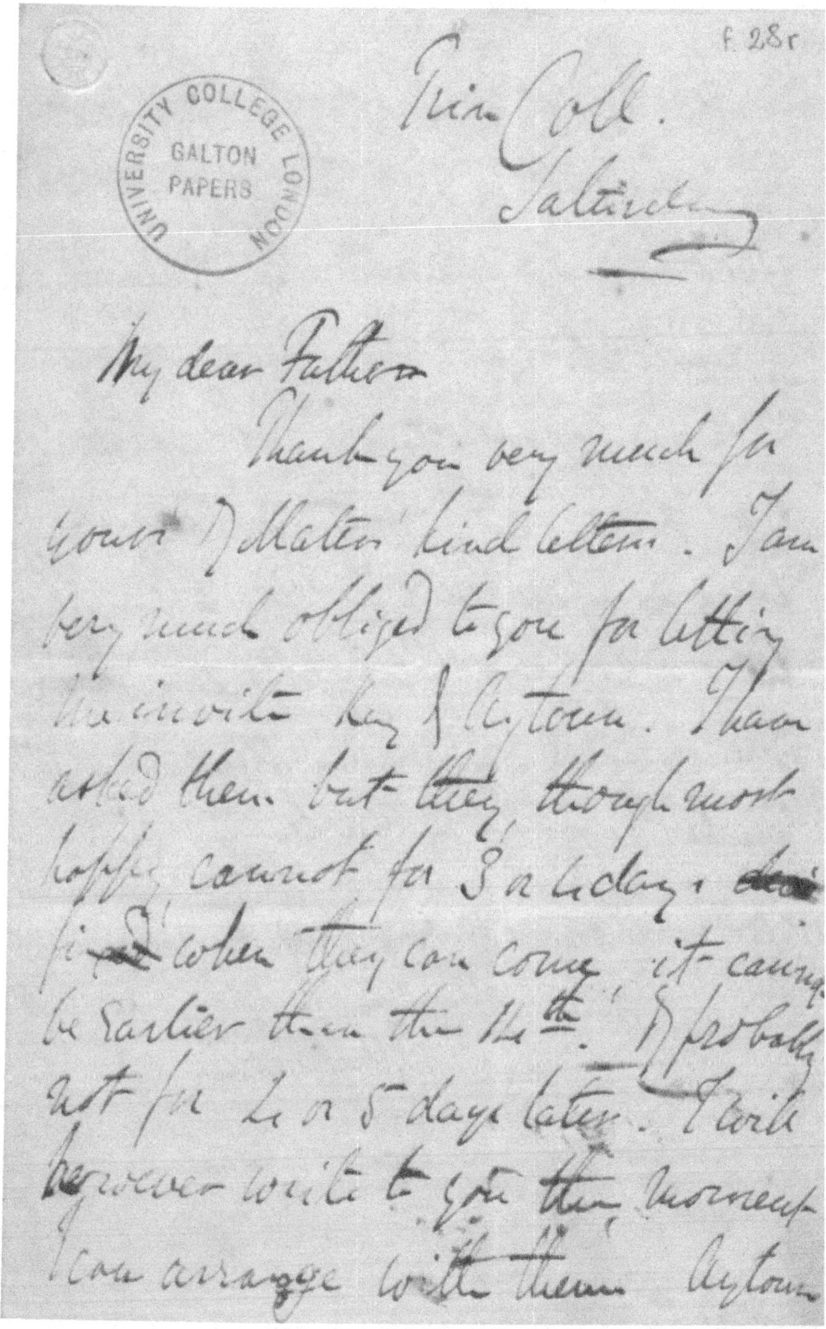

f. 28r

Trin Coll.
Saturday

My dear Father
Thank you very much for your & Mater's kind letters. I am very much obliged to you for letting me invite Key & Aytoun. I have asked them but they though most happy cannot for 3 a holiday; & when they can come it cannot be earlier than the 16th & probably not for 4 or 5 days later. I will however write to you the moment I can arrange with them. Aytoun

Figure 3.55: 1842-12-03. FG to STG page 1.

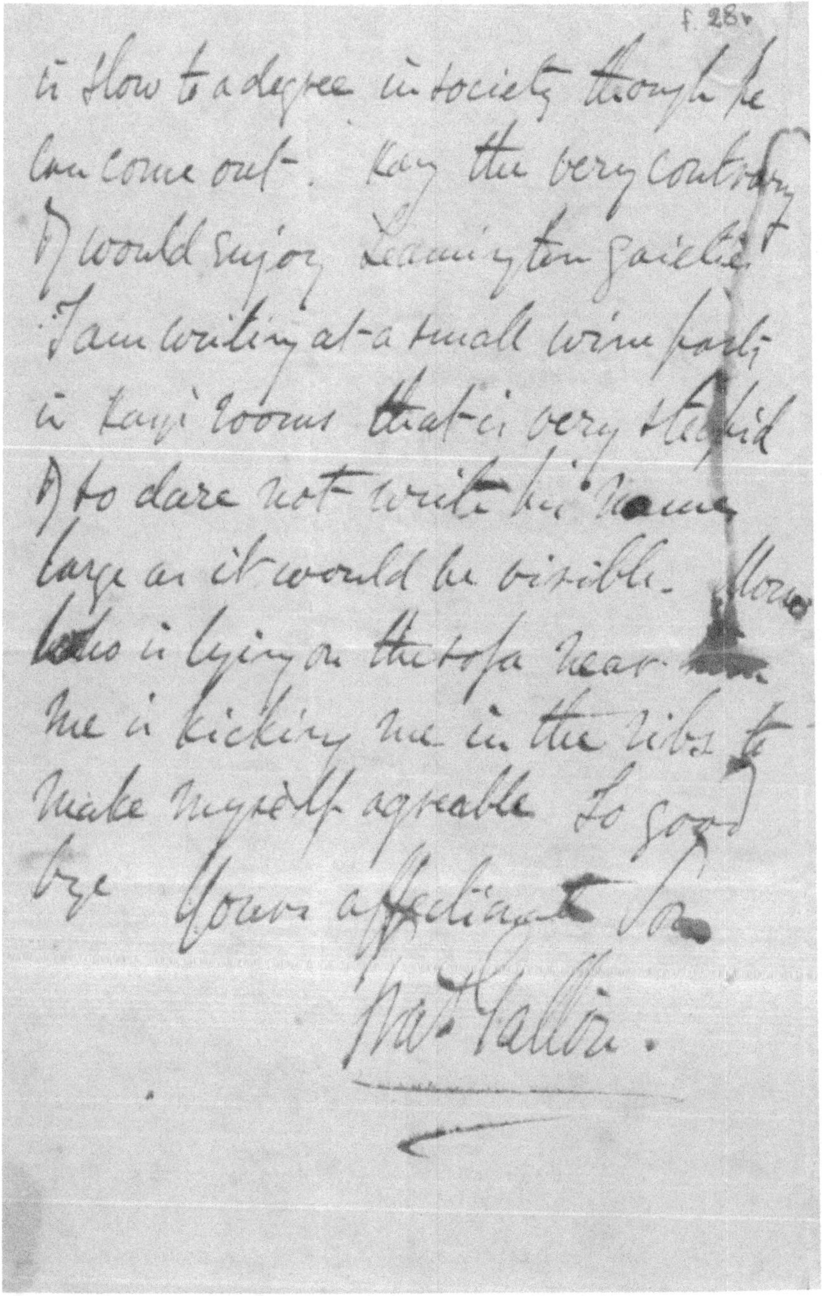

f. 28v

is slow to a degree in society though he has come out. Nay the very contrary I would enjoy Leamington gaieties I am writing at a small wine party in Kay's rooms that is very stupid & so dare not write his name large as it would be visible. Moore who is lying on the sofa near me is kicking me in the ribs to make myself agreeable. So good bye Yours affectionate Son

Fras Galton.

Figure 3.56: 1842-12-03. FG to STG page 2.

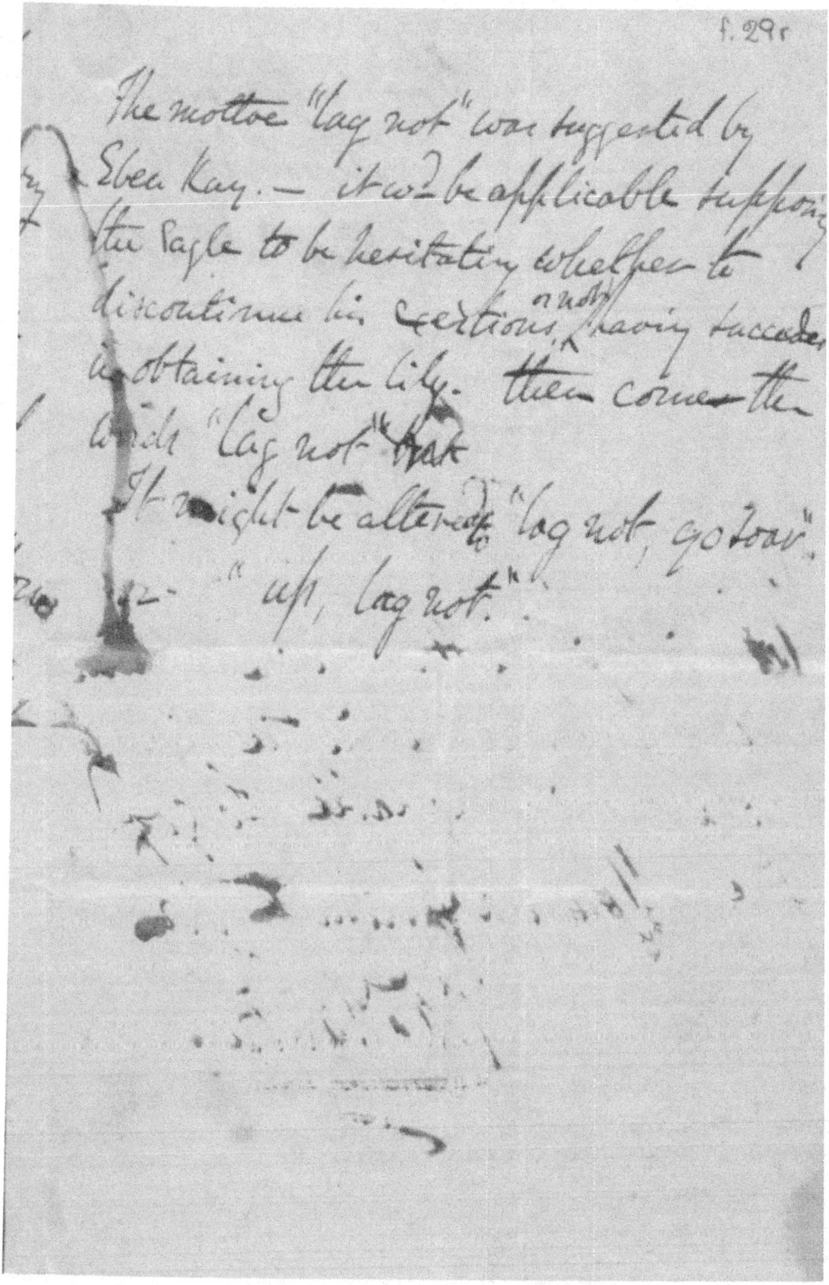
f. 29r

The mottoe "lag not" was suggested by
Eben Kay.— it wd be applicable supposing
the Eagle to be hesitating whether to
discontinue his exertions, and not having succeeded
in obtaining the lily. then comes the
words "lag not" that
It might be altered to "lag not, go on"
"up, lag not".

Figure 3.57: 1842-12-03. FG to STG page 3.

Fras. Galton
Cambridge
3 Decr 1842.

Figure 3.58: 1842-12-03. FG to STG page 4.

1842/12/22–31. Diary of Samuel Tertius Galton

December 22nd. Thursday. Leamington.

... Francis arrived before dinner from London but did not look well; he and Emma went to a party at the Granvilles in the evening.

December 23rd. Friday. Leamington.

... In the evening I had a conversation with Francis relative to his prospects.

December 26th. Monday. Leamington.

... Mr. Brown, a college friend of Francis, lunched with us.

December 27th. Tuesday. Leamington.

... Francis and Emma dined with Miss Phillips.

December 31st. Saturday. Leamington.

I went with Violetta to Claverdon where we were afterwards joined by Irwin Maling, Darwin, Erasmus and Francis; the three latter dined with us.

Chapter 4

1843

1843/01/01–04. FG Diary

1. Sunday.
Dr. Kaye Bishop of Lincoln
preached.[1]

2. Monday.
The Dean's last tea with us.

3. Tuesday.
Dinner party at home. $\frac{1}{4}7$
Mr. & Mrs. Hopkins dined with us.[2]

4. Wednesday.
Dinner at Mr. Russell's.[3]
Subscription Ball.

1842/01/04-05. Diary of Samuel Tertius Galton

January 4th. Wednesday. Leamington.
... Erasmus and Francis dined with Mr. Russell.

January 5th. Thursday. Leamington.
... Erasmus, Bessy and Francis went to Birmingham to stay a few days with Lucy. ...

[1] John Kaye (1783–1853).
[2] Hopkins, William. See page 484.
[3] Possibly Russell, James, see page 490.

1843/01/05–10. FG Diary

5. Thursday.
Went to Birmingham to Selly Hall.[4]

6. Friday.
Prior's Ball (cant go)
Lucy children acted Cinderella
in French,

7. Saturday.
Called at Moore Green.[5]

8. Sunday.
Went to Moseley. Miss Taylor
at church.

9. Monday
Snow & blowy.

10. Tuesday.
Mrs Young's Evg party.
Tableaux. Returned
from Selly Hall, having called
at the Hodgsons.[6]
To Blakeley £49.[7]

1842/01/10–11. Diary of Samuel Tertius Galton

January 10th. Tuesday. Leamington.

Cold sleety day. Bessy and Francis returned from Selby Hall. Francis and Emma went in the evening to a party at Mrs. Young's.

January 11th. Wednesday. Leamington.

Francis and Emma went to Edstone and to the Stratford hunt ball. ...

---

4 To his sister Lucy Moilliet.
5 Seat of the Taylor family, in Kings Norton, Worcestershire.
6 Hodgson, Joseph. See page 483.
7 Appendix. His classical tutor at Cambridge.

1843/01/11–12. FG Diary

11. Wednesday.
Stratford Ball, dined & slept at Edstone.[8] Emma P. came out.

12. Thursday
Returned to Leamington by Claverdon[9]—very seedy.

1842/01/12. Diary of Samuel Tertius Galton

January 12th. Thursday. Leamington.

Violetta went to Claverdon and brought back Francis and Emma from Edstone.

1843/01/13–18. FG Diary

13. Friday.
Dinner at home.
Barometer 27.9°windy & snowy.

14. Saturday.
Walked with Mr. Hopkins[10] called on the Rylands. Snowy & unpleasant.

16. Monday.
The Hopkins went back.

17. Tuesday.
Warneford ball, dinner at home $\frac{1}{4}$7

---

[8] His brother Darwin Galton's country estate.
[9] The Galton estate since 1824, near Warwick.
[10] Hopkins, William. See page 484.

18. Wednesday
Miss Foley, dinner $6\frac{1}{2}$
Mrs. Madock's ball, very
good Roman punch.

1842/01/18–19. Diary of Samuel Tertius Galton

January 18th. Wednesday. Leamington.

... Emma, Francis and myself dined at Miss Foleys and the two former went afterwards to a party at Mrs. Madocks.

January 19th. Thursday. Leamington.

... Bessy, Francis and myself dine with Sir J. Cave and met the Watts, Russells, Foxes, Miss Granville and Captain Henderson.

1843/01/19–20. FG Diary

19. Thursday.
Sir John Cave, dinner $\frac{1}{4}7$.
We practised the Lancers.

20. Friday.
Subscription ball.
Dinner Mrs. Cameron. 7.

1842/01/20. Diary of Samuel Tertius Galton

January 20th. Friday. Leamington.

... Francis dined with the Camerons; he and Emma went to the Ball in the evening. ...

1843/01/22–31. FG Diary

22. Sunday.
Mr. Wilson's dray got overturned
coming home from a dinner Coventry
1 horse killed—Mrs. Dean died.

23. Monday.
Ball at home, about 200,
had a reel & the Lancers.
Walked to Warwick & Guys Cliff
with Brown.[11] Miss Hodgson came
to stay with us.

24. Tuesday.
Go to Hadzor,[12] leave Birm: at $3\frac{1}{2}$.
Worcester ball. Didn't go.
Lady Thompson, evening party
don't go.

25. Wednesday.
Miss Hodgson left us. Miss Taylor came.
Dine Dr. Jephson $6\frac{1}{2}$[13]
Queen's Ball at Birm:
Taught the Jephsons the
game of Nimikin [Storitch].[14]

26. Thursday.
Thursby ball (recalled)
Countess of Kilmorey ball
Capital singing.
Dinner at home.
Dr. Jephson was spilt out of his carriage.

27. Friday.
Emma Phillips came to stay
here.[15] Walked with Ramsay
who took me to see Mr. Jones
mesmerize a patient before cutting
his leg off.

28. Saturday.
Miss Judith Hodgson tea-ed with

---

[11] Laurence Lawson Brown, who later emigrated to New Zealand.
[12] To visit his uncle Howard Galton.
[13] Jephson, Dr. Henry. See page 485.
[14] Not traced.
[15] A sister of Darwin Galton's wife Mary.

us.[16] Darwin asked me to go to
Edstone & rabbit shoot, could
not get a lift there, didn't go.
To Barclay [£]22.7
Mortlock & co. [£]50.[17]

29. Sunday.
Walked with Mater.

31. Tuesday.
Dinner Smiths Holly Walk
(cant go)
Returned to Cambridge.
Wray on the coach.[18] Chatted with
E. Kay.[19]

1843/01/31. Diary of Samuel Tertius Galton

January 31st. Tuesday. Leamington.
Fine morning. Francis started at 9 a.m. for Cambridge. ...

1843/02/01–17. FG Diary

[February]

1. Wednesday
Mr. Stacey dinner $6\frac{1}{2}$, cant
go.

3. Friday.
Mrs. Bishopps ball.
Doveton's Evg party
Elected President of His.

6. Monday.
Dine with L L Brown.

---

[16] Not traced. She was not the daughter of Dr. Joseph Hodgson, who was Mary Anne.
[17] Appendix.
[18] Possibly Henry Wray (?-1880) of Trinity.
[19] Kay, Edwin Ebenezer. See page 485.

7. Tuesday.
Supper party in my rooms
St. Piot (Dalyell's friend)[20]
came he sang capital
French songs.

8. Wednesday.
Wined in my own rooms
to Committee of the Historical.

9. Thursday.
Wine with L. L. Brown.

10 Friday.
(Manners Sutton's dinner)
Wined with Aytoun.[21]

11. Saturday.
Wine Brodie. Boulton[22] was
making some gas experiments.

12. Sunday.
Walked with Boulton along
the river.
Breakfast E. Kay.[23]

13. Monday.
Was proposed for Library Committee
at the union. Peacocke made
a great row [24]
Wine Alderson.[25]

14. Tuesday.

---

[20] Dalyell, Robert Alexander Osborne, see page 482. "St. Piot" could not be traced.

[21] Aytoun, Roger Sinclair. See page 480.

[22] Boulton, Matthew Piers Watt. See page 480.

[23] Kay, Edwin Ebenezer. See page 485.

[24] George Montague Warren Peacocke (1822?-1879). He changed his name to Sandford in 1866.

[25] Edward Samuel Alderson (1822-1903).

Was elected. Gibbs president, King secr.
Ogier Treasurer,[26] Stewart, Shakespear[27]
& Tooke the other Library Committee.
Peacocke made a shameful disturbance,
groaning at Gibbs &c.

15. Wednesday.
Curtis lectured on Shorthand
at the Union. Peacocke tried to
groan him down. Craufurd[28]
went to him and told him his conduct
was blackguardly. Thorp in the chair.

16. Thursday.
Peacocke licked Craufurd
with a Life Preserver. I had
a supper party in my room which
was broken up at the news.

1843/02/16. Diary of Samuel Tertius Galton

February 16th. Thursday. Leamington.

Francis attained his majority on this his birthday. ...

1843/02/17–18. FG Diary

17. Friday.
Lunch J. Kay.

18. Saturday.
Breakfast with Bristed.[29]
Lunched Craufurd.
Peacocke rusticated sine die.
Craufurd—one term.

---

[26] John Creuzé Hingeston Ogier (1821?-1913), who later emigrated to Australia.

[27] John Joseph Arthur Shakespear (1820?-1881).

[28] Edward Henry John Craufurd (1816—1887).

[29] Bristed, Charles Astor. See page 482.

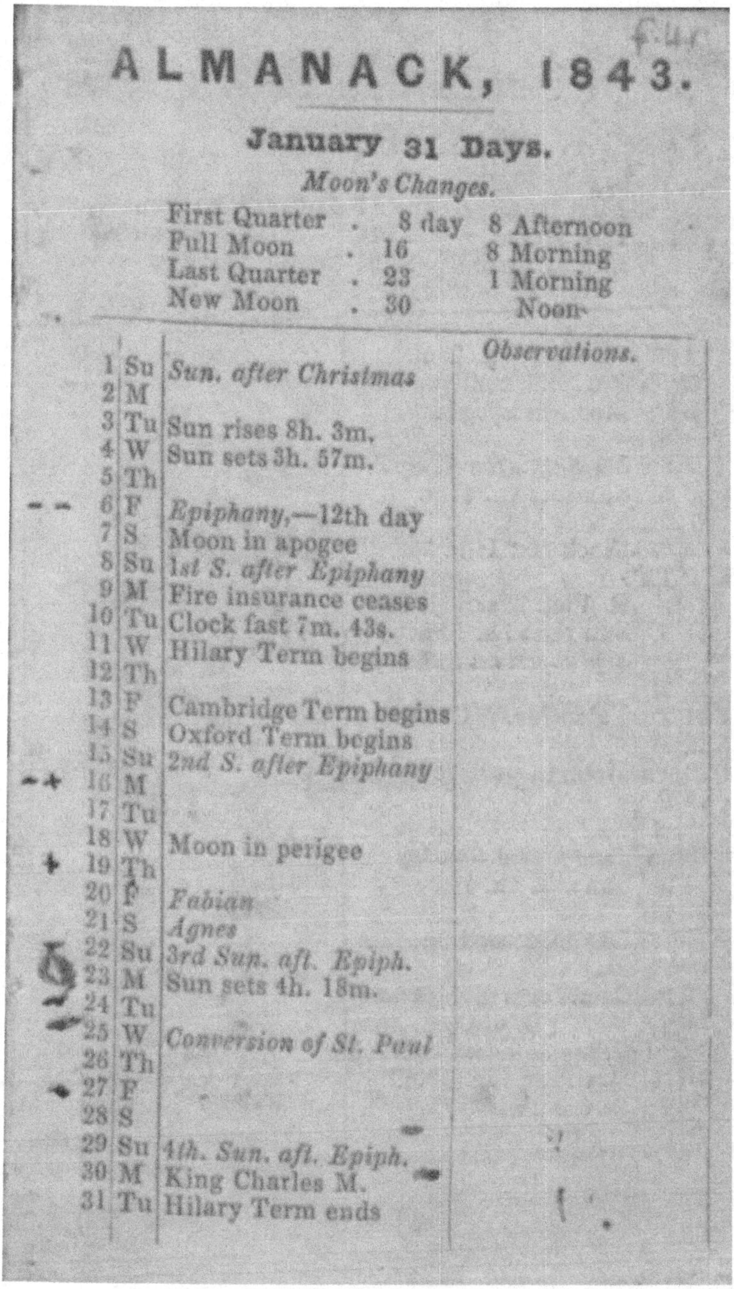

ALMANACK, 1843.

January 31 Days.

Moon's Changes.

First Quarter	.	8 day	8 Afternoon
Full Moon	.	16	8 Morning
Last Quarter	.	23	1 Morning
New Moon	.	30	Noon

			Observations.
1	Su	*Sun. after Christmas*	
2	M		
3	Tu	Sun rises 8h. 3m.	
4	W	Sun sets 3h. 57m.	
5	Th		
6	F	*Epiphany*,—12th day	
7	S	Moon in apogee	
8	Su	*1st S. after Epiphany*	
9	M	Fire insurance ceases	
10	Tu	Clock fast 7m. 43s.	
11	W	Hilary Term begins	
12	Th		
13	F	Cambridge Term begins	
14	S	Oxford Term begins	
15	Su	*2nd S. after Epiphany*	
16	M		
17	Tu		
18	W	Moon in perigee	
19	Th		
20	F	*Fabian*	
21	S	*Agnes*	
22	Su	*3rd Sun. aft. Epiph.*	
23	M	Sun sets 4h. 18m.	
24	Tu		
25	W	*Conversion of St. Paul*	
26	Th		
27	F		
28	S		
29	Su	*4th. Sun. aft. Epiph.*	
30	M	King Charles M.	
31	Tu	Hilary Term ends	

Figure 4.1: Coded Almanack for 1843: January.

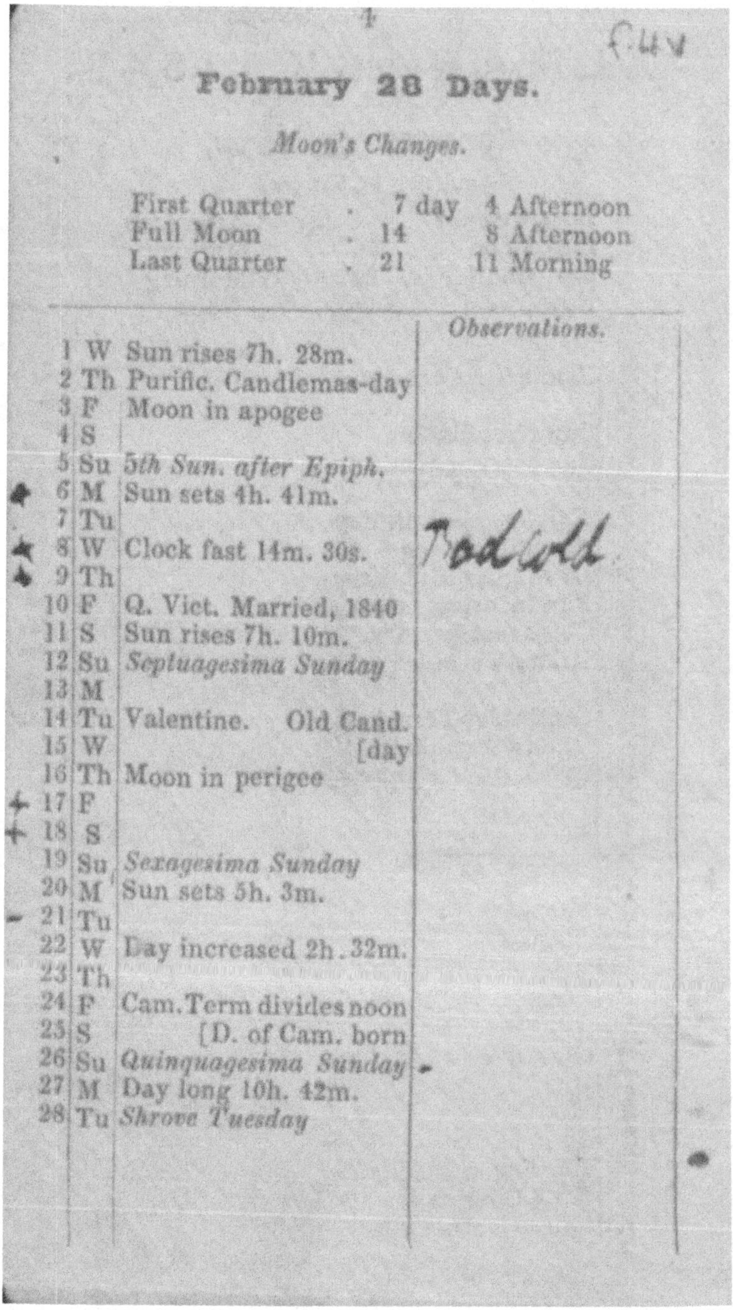

February 28 Days.

Moon's Changes.

First Quarter	. 7 day	4 Afternoon
Full Moon	. 14	8 Afternoon
Last Quarter	. 21	11 Morning

			Observations.
1	W	Sun rises 7h. 28m.	
2	Th	Purific. Candlemas-day	
3	F	Moon in apogee	
4	S		
5	Su	*5th Sun. after Epiph.*	
6	M	Sun sets 4h. 41m.	
7	Tu		
8	W	Clock fast 14m. 30s.	
9	Th		
10	F	Q. Vict. Married, 1840	
11	S	Sun rises 7h. 10m.	
12	Su	*Septuagesima Sunday*	
13	M		
14	Tu	Valentine. Old Cand.	
15	W	[day	
16	Th	Moon in perigee	
17	F		
18	S		
19	Su	*Sexagesima Sunday*	
20	M	Sun sets 5h. 3m.	
21	Tu		
22	W	Day increased 2h. 32m.	
23	Th		
24	F	Cam. Term divides noon	
25	S	[D. of Cam. born	
26	Su	*Quinquagesima Sunday*	
27	M	Day long 10h. 42m.	
28	Tu	*Shrove Tuesday*	

Figure 4.2: Coded Almanack for 1843: February.

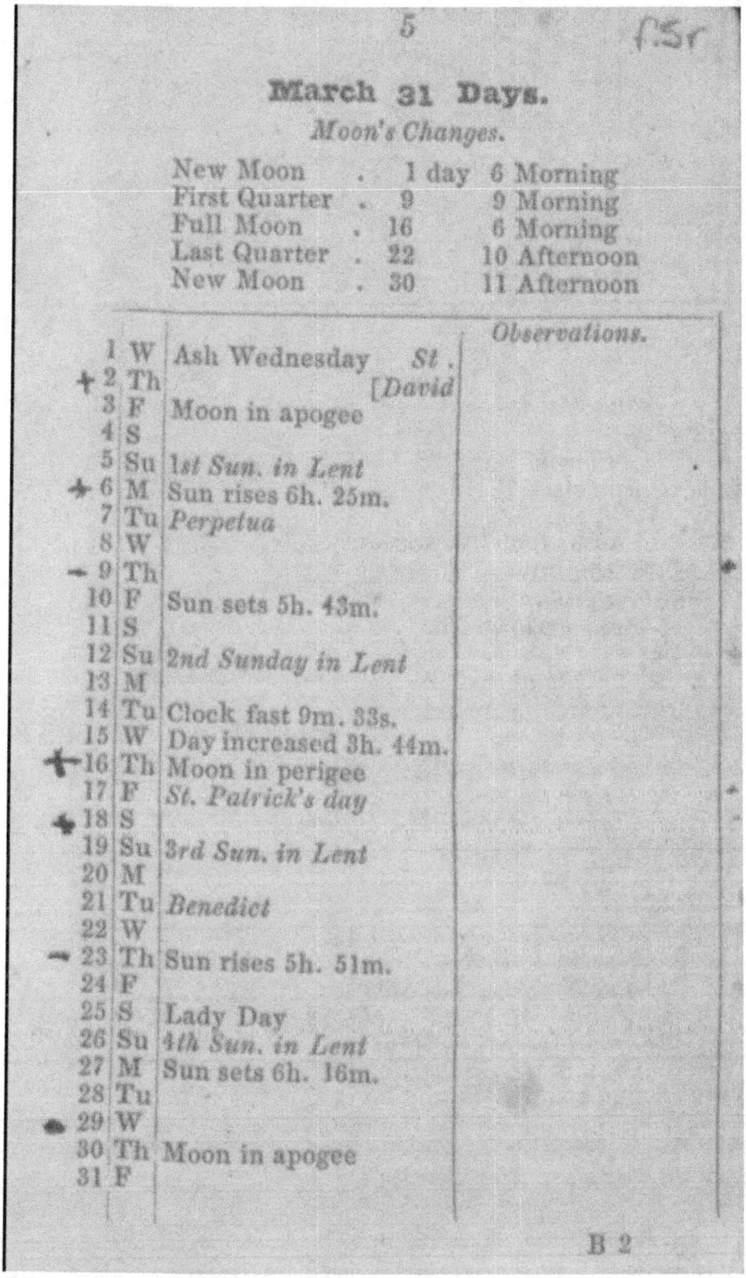

5 f.5r

March 31 Days.

Moon's Changes.

New Moon	.	1 day	6 Morning
First Quarter	.	9	9 Morning
Full Moon	.	16	6 Morning
Last Quarter	.	22	10 Afternoon
New Moon	.	30	11 Afternoon

			Observations.
1	W	Ash Wednesday St.	
2	Th	[David	
3	F	Moon in apogee	
4	S		
5	Su	*1st Sun. in Lent*	
6	M	Sun rises 6h. 25m.	
7	Tu	*Perpetua*	
8	W		
9	Th		
10	F	Sun sets 5h. 43m.	
11	S		
12	Su	*2nd Sunday in Lent*	
13	M		
14	Tu	Clock fast 9m. 33s.	
15	W	Day increased 3h. 44m.	
16	Th	Moon in perigee	
17	F	*St. Patrick's day*	
18	S		
19	Su	*3rd Sun. in Lent*	
20	M		
21	Tu	*Benedict*	
22	W		
23	Th	Sun rises 5h. 51m.	
24	F		
25	S	Lady Day	
26	Su	*4th Sun. in Lent*	
27	M	Sun sets 6h. 16m.	
28	Tu		
29	W		
30	Th	Moon in apogee	
31	F		

B 2

Figure 4.3: Coded Almanack for 1843: March.

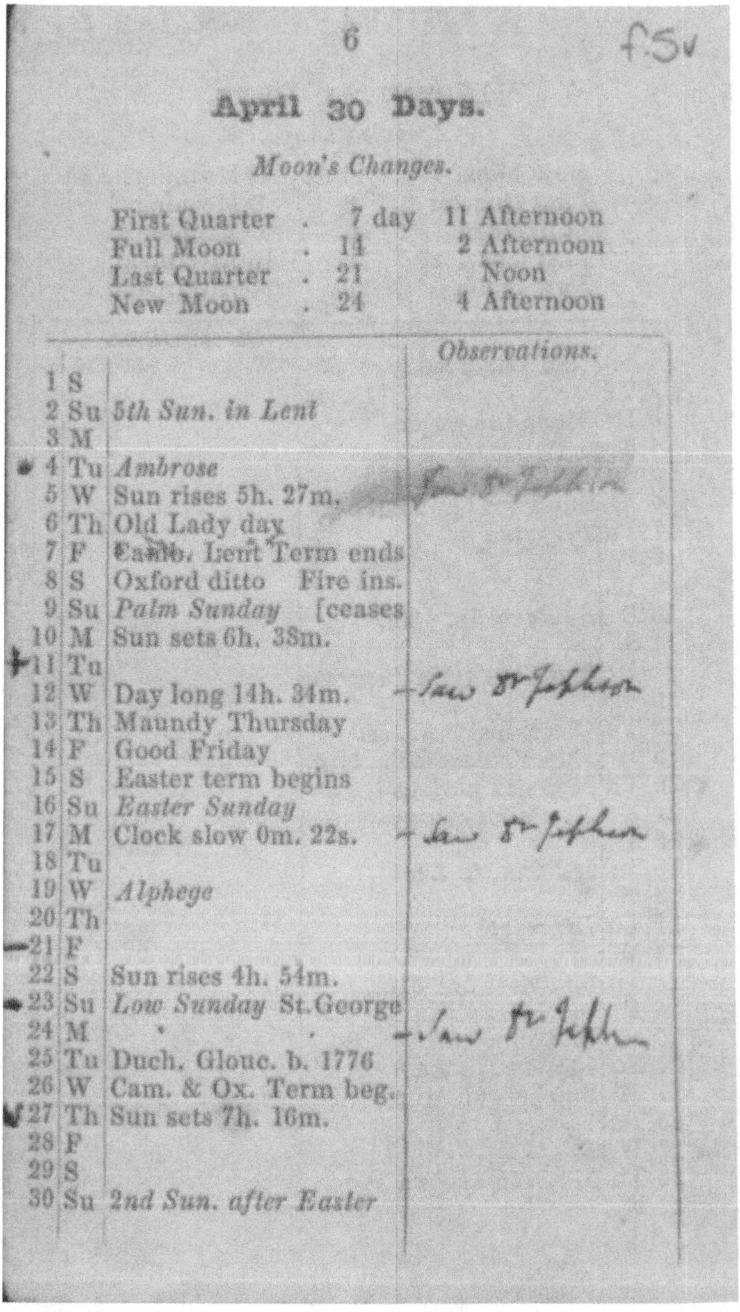

6 f.5v

April 30 Days.

Moon's Changes.

First Quarter	.	7 day	11 Afternoon
Full Moon	.	14	2 Afternoon
Last Quarter	.	21	Noon
New Moon	.	24	4 Afternoon

			Observations.
1	S		
2	Su	*5th Sun. in Lent*	
3	M		
4	Tu	*Ambrose*	
5	W	Sun rises 5h. 27m.	Saw Dr Jephson
6	Th	Old Lady day	
7	F	Camb. Lent Term ends	
8	S	Oxford ditto Fire ins.	
9	Su	*Palm Sunday* [ceases	
10	M	Sun sets 6h. 38m.	
11	Tu		
12	W	Day long 14h. 34m.	– Saw Dr Jephson
13	Th	Maundy Thursday	
14	F	Good Friday	
15	S	Easter term begins	
16	Su	*Easter Sunday*	
17	M	Clock slow 0m. 22s.	– Saw Dr Jephson
18	Tu		
19	W	*Alphege*	
20	Th		
21	F		
22	S	Sun rises 4h. 54m.	
23	Su	*Low Sunday* St. George	
24	M		– Saw Dr Jephson
25	Tu	Duch. Glouc. b. 1776	
26	W	Cam. & Ox. Term beg.	
27	Th	Sun sets 7h. 16m.	
28	F		
29	S		
30	Su	*2nd Sun. after Easter*	

Figure 4.4: Coded Almanack for 1843: April.

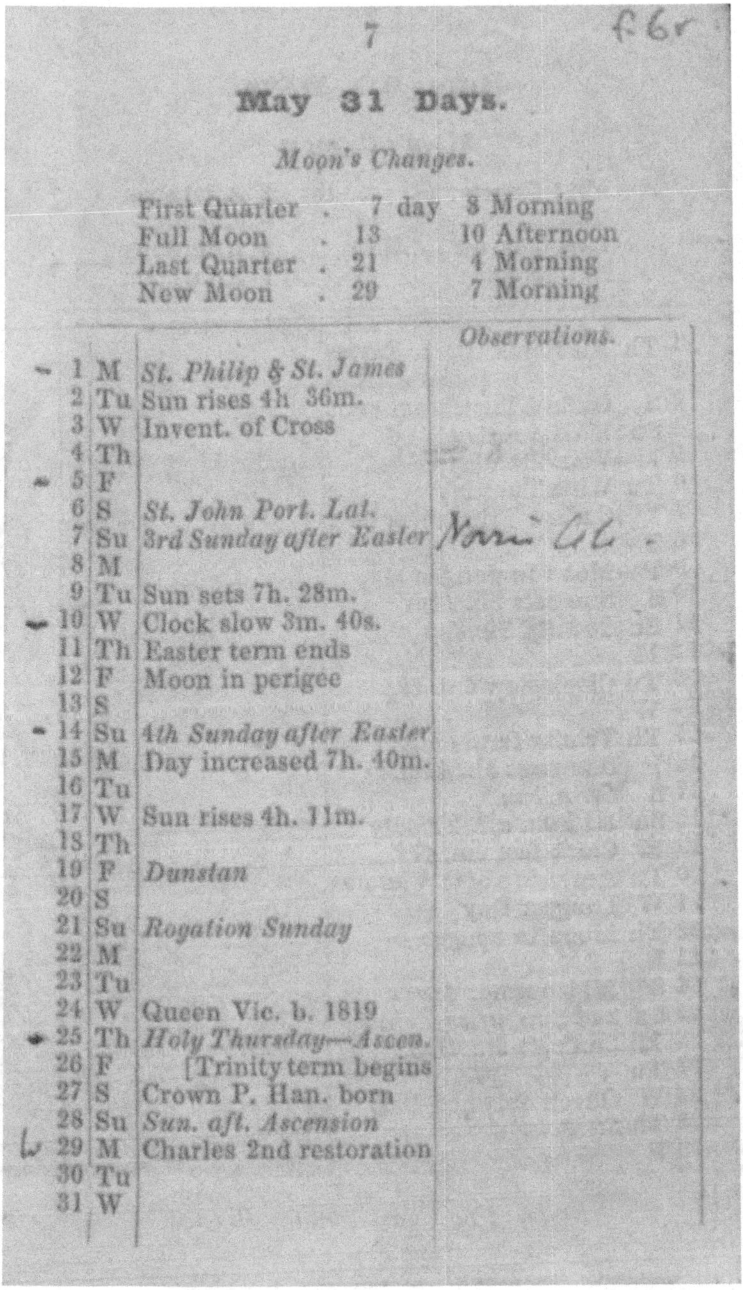

7

May 31 Days.

Moon's Changes.

First Quarter	. 7 day		8 Morning
Full Moon	. 13		10 Afternoon
Last Quarter	. 21		4 Morning
New Moon	. 29		7 Morning

			Observations.
1	M	*St. Philip & St. James*	
2	Tu	Sun rises 4h 36m.	
3	W	Invent. of Cross	
4	Th		
5	F		
6	S	*St. John Port. Lat.*	
7	Su	*3rd Sunday after Easter*	
8	M		
9	Tu	Sun sets 7h. 28m.	
10	W	Clock slow 3m. 40s.	
11	Th	Easter term ends	
12	F	Moon in perigee	
13	S		
14	Su	*4th Sunday after Easter*	
15	M	Day increased 7h. 40m.	
16	Tu		
17	W	Sun rises 4h. 11m.	
18	Th		
19	F	*Dunstan*	
20	S		
21	Su	*Rogation Sunday*	
22	M		
23	Tu		
24	W	Queen Vic. b. 1819	
25	Th	*Holy Thursday—Ascen.*	
26	F	[Trinity term begins	
27	S	Crown P. Han. born	
28	Su	*Sun. aft. Ascension*	
29	M	Charles 2nd restoration	
30	Tu		
31	W		

Figure 4.5: Coded Almanack for 1843: May.

Figure 4.6: Charles Astor Bristed.

1843/02/19. Francis Galton to Samuel Tertius Galton

Sunday.

My dear Father

Thank you very much for all three of your letters, which were certainly very amusing. All my time has lately been taken up by canvassing and afterwards by a most unfortunate collision between two of my friends with one of whom I was much interested & in some degree involved. Peacocke & Craufurd are the two men. The first you have heard me mention before. The last is a son I believe of Sir Thos Craufurd and was a fellow commoner.[30] Peacocke is a man whose whole object was to make a very extended acquaintance, in which he certainly succeeded, but at the same time was very unpopular, from being a pushing sort of man and often mixed in quarrels, and a very noisy arguer at the Union. He was a great friend of Theodore's at one time who afterwards was rather offended at him. To proceed. On Monday night the candidates for the Union officers were proposed. Peacocke proposed me & Craufurd seconded—in some private business that followed Peacocke made a great disturbance, calling out groaning &c &c. in which he urged on 4 or 5 Magdalene men to support him. I on meeting him that evening assured him that had I been president I should have fined or expelled him but he seemed to look upon the whole matter as a joke, and assured me that with 4 or 5 supporters he would break the Union lamps and upset anything like order. The reason of his anger was that a different man was elected president to the one he supported. The next day he apologised for the disturbance he had made. When that very evening on Gibbs[31] the new President taking the chair he was so shamefully uproarious that nothing could go on. At one time I heard him cry out 3 groans for the President which he & his men gave. I then went to the President & requested him to censure or expel him, for the whole Union was in an uproar. The president shortly after seeing one of his supporters Childe[32] crying out fined him a sovereign. Peacocke rushed across the room crying out "infamous," and was neither fined nor expelled as he ought to have been. I then spoke

[30] He was a son of the Scot John Craufurd of Auchenames, and later became a radical politician.
[31] Frederick Waymouth Gibbs (1821-1898).
[32] James Gay Child (1819-1870) of Magdalene.

to some other members of the Historical about his immediate expulsion out of that society in which they cordially agreed but we determined to talk it over the next day Wednesday. On Wednesday Morning I met Peacocke who told me he was highly vexed at his conduct the night before. When by all that is shameful that very night at a lecture which was given in the Union & Mr Thorp[33] the tutor of Trinity kindly took the chair (he had been when as undergraduate a President) Peacocke was more uproarious than ever urging on several Magdalene men who stood behind the furthest benches and kept up continually stamping so that nothing could go on. Thorp threatened twice to leave the chair and was going to do so, when Craufurd jumped up and rushed into the midst of them confronted Peacocke and told him that his conduct was disgraceful and blackguardly as it had been the night before, he then turned round and said his observations applied to all who had assisted in the row. When turned, Craufurd felt a push on the shoulder of which he took no heed but turning again repeated his observations to Peacocke. He then spoke to all the men who had left their seats (about 200) & were crowding round, and said is it your wish these men should be turned out, they all cried turn them out, turn them out. Cries of Chair Chair recalled the men to their places & Peacocke & his associates immediately left the room. I immediately drew up a requisition to Peacocke to leave the Historical which was signed by all who saw it about 17 and then began to take steps for expelling him the Union. When going to Craufurd's room I found him half mad hearing that Peacocke had spread a report ~~in the Union~~ that he had struck Craufurd in the Union who was too cowardly to return it. Craufurd then put a horsewhip in his pocket and went everywhere in search of Peacocke, but could not find him. Late in the evening he returned to his rooms with two friends where he found Peacocke with Childe & Hughes[34] of Magd. who said I have heard that you have been looking for me all day, here I am. Craufurd said he wished to speak with him by himself. On his demurring he gave his word that he need be under no bodily fear. So they went down into the cloisters. On arriving there Craufurd asked him if it was true that he had spread the above report. Peacocke replied that he had struck him in the Union. Craufurd drew out his horsewhip and

[33] Thomas Thorp (1797-1877).
[34] Robert Edgar Hughes (1822-1863).

held it over him and said consider yourself horsewhipped. Peacocke said you have not struck me. Craufurd dropped the whip on his shoulders. Peacocke drew out a life-preserver and struck Craufurd ferociously over the temples which quite confused him; he however closed in when Childe and Hughes actually pinioned Craufurd & whilst Craufurd's two friends were trying to tear them off Peacocke deliberately hit Craufurd several times on the head. Peacocke turned away & said I have witnesses that you first struck me. He since owned that the whole affair was prearranged. All Craufurd's friends were of course almost mad. It was beneath his dignity to challenge him. To skip over all the different plans that were proposed & laid aside it has ended in laying the matter before the College Authorities who have rusticated Peacocke "sine die" which is the same as expelling him for they take his name off the boards[35] & they have not allowed Craufurd to reside this term which is almost a nominal punishment as he is a bachelor scholar, and in no way to be injured by such sentence. I need not add that Peacocke is universally cut and I understand that he has threatened being revenged on me.

Craufurd was a good deal hurt. A large committee of men of whom I am one are always together either at Craufurd's or some others' rooms. He breakfasts with me to-morrow before going down & the whole of Trinity will probably see him off. The motion for ~~his~~ Peacocke's expulsion comes on to-morrow evening.[36] Denman[37] son of Lord Denman & Senior Classic takes the chair.

A printed statement will be published as soon as the Magdalene men have been punished & I will send you one.

The tests for arsenic are very easily applied and quite cheaply, but the 4 pounds is I believe the Chemist's fee.

Your affectionate son,
Fras. Galton

I fear the shortness of the above statement will not give you a very clear notion of the way matters stand.

---

35 He migrated to Oxford, receiving a BA there in 1844, and then migrated back to Magdalene College at Cambridge, receiving an MA in 1849.

36 Bristed 1852, 122 states that a motion to expel Peacocke from the Union could not carry the required 2/3 majority, but Galton states in his diary that a 3/4 majority was required.

37 George Denman (1819-1896).

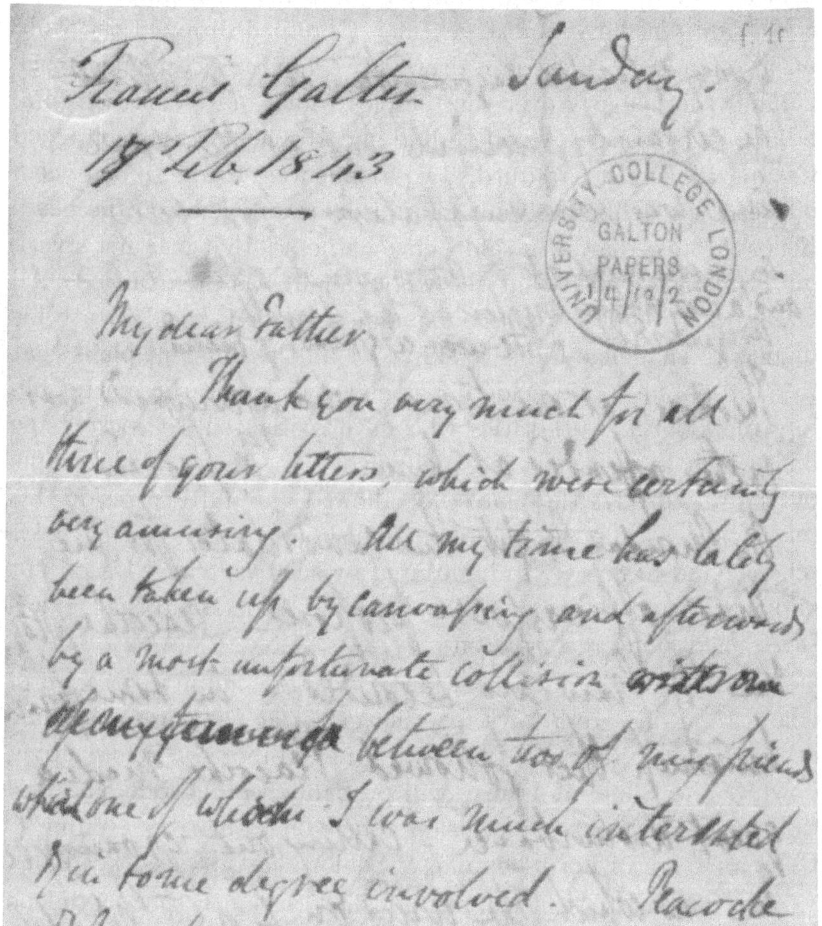

Francis Galton
19 Feb 1843
Sunday.

My dear Father
Thank you very much for all three of your letters, which were certainly very amusing. All my time has lately been taken up by canvassing and afterwards by a most unfortunate collision [illegible] between two of my friends with one of whom I was much interested & in some degree involved. Peacocke & Crawford are the two men. The first you have heard me mention before. The last is a son I believe of Sir Thos Crawford and was a fellow Commoner. Peacocke is a man whose whole object was to make

Figure 4.7: 1843-02-19. FG to STG page 1.

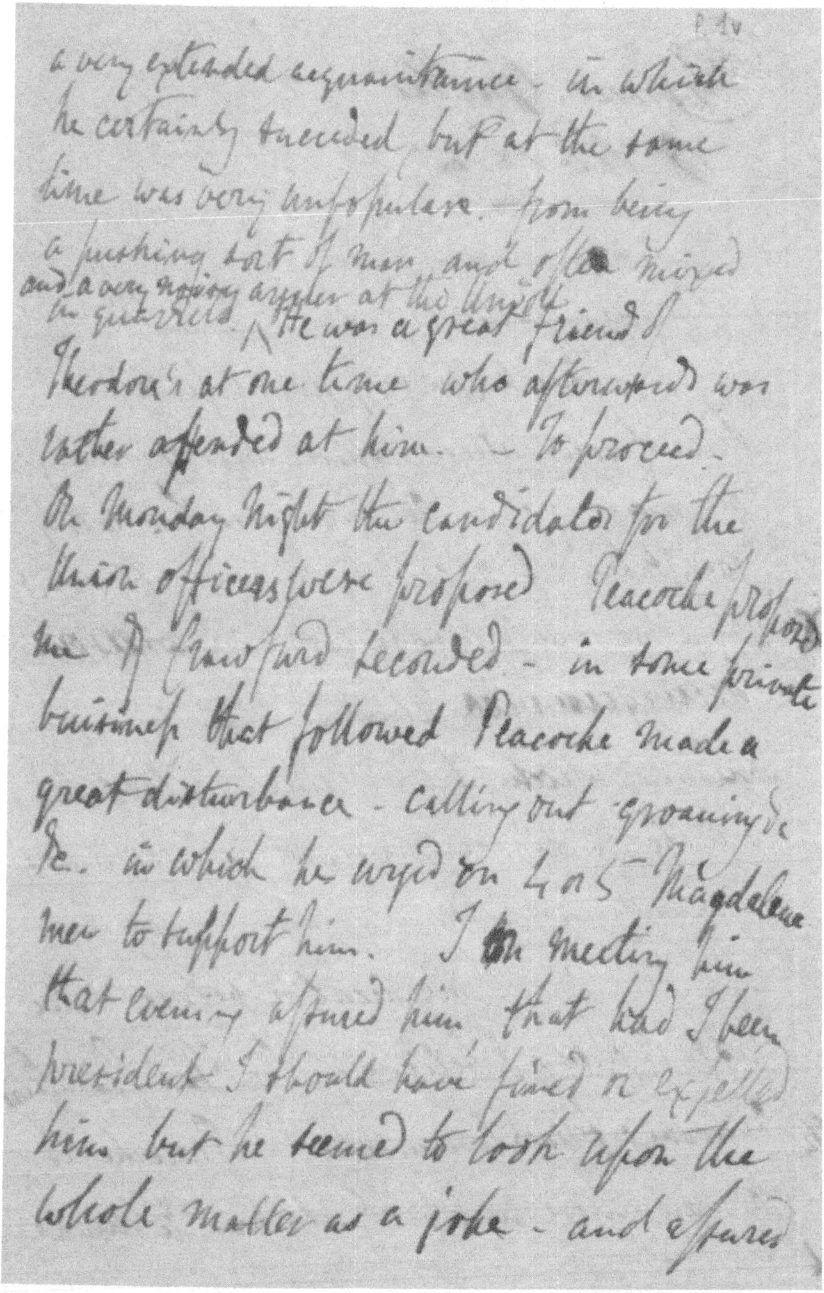
f. 1v

a very extended acquaintance - in which
he certainly succeeded, but at the same
time was very unpopular - from being
a pushing sort of man, and often mixed
and a very noisy arguer at the Union
in quarrels He was a great friend of
Theodore's at one time who afterward was
rather offended at him. - To proceed -
On Monday night the candidates for the
Union offices were proposed Peacocke proposed
me & Crawford seconded - in some private
business that followed Peacocke made a
great disturbance - calling out groaning &c
&c. in which he urged on 4 or 5 Magdalene
men to support him. I on meeting him
that evening assured him that had I been
president I should have fined or expelled
him but he seemed to look upon the
whole matter as a joke - and assured

Figure 4.8: 1843-02-19. FG to STG page 2.

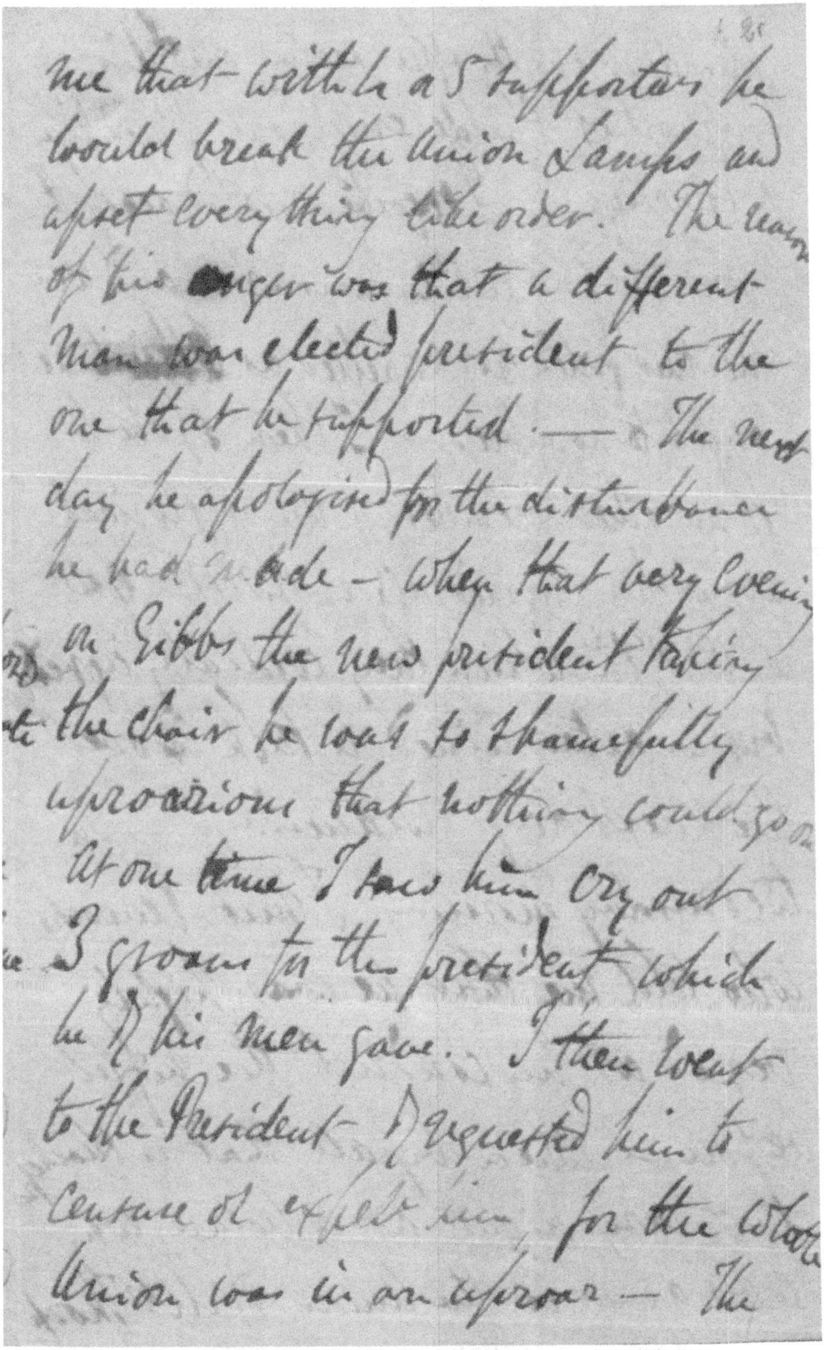

me that with his 5 supporters he
would break the Union Lamps and
upset every thing like order. The reason
of his anger was that a different
man was elected president to the
one that he supported. — The next
day he apologized for the disturbance
he had made — when that very evening
on Gibbs the new president taking
the chair he was so shamefully
uproarious that nothing could go on
At one time I saw him cry out
3 groans for the president which
he & his men gave. I then went
to the President & requested him to
censure or expel him for the whole
Union was in an uproar — The

Figure 4.9: 1843-02-19. FG to STG page 3.

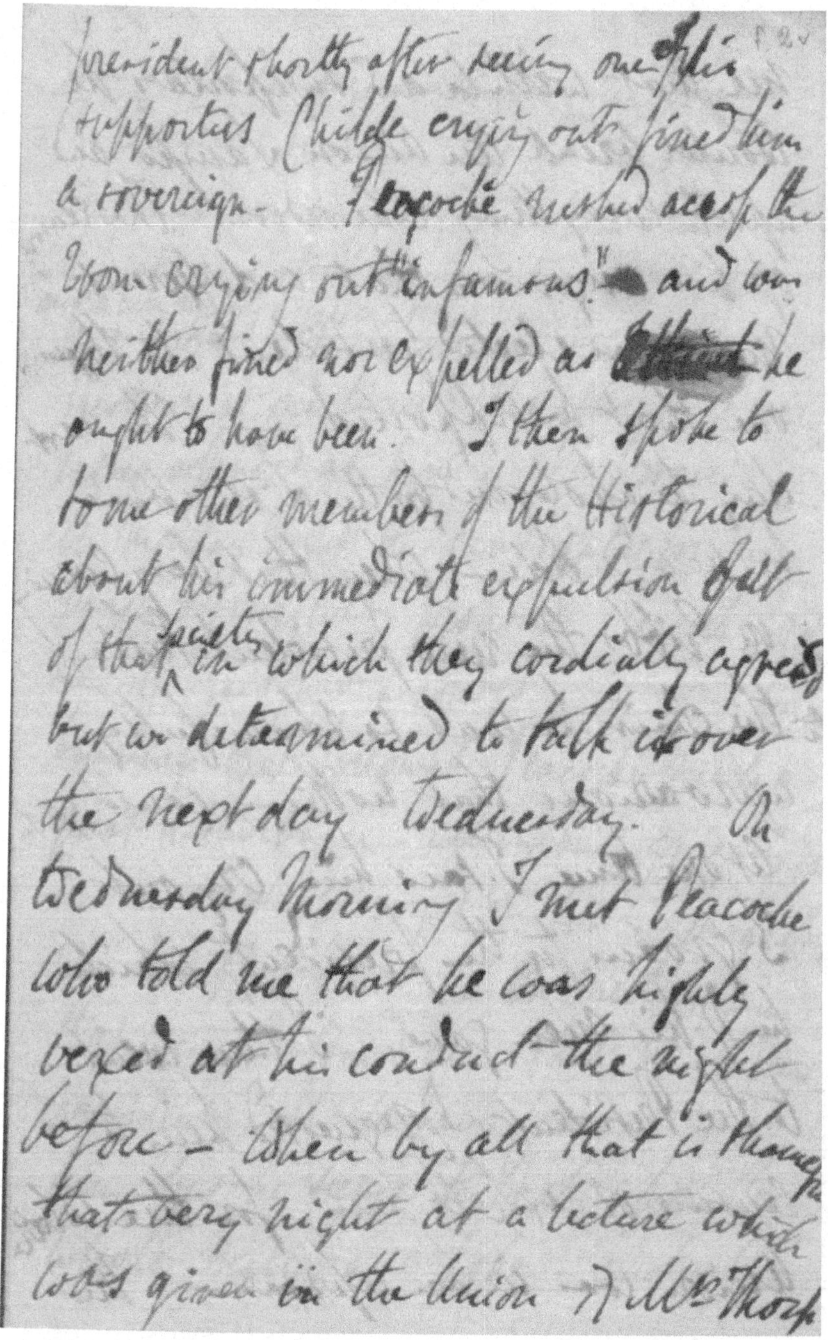
president shortly after seeing one of his
supporters Childe crying out fined him
a sovereign - Peacocke rushed across the
room crying out "infamous" and was
neither fined nor expelled as he
ought to have been. I then spoke to
some other members of the Historical
about his immediate expulsion out
of that society in which they cordially agreed
but we determined to talk it over
the next day Wednesday. On
Wednesday morning I met Peacocke
who told me that he was highly
vexed at his conduct the night
before - when by all that is monstrous
that very night at a lecture which
was given in the Union by Mr Thorp

Figure 4.10: 1843-02-19. FG to STG page 4.

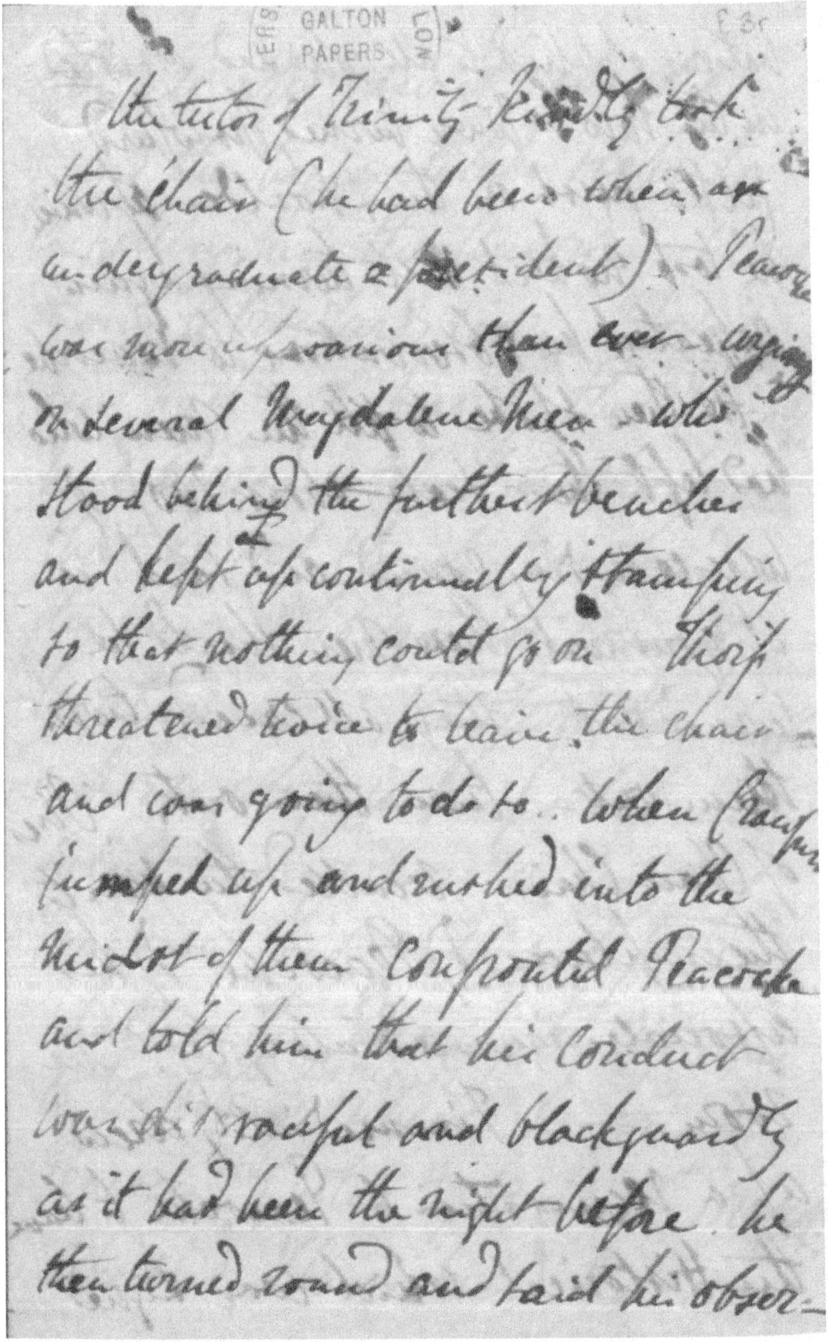

the tutor of Trinity kindly took
the chair (he had been when an
undergraduate a president). Peacock
was more uproarious than ever urging
on several Magdalene men who
stood behind the furthest benches
and kept up continually stamping
so that nothing could go on. Thorp
threatened twice to leave the chair
and was going to do so. When Crompton
jumped up and rushed into the
midst of them confronted Peacock
and told him that his conduct
was disgraceful and blackguardly
as it had been the night before. he
then turned round and said his obser-

Figure 4.11: 1843-02-19. FG to STG page 5.

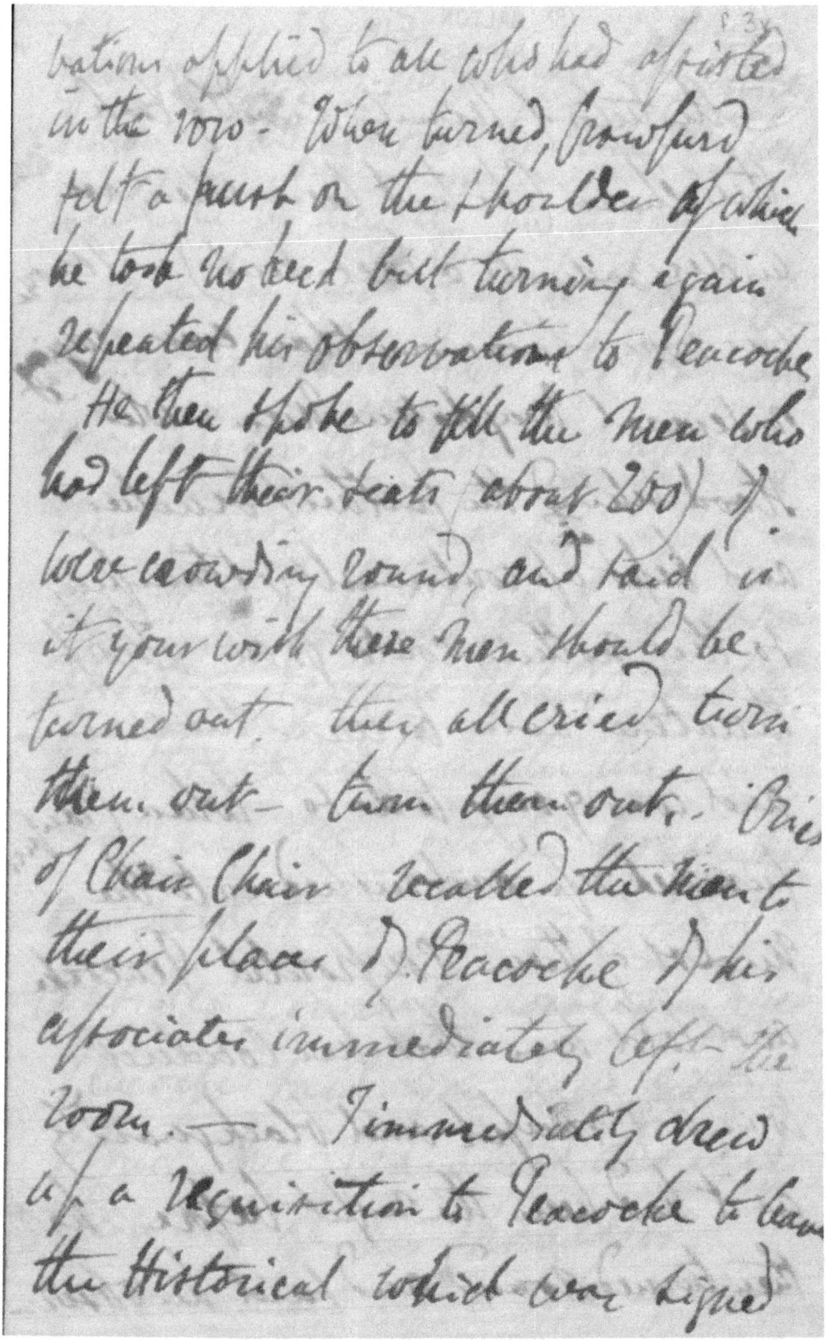
bations applied to all who had assisted
in the room - When turned, Crawford
felt a push on the shoulder of which
he took no heed but turning again
repeated his observations to Peacocke
He then spoke to tell the men who
had left their seats (about 200) &
were crowding round, and said is
it your wish these men should be
turned out, they all cried turn
them out - turn them out. - Cries
of Chair Chair recalled the men to
their places & Peacocke & his
associates immediately left the
room. — I immediately drew
up a Requisition to Peacocke to leave
the Historical which was signed

Figure 4.12: 1843-02-19. FG to STG page 6.

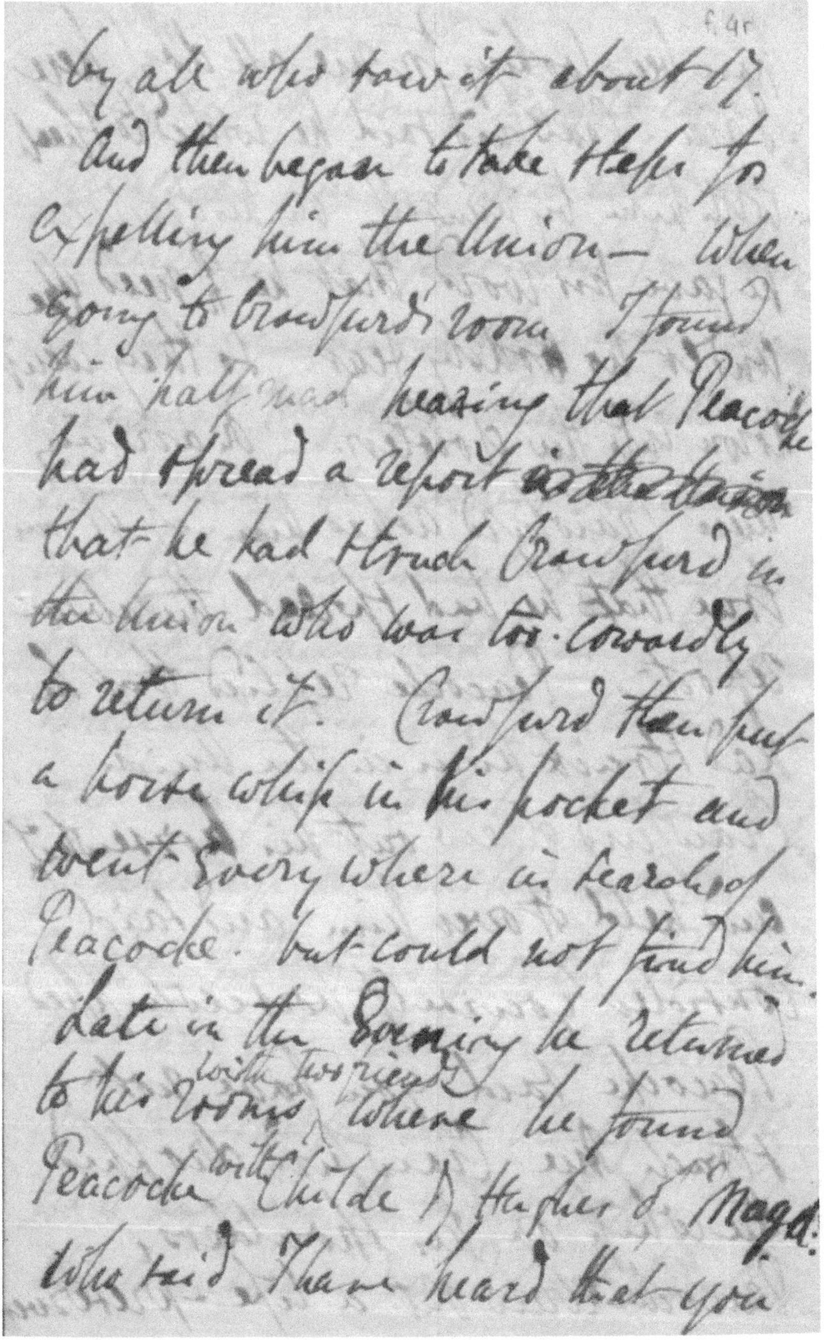

f. 4r

by all who knew it about 17.
And then began to take steps for
expelling him the Union— When
going to Crawfurd's room I found
him half mad hearing that Peacocke
had spread a report in the Union
that he had struck Crawfurd in
the Union who was too cowardly
to return it. Crawfurd then put
a horse whip in his pocket and
went every where in search of
Peacocke. but could not find him.
Late in the evening he returned
to his rooms with two friends where he found
Peacocke with Childe & Hughes & Magd:
who said I have heard that you

Figure 4.13: 1843-02-19. FG to STG page 7.

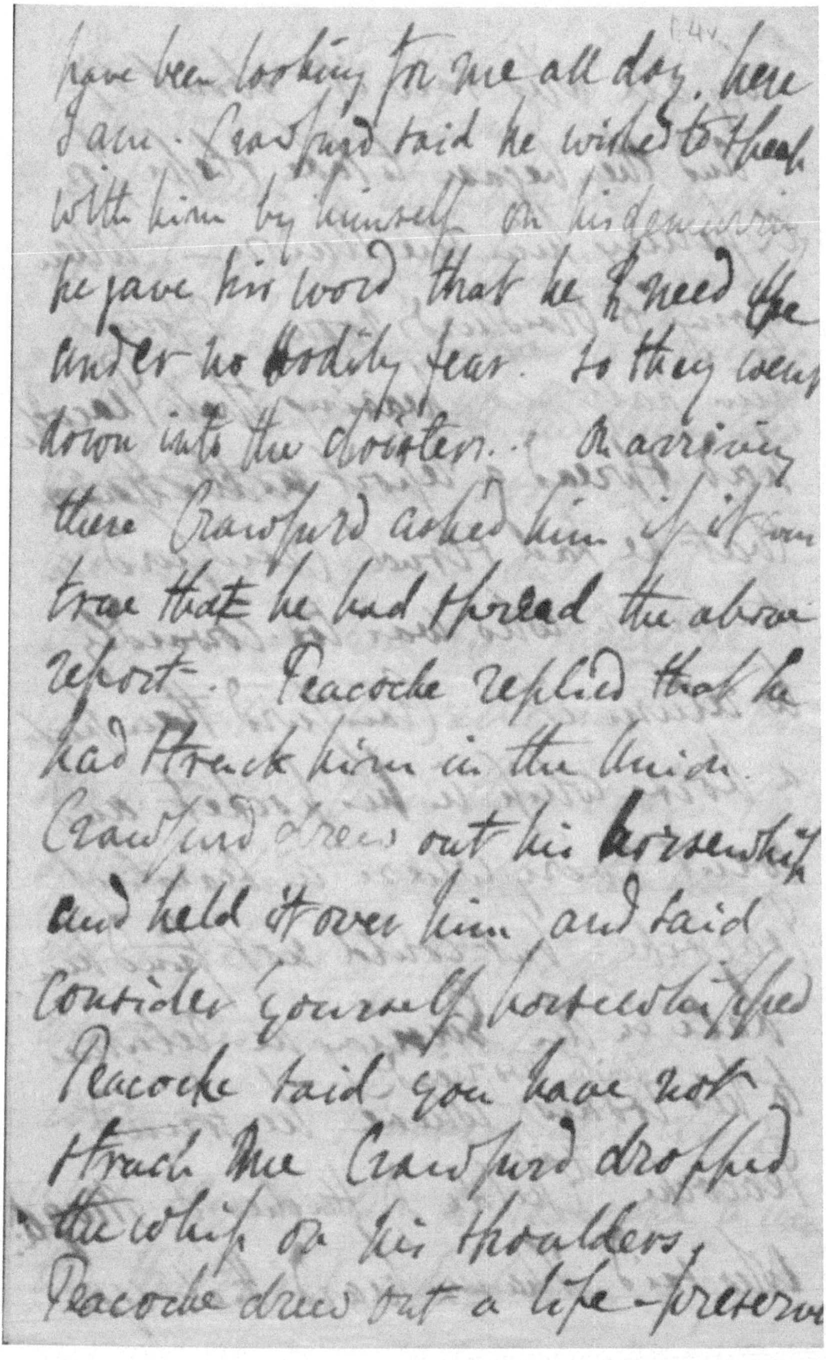
f.4v
have been looking for me all day, here
I am. Crawfurd said he wished to speak
with him by himself on his demurring
he gave his word that he should be
under no bodily fear. So they went
down into the cloisters. On arriving
there Crawfurd asked him if it was
true that he had spread the above
report. Peacocke replied that he
had struck him in the Union.
Crawfurd drew out his horsewhip
and held it over him and said
Consider yourself horsewhipped
Peacocke said you have not
struck me Crawfurd dropped
the whip on his shoulders,
Peacocke drew out a life-preserver

Figure 4.14: 1843-02-19. FG to STG page 8.

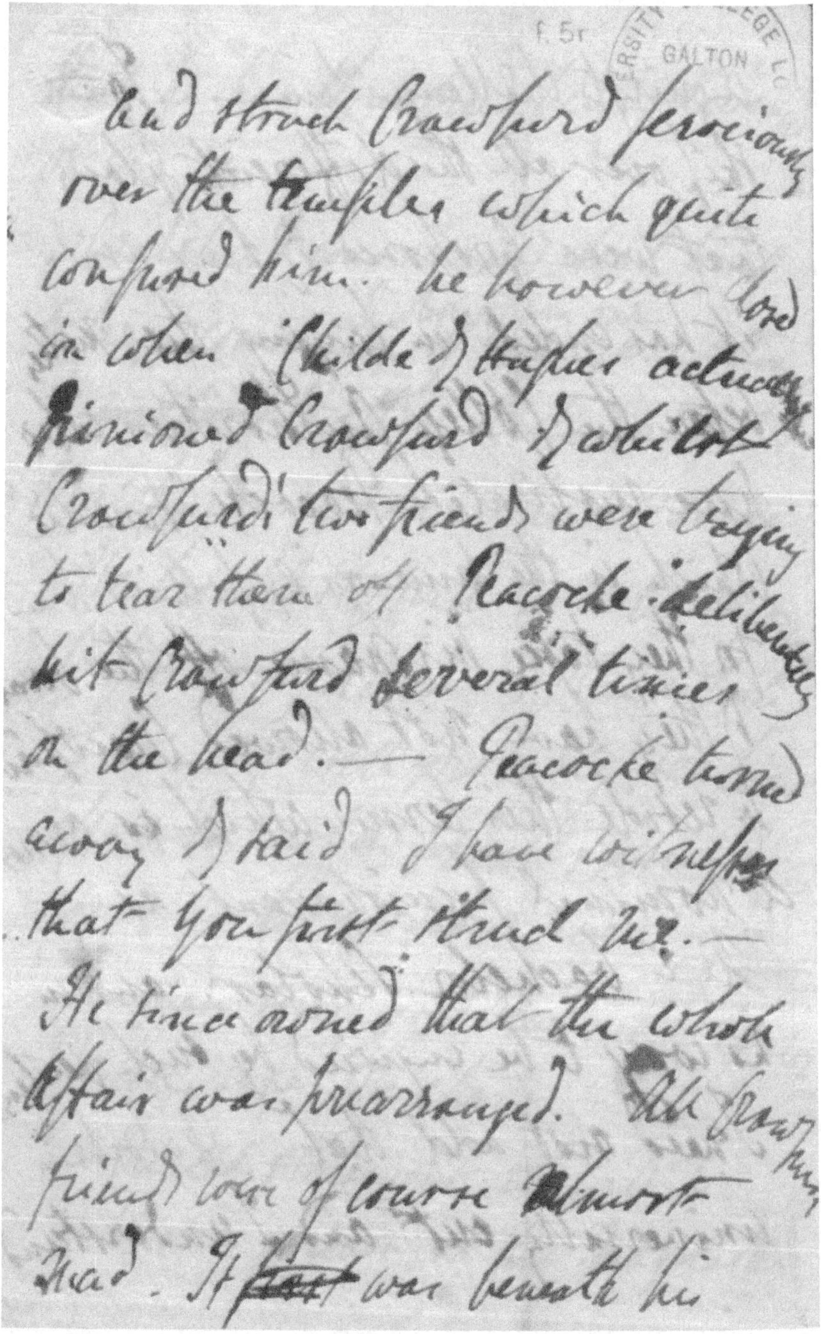
and struck Crawford ferociously
over the temples which quite
confused him. he however closed
in when Childe & Staples actually
pinioned Crawford & whilst
Crawford's two friends were trying
to tear them off Peacocke deliberately
hit Crawford several times
on the head. — Peacocke turned
away & said I have witnesses
that you first struck me. —
He since owned that the whole
affair was prearranged. All Crawfords
friends were of course almost
mad. It ~~[illegible]~~ was beneath his

Figure 4.15: 1843-02-19. FG to STG page 9.

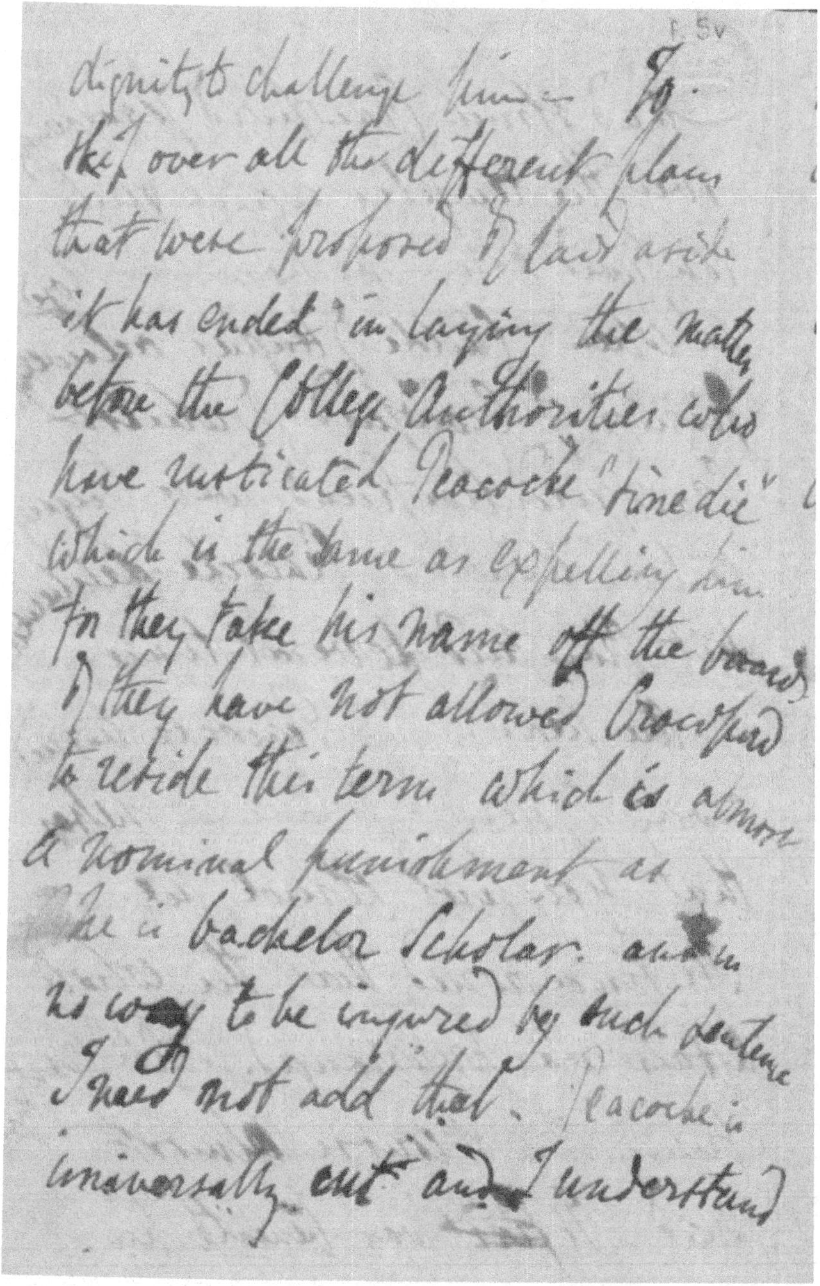
dignity to challenge him. To
step over all the different plans
that were proposed & laid aside
it has ended in laying the matter
before the College Authorities who
have rusticated Peacocke "sine die"
which is the same as expelling him
for they take his name off the boards
& they have not allowed Crawford
to reside this term which is almost
a nominal punishment as
he is bachelor Scholar. and in
no way to be injured by such sentence
I need not add that. Peacocke is
universally cut and I understand

Figure 4.16: 1843-02-19. FG to STG page 10.

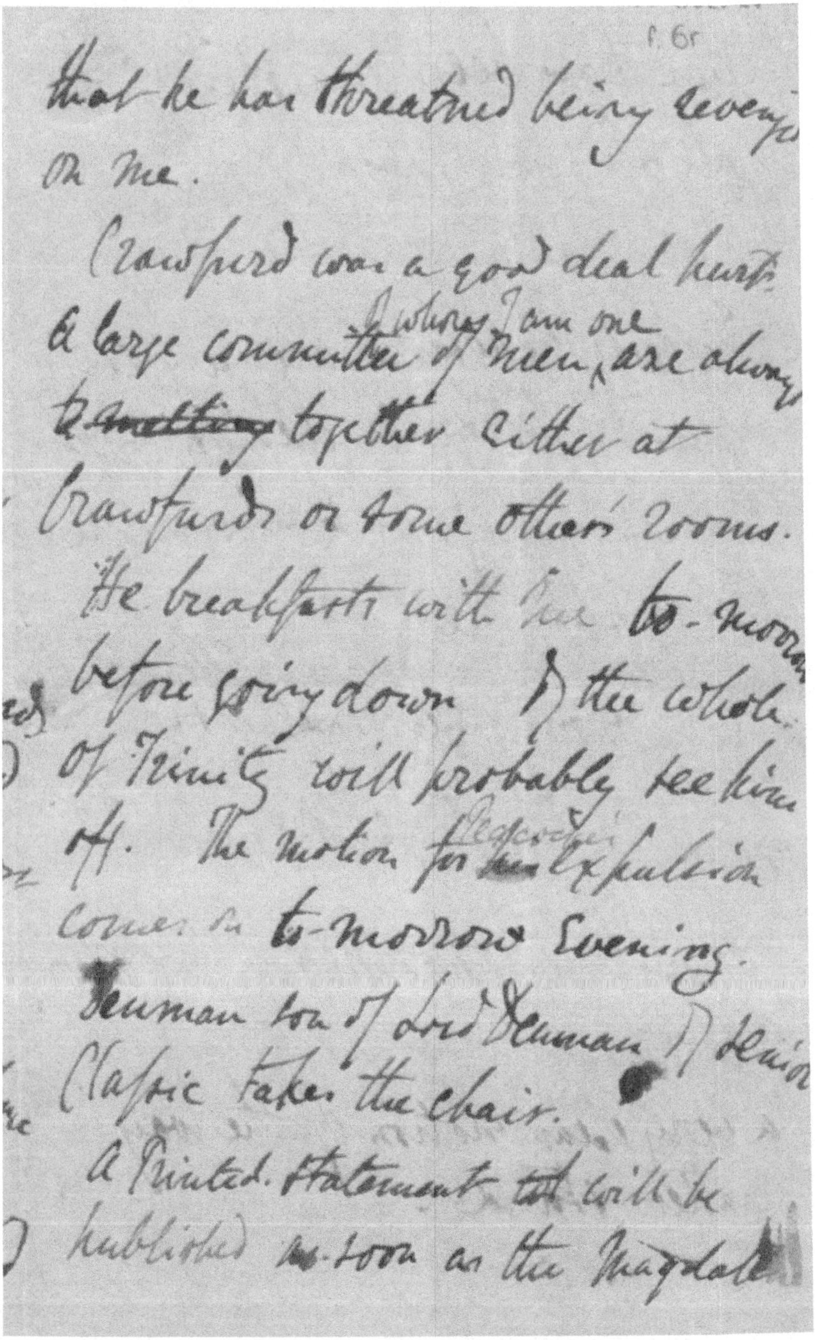
that he has threatened being revenged
on me.
Crawford was a good deal hurt.
A large committee of men, whereof I am one, are always
~~meeting~~ together either at
Crawfords or some others' rooms.
He breakfasts with me to-morrow
before going down & the whole
of Trinity will probably see him
off. The motion for [illegible] expulsion
comes on to-morrow evening.
Denman son of Lord Denman & Senior
Classic takes the chair.
A Printed statement ~~[illegible]~~ will be
published ~~as~~ soon as the Magdale

Figure 4.17: 1843-02-19. FG to STG page 11.

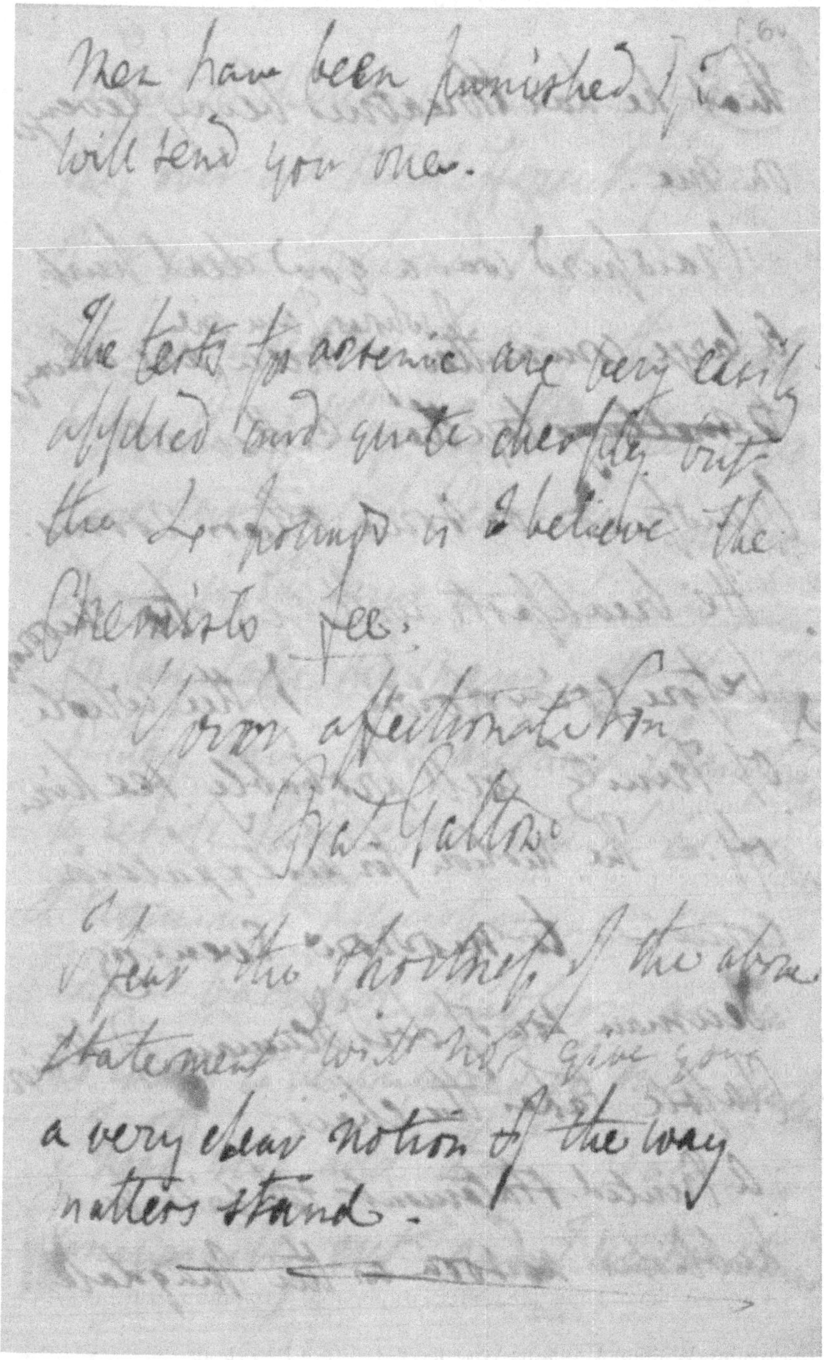

these have been furnished I will send you one.

The tests for arsenic are very easily applied and quite cheaply, but the 2 pounds is I believe the Chemists fee.

Your affectionate Son
Frank Galton

I fear the shortness of the above statement will not give you a very clear notion of the way matters stand.

Figure 4.18: 1843-02-19. FG to STG page 12.

1843/02/19–28. FG Diary

19. Sunday.
Breakfast E. Kay.[38]
Walked out with Boulton.[39]

20. Monday.
Breakfast party in my
rooms. Special meeting
at Union in committee to consider about
the expulsion of Peacocke.
Lunch E Kaye.[40] Wine King & Hutton.[41]

21. Tuesday.
Poll gave more than $\frac{3}{4}$ths
agst. Peacocke. meeting of
the house.
Breakfast Craufurd at Walton's rooms.[42]
Saw him off in coach.

22. Wednesday.
For the accepting of the report of
the Committee 236, agst 103. Not carried
since $\frac{3}{4}$ths were necessary.
Wine party in my rooms.
Supper party at Armitage.[43]

23. Thursday.
Wine Pitcairn.[44]

28. [Tuesday] Bad cold.[45]

---

38 Kay, Edwin Ebenezer. See page 485.
39 Boulton, Matthew Piers Watt, see page 480.
40 Kay, Edwin Ebenezer. See page 485.
41 George William King (1822–1881); possibly Crompton Hutton (1822–1910).
42 Possibly William Williams Walton (?–1859) of Peterhouse.
43 Possibly James Armitage (1818–1852).
44 Possibly James Pelham Pitcairn (1821–1892) of Christ's.
45 Almanac.

1843/03/05–11. FG Diary

5. Sunday.
Walked with Hallam[46] & surveyed
ground for gymnasium.
Wine Dalyell.[47]
Evening Brown (not going)
Boulton[48] gave me engraving of
Raphael [...]

6. Monday.
Concert. Parry & Miss Beaufort
were down I did not go.
Saw Boulton off in Coach.

7. Tuesday
Breakfast Keary[49]
Dine E. Kay.[50]

8. Wednesday.
Dine Housman.[51]
Brought forward 2 motions
at the Union.

1843/03/11. Francis Galton to Samuel Tertius Galton

Saturday Evening

My Dear Father

I would most strongly support Baker[52] if I were possessed of a vote at the Hospital as I am convinced that he is a person of more than common abilities—for besides the fact that his diagnoses have always been exceedingly correctly there are many other reasons why I feel confirmed in the above opinion. For example in

[46] Hallam, Henry Fitzmaurice. See page 483.
[47] Dalyell, Robert Alexander Osborne. See page 482.
[48] Boulton, Matthew Piers Watt. See page 480.
[49] Henry Keary (1821–1852), later a master at Harrow.
[50] Kay, Edwin Ebenezer. See page 485.
[51] Possibly George Vernon Housman (1820–1887).
[52] Alfred Baker (1815-1893).

most hospitals the house-surgeon is ~~generally~~ looked down upon and snubbed by the other medical officers, is looked upon as an intruder in their consultations & bears but little weight along with his opinions, but the reverse was the case with Baker. He was always asked for his advice not only in regular consultation but even in ordinary cases of slight difficulty. In every visit that I have made to the hospital I have been struck by seeing how completely he is the manager there. Nothing seems done without his advice. Even Hodgson[53] has given way to him in many instances & indeed promised to support him in case of an election. Baker has lately given clinical lectures & has avowed his determination of using every means in his power to have regular ones instituted. Their advantage is fully acknowledged, surely then if he were elected surgeon their present deficiency w^{d} be most effectively remedied. He is also a man who has been all his life in the hospital & who would, I do not doubt, have a strong interest in its welfare. Lastly he is at present in his practice and w^{d} therefore better himself all the more vigorously. The only objection to him that I have ever heard is that his manners are not gentlemanly, I admit that he is somewhat snobbish, but he is vastly less so than when I was a pupil at the hospital, and even then he was the greatest imaginable favourite with the patients who are not nice discriminators of manners.

Amphlett[54] has by no means a high reputation as a practitioner though I understand he is a very gentlemanly person.

Crompton, Elkington & Gutteridge[55] are I believe about equal in their attainments & superior to Amphlett. But I should most strongly recommend you to vote for Baker.

I saw something about Hodgson's row in the paper, but the case there appeared to be that he had acknowledged that it was he who had informed Ledsam[56] and when accused of it by Wood he apologised much in the same way that Ledsam had done. It seemed very strange but I could not quite make it out.

Mathew Boulton[57] has gone down, he gave me a beautiful engraving before he left and asked me to stay with him a few days in the Easter Vacation which I accepted conditionally. I should much like to go. I am quite sorry that he has left, for we became

---

53 Hodgson, Joseph. See page 483.
54 Samuel Holmden Amphlett (1812-1857)
55 Dickinson Webster Crompton, Francis Elkington and Thomas Gutteridge.
56 John Joseph Ledsam.
57 Boulton, Matthew Piers Watt. See page 480.

great friends, he has quite altered all his shyness is gone and he looks the very perfection of a head of a family.

Your affectionate son
Fras Galton.

1843/03. Francis Galton to Samuel Tertius Galton

My dear Father

Thank you very much for the vote you have kindly given to Baker & which I feel the greatest confidence not to be misplaced. He has written two letters to me the first of which arrived the same evening that I received your last asking me to do anything that I could ~~for him~~ and a second since expressing thoroughly his gratitude for the kind way in which you had written to him, but at the same time quite despairing of success on this occasion, but very anxious about the next vacancy which he says is rumoured as likely to take place shortly (Vaux's I suppose).

He says his great object ever since he chose the profession has been to become surgeon at the Hospital and this wish has led him to exert himself far more than was required ~~from him~~ in which fact I can most strongly add my testimony.

Under the present system of electing Officers his chance is obviously much diminished since the great majority of voters have but little power of judging of relative qualifications and accordingly give their vote and interest to their personal friends. Men therefore like Crompton & Amphlett who have been in general practice & making acquaintances for some years must have a far superior chance over another who has spent an equal length of time drudging in the retirement of an Hospital.

I cannot tell you how much obliged I should feel to you if you could in any way get him votes at least for the next election in doing which I should think you would be quite justified from Hodgson's strong testimony to his character and the weight of his testimonials.

Do you think it possible that the law of Election might be altered so that it would lie in the hands of the most proper persons the medical officers? Please let me know your opinion.

Your affectionate Son
Fras. Galton

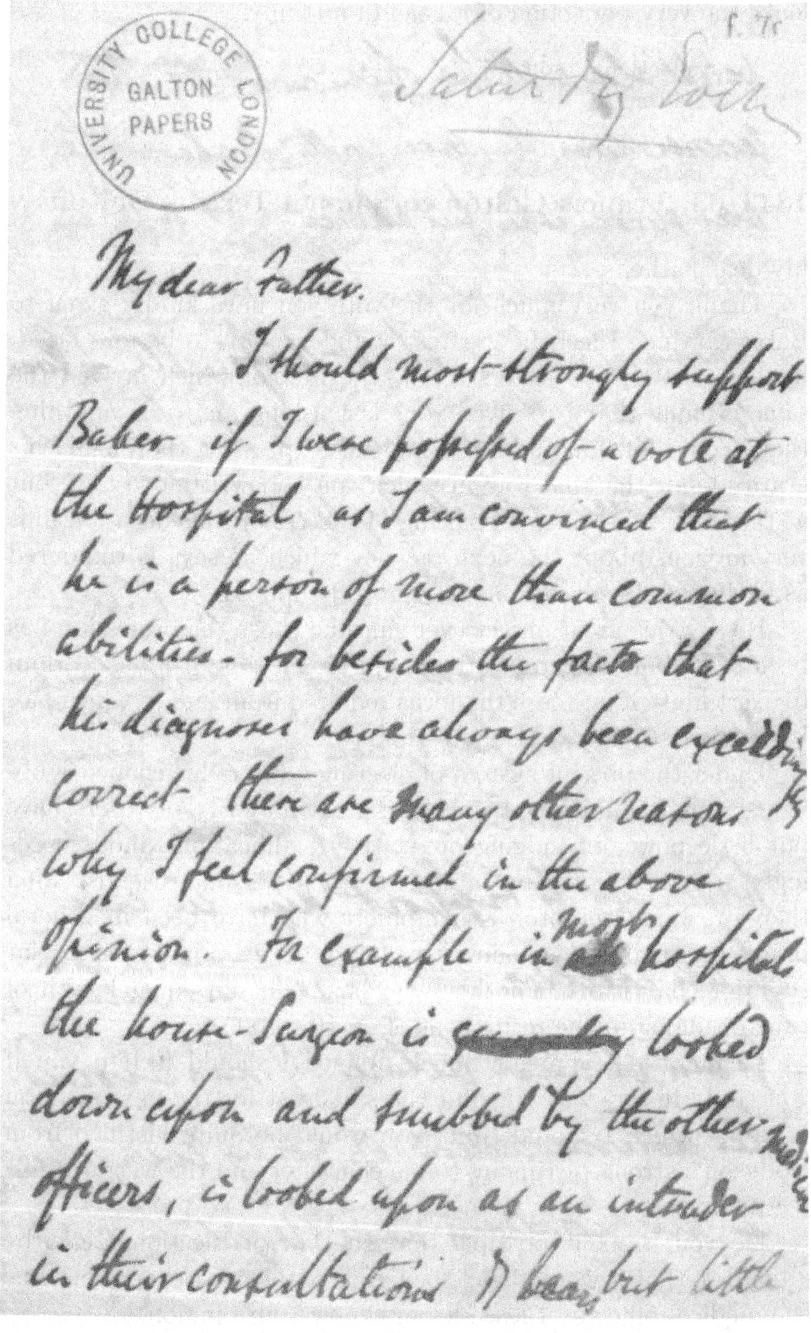

f. 75

Saturday Even

My dear Father.

I should most strongly support Baker if I were possessed of a vote at the Hospital as I am convinced that he is a person of more than common abilities – for besides the facts that his diagnoses have always been exceedingly correct there are many other reasons why I feel confirmed in the above opinion. For example in most hospitals the house-Surgeon is looked down upon and snubbed by the other medical officers, is looked upon as an intruder in their consultations & bears but little

Figure 4.19: 1843-03-11. FG to STG page 1.

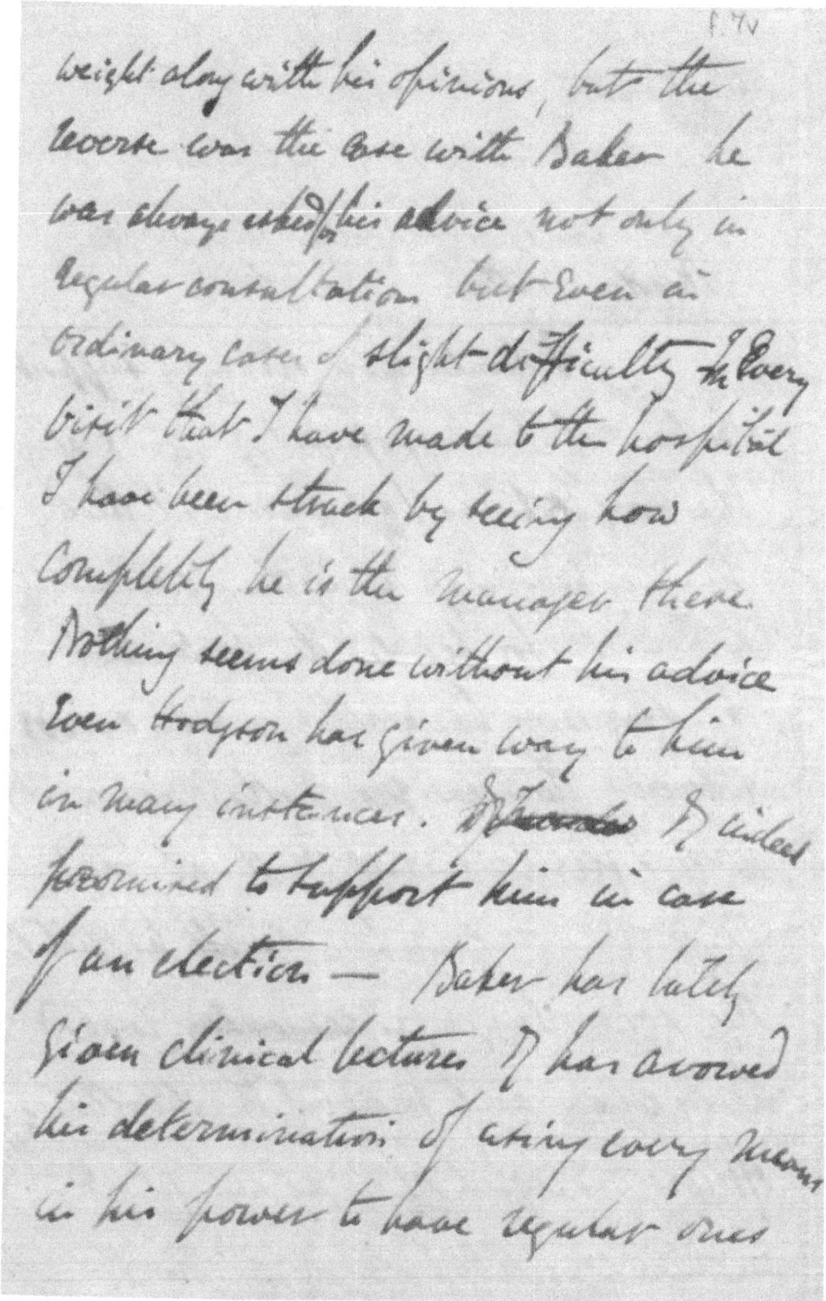
weight along with his opinions, but the
reverse was the case with Baker he
was always asked for his advice not only in
regular consultations but even in
ordinary cases of slight difficulty & Every
visit that I have made to the hospital
I have been struck by seeing how
completely he is the manager there.
Nothing seems done without his advice
Even Hodgson has given way to him
in many instances. & [illegible]
promised to support him in case
of an election — Baker has lately
given clinical lectures & has avowed
his determination of using every means
in his power to have regular ones

Figure 4.20: 1843-03-11. FG to STG page 2.

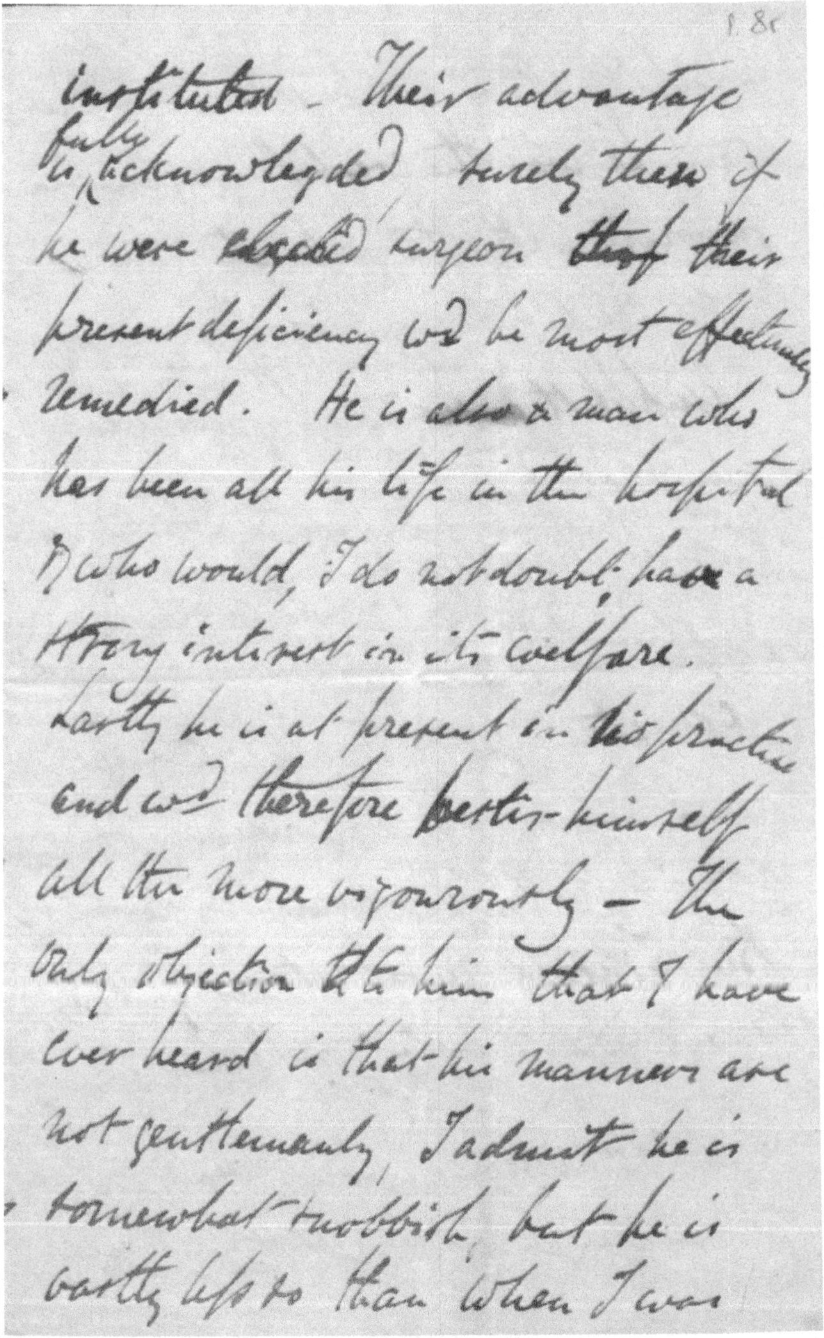

institution - Their advantage
is fully acknowledged surely then if
he were elected surgeon their
present deficiency wd be most effectually
remedied. He is also a man who
has been all his life in the hospital
& who would, I do not doubt, have a
strong interest in its welfare.
Lastly he is at present in no practice
and wd therefore exert himself
all the more vigourously - The
only objection th to him that I have
ever heard is that his manners are
not gentlemanly, I admit he is
somewhat snobbish, but he is
vastly less so than when I was

Figure 4.21: 1843-03-11. FG to STG page 3.

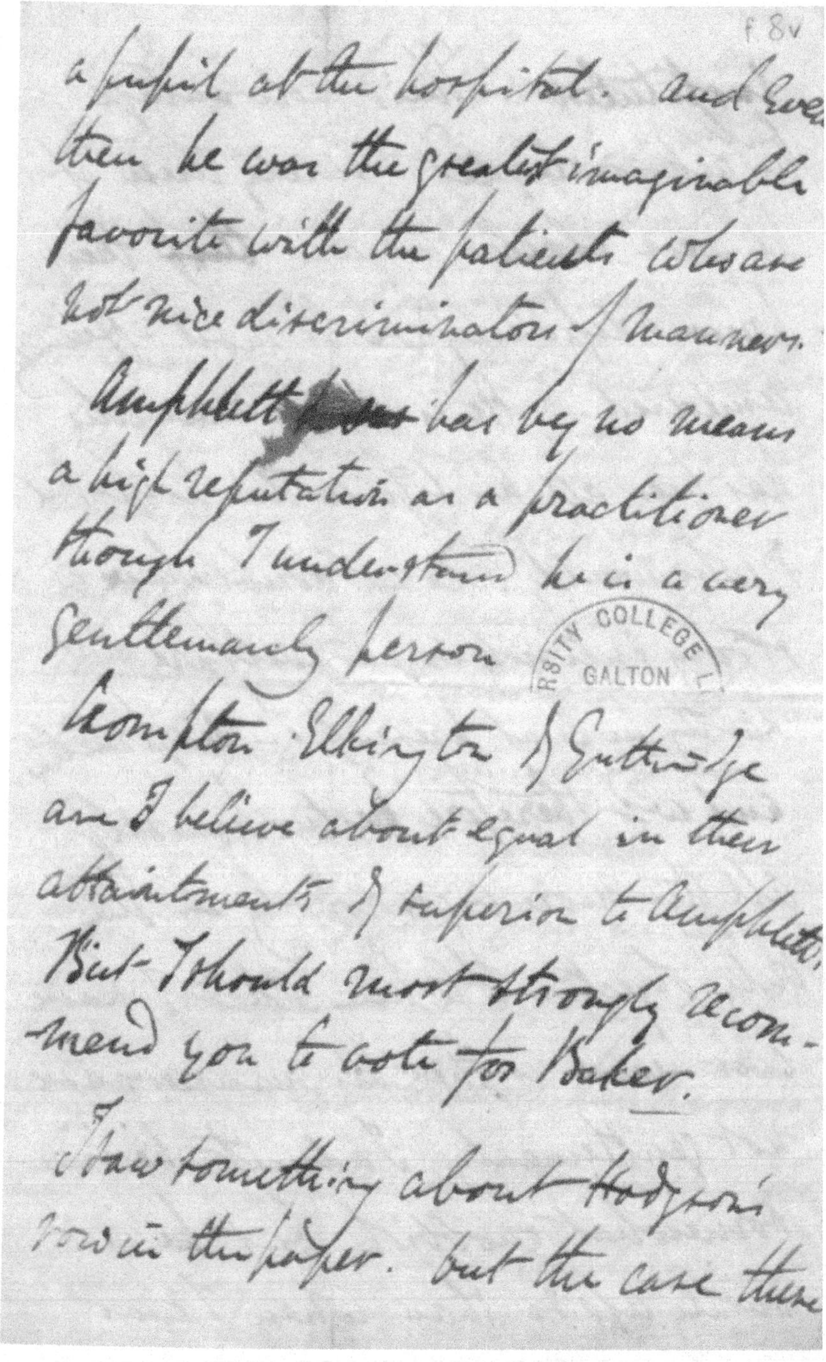

f. 8v

a pupil at the hospital. and even
then he was the greatest imaginable
favorite with the patients who are
not nice discriminators of manners.
Amphlett has by no means
a high reputation as a practitioner
though I understand he is a very
gentlemanly person
Crompton Elkington & Guthridge
are I believe about equal in their
attainments & superior to Amphlett,
But I should most strongly recom-
-mend you to vote for Baker.
I saw something about Hodgson's
son in the paper. but the case there

Figure 4.22: 1843-03-11. FG to STG page 4.

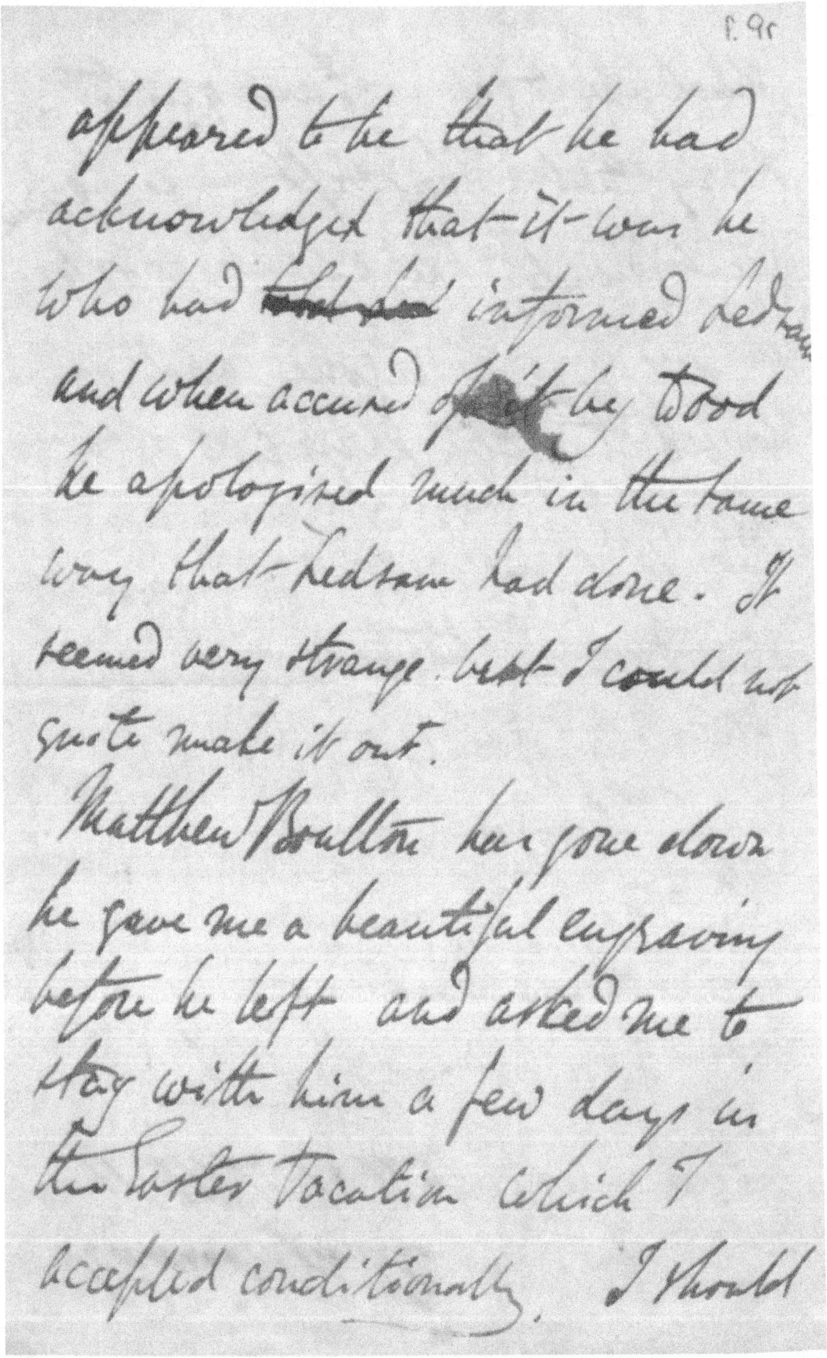

f. 95

appeared to be that he had
acknowledged that it was he
who had ~~[illegible]~~ informed Ledsam
and when accused of it by Wood
he apologized much in the same
way that Ledsam had done. It
seemed very strange but I could not
quite make it out.
Matthew Boulton has gone down
he gave me a beautiful engraving
before he left and asked me to
stay with him a few days in
the Easter vacation which I
accepted conditionally. I should

Figure 4.23: 1843-03-11. FG to STG page 5.

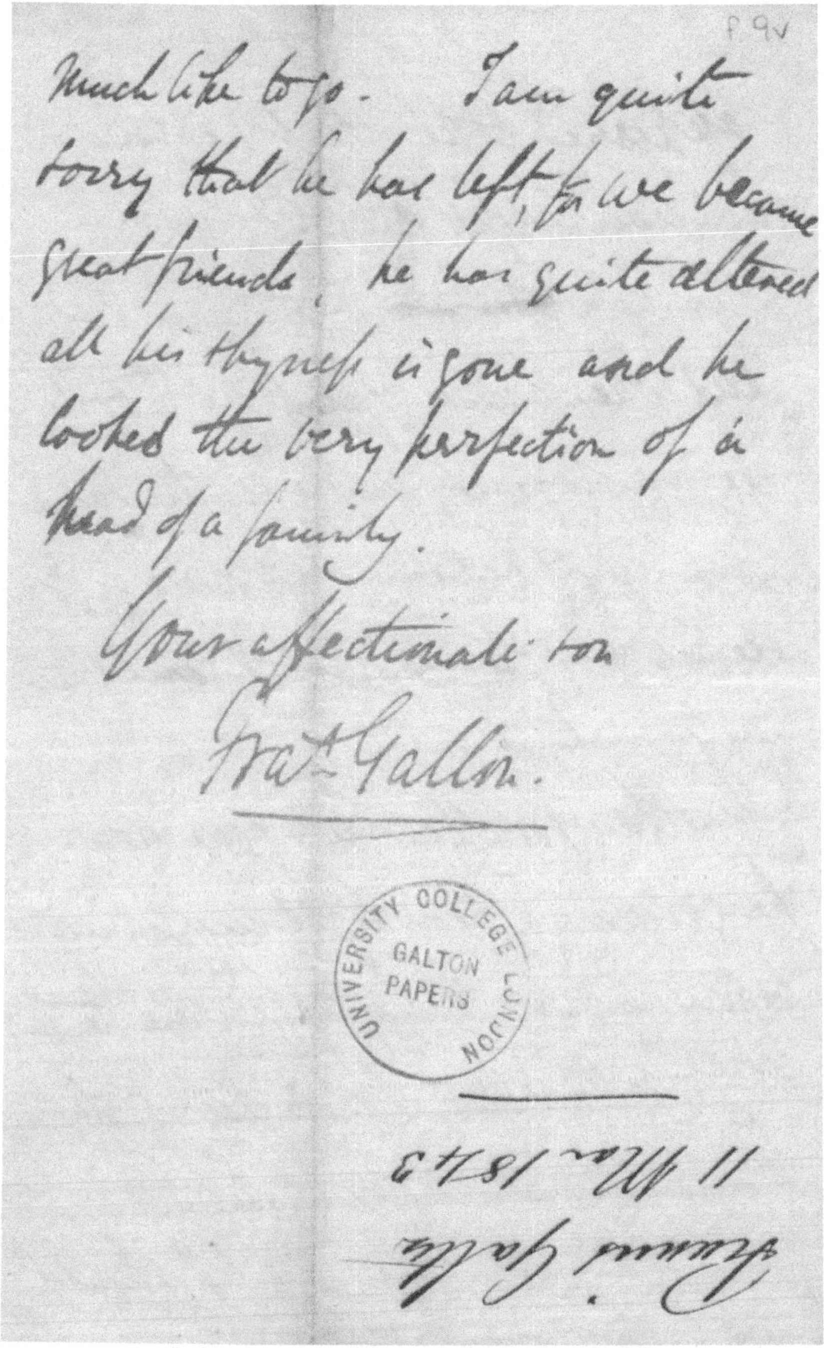

f. 9v

much like to go. I am quite sorry that he has left, for we became great friends, he has quite altered all his shyness is gone and he looked the very perfection of a head of a family.

Your affectionate son

Fra^s Galton.

11 Mar 1843

Francis Galton

Figure 4.24: 1843-03-11. FG to STG page 6.

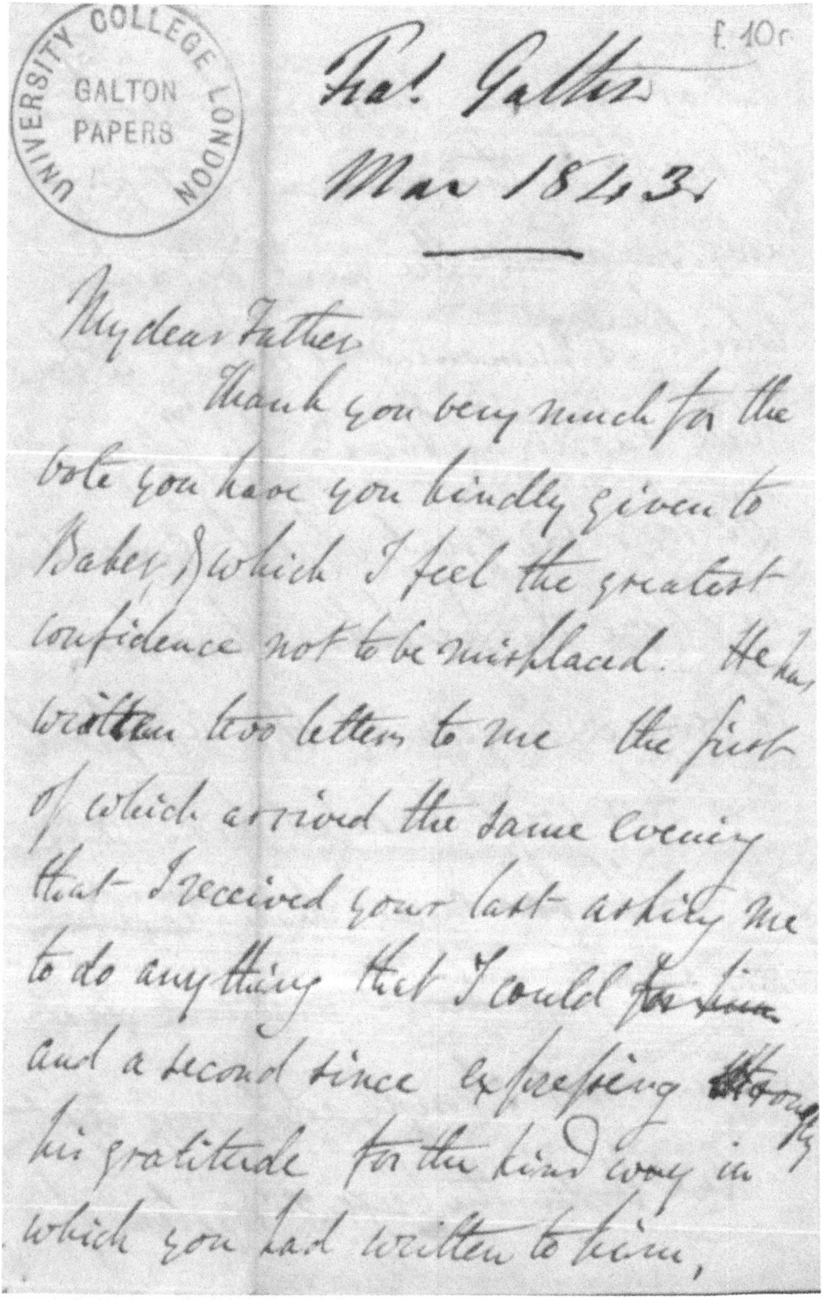

Frat. Galton
Mar 1843.

My dear Father
Thank you very much for the vote you have so kindly given to Baker, & which I feel the greatest confidence not to be misplaced. He has written two letters to me the first of which arrived the same evening that I received your last asking me to do anything that I could for him and a second since expressing strongly his gratitude for the kind way in which you had written to him,

Figure 4.25: 1843-03. FG to STG page 1.

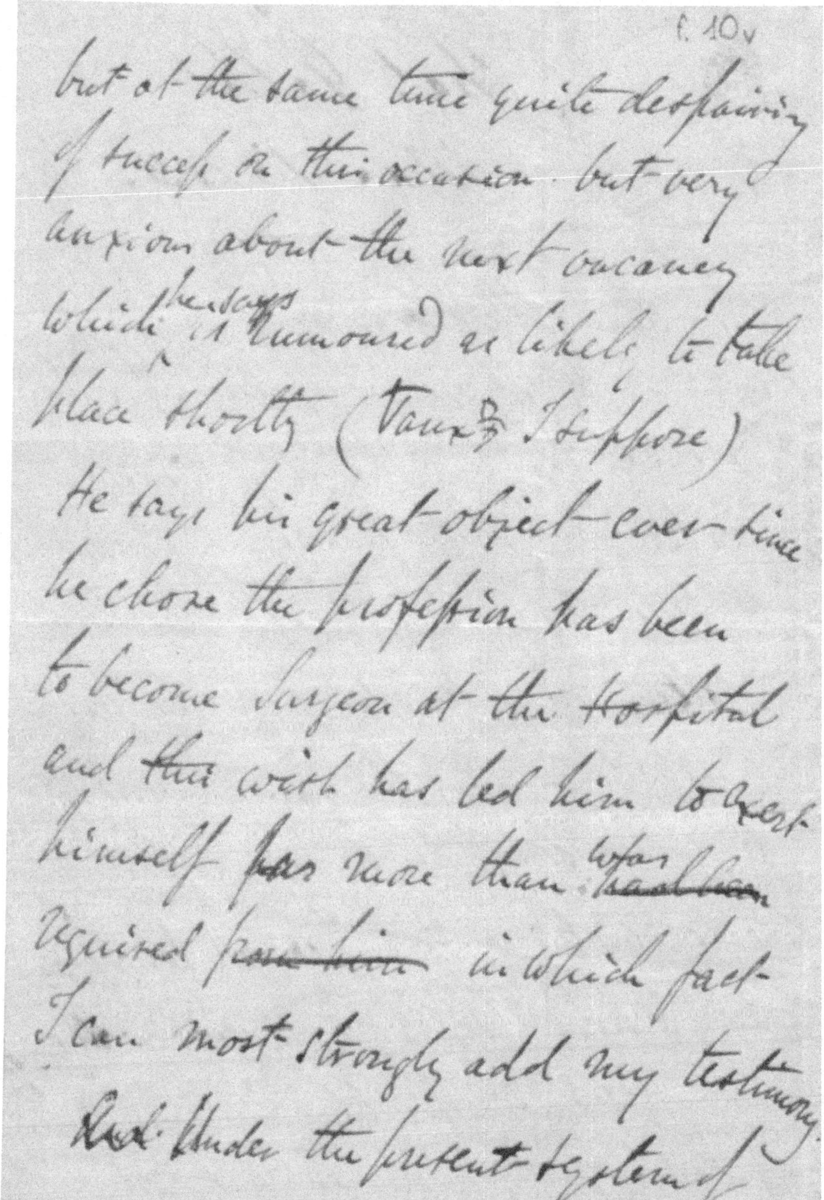

f. 10v

but at the same time quite despairing
of success on this occasion but very
anxious about the next vacancy
which he says is rumoured as likely to take
place shortly (Vaux's I suppose)
He says his great object ever since
he chose the profession has been
to become Surgeon at the Hospital
and ~~this~~ wish has led him to exert
himself far more than was ~~had been~~
required ~~from him~~ in which fact
I can most strongly add my testimony
~~And~~ Under the present system of

Figure 4.26: 1843-03. FG to STG page 2.

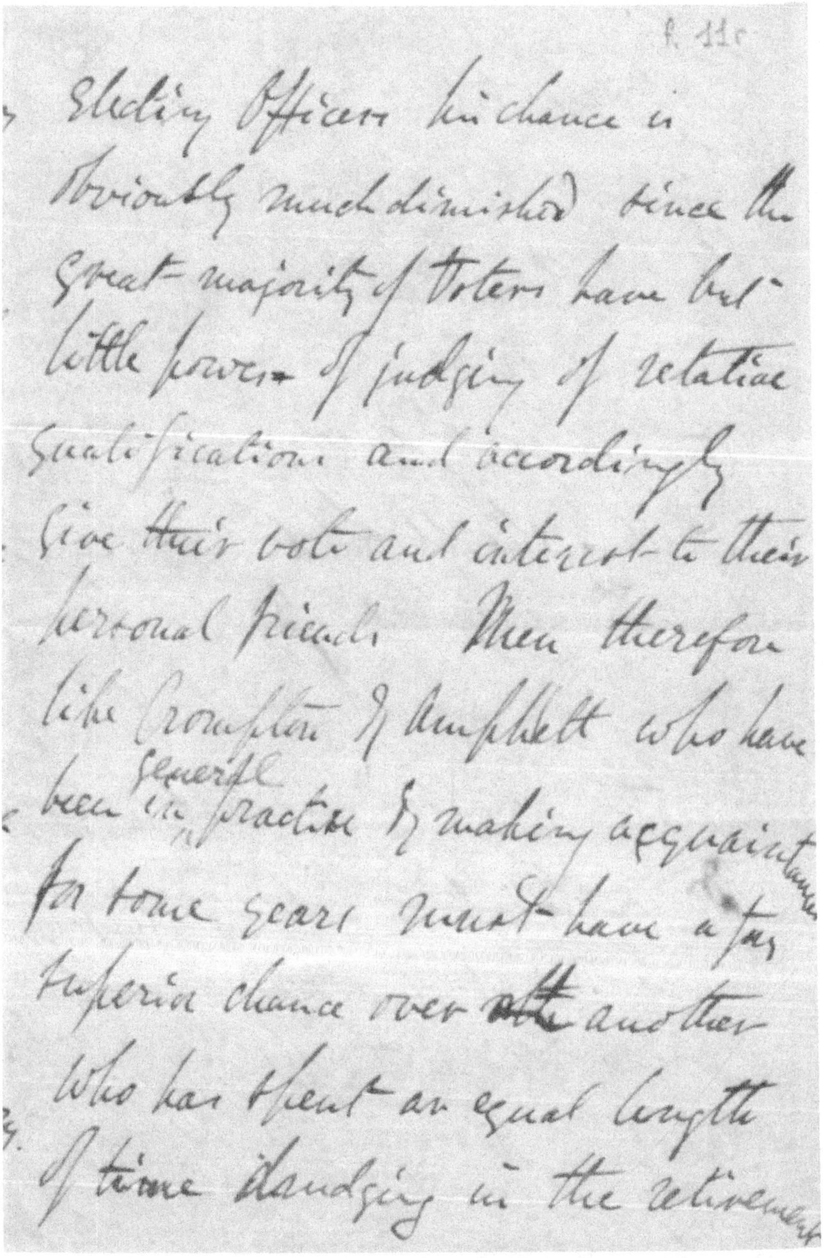
f. 11r

Electing Officers his chance is
obviously much diminished since the
great majority of Voters have but
little power of judging of relative
qualifications and accordingly
give their vote and interest to their
personal friends. Men therefore
like Crompton & Campbell who have
been generally in practice & making acquaintance
for some years must have a far
superior chance over ~~all~~ another
who has spent an equal length
of time standing in the retirement

Figure 4.27: 1843-03. FG to STG page 3.

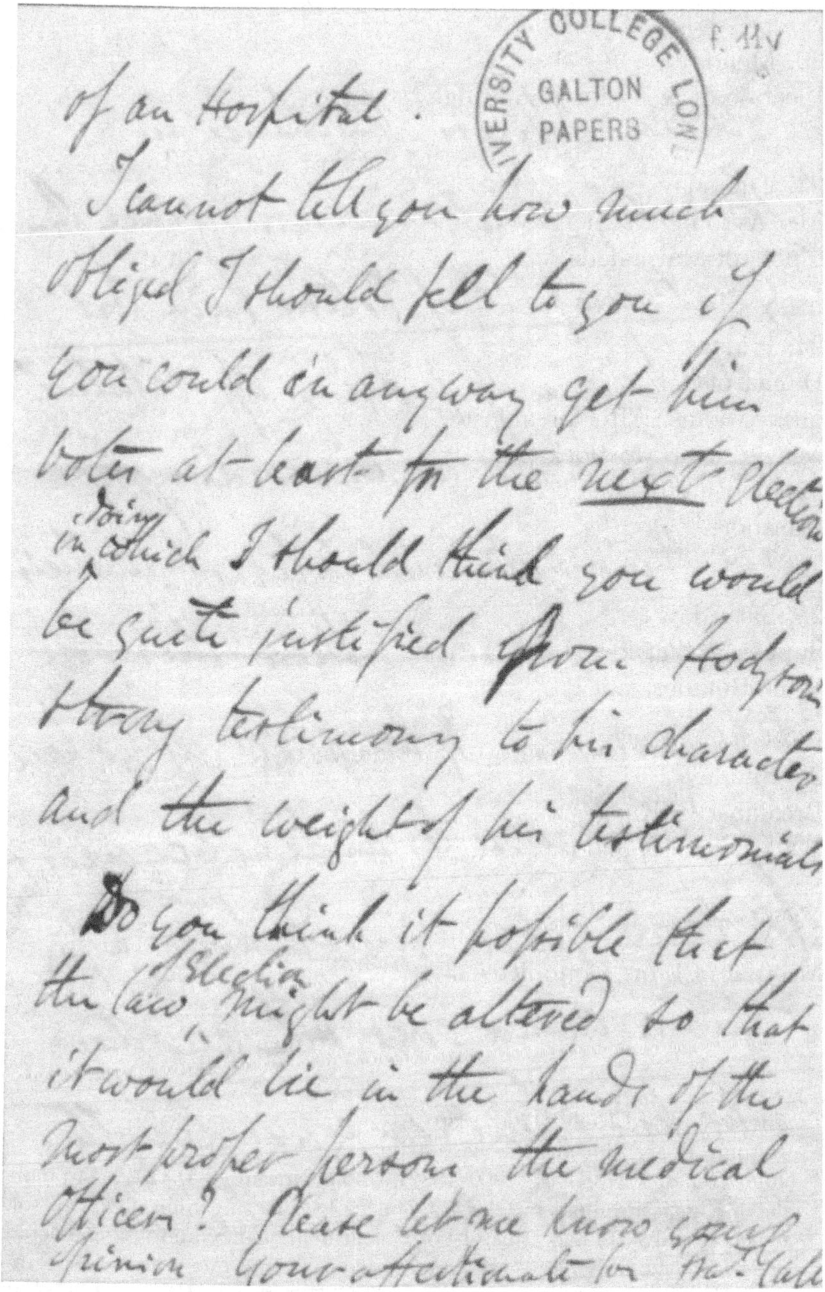

of an Hospital.
I cannot tell you how much obliged I should feel to you if you could in anyway get him votes at least for the next Election in doing which I should think you would be quite justified from Hodgson's strong testimony to his character and the weight of his testimonials.
Do you think it possible that the law of Election might be altered so that it would lie in the hands of the most proper person the medical Officer? Please let me know your opinion Your affectionate bro Fra. Galt

Figure 4.28: 1843-03. FG to STG page 4.

1843/03/12–27. FG Diary

20. Monday.
Election of Kelly for Cambridge.[58]

23. Thursday.
Mr. Aytoun[59] sent me
6 pots of Marmalade.

24. Friday.
Dinner of [...]
in my rooms. After breakfast
went to hear Mortlock tried
for shooting his uncle 21 years
transport.[60]

25. Saturday.
Supper & chat in my rooms.
Wine Boudier. not go.[61]

26. Sunday.
Breakfast E. Kay.[62]
Walked with Hallam.[63]

27. Monday.
Meeting of Libr: Committee.

---

[58] Fitzroy Edward Kelly (1796–1880).
[59] Aytoun, Roger Sinclair. See page 480.
[60] John Frederick Mortlock (1809–1882), who in November 1842 tried to murder his uncle Edmund in Christ's College by shooting him with a pistol, unsuccessfully. He had a history of violent delinquency and was convicted of "grievious bodily harm", then sentenced to twenty-one years transportation. He illegally returned to England from New South Wales in 1857, but was rearrested, jailed and sent back to Western Australia.
[61] George John Boudier (1820-1899) of King's.
[62] Kay, Edwin Ebenezer. See page 485.
[63] Hallam, Henry Fitzmaurice. See page 483.

1843/04/05. FG Diary

5. [Wednesday.]
Saw Dr. Jephson.[64]

1843/04/05. Francis Galton to Samuel Tertius Galton

Trin. Col.

My dear Father

I am having the greatest fun imaginable in getting up an "English Epigram Society" which is to meet 3 times a term the members are to send in their epigrams anonymously and they are to be read by some one chosen by lot. The subject is to be chosen out of those proposed by a majority. I have got the first man in the University, among the rising men, to join it two young Fellows of Trinity & bachelors &c. so I expect that some of the Epigrams will be first rate. The society consists of 12. All the men I have spoken to have jumped at the idea, and I have great hopes of its working exceedingly well.

I think I shall be able to come down this week when I would coach it to Hatton & then walk on. I send you a poem I have just sent it for the Camden medal & fear it will not interest you much as it is all in relation to the present great controversy as to whether man has a conscience (innate I mean) or not. Paley & Locke & many Greek philosophers as you know against it, Plato & Bishop Butler & some German metaphysicians & Whewell on the other side. Stewart seems to be for it but does not give a decided opinion one way or the other.

The mottoes I have chosen explain the point of the whole I take the paraphrase of the one from Plato to be "They have ever in their soul a specimen of the Divine nature lasting and bright as silver or gold."

I was obliged to print it before sending it in.

I leave it on the honor of the Family that it be not shown to any besides themselves I mean my Father & Mother Bessy Adèle Emma, nor in any way to be spoken of to others.

Your affectionate son
Fras Galton.

[64] Almanac. Jephson, Dr. Henry, see page 485.

1842/04/10. Diary of Samuel Tertius Galton

April 10th. Monday. Claverdon.

Violetta, Bessy, Francis and myself lunched at Edstone and Adèle drove the latter back in her pony carriage to Claverdon. ...

1843/04/12–17. FG Diary

12. W[ednesday].
Saw Dr. Jephson.[65]
12. Wednesday.
Went to Leamington & saw Dr. Jephson.
Rx. Decoct Taraxaci[66] 3x
Magnesia sulph. 3 [...][67]
Acid Sulph Dil 3 [...][68]
Pais sext vis die

17. M[onday].
Saw Dr. Jephson.[69]

1842/04/17–24. Diary of Samuel Tertius Galton

April 17th. Monday. Claverdon.

... Bessy, Emma and Francis went to Leamington, the former to prepare for her journey to London tomorrow ... and the latter to consult Dr. Hodgson.[70] ...

April 19th. Wednesday. Claverdon.

Francis, Emma, Irwin and myself went to Leamington and found the Huberts at the Regent.

April 24th. Monday. Claverdon.

Francis left us this morning to dine with Dr. Luard on his way to Cambridge tomorrow from Leamington. ...

---

65 Almanac. Jephson, Dr. Henry, see page 485
66 Dandelion.
67 Epsom salt.
68 Diluted sulfuric acid.
69 Almanac.
70 Hodgson, Joseph. See page 483.

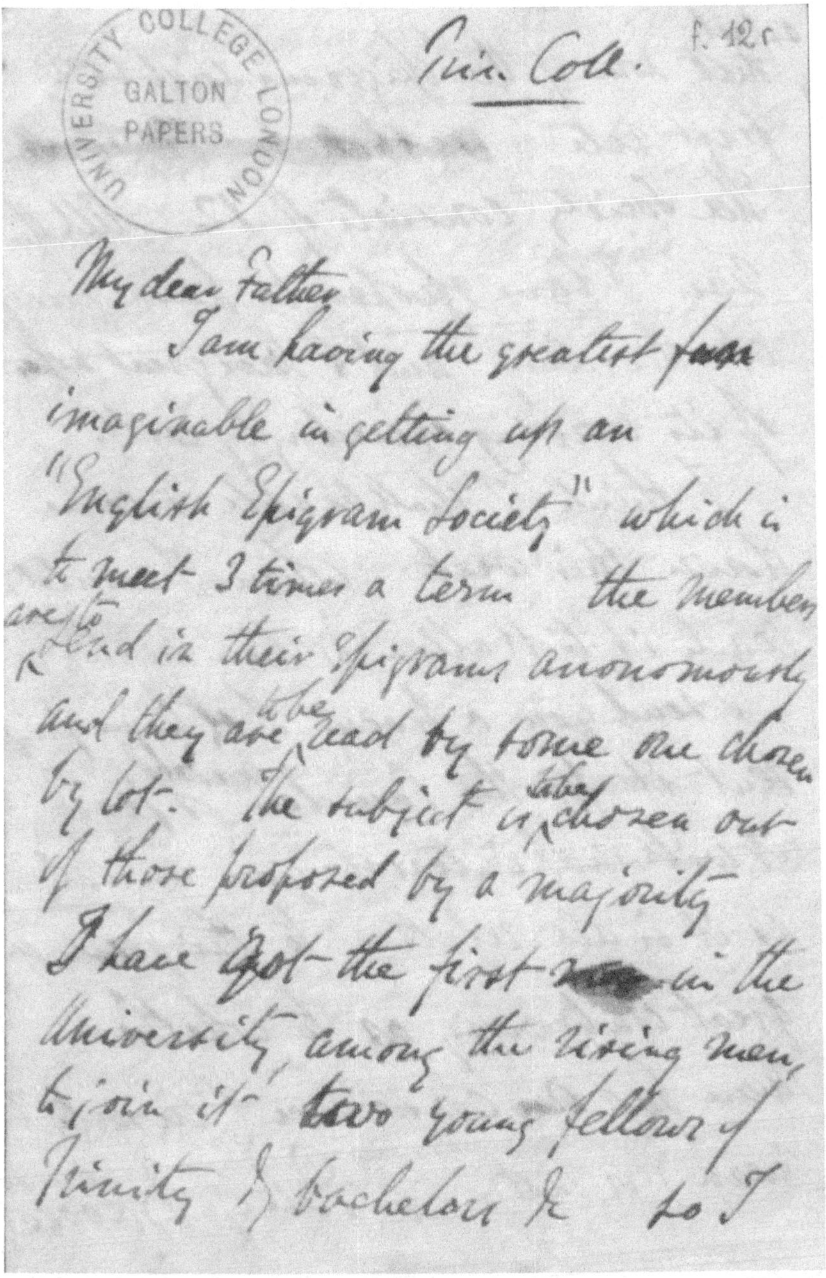

Trin. Coll.

f. 12r

My dear Father
I am having the greatest fuss imaginable in getting up an "English Epigram Society" which is to meet 3 times a term the members are to send in their Epigrams anonymously and they are to be read by some one chosen by lot. The subject is taken chosen out of those proposed by a majority. I have got the first men in the University among the rising men, to join it two young fellows of Trinity & bachelors &c so I

Figure 4.29: 1843-04-05. FG to STG page 1.

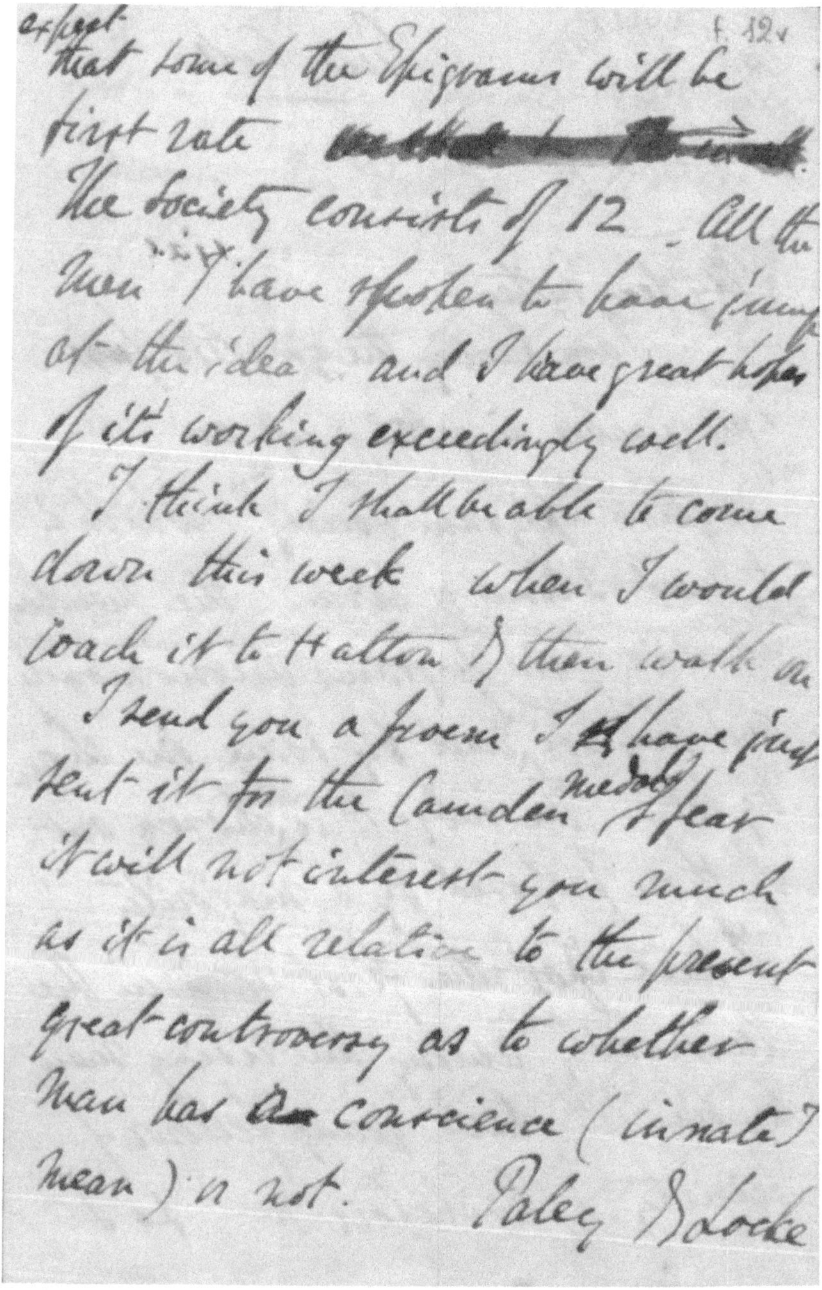

f. 12v

expect
that some of the Epigrams will be
first rate [redacted]
The society consists of 12 – All the
men I have spoken to have jump[ed]
at the idea and I have great hopes
of its working exceedingly well.
I think I shall be able to come
down this week when I would
coach it to Halton & then walk on
I send you a poem I have just
sent it for the Camden medal I fear
it will not interest you much
as it is all relation to the present
great controversy as to whether
man has a conscience (innate I
mean) or not. Paley & Locke

Figure 4.30: 1843-04-05. FG to STG page 2.

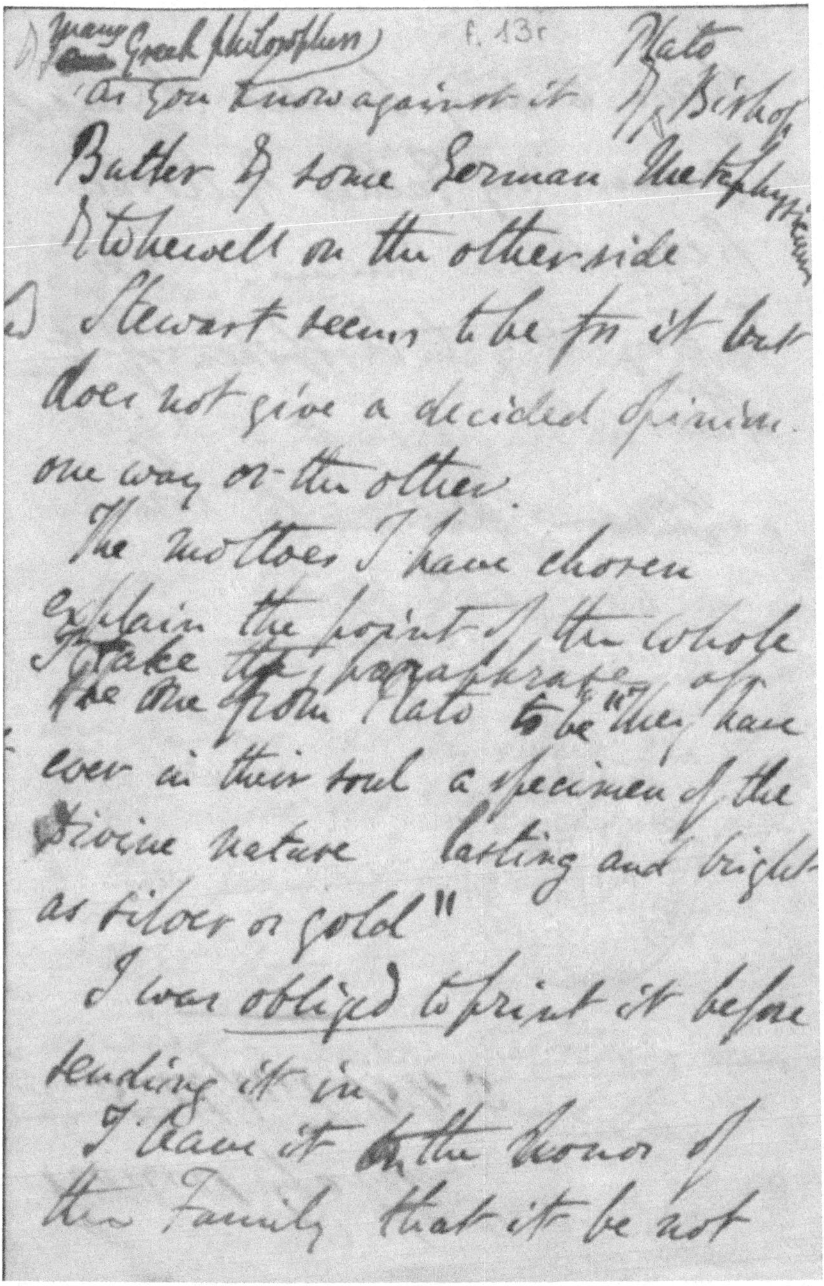

many Great Philosophers) f. 13r Plato
as you know against it & Bishop
Butler & some German Metaphysicians
& Whewell on the other side
& Stewart seems to be for it but
does not give a decided opinion
one way or the other.
The mottoes I have chosen
explain the point of the whole
I take the paraphrase of
the one from Plato to be "they have
ever in their soul a specimen of the
divine nature lasting and bright
as silver or gold"
I was obliged to print it before
sending it in
I leave it to the honor of
the Family that it be not

Figure 4.31: 1843-04-05. FG to STG page 3.

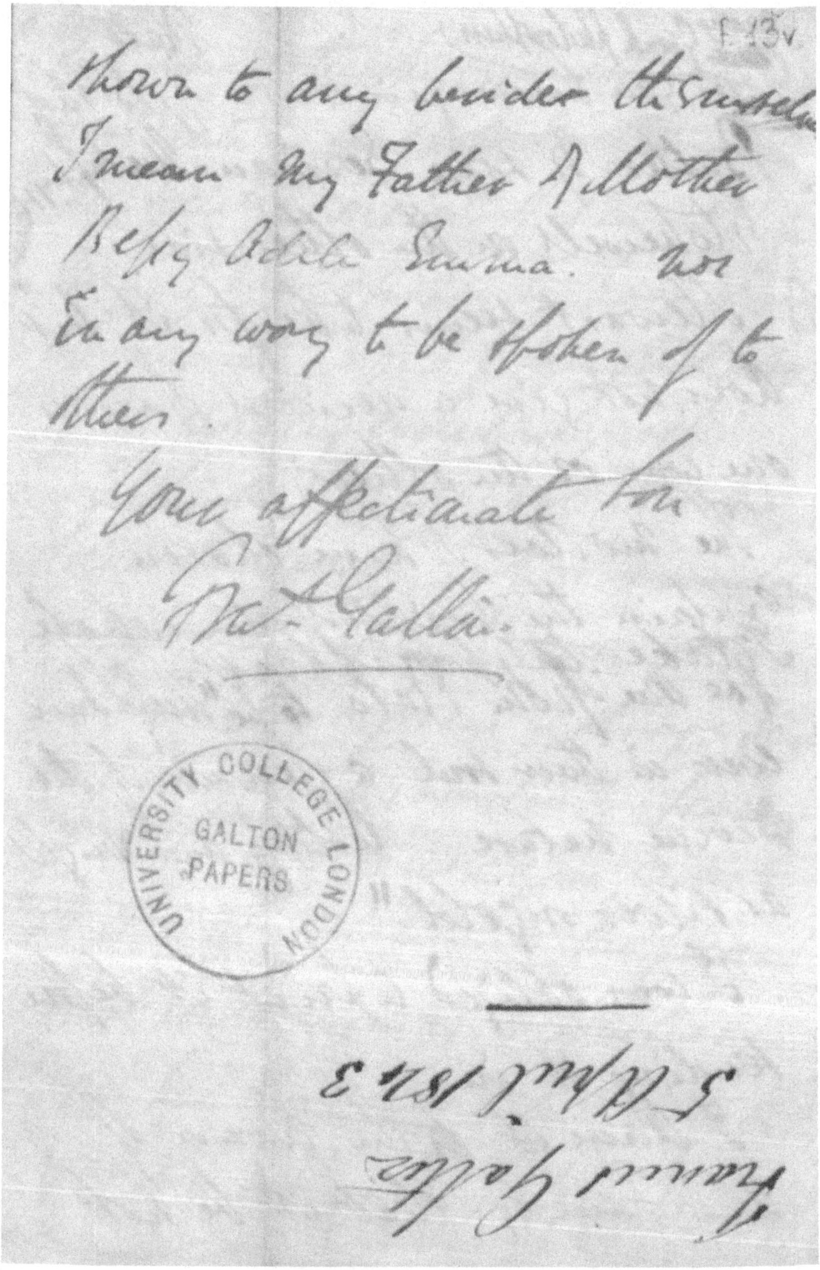
f 13v

shown to any besides themselves
I mean my Father & Mother
Bessy Adele Emma. nor
in any way to be spoken of to
others.

Your affectionate son
Frat Galton

Samuel Galton
5 April 1843

Figure 4.32: 1843-04-05. FG to STG page 4.

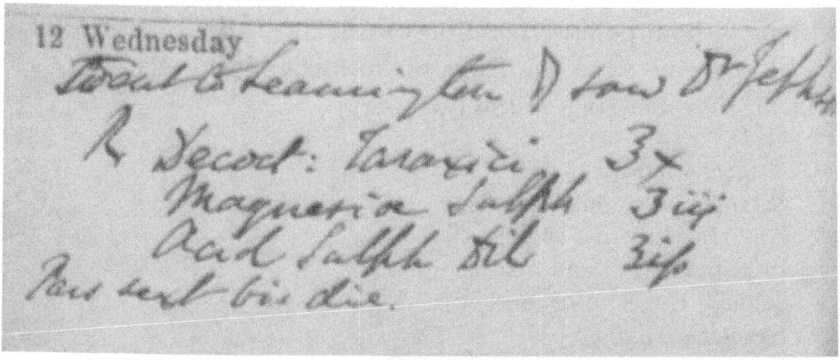

Figure 4.33: 1843-04-12: Jephson's Prescription of Dandelion Extract.

1843/04/24–30. FG Diary

24. M[onday].
Saw Dr. Jephson.[71]

1843/05 circa. Francis Galton to Samuel Tertius Galton

Trinity

My dear Father,

I do not think I can get any Weimar introduction from my London Medical friends though they may have some acquaintances in Jena. Jena is I find a stage from Weimar. I am making all enquiries I can, and from all I can gather Weimar is decidedly the place for us.

The Historical Society flourishes. I speechified there the last meeting. The Epigram Society appears most prosperous we had its first meeting last night. A great many were sent in, and 5 chosen out of them by ballot. Then we have subscribed for a superb manuscript book with AMOENITATES CANTABRIGIENSES on it in which such epigrams as are chosen are inserted. I have not time to write out for you the five in question but they were very fair. I ought to say that we take the word epigram in its most general sense that is any poem of any character on a given subject,

[71] Almanac.

with or without point. The subject was via trita via tuta (the worn way is the safe way) the Duke of Norfolk's motto. The verses I sent in, they were one of the five, were

A plucky lad was he,
Who fastened quills together
And tried to cross the sea
In spite of wind and weather
Though better to have wept
In silence Minos' ire
Facts only prove he leapt
From frying pan to fire.

Shareholders save your load,
Save money save material,
Go keep the turnpike road
And sell your steam aerial.

Your affectionate Son
Fras. Galton.

1843/05/01–27. FG Diary

7. Sunday
Norris GG[72]

19. Friday.
First meeting of the Epigram Society. Members Bristed[73] G.W. Clarke,[74] Douglass,[75] Galton Johnson, J. Kay,[76] Keary, Maine[77] Rendall, Richards.

23. Tuesday.
Supper $10\frac{1}{2}$.

---

[72] Almanac.
[73] Bristed, Charles Astor. See page 482.
[74] William George Clark (1821-1878).
[75] Possibly Charles Edward Douglass (?–1904).
[76] Kay, Joseph. See page 486.
[77] Maine, Henry James Sumner. See page 486.

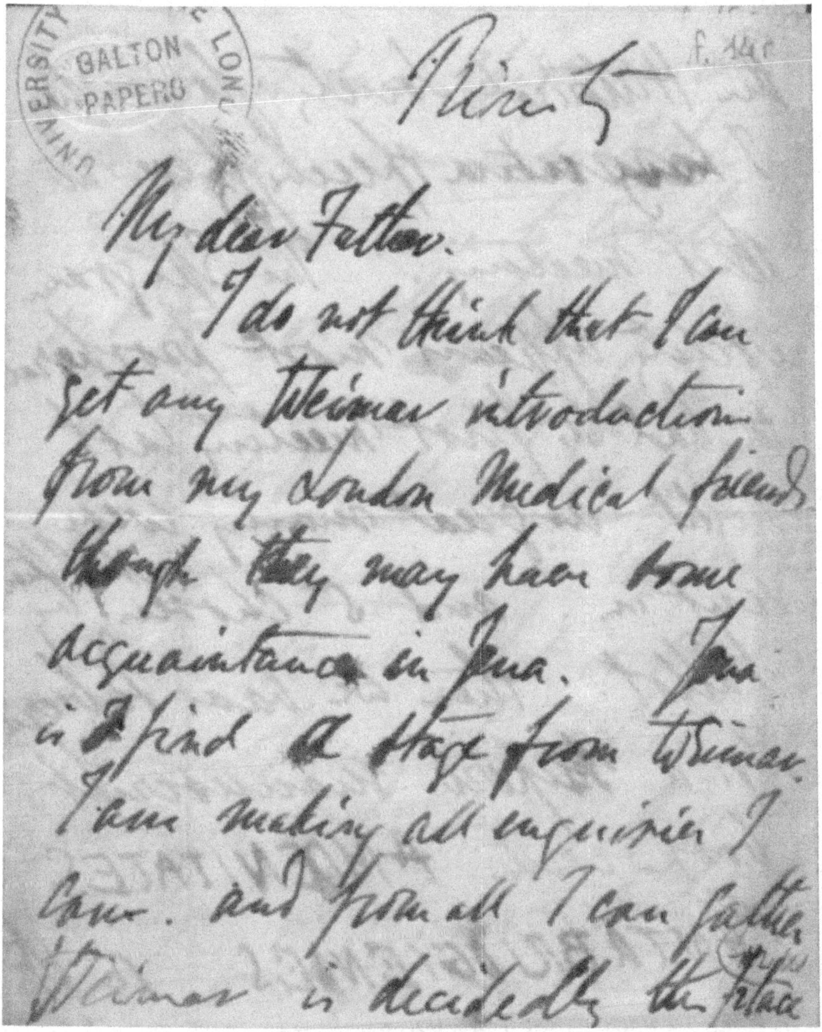

Private

My dear Father.
I do not think that I can get any Weimar introduction from my London Medical friends though they may have some acquaintance in Jena. Jena is I find a stage from Weimar. I am making all enquiries I can. and from all I can gather Weimar is decidedly the place

Figure 4.34: 1843-05. FG to STG page 1.

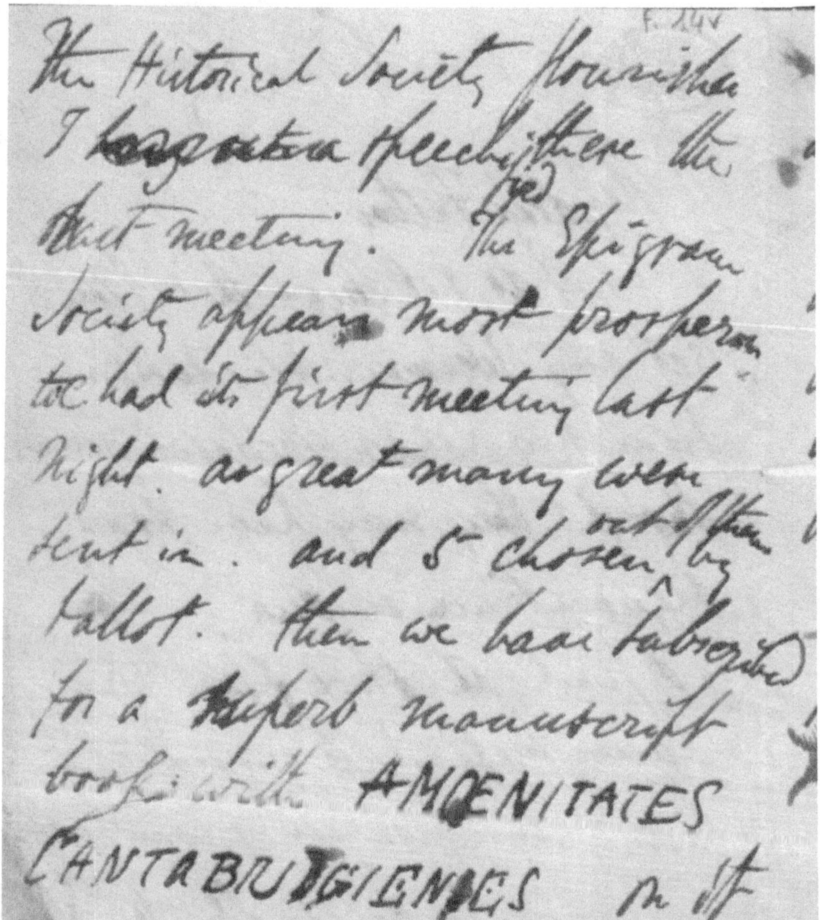

f. 244v

the Historical Society flourishes
I extra speech there the
first meeting. The Epigram
Society appears most prosperous
it had its first meeting last
night. a great many were
sent in. and 5 chosen out of them by
ballot. then we have subscribed
for a superb manuscript
book with AMOENITATES
CANTABRIGIENSES on it

Figure 4.35: 1843-05. FG to STG page 2.

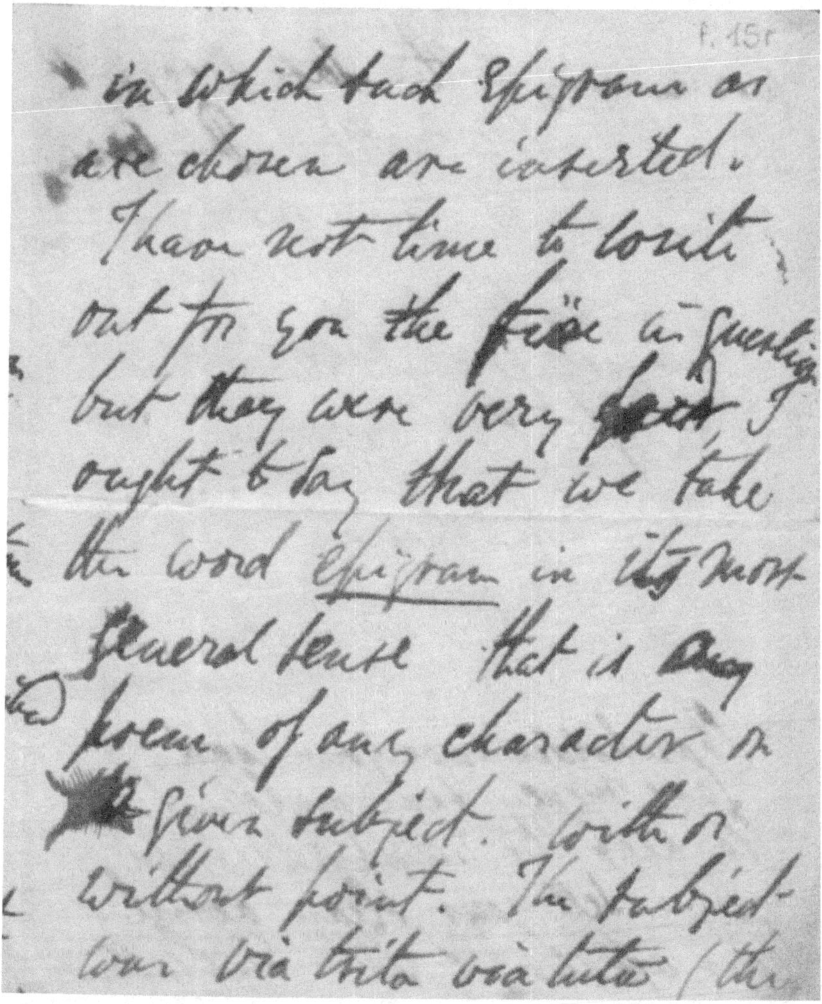
in which such Epigrams as
are chosen are inserted.
I have not time to write
out for you the five in question
but they were very good I
ought to say that we take
the word Epigram in its most
general sense that is any
poem of any character on
a given subject. with or
without point. The subject
was Via trita via tuta (the

Figure 4.36: 1843-05. FG to STG page 3.

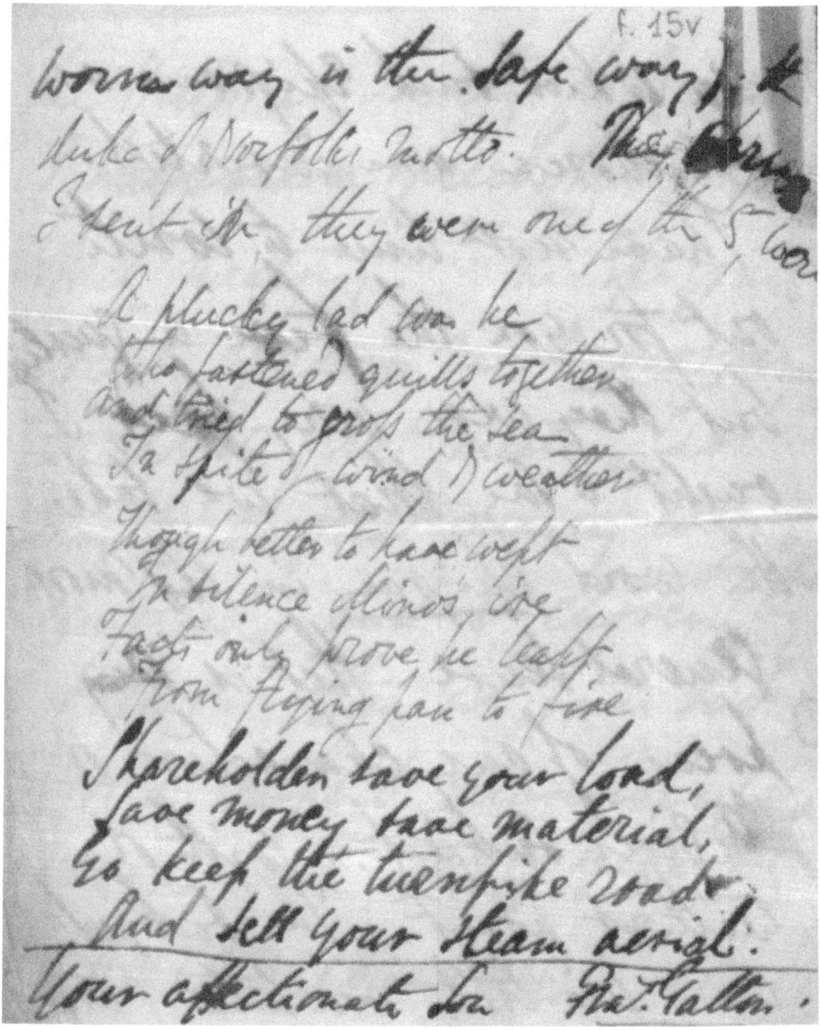

f. 15v

worse way is the safe way, the
duke of Norfolk's motto. They [illegible]
spent it, they were one of the 5 wor[illegible]

A plucky lad was he
Who fastened quills together
And tried to cross the sea
In spite of wind & weather
Though better to have wept
In silence Minos' ire
Facts only prove he leapt
From frying pan to fire
Shareholders save your land,
Save money save material,
Go keep the turnpike road
And sell your steam aerial.

Your affectionate Son Fras. Galton.

Figure 4.37: 1843-05. FG to STG page 4.

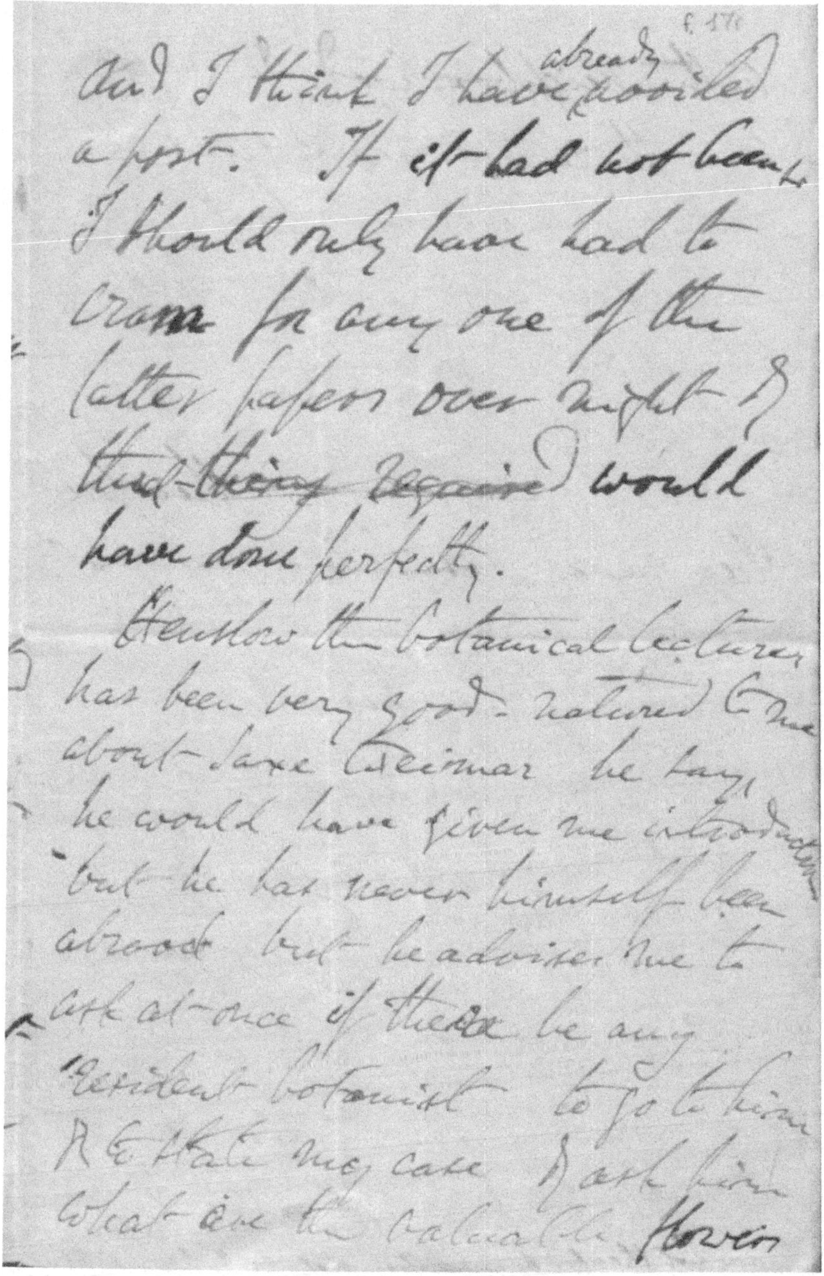

And I think I have already spoiled
a post. If it had not been so
I should only have had to
cram for any one of the
latter papers over night &
~~that thing required~~ would
have done perfectly.

Henslow the botanical lecturer
has been very good-natured to me
about Saxe Weimar he says
he would have given me introductions
but he has never himself been
abroad but he advises me to
ask at once if there be any
resident botanist to go to him
& state my case & ask him
what are the valuable flowers

Figure 4.38: 1843-05. FG to STG page 3.

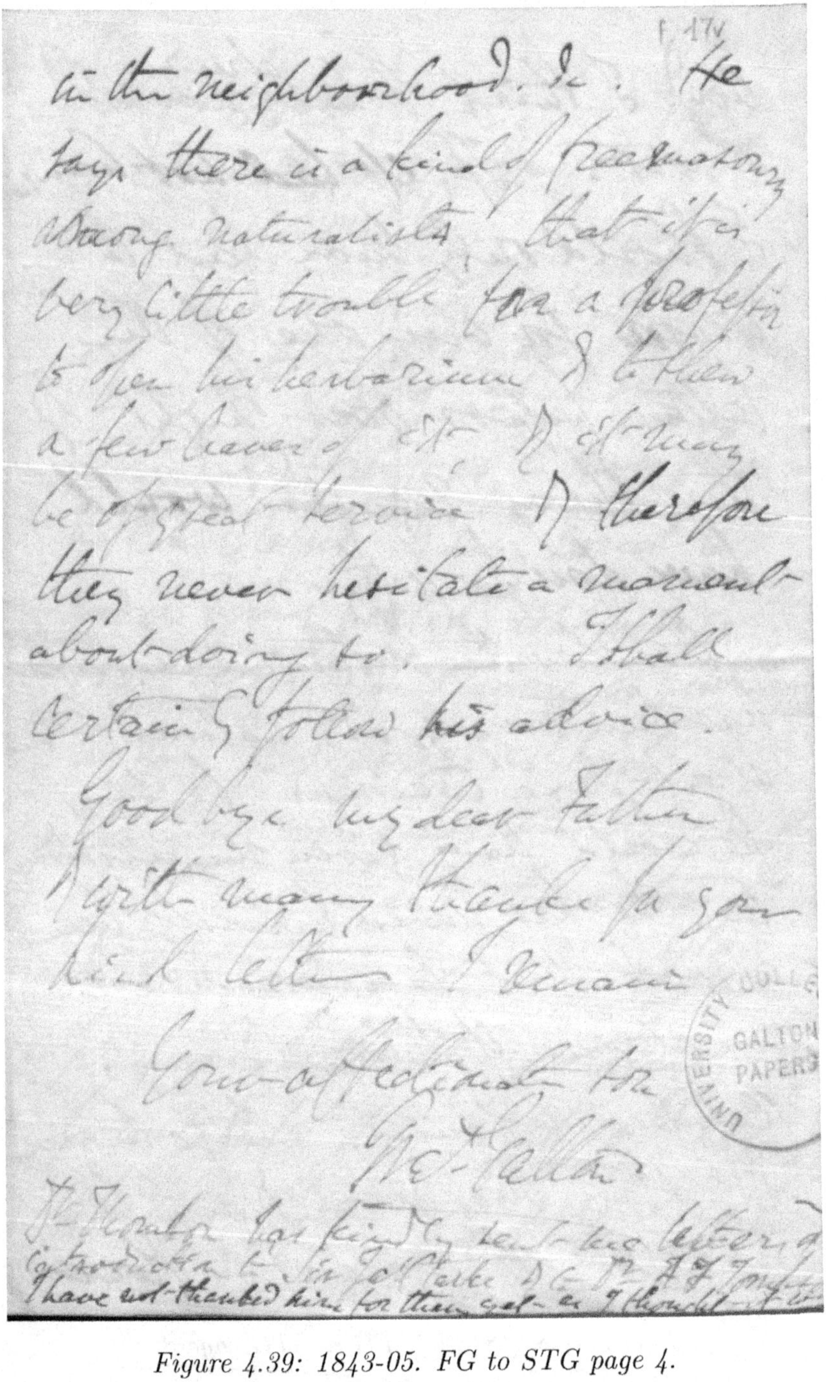
f. 174

in the neighbourhood. &c. He
says there is a kind of freemasonry
among naturalists that it is
very little trouble for a professor
to open his herbarium & to shew
a few leaves of &c. & it may
be of great service & therefore
they never hesitate a moment
about doing so. I shall
certainly follow his advice.
Good bye my dear Father
& with many thanks for your
kind letter I remain
Your affectionate son
F. Galton

Dr Thornton has kindly sent me letters of
introduction to Sir J. Clarke D.G. Dr A. J. ...
I have not thanked him for them yet – as I thought it ...

Figure 4.39: 1843-05. FG to STG page 4.

Figure 4.40: King's College Chapel, Cambridge, May 1843, by FG.

25. Thursday.
Newman wine

27. Saturday.
Epigram.

1843/05 circa. Francis Galton to Samuel Tertius Galton

Wed^{y} Evening

My dear Father

Thank you very much for your kind present to me which will be very acceptable as I do not doubt but that my journey will be somewhat expensive. I will be with you on Monday Evening, as I propose to leave Cambridge for Claverdon by the Eagle on that day. Then I was thinking of staying with you there until Wed^{y} or Thursday & then joining Emma in London & starting off with her on Monday.

I cannot write a longer letter now as we are being plagued with an Examination, in which however I am not trying to do much as I am quite indifferent as to my place in the Classes & only want to avoid being posted which is a bore because the name is published in the newspapers as such & relations are generally ignorant of the nature and character of these examinations. I have prepared literally nothing but trusted to the light of nature which has been very useful so far and I think I have already avoided a post. If it had not been so I should only have had to cram for any one of the latter papers over night & that ~~thing required~~ would have done perfectly.

Henslow the botanical lecturer has been very good-natured to me about Saxe Weimar he says he would have given me introductions but he has never himself been abroad but he advises me to ask at once if there be any resident botanist to go to him & to state my case & ask him what are the valuable flowers in the neighbourhood, &c. &c. He says there is a kind of freemasonry among naturalists, that it is very little trouble for a professor to open his herbarium & to spew a few leaves of it, & it may be of great service & therefore they never hesitate a moment about doing so. I shall certainly follow his advice.

Goodbye my dear Father & with many thanks for your kind letters I remain

Your affectionate son
Fras. Galton

Dr. Thompson has kindly sent me letters of introduction to Sir Jas. Clarke & to Dr. A. J. [Thompson]. I have not thanked him for them yet as I thought it was the better to delay doing so till I had used them.

1843/06/05. FG Diary

5. Monday.
Went to Claverdon.[78]

1843/06/05–07. Diary of Samuel Tertius Galton

June 5th. Monday. Claverdon.

... Darwin and Mary dined with us and Francis on his arrival from Cambridge walked from Warwick in the evening. Much heavy rain.

[78] Almanac.

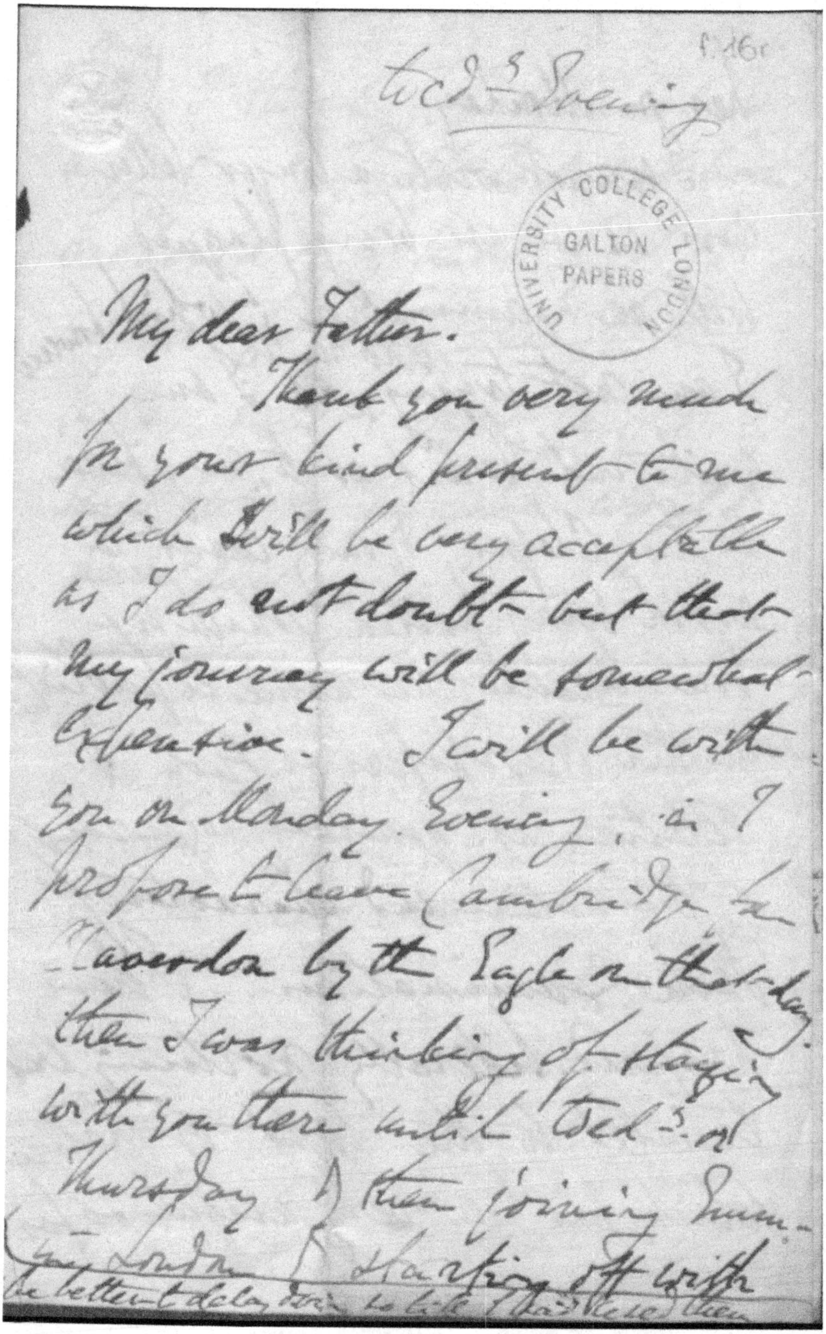

Wed^s Evening

f. 16c

My dear Father.
Thank you very much
for your kind present to me
which I will be very acceptable
as I do not doubt but that
my journey will be somewhat
expensive. I will be with
you on Monday Evening, as I
propose to leave Cambridge for
Claverdon by the Eagle on that day.
then I was thinking of staying
with you there until Wed^s or
Thursday & then joining [illegible]
[illegible] London & starting off with
be better to delay [illegible]

Figure 4.41: 1843-05. FG to STG page 1.

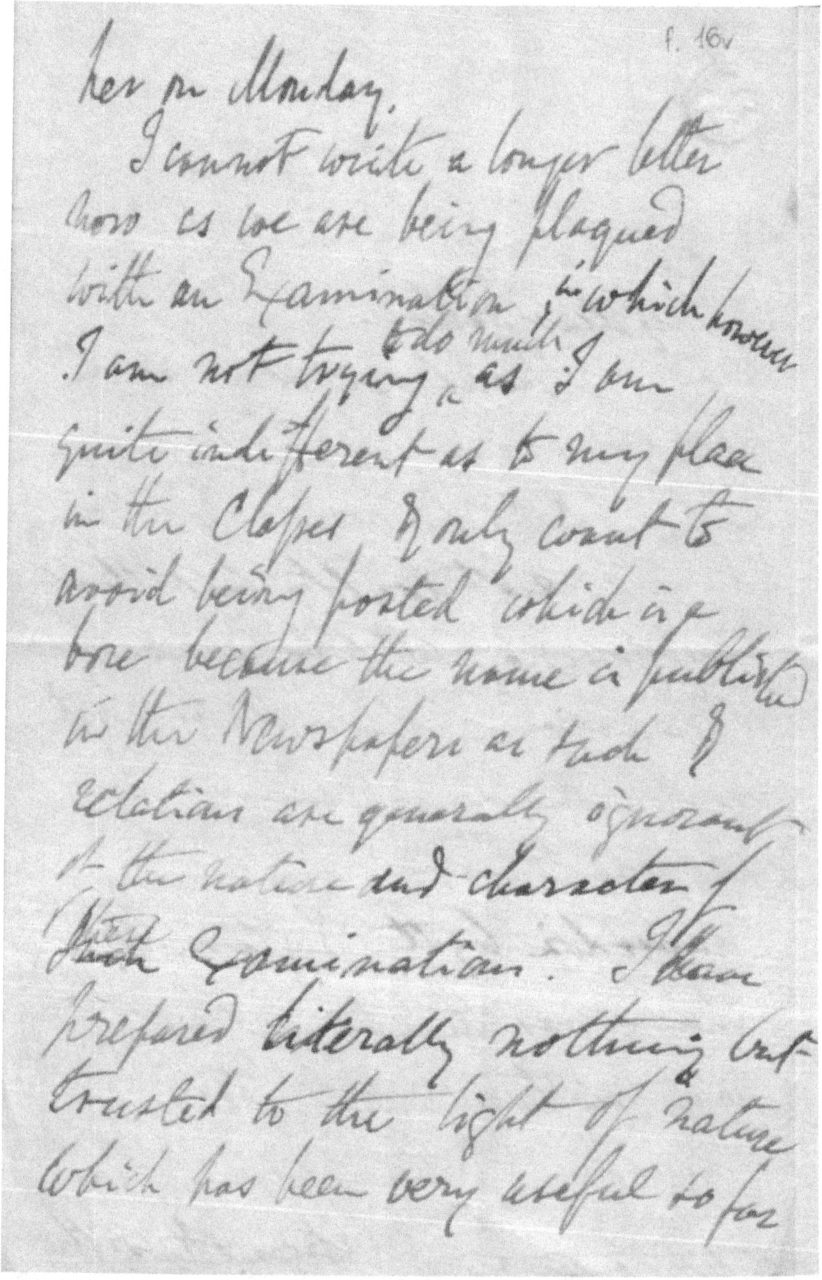
f. 16v

her on Monday.
I cannot write a longer letter
now as we are being plagued
with an Examination in which however
I am not trying to do much as I am
quite indifferent as to my place
in the Classes & only want to
avoid being posted which is a
bore because the name is published
in the Newspapers as such &
relations are generally ignorant
of the nature and character of
such Examinations. I have
prepared literally nothing but
trusted to the light of nature
which has been very useful so far

Figure 4.42: 1843-05. FG to STG page 2.

June 6th. Tuesday. Claverdon.
... I gave Francis credit for £200.

June 7th. Wednesday. Claverdon.
I took Mary, Bessy, Adèle and Francis to Leamington when the latter proceeded to London to join Emma on their way to Germany. ...

1843/06/07. FG Diary

7. Wednesday.
Went to London[79]

1843/06/10. Francis Galton to Samuel Tertius Galton

1. St Alban's Place
St James'

My dear Father,

Emma and I are bustling about in every direction to complete our arrangements. My occupations have been as follows. I got into London at $6\frac{1}{2}$ found Brown at Lanc's, where I am staying & then sallied forth with him to see the farce of Fortunio at Drury Lane which is certainly most absurd & contains more puns than it has hitherto fallen to my lot to listen to even from yourself. Thursday morning I called on Emma, on Partridge,[80] got passport & dined at the Huberts, this morning breakfast with the Charles Buxtons whom Emma had met with at the Huberts & who kindly asked her & myself to breakfast with them. We then called on the Horners.[81] Frances H. was very energetic concerning the late bills in parliament. I must take an opportunity of recommending Irwin Maling to her notice. They have given us an introduction at Dresden or Munich, I forget which. We then returned to Portman Square to await the appointed call of Denham Cookes. He has the funniest head I ever saw, is exceptionally agreeable & at his ease, nobody except his lawyer knows where he lives when under cover to whom all communications are addressed. His hair is yellowish

[79] Almanac.
[80] Partridge, Richard. See page 490.
[81] Horner, Leonard. See page 484.

red. His face something like this [sketch]. He told us a great deal. Mentioned that Colonel Todd lives & is very popular at Dresden, gave us an introduction to him. Emma says we knew him slightly at Leamington. Next we went to the Gurneys & I dined again with the Huberts. Tomorrow I breakfast with the Kays & dine with Partridge.

On carefully considering over our plans, as we have introductions in abundance to Dresden & Munich & none to Saxe Weimar; as moreover there is a picture gallery in Dresden & Munich & none at Saxe Weimar & in [time?] as every German gentleman leaves Saxe Weimar during the summer, we have resolved upon changing our plans & going to Hamburg & then to Dresden there to stay half our time until we can get on moderately well in German & then to go to Munich, through Saxe Weimar & by which time Lord Erskine will have fully recovered the death of his wife & stay there the other half of our time and then return homeward. By the way, Mr Chas. Barclay has given us another introduction to Lord Erskine, who is under great pecuniary obligation to him.

Partridge has given me another letter of introduction to Sir John Clarke on whom I call tomorrow.

We start at 2 on Wed. morning.

You affectionate son.
Fras Galton.

1843/06/13. FG Diary

13. Tuesday.
Set off with Emma
to Hamburg.

1843/08/11–16. FG Diary

[August]

11. Friday.
Went with Dr. Schriker[82] [in a]

[82] Possibly Dr. August Schriker.

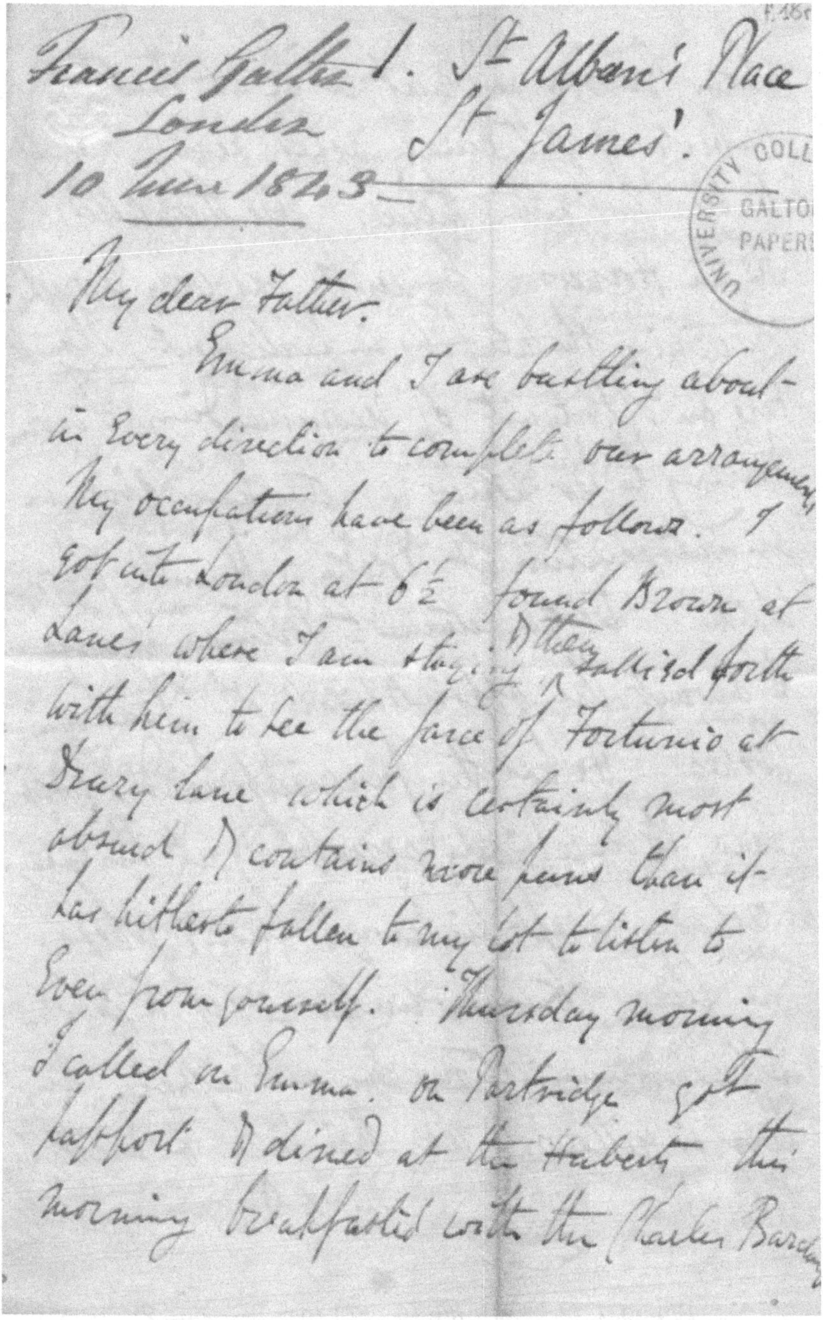
f.18r

Francis Galton 1. St Alban's Place
London St James'.
10 June 1843

My dear Father.

Emma and I are bustling about in every direction to complete our arrangements. My occupations have been as follows. I got into London at 6½ found Brown at Lane's where I am staying & then sallied forth with him to see the farce of Fortunio at Drury Lane which is certainly most absurd & contains more puns than it has hitherto fallen to my lot to listen to even from yourself. Thursday morning I called on Emma, on Partridge got passport & dined at the Harberts this morning breakfasted with the Charles Barclays

Figure 4.43: 1843-06-10. FG to STG page 1.

whom Emma had met with at the
Huberts & who had kindly asked her & myself
to breakfast with them. We then called
on the Horners. Frances H. was very energetic
concerning the late bills in parliament. I must
take an opportunity of recommending Iron
healing to her notice. they have given us
an introduction at Dresden or Munich, I forget
which. We then returned to Portman Square
to await the appointed call of Danhorn
Cookes. He has the funniest head I ever
saw, & is excessively agreable & at his ease,
nobody except his lawyer knows where
he lives. under cover to whom
all communications are addressed. his
hair is yellowish red & his face something
like this  . He told us a great deal

Figure 4.44: 1843-06-10. FG to STG page 2.

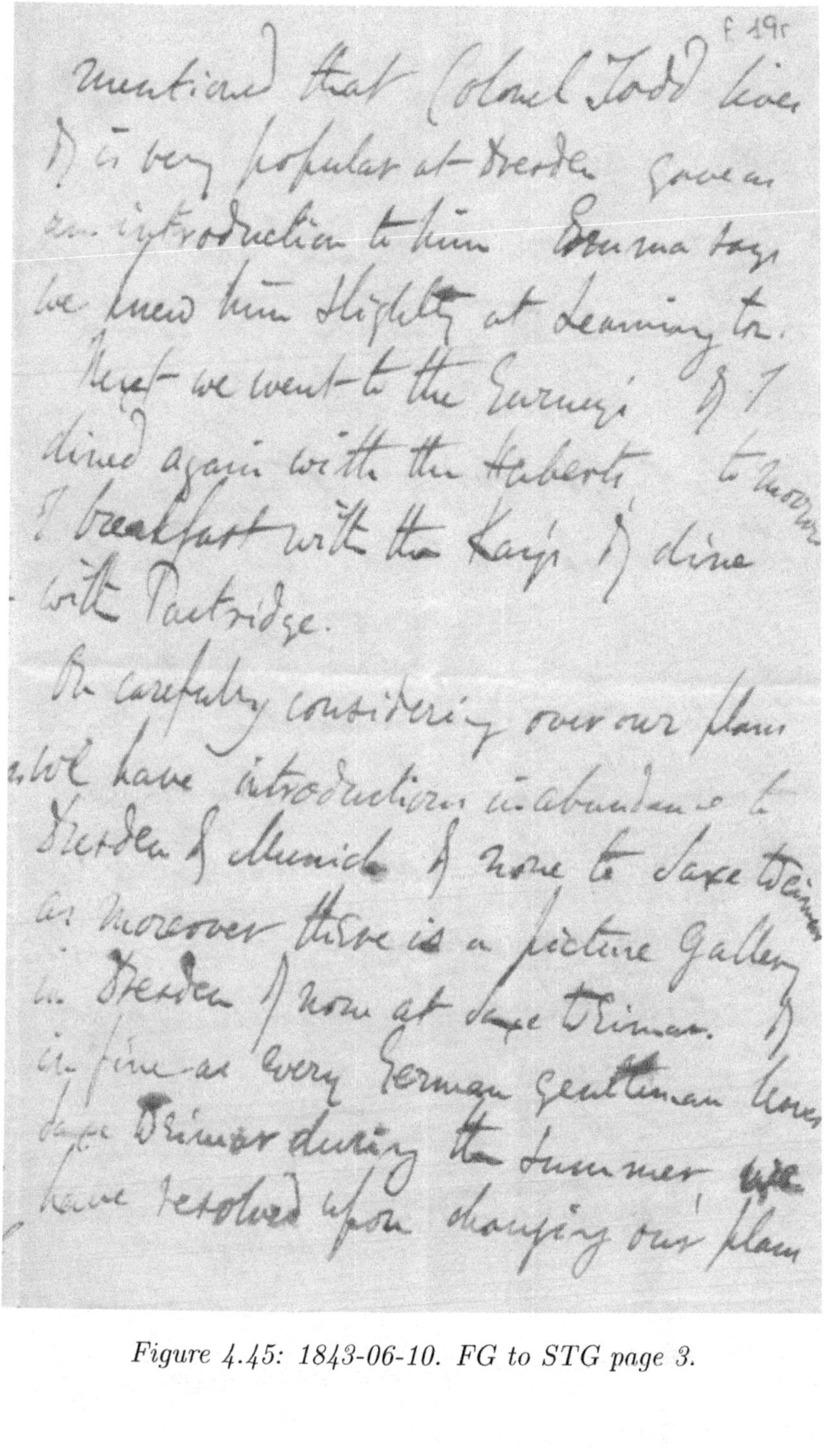

f 19r

mentioned that Colonel Todd knew
& is very popular at Dresden gave us
an introduction to him Emma says
we knew him slightly at Leamington.
Next we went to the Gurneys & I
dined again with the Herberts to morrow
I breakfast with the Kays & dine
with Partridge.
On carefully considering over our plans
we have introductions in abundance to
Dresden & Munich & none to Saxe Weimar
as moreover there is a picture Gallery
in Dresden & none at Saxe Weimar. &
in fine as every German gentleman leaves
Saxe Weimar during the summer
have resolved upon changing our plans

Figure 4.45: 1843-06-10. FG to STG page 3.

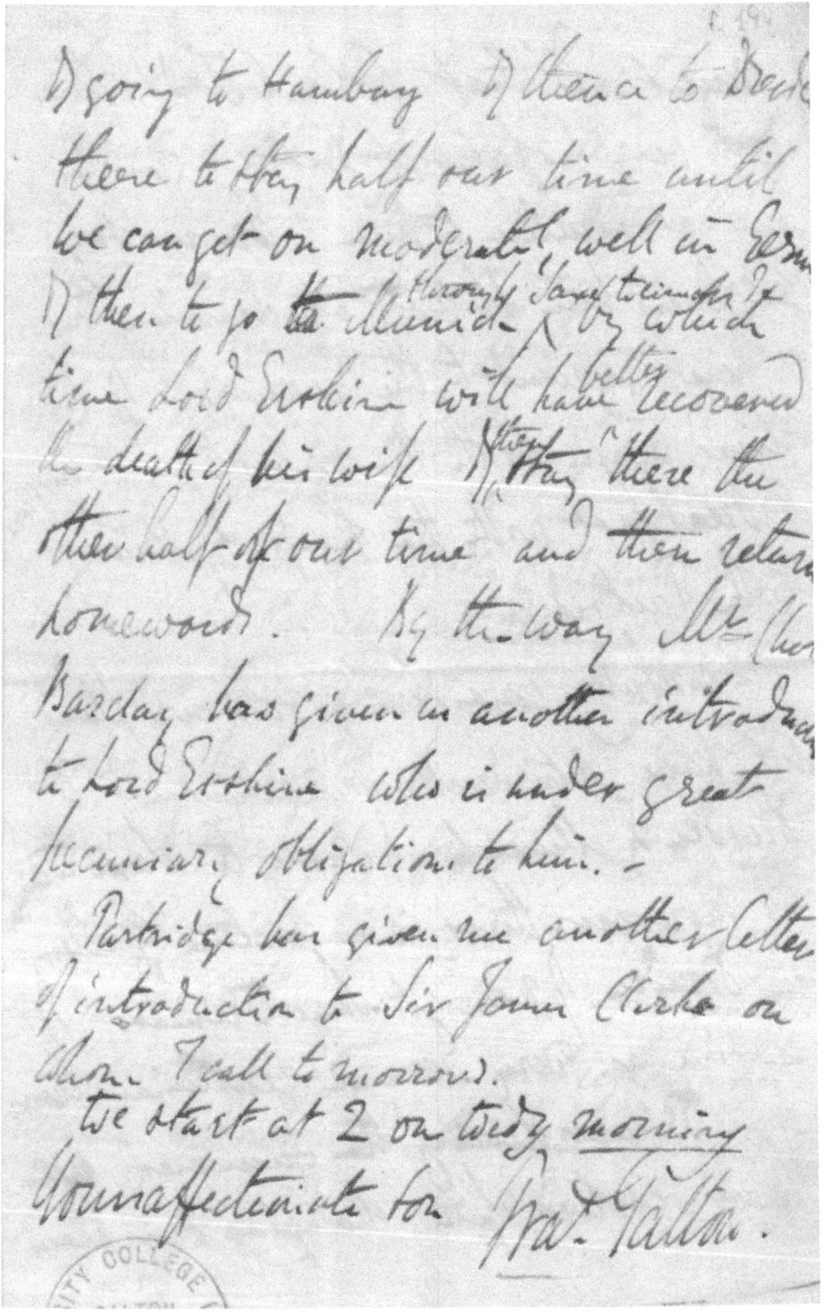

& going to Hamburg & thence to Dresden
there to stay half our time until
we can get on moderately well in German
& then to go to Munich through Saxony to Vienna & by which
time Lord Erskine will have better recovered
the death of his wife & then stay there the
other half of our time and then return
homewards. By the way Mr Chas
Barclay has given us another introduction
to Lord Erskine who is under great
pecuniary obligations to him. —
Partridge has given me another letter
of introduction to Sir James Clarke on
whom I call tomorrow.
We start at 2 on Wedy morning
Your affectionate son Fras. Galton.

Figure 4.46: 1843-06-10. FG to STG page 4.

particularly small hospital.

14. Monday.
Went with Dr. [Leis] over the blind institute, found Dr. Todd[83] & his brother Mr. Todd of Dublin in the Gallery they all 3 tea[-ed] with us.
16. Wednesday.
Wrote to Dr. Seiler to say that I had not given him earlier my letters of introduction from Dr. Hodgson[84] & from Dr. Hooker (?) of Berlin as I heard that his health was bad.[85]

1843/08/16. Francis Galton to Samuel Tertius Galton

My dear Father Mother Bessy & Delly.

We are in full preparations for leaving Dresden early the day after tomorrow. We go by steamer to Tetschen to the house of Mr. Noel[86] Mr. Woodbridge's friend, he called upon us the other day whilst I was snoozing in bed at 9, and was very good natured to Pem & asked us to stay in his house on our way to Prague. Accordingly we go to him on Thursday and stay until Saturday, which time he wrote to say would suit him. Coombe the phrenologist is I believe staying at Tetschen—at least Mr Noel came to Dresden to meet him & afterwards returned with him in the steamer home. Mrs. Noel is a Bohemian lady of very good birth sister to a Count Thun who is a great man in these parts. Her face is not comely but the bumps on her head are undeniably good, Mr. Noel himself being mad on that subject. Emma suggested that Lecky who was second in the late duel might be my old schoolfellow at Boulogne, have you any means of ascertaining whether such be the case? I do not remember him personally, though I remember passing a Sunday at Colonel Lecky's. The Hallams[87] go on Monday but we have made arrangements for meeting them two or three times on

[83] Robert Bentley Todd (1809–1860).
[84] Hodgson, Joseph. See page 483.
[85] Dr. Burkhard Wilhelm Seiler (1779–1843), who died a few months later, on September 27.
[86] Robert Ralph Noel, the expatriate phrenologist.
[87] Hallam, Henry Fitzmaurice. See page 483.

our way southward and then they will stay 10 days at Munich. Can you find out by means of Mrs. Young or by any other way what Miss Hallam's age is, as Emma and I have sore contention on that subject, Emma declaring that she is not more than 21 and I maintaining her to be older.[88] We have no way of finding it out here as Pemmy says that it is not etiquette for ladies to ask each other their ages. I am glad that Miss Hawkes and Nurse Ricketts look gracious on young Gumbly, but in which ward is he? Nurse S. used to be a great friend of mine & had the nicest [w...] but she ran away with and married a patient. There are scarcely any medical institutions here but doctors abound. I have been over what there are, but they are exceedingly small & close and abominable. Yesterday as I was walking in the picture gallery who should I meet but Dr Todd & his brother who looked like Mr [Brew] turned dissenting minister, we had a long talk over King's College matters & he had tea with us afterwards & I got a Dresden Doctor to meet him. He went off this morning southwards. Atkinson & Rhodes have already gone 4 days & Oakes went this morning in the same diligence with Dr Todd, so we are utterly inconsolable. Mr. Beech of Leamington was in the picture gallery this morning where Emma recognised him, his altitude is very conspicuous in Germany, where the average run of Gentlemen come up to Emma's shoulder. He talks about buying copies of some of the pictures here. We had a very agreeable dinner party at a Mr Smith who we discovered knew Uncle & Aunt Howard of whom we had had a letter from Mr [Deveux]. They live about 14 miles off in German style, manage a small farm with two donkeys & certain geese. They have also some corn which had just been cut when we were there & in consequence the labourers had presented the usual harvest [home] wreath of flowers worked with ribbons & spangled with gold leaf which is hung up in the hall until the ensuing harvest. After dinner (at 2) we adjourned to the balcony on the rails of which the gentlemen sat smoking cigars [the] while they talked to the ladies. The gentlemen do not sit round the table after the ladies go out but they walk in to the Drawing Room arm & arm together just as the left it. I am very glad Mrs. Phillips enjoyed her Irish Tour so much but in moving about so unprotected from place to place she must have run great risk from the susceptible Irishmen—perhaps she imitated the conduct of the lady in Moore's

[88] Hallam, Julia Maria Frances, see page 483. She was 25 in 1843.

Figure 4.47: 1843-08-16 FG to STG: Carrier pigeon bearing portrait.

Irish Melodies whom Darwin used to be so fond of quoting & who repelled with indignation the knight who said

Lady dost thou not fear to stray
So lone & lovely in this bleak way
Are the sons of Erin so good or so cold
As not to tempted by woman or gold?

I can fancy her making the reply according to the continuation of the song "Sir Knight, I feel not the least alarm &c."

I am very much amused by your description of Mrs P's pictures and of their youthful appearance. What a happy thing it is that now-a-days courtship does not proceed in the chivalrous manner, when a carrier pigeon brought the lady's picture & the gentleman who had not see the original straightaway fell in love as there might be some misunderstanding in the present case. [See Figure 4.47, p. 313.]

I am very glad that Douglas is so much better. Everybody talks of the Gibraltar fever as so common & so dangerous among the troops stationed there at present. I hope he will quite get over the effects of it.

Good bye your very affectionate son, brother
Fras. Galton.

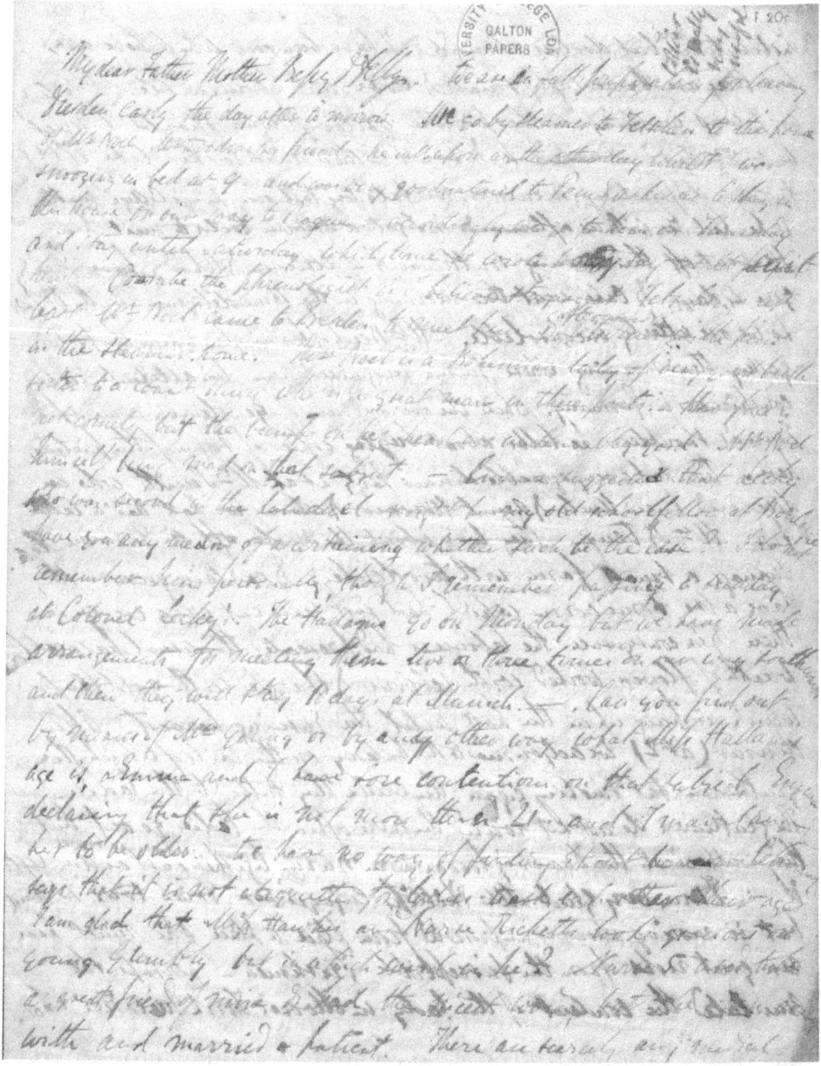

Figure 4.48: 1843-08-16. FG to STG page 1.

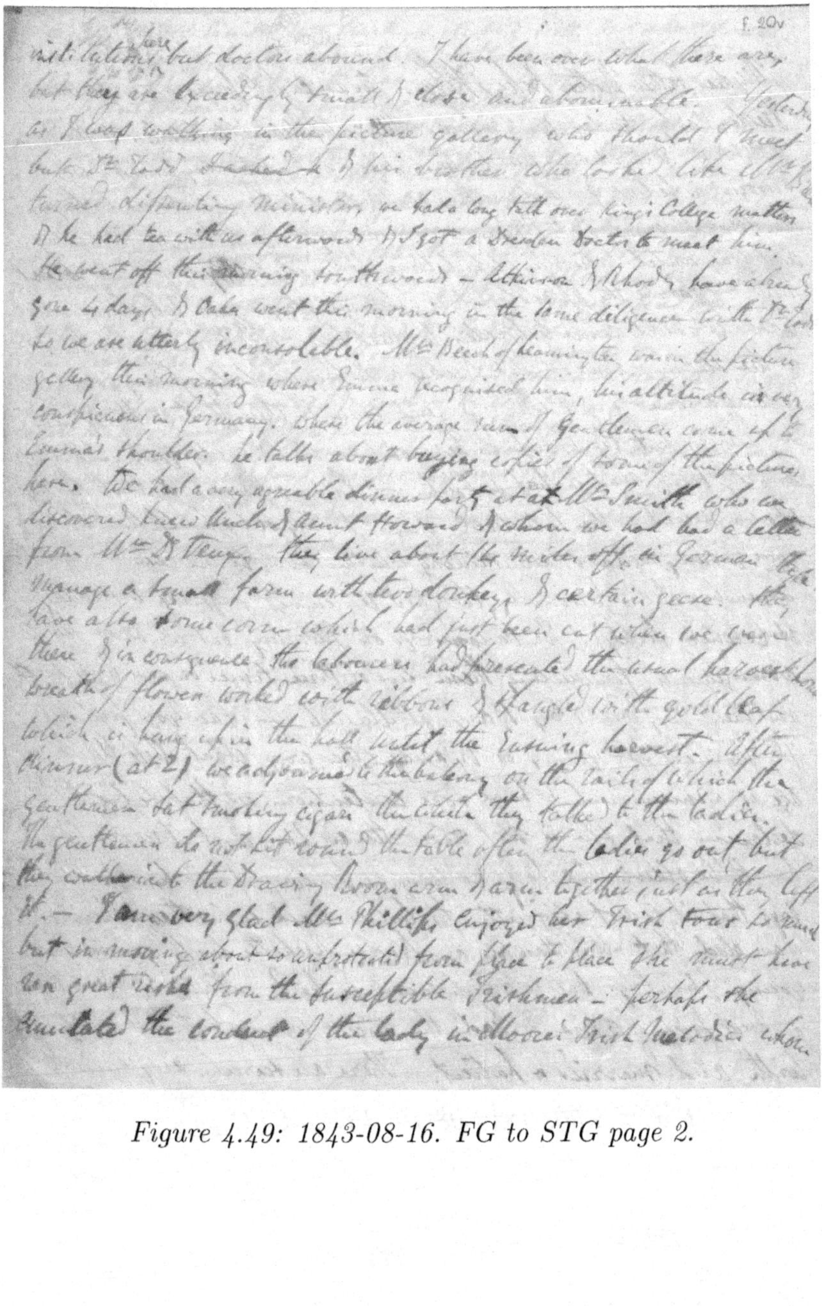

f. 20v

institutions here but doctors abound. I have been over what there are, but they are exceedingly small & close and abominable. Yesterday as I was walking in the picture gallery who should I meet but Dr Todd [illegible] & his brother who looks like Mr B turned Dissenting minister, we had a long talk over King's College matters & he had tea with us afterwards & I got a Dresden Doctor to meet him. He went off this morning southward — Atkinson & Rhodes have already gone 4 days & Oake went this morning in the same diligence with Dr Todd so we are utterly inconsolable. Mr Beech of Leamington was in the picture gallery this morning where Emma recognised him, his altitude is very conspicuous in Germany, where the average run of Gentlemen come up to Emma's shoulder. he talks about buying copies of some of the pictures here. We had a very agreeable dinner party at Mr Smith who we discovered knew Uncle & Aunt Howard & whom we had had a letter from Mr D Teague. they live about the miles off in German they manage a small farm with two donkeys & certain geese. they have also some corn which had just been cut when we were there & in consequence the labourers had presented the usual harvest wreath of flowers worked with ribbons & spangled with gold leaf which is hung up in the hall until the ensuing harvest. After dinner (at 2) we adjourned to the balcony on the rails of which the gentlemen sat smoking cigars while they talked to the ladies. The gentlemen do not sit round the table after the ladies go out but they walked into the Drawing Room arm & arm together just as they left it. — I am very glad Mrs Phillips enjoyed her Irish tour so much but in moving about so unprotected from place to place she must have run great risks from the susceptible Irishmen — perhaps she emulated the conduct of the lady in Moore's Irish Melodies whom

Figure 4.49: 1843-08-16. FG to STG page 2.

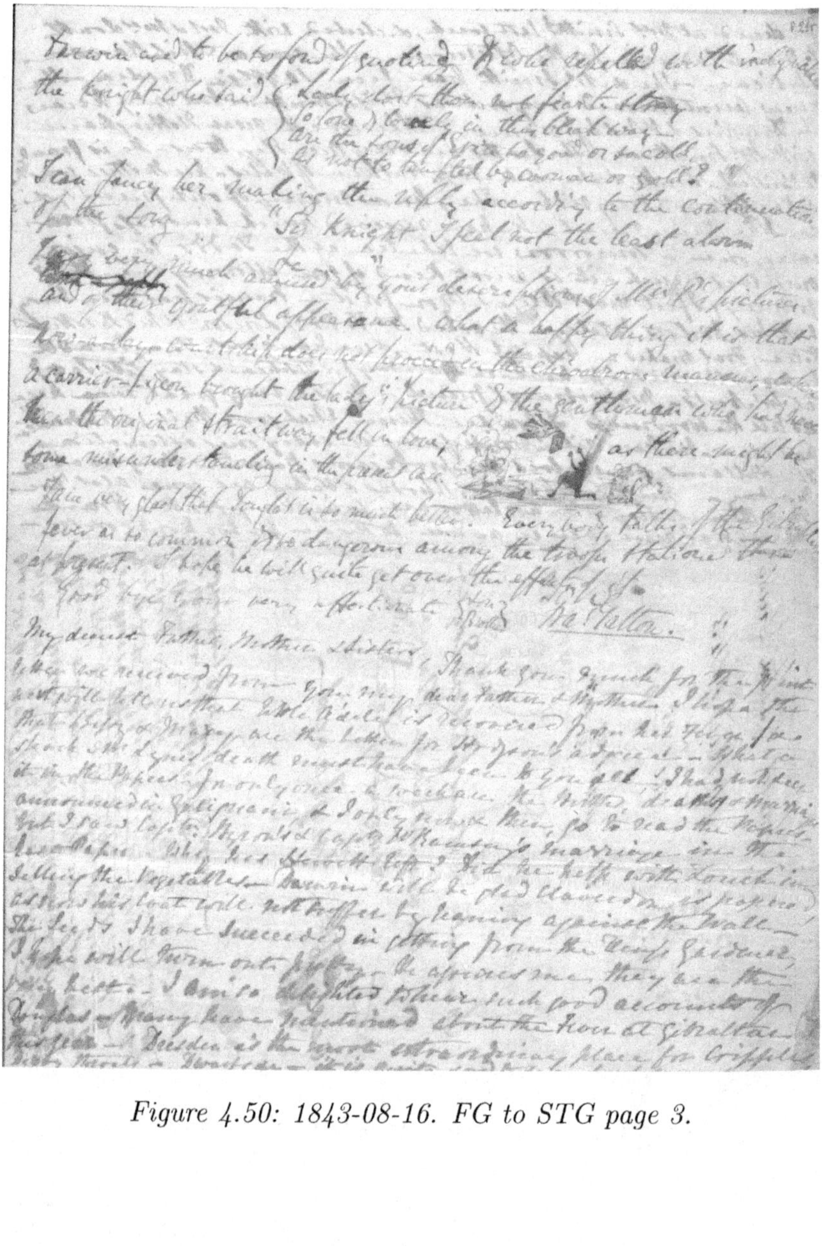

Figure 4.50: 1843-08-16. FG to STG page 3.

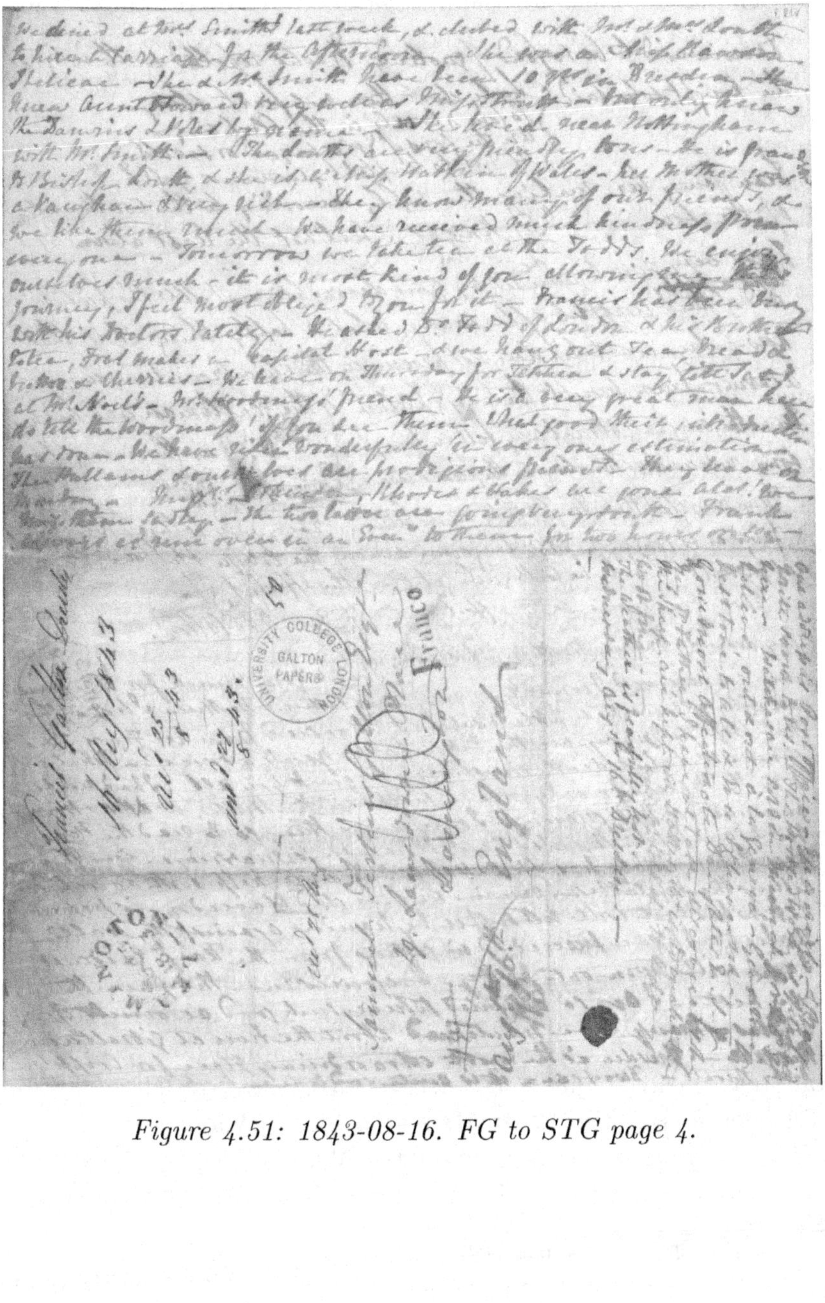

Figure 4.51: 1843-08-16. FG to STG page 4.

1843/08/16. Emma Sophia Galton to Samuel Tertius Galton

My dearest Father, Mother & sisters,

Thank you much for the first letter we received from you, my dear Father and Mother. I hope the rest will tell us that little Adèle is recovered from her fever & that Bessy & Mary are the better for Hodgson's advice.[89] What a shock Mr Lynes' death must have been to you all. I had not seen it in the papers for only once a week are the Births Deaths & Marriages announced in Galignani[90] & I only now and then go to read the papers but I saw Capt Byron's & Capt W. Ramsay's marriage in the last paper. Why has Hewitt left? Did he help with [Louch] in selling the vegetables. Darwin will be glad Claverdon is papered as now his coat will not suffer by leaning against the wall. The seeds I have succeeded in getting from the King's Gardens I hope will turn out pretty. He assures me they are the very best. I am so delighted to hear such good accounts of Douglas. Many have mentioned about the fever at Gibraltar this year. Dresden is the most extraordinary place for cripples [Dirty] throats, Dwarfs &c. &c. it is quite sad to see such numbers.

We dined at Mrs Smith's last week & clubbed with Mr & Mrs Louth to hire a carriage for the afternoon. She was a Miss Rawson I believe. She & Mr. Smith have been 10 yrs in Dresden. She knew Aunt Howard very well as Miss Strutt – but only knew the Darwins & Poles by name. She lived near Nottingham with Mr Smith. The Louths are very friendly to us. He is grandson to Bishop Louth & she is a Miss Watkins of Wales. He mother was a Vaughan & very rich. They know many of our friends & we like them much. We have received much kindness from everyone. Tomorrow we take tea at the Todds. We enjoy ourselves much—it is most kind of you allowing me this journey, I feel most obliged to you for it. Francis has been busy with his doctors lately. He asked Dr Todd of London his brother to tea. Fras. makes a capital host & we hand out tea, bread & butter & cherries. We leave on Thursday for Tetschen & stay till Sat at Mr Noel's—Mr. [Woodward's] friend. He is a very great man here do tell the [Woodwards] if you see them what good their introduction has done. We have risen wonderfully

---

89 Hodgson, Joseph. See page 483.

90 *Galignani's Messenger*, a progressive newspaper published on the continent.

in everyone's estimation. The Hallams & ourselves are prodigious friends. They leave on Monday. Messrs. Atkinson, Rhodes & [Blakes] are gone alas! We miss them sadly. The two latter are going very south. Frank always c^d run over in an even. to them for two hours or so.

Our address is Post Office Munich.[91] Pray write soon to us. The Worsted Work is splendid here. Watches are sold in Munich for £6 I believe, out & out à la Geneva. Give my best love to all & ever believe me Your most affectionate daughter. E.S.G.

My dr. Father, I have at last discovered why the sheep are kept in stables, it is, to make the wool fine. I saw Mr Beech today. The weather is fearfully hot.

Wednesday Augst. 16th Dresden.

1843/09/11–12. FG Diary

[September]

11. Monday.
Called on Mons. Le Sage who is secretary
to the Duke du Leuchtenberg[92] with
a note from Mr. Noel of Rosawitz.[93]
he was lame & said he would send
his brother in law to lionise me.

12. Tuesday
Captain N. Neumayer
Schönfield Gasse No 4 called.

1843/10/14. Dr. Henry Jephson to FG

Dr. Jephson will have much pleasure in Dining with Mr. Galton on Saturday the 14th Inst. at 6 o'clock.

Beech Lawn Saturday Evg.[94]

---

[91] Cross-written.

[92] Maximilian de Beauharnais (1817-1852).

[93] Robert Ralph Noel.

[94] Pencil annotation: 1843. The 14th only fell upon a Saturday in January and October that year. We know (see below) that Galton saw the Jephsons the weekend that he returned from his tour to the continent, hence the dating.

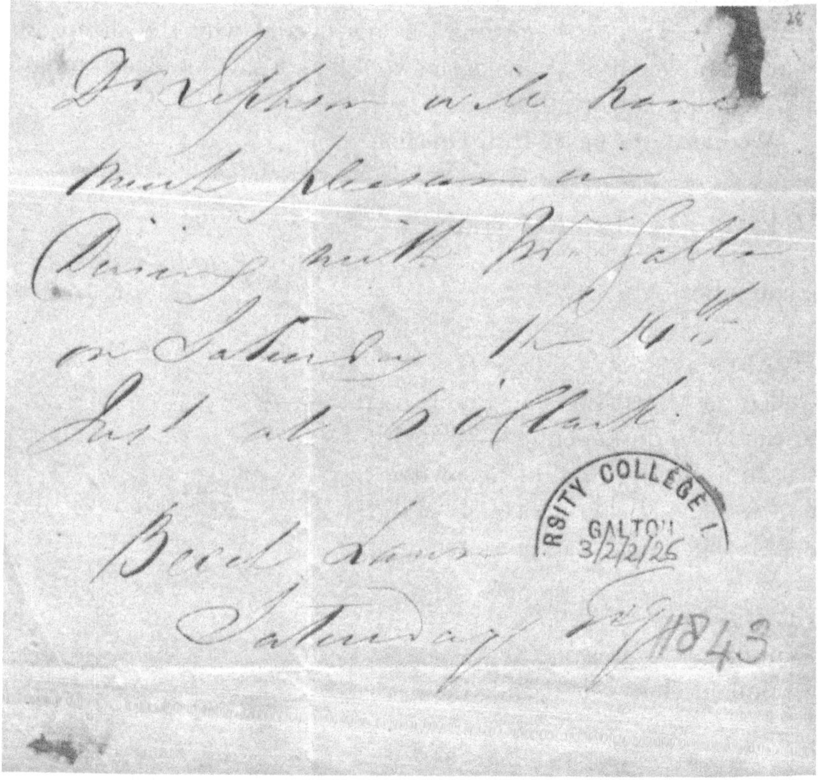

Dr Jephson will have much pleasure in Dining with Mr Galton on Saturday the 14th Inst at 6 o'clock.

Bird Lane

Saturday Oct 1843

Figure 4.52: 1843-10-14. Dr Henry Jephson to FG.

1843/10/14–20. Diary of Samuel Tertius Galton

October 14th. Saturday. Leamington.

... Emma returned at about 4 p.m. from her continental tour having had a narrow escape from a stormy passage. She and Francis embarked from Ostend on Thursday 1 a.m. and did not reach London till Friday at 2 p.m., being a passage of 37 hours; they were driven upon a sand bank and between some rocks in the dark. ...

October 16th. Monday. Leamington.

... Bessy and Francis arrived from Selly Hall. ... Francis gave Mrs. Jephson a present from Munich.[95] ...

October 17th. Tuesday. Leamington.

Much rain in the night. Violetta, Emma, Francis and myself went to Claverdon. ...

October 20th. Friday. Leamington.

Francis left us for Cambridge at 9 a.m. ...

1843/10/20. Francis Galton to Samuel Tertius Galton

My dear Father

The best thing in the world that you can possibly do is to come up on Tuesday by the Eagle with Emma & Bessy & to take up your quarters with me & see her majesty with Cantabs in all their glory. You must come, Emma can sleep in my bed. You in my sitting room, and I where I can in some other Cantab's room, then the sitting room will be cleaned at breakfast &c. & we can see every thing—this is the worst view of your accommodation, it is possible that I can get the loan of a set of rooms, but as a supper party is waiting for me I have not time to explain.

I should have some difficulty in escorting Emma & Bessy myself alone, but if you cannot come I will do all I can if you can send one of [the] sisters by themselves. Perhaps Darwin would like the jaunt if it is too much for you. If so I hope he will come.

Unless there is a gentleman to escort the lady besides myself I am afraid that a great deal will be left unseen by both as the ladies

---

[95] Wife of Jephson, Dr. Henry, see page 485.

& the undergraduates are often separated & the lady must remain under escort & so neither can see the [sights]. But remember if neither you nor Darwin can come Pemmy must as it is better she should see something than nothing.

Your affectionate Son
F. G.

Please write by return of post.

1843/10/30. Francis Galton to Samuel Tertius Galton

Trin Coll
Monday Morning

My dear Father

To the best of my belief Cheshire would make a very fit surgeon to the Eye Infirmary.

Do you know Mr. Cheshire well? His note seems rather to demand than to petition your vote.

I write in haste to secure the morning post & will write again this evening. I have much to say about medicals.

Your very affectionate Son
Fras. Galton

All the articles were left purposely except the watch key, please ask Mater to keep them all in custody—how is her face?

1843/10/31. Francis Galton to Samuel Tertius Galton

Trin Coll

My dear Father,

I have been talking with Dr Haviland about the lectures I have to attend. He tells me that if I put a "spurt" on and go to 4 lectures a day that I shall be able to finish with Cambridge by the end of next term, he absolves me from hospital practice in Cambridge & accordingly I shall be able to practice (if I like) next June 2 years. I have without hesitation adopted his plan. I must pass an examination by each of the Professors separately & the examination takes place at the close of their lectures. Whether I shall feel myself strong enough to go in for all four or any of them

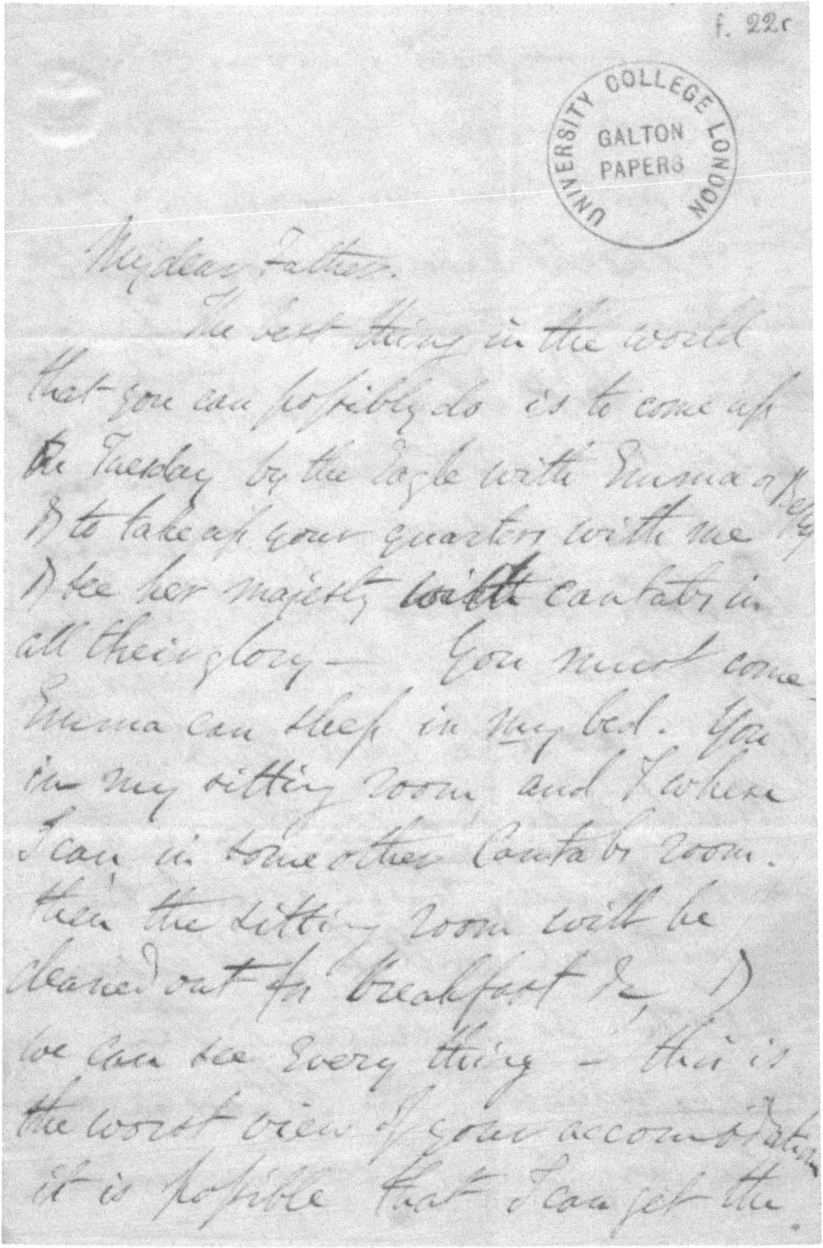
f. 22c

UNIVERSITY COLLEGE LONDON GALTON PAPERS

My dear Father.
The best thing in the world that you can possibly do is to come up on Tuesday by the Eagle with Emma & Bes & to take up your quarters with me & see her majesty with Cantabs in all their glory— You must come. Emma can sleep in my bed. You in my sitting room and I where I can in some other Cantab room. then the sitting room will be cleared out for breakfast &c, & we can see every thing — this is the worst view of your accommodation it is possible that I can get the

Figure 4.53: 1843-10-20. FG to STG page 1.

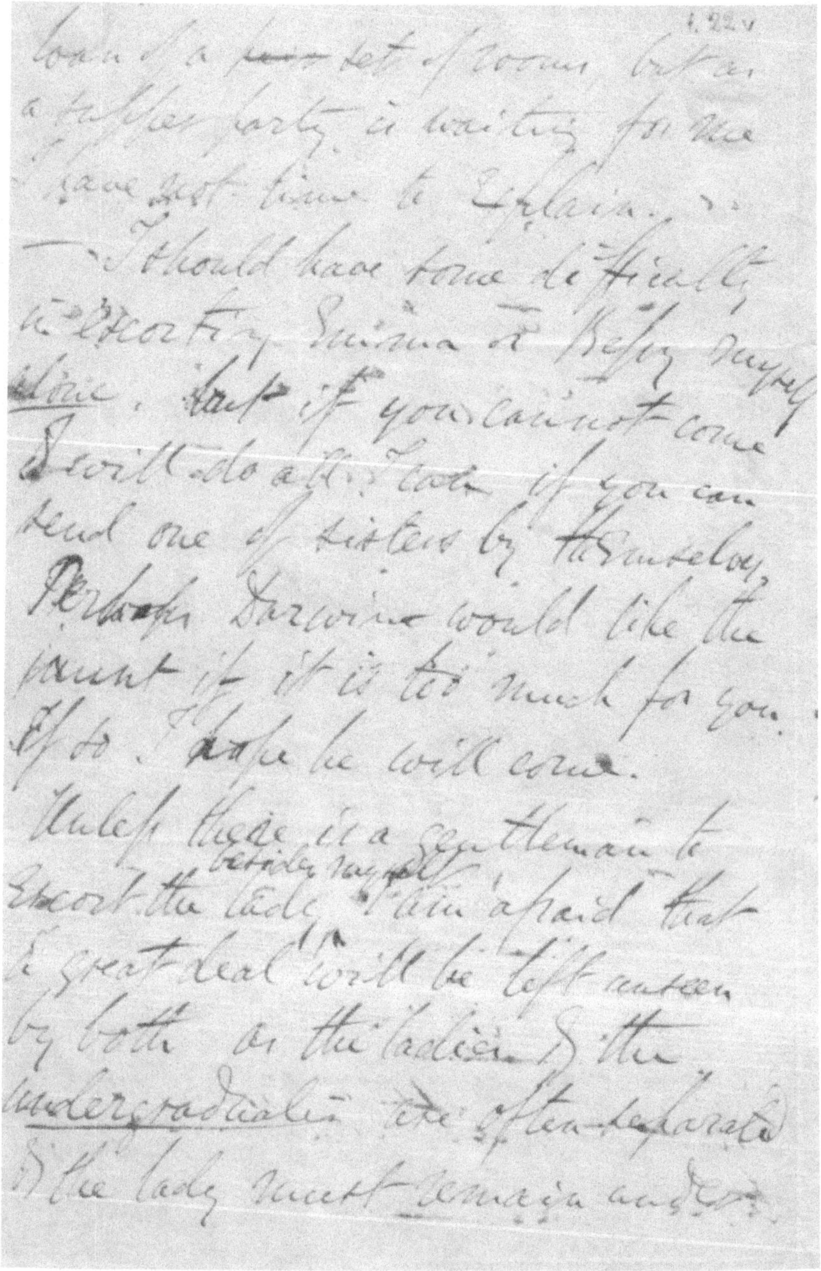

f. 22 v

board of a ~~pair~~ set of rooms, but as
a tea party is waiting for me
I have not time to explain. —
I should have some difficulty
in escorting Emma & Betsy myself
alone. but if you cannot come
I will do all I can. if you can
send one of sisters by themselves,
Perhaps Darwin would like the
jaunt if it is too much for you.
If so I hope he will come.
Unless there is a gentleman to
escort the lady besides myself I am afraid that
a great deal will be left unseen
by both or the ladies & the
undergraduates are often separated
& the lady must remain under

Figure 4.54: 1843-10-20. FG to STG page 2.

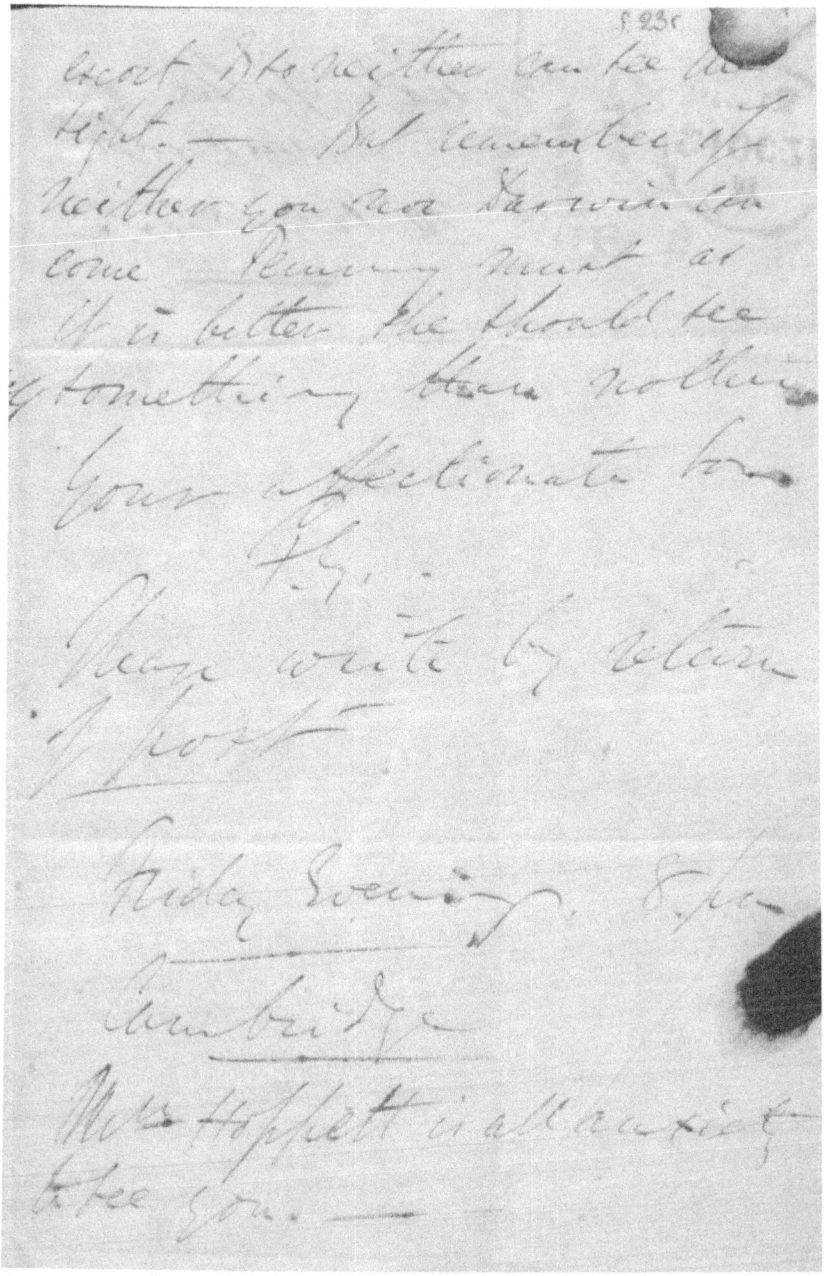
least 3 to neither can see the
light. — But remember if
neither you nor Darwin can
come Jemmy must as
it is better she should see
something than nothing
Your affectionate son
F.G.
Please write by return
of post
Friday Evening. 8 pm
Cambridge
Mrs Hoppett is all anxiety
to see you. —

Figure 4.55: 1843-10-20. FG to STG page 3.

Figure 4.56: 1843-10-20. FG to STG page 4.

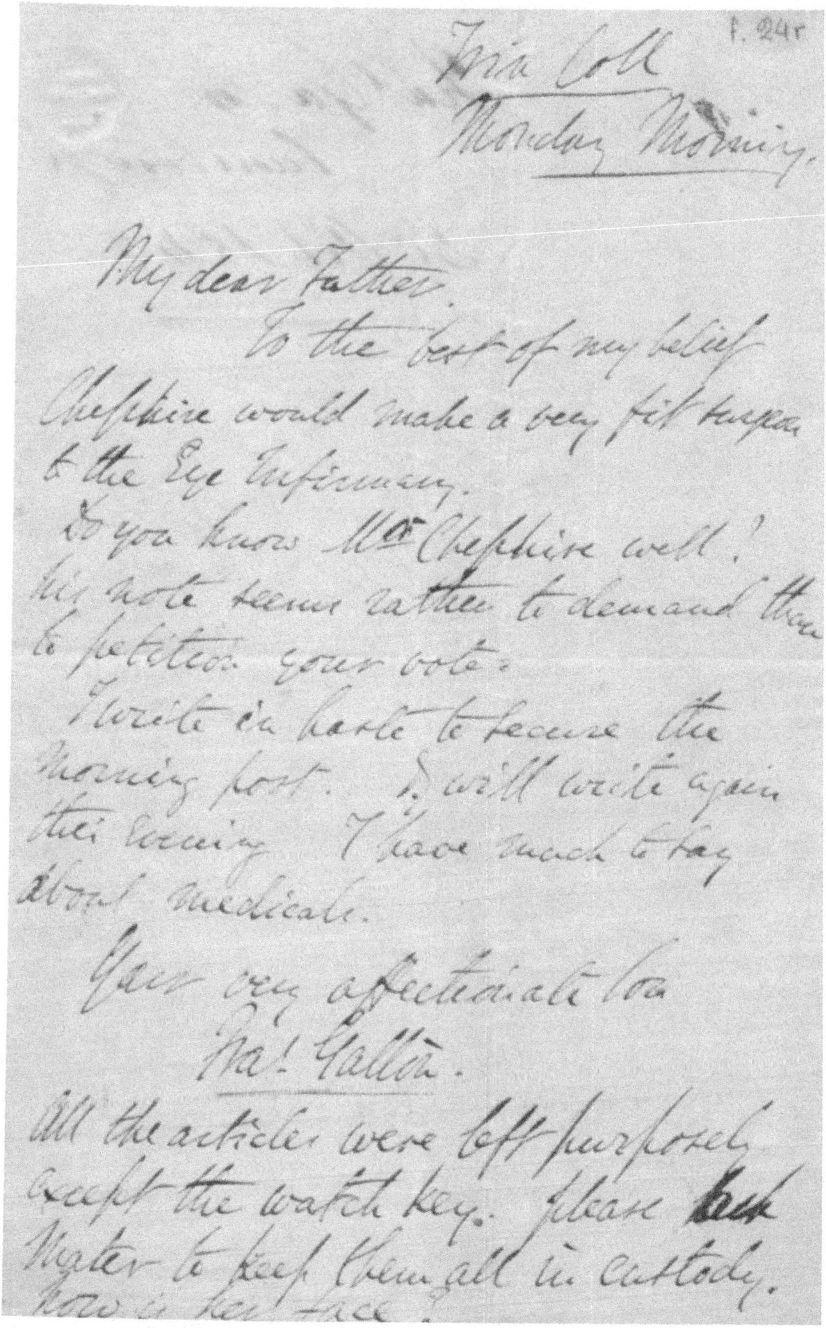

Trin Coll
Monday Morning.

My dear Father.
To the best of my belief Chesshire would make a very fit surgeon to the Eye Infirmary.
Do you know Mr Chesshire well? his note seems rather to demand than to petition your vote.
I write in haste to secure the morning post. I will write again this evening. I have much to say about medicals.
Your very affectionate Son
Fra: Galton.
All the articles were left purposely except the watch key. Please ask Mater to keep them all in custody. How is her face?

Figure 4.57: 1843-10-30. FG to STG page 1.

Figure 4.58: 1843-10-30. FG to STG page 2.

at the end of the course I attend, I do not know, if not I shall do so next year. I am working ferociously at present. First for my degree & secondly on the 4 different subjects Anatomy Practice of Medicine Chemistry & Forensic Medicine. I did at one time know these same subjects well enough but they slip in an extraordinary degree out of the mind. I am writing in the Union & waiting for some motions to come on that I was pledged last term to open. I wrote you a note by this morning's post. My gown was not among the number thrown down for the Queen to walk upon & caught up before the maids of honor (bless their pretty feet) could do so, as I had no inclination to assimilate my loyalty to that of the aldermen of Southampton.

Goodbye your affectionate son
Fras. Galton.

I have called on & hung out to Ingelby.[96] He is not very polished but does ...

I open my note because I was detained speechifying till too late for the post.

If you have occasion to set up in Claverdon or elsewhere any new cupboards, I have just seen the jolliest dodge imaginable to supersede the old plan of bolting the left door into which the [diagram] right door locks & which it is always a bore to open breaking finger nails & very troublesome. The plan is this. Into the wood work above and below the meeting of the doors an affair of this sort is placed [diagrams] A rod AB moving on a centre C pressed upward on the side B by a spring S, & carries two projections, one at either end of which the right hand one is rounded. Hence when the right door is shut it presses down B & the projection at A sticks up & when both the doors are shut together the projection at A both above and below stick into holes at the top & bottom of the left hand door & in fact bolt it.

It acts perfectly is very cheap, scarcely any more expensive than the ordinary bolts & is more secure as the latches[?] have been lifted up by boring a very small hole through the wood work & working them up with an instrument & afterwards filling up the hole with colored putty—these cannot be picked as the bolts cannot be depressed until the right hand door is unlocked & opened.

---

96 The solicitor Clement Ingleby of Ingleby Wragge and Cope, Bennet's Hill, Birmingham.

Do give me an order to get you a pair if you have any more cupboards.

Poor Pemmy and her face.

1843/10/31. FG Diary

31. Tuesday.
Sat after hall in F. Buxton's rooms.
Wined also with Hutton.

1843/11/01–1843/11/30. FG Diary

1. Wednesday.
Dine with Pinder.

2. Thursday.
Breakfast Law.
Epigram Club in Bristed's rooms.[97]

3. Friday.
Breakfast in my rooms.
Mrs. Hopkins Ev. party $8\frac{1}{4}$[98]
Mrs. Hopkins ill $\therefore$ not go.
Historical Society met first
time in Black Bear.[99]

4. Saturday.
Breakfast in my rooms.
Put up placards every where for
"Aid to the National Fund" by the Bachelors
and Undergrads of Cambridge. Joe Kay[100]
and I concocted the plan late last night.
Men talked favourably of it.
C. Buxton chatted in my rooms eating
pie—discussion on nature of beauty.

---

97 Bristed, Charles Astor. See page 482.
98 Hopkins, William. See page 484.
99 A Public House in the town.
100 Kay, Joseph. See page 486.

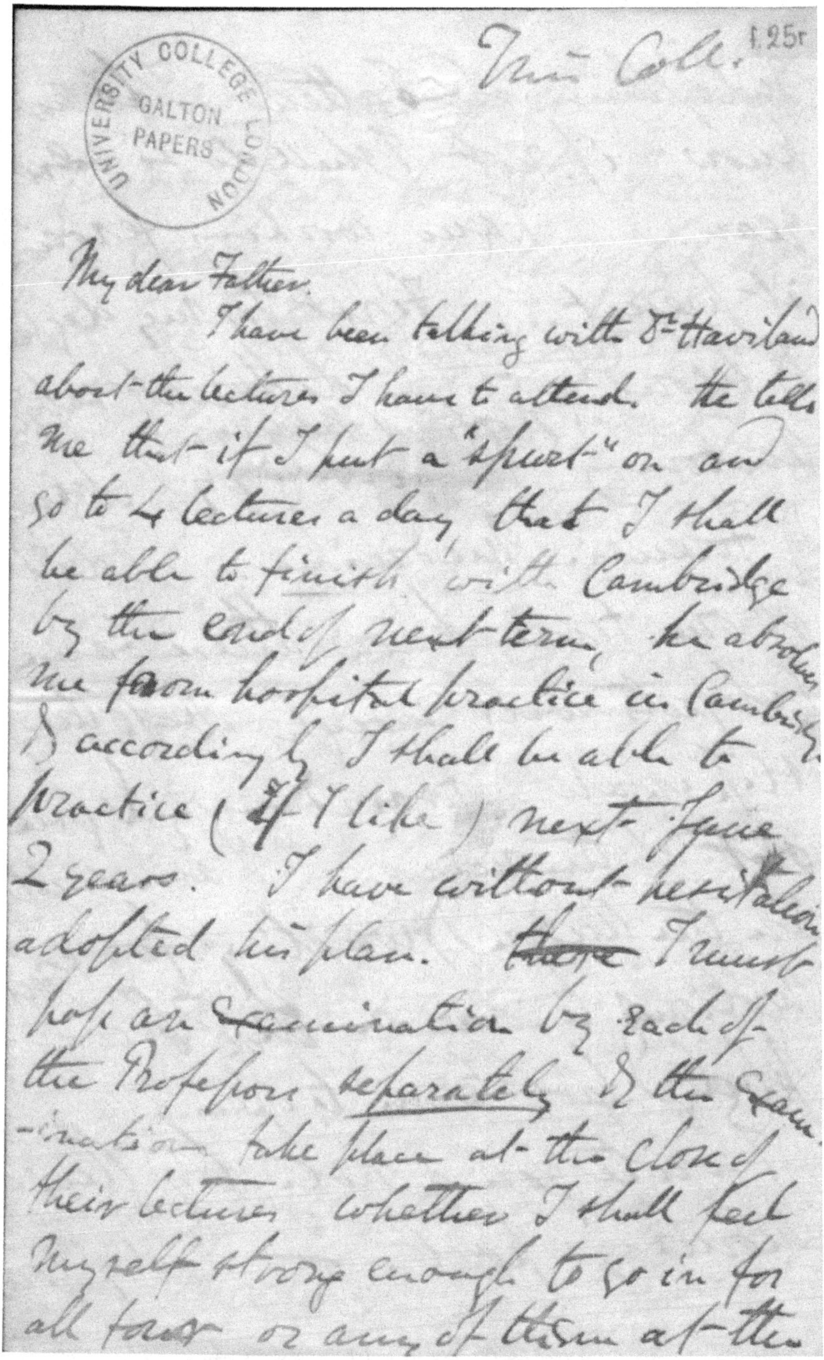

UNIVERSITY COLLEGE LONDON GALTON PAPERS

Trin: Coll. 125r

My dear Father.

I have been talking with Dr Haviland about the lectures I have to attend. He tells me that if I put a "spurt" on and go to 4 lectures a day that I shall be able to finish with Cambridge by the end of next term, he absolves me from hospital practice in Cambridge & accordingly I shall be able to practice (if I like) next June 2 years. I have without hesitation adopted his plan. ~~there~~ I must pass an examination by each of the Professors separately & the examinations take place at the close of their lectures whether I shall feel myself strong enough to go in for all four or any of them at the

Figure 4.59: 1843-10-31. FG to STG page 1.

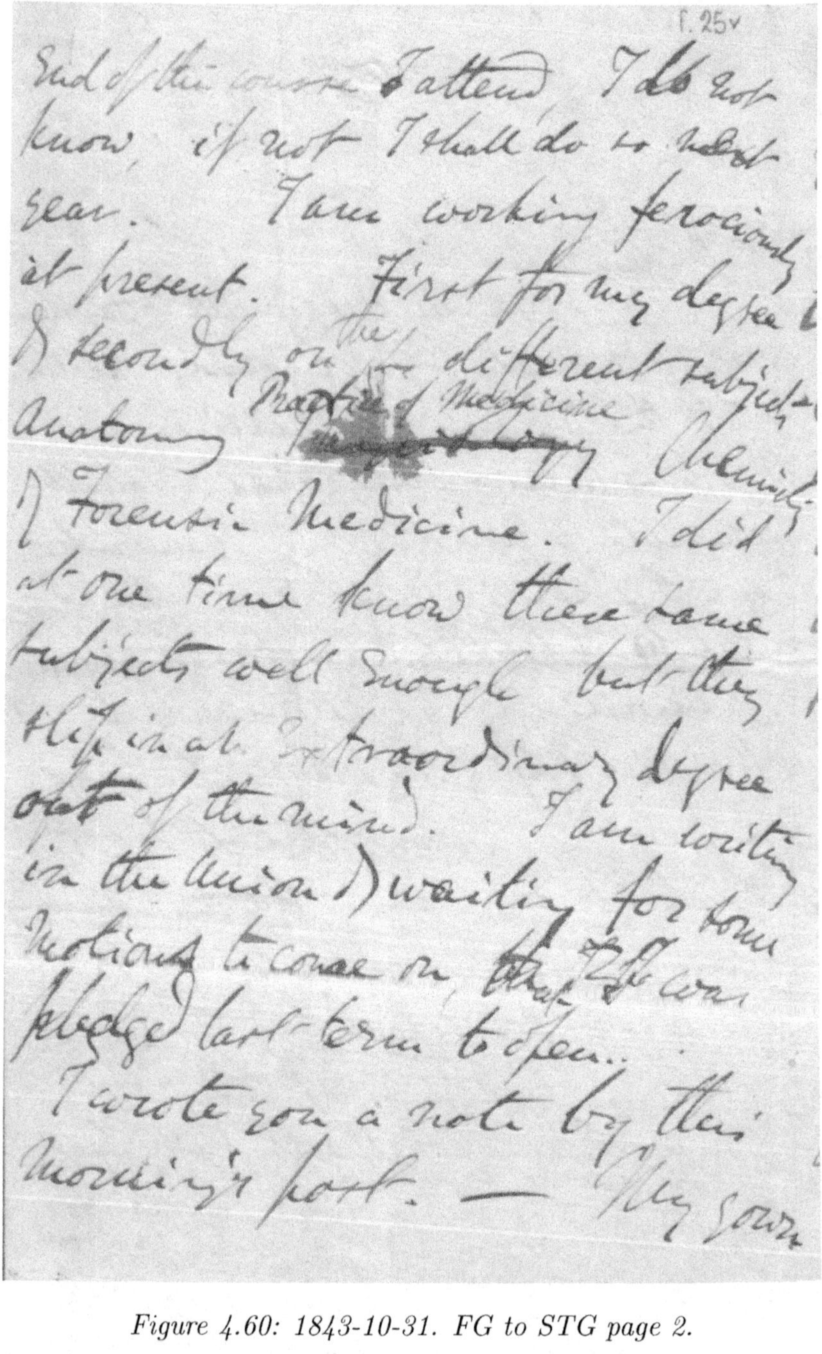

f. 25v

End of the course I attend I do not know, if not I shall do so next year. I am working ferociously at present. First for my degree & secondly on the different subjects Anatomy, Practice of Medicine ~~Physiology~~ Chemistry & Forensic Medicine. I did at one time know these same subjects well enough but they slip in an extraordinary degree out of the mind. I am writing in the Union & waiting for some motions to come on that I was pledged last term to open.. I wrote you a note by this morning's post. — Ever yours

Figure 4.60: 1843-10-31. FG to STG page 2.

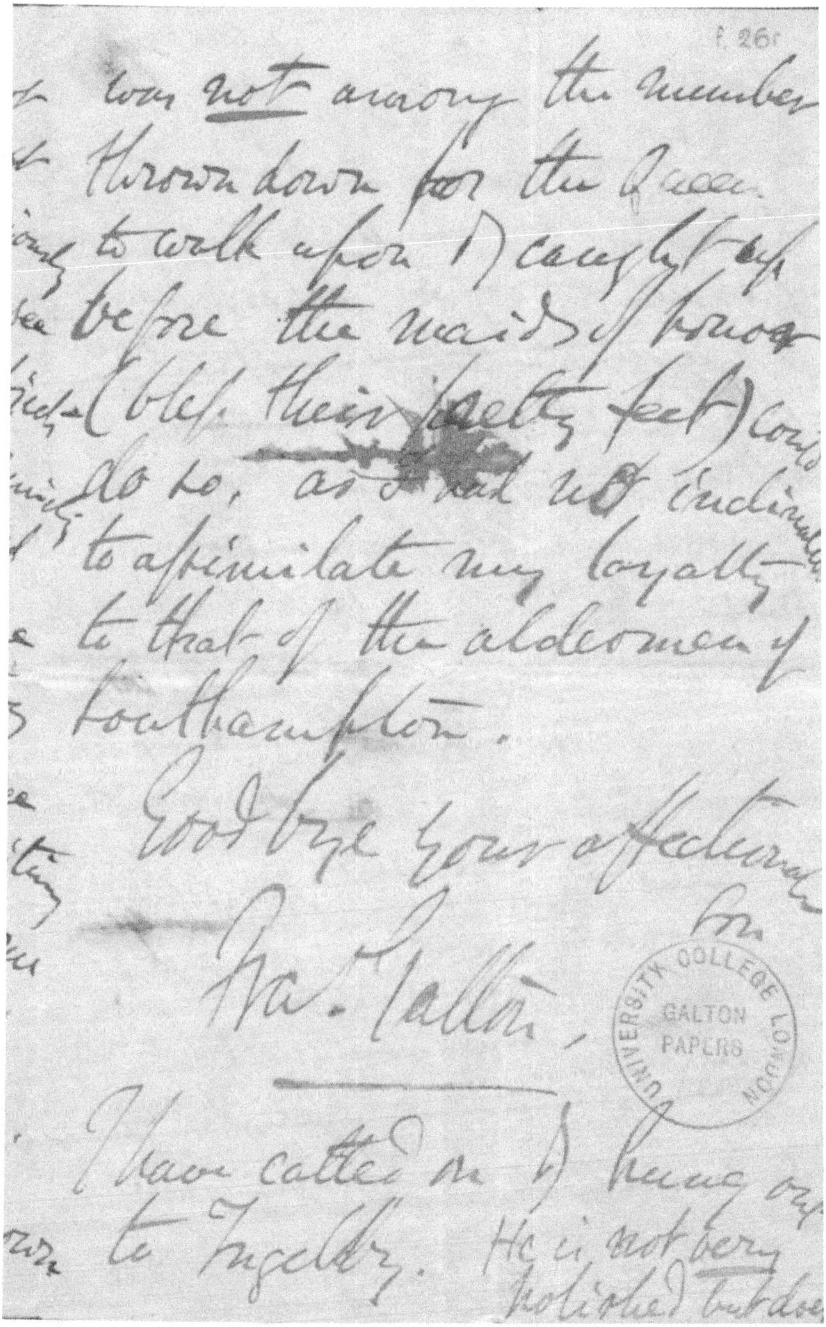

was not among the number
thrown down for the Queen
to walk upon & caught up
before the maids of honour
(bless their pretty feet) could
do so, as I had not inclination
to assimilate my loyalty
to that of the aldermen of
Southampton.
Good bye your affectionate
son
Fras. Galton.

I have called on [illegible]
to Ingelby. He is not very
polished but does

Figure 4.61: 1843-10-31. FG to STG page 3.

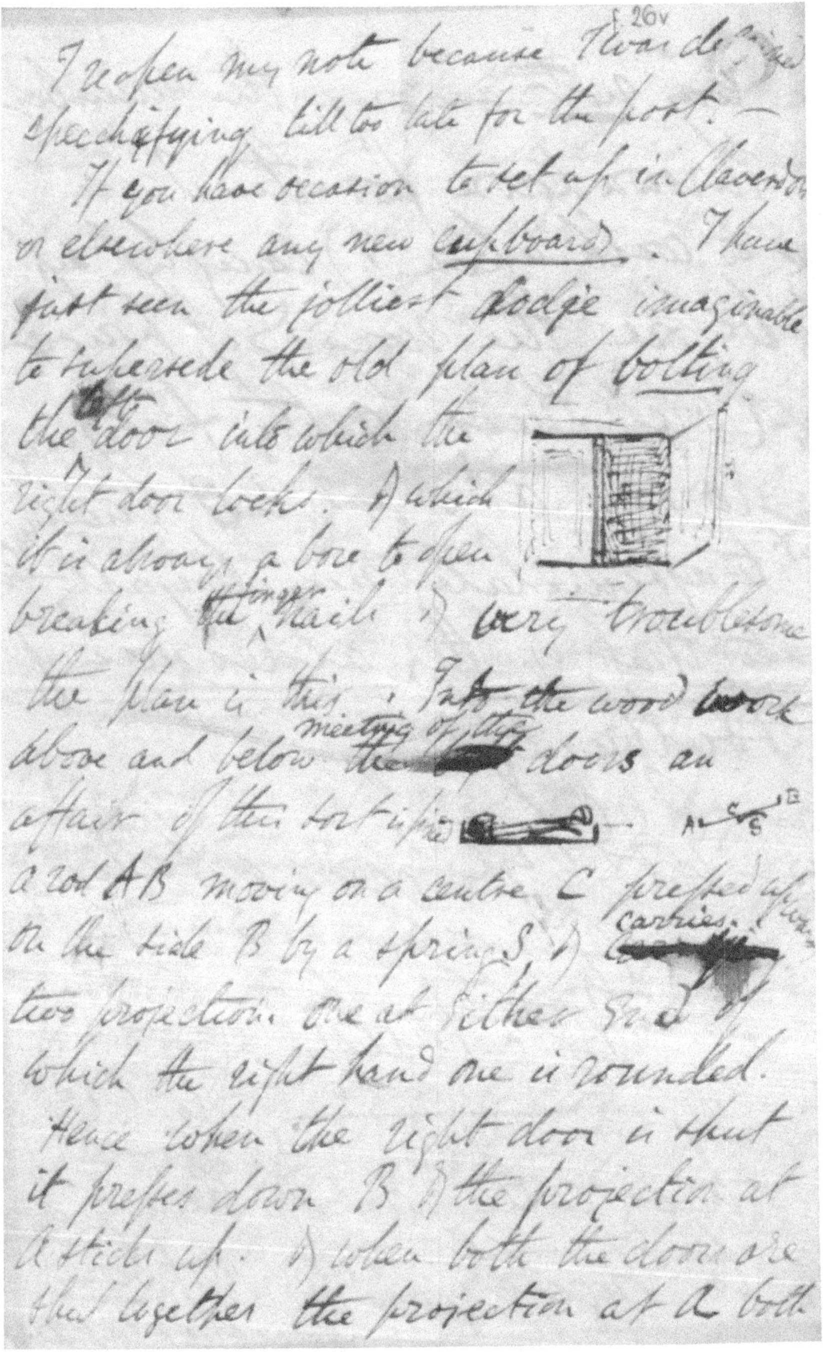

f. 26v

I reopen my note because I was de[tained] speechifying till too late for the post.—

If you have occasion to set up in Clarendon or elsewhere any new cupboard. I have just seen the jolliest dodge imaginable to supersede the old plan of bolting the left door into which the right door locks. & which it is always a bore to open breaking the finger nails & very troublesome the plan is this. Into the wood work above and below the meeting of the doors an affair of this sort is fixed

a rod AB moving on a centre C pressed upwards on the side B by a spring S & carries two projections one at either end of which the right hand one is rounded. Hence when the right door is shut it presses down B & the projection at A sticks up. & when both the doors are shut together the projection at A both

Figure 4.62: 1843-10-31. FG to STG page 4.

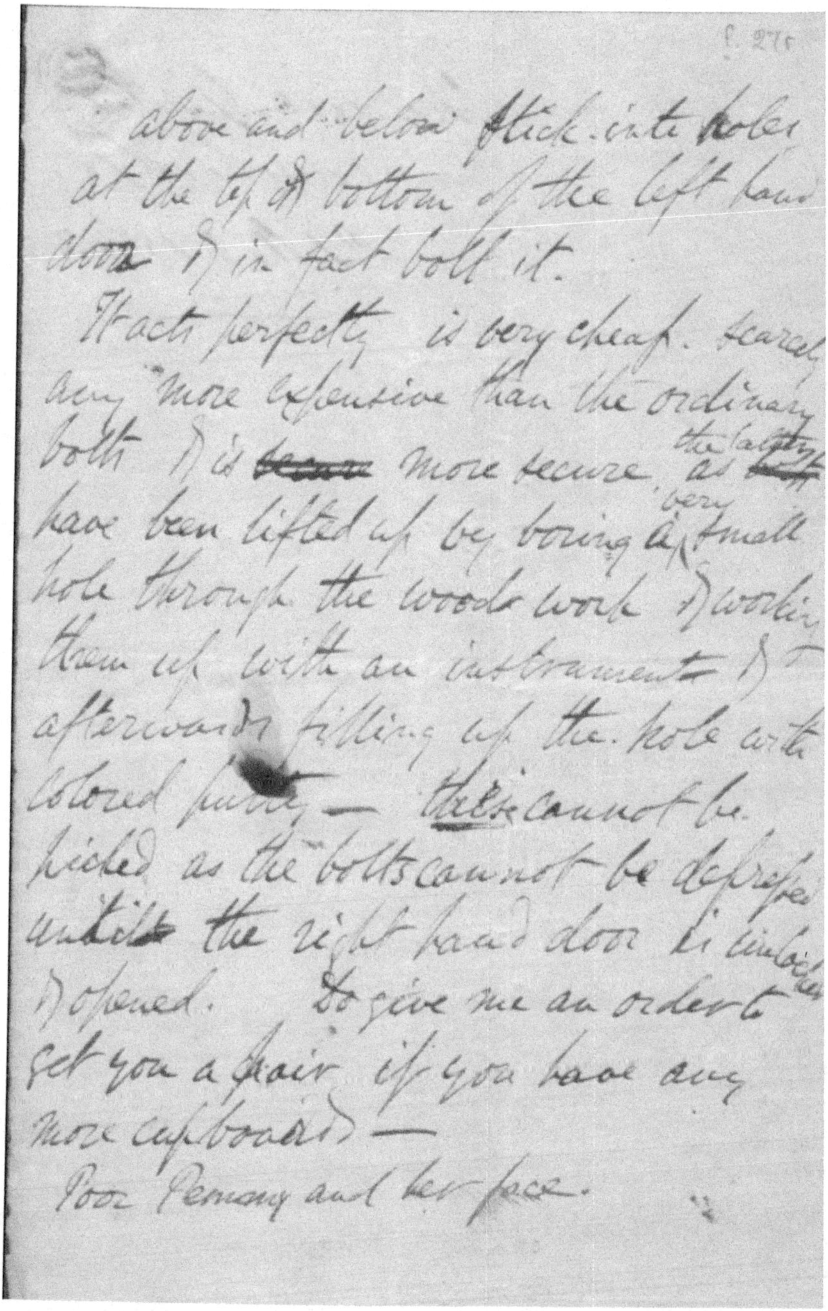
above and below stick into holes
at the top & bottom of the left hand
door & in fact bolt it.
It acts perfectly is very cheap. scarcely
any more expensive than the ordinary
bolt & is more secure, as the latches
have been lifted up by boring a very small
hole through the wood work & working
them up with an instrument & then
afterwards filling up the hole with
colored putty — these cannot be
picked as the bolts cannot be depressed
until the right hand door is unlocked
& opened. Do give me an order to
get you a pair if you have any
more cupboards —
Poor Pemony and her face.

Figure 4.63: 1843-10-31. FG to STG page 5.

Francis Galton
Cambridge
31 Oct. 1843

Figure 4.64: 1843-10-31. FG to STG page 6.

5. Sunday.
Breakfast Dalyell[101]
also with Tooke, T. Taylor[102]
and other apostles there, much talk
about Tennyson, especially the Vision
of Sin. "The awful rose of dawn" suggested to
be a man's moral feelings?

6. Monday.
Wine Sykes.
Subscriptions for National Fund up to £80
not very successful.
meeting at Union for curtailment
of expenses.

7. Tuesday.
Breakfast Douglas 9
Dine Aytoun[103] $6\frac{1}{2}$
Wine Simpson John's.

8. Wednesday.
Wine Hallam.[104]
Breakfast Armitage $9\frac{1}{2}$

9. Thursday.
Wine Pitcairn.
" [Hooke] cant go.

10. Friday.
Wine [Western].

11. Saturday.
Breakfast J. Kay 9.[105]

13. Monday.

---

[101] Dalyell, Robert Alexander Osborne. See page 482.
[102] Thomas Hammond Tooke (1813-1888); Tom Taylor (1817-1880), later editor of Punch.
[103] Aytoun, Roger Sinclair. See page 480.
[104] Hallam, Henry Fitzmaurice. See page 483.
[105] Kay, Joseph. See page 486.

Mrs. Drake Evg. party

15. Wednesday.
Ball. Hopkins.[106]

16. Thursday.
Wine [Airlie]

17. Friday.
Ball Dr. Haviland.

18. Saturday.
Cd to cash

In father's hands	£38	14	1
Mortlock's	50		
In hand	14		
	£102	14	1

20. Monday.
Breakfast Gibbs.
Stomachish.

21. Tuesday.
Wine Wood.

23. Thursday.
Wine Pepys.

24. Friday.
Ball Mrs. Wilkins.

26. Sunday.
Breakfast in my rooms
Wine Dalyell[107] not go

27. Monday.
Wine Yeoman
Breakfast Keary not go

[106] Hopkins, William. See page 484.
[107] Dalyell, Robert Alexander Osborne. See page 482.

28. Tuesday.
Macaroni & Cheese E. Kay[108]
Paid subscription £12.12 to British
& Foreign Institute, afraid it is a humbug

29. Wednesday.
Dine Dr. Fisher 6pm
Evg. party Marchesa di Spineto[109]

30. Thursday.
Wine Hallam.[110]

That L. L. Brown will be
plucked
I bet my picture & frame
agst £1.10

Paid in Mortlock to credit
of National Fund
£59.9.6
35.4.0

1843/11. Francis Galton to Samuel Tertius Galton

My dear Father.

Thank you very much for the Bristol water Dr Fisher[111] was most overcome. Did ~~you~~ Emma receive a letter I wrote 3 or so days before my last one, talking about the British & Foreign Institute, I have delayed writing hoping to hear something positive or negative from you in reference to its merits. However I could not well delay longer & am already elected & propose entering myself as a country member which will cost 6 ~~pounds~~ guineas viz. for entrance & 2 yearly subscription.

I am very anxious to begin pure mediculizing again. I find that I shall have to stay in Cambridge till about the 1st of June &

[108] Kay, Edwin Ebenezer. See page 485.
[109] c.1774 – 1849. Married to Elizabeth Campbell.
[110] Hallam, Henry Fitzmaurice. See page 483.
[111] William Webster Fisher.

then have completely done with the place, excepting some further examinations. I fully concur in all that Dr. Hodgson[112] said about St George's and shall like very much to pursue the medical practice there. I am not quite sure what would be the next thing to do immediately after leaving Cambridge as the medical sessions are not then going on. I however rather think that London w^d^ be the best place as at all events I could learn Botany and Materia Medica, together with some degree of hospital practice. If not I could dissect in Paris though after all minute anatomy is really useless to a practitioner who does not operate, and I think I know quite enough of general anatomy. I propose taking rooms in King's College next October as I think it much nicer than being out in lodgings, for one cannot have preparations and all that so well in the latter and besides it is more collegiate. I shall be obliged to leave my rooms next term and am now looking out for lodging in the town. How is my mother's face? I wish it would get well. I wrote to ask her in my last letter to Emma if she could have Charley Buxton for a week in Leamington, as she was so kind as to give me a carte blanche. Would somewhere about the 16th suit? He would give his ears for dancing & so I should prefer that he should come down for a dancing season. Is Emma with Miss Taylor? How is Mr. Gurney? I think that I shall only be able to be down for about 10 days or a week, during the vacation, so if Mater could tell me what time would suit her best I would make arrangements at once with Charles Buxton.

your affectionate Son Fras. Galton

1843/12/01–05. FG Diary

[December]

1. Friday.
Dine Hopkins at 6[113]
Ball Purchas's

2. Saturday.
Wine Johnson Kings.

---

[112] Hodgson, Joseph. See page 483.
[113] Hopkins, William. See page 484.

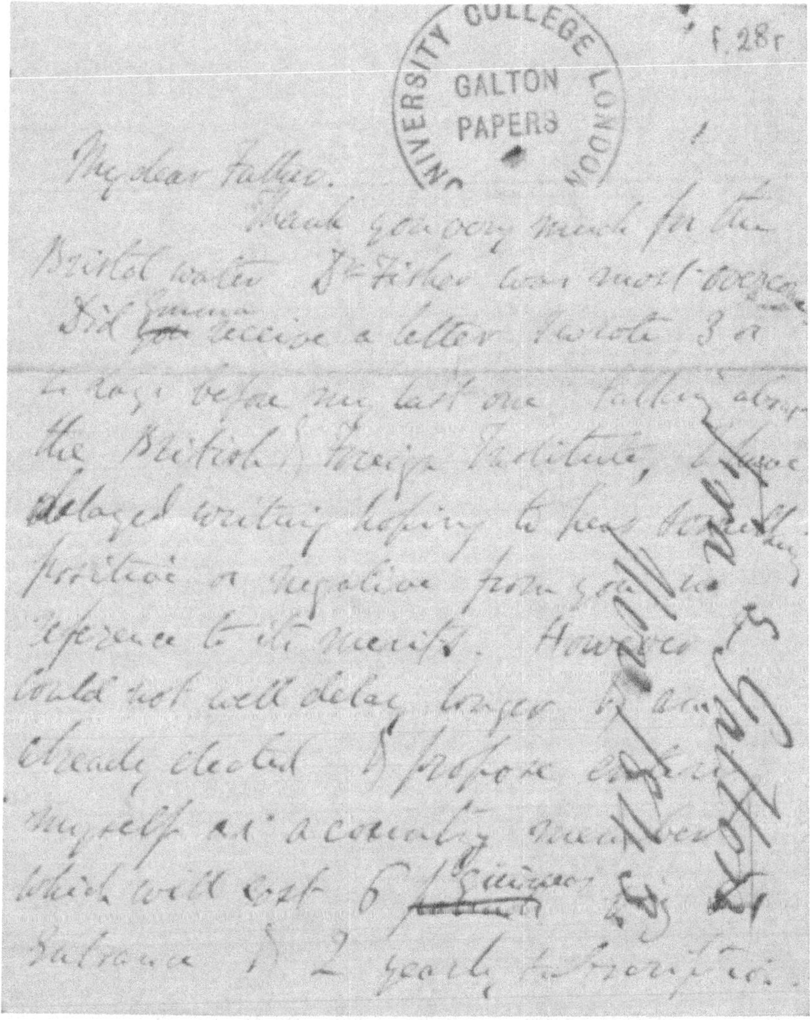

f. 28r

1

My dear Father.

Thank you very much for the Bristol water. Dr Fisher was most energe[tic]. Did mamma receive a letter I wrote 3 or 4 days before my last one talking about the British & Foreign Institute, I have delayed writing hoping to hear something positive or negative from you in reference to its merits. However I could not well delay longer & am already elected & propose entering myself as a country member which will cost 6 guineas entrance & 2 yearly subscription.

Figure 4.65: 1843-11. FG to STG page 1.

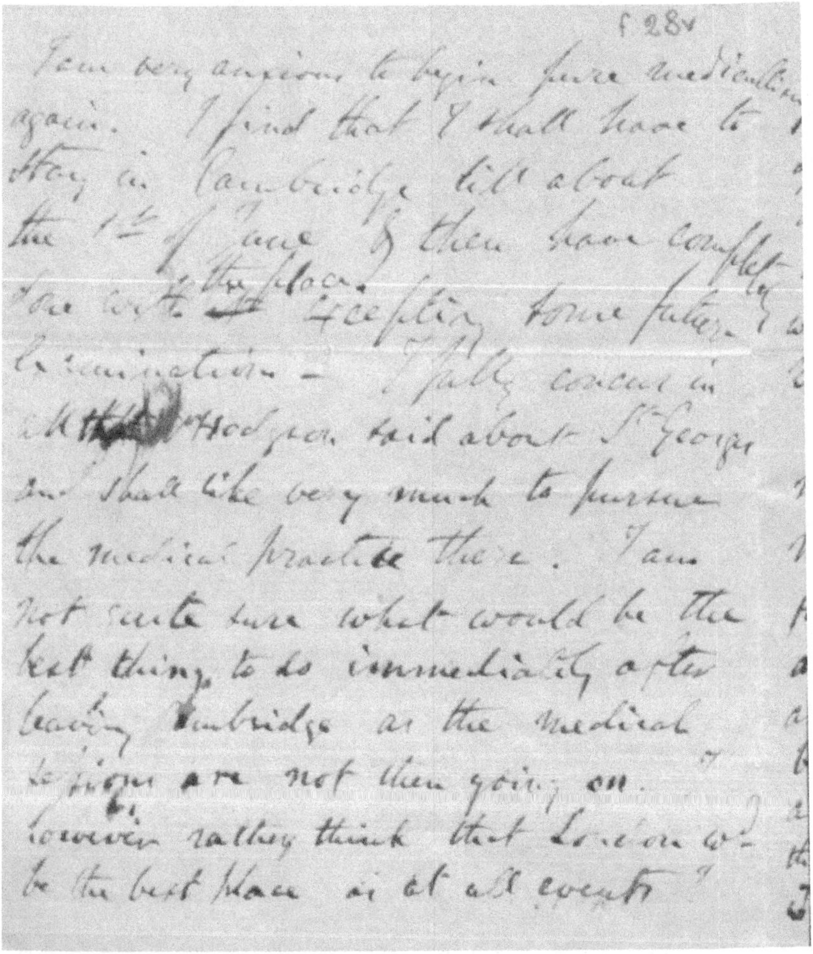
f 28v

I am very anxious to begin pure medicine
again. I find that I shall have to
stay in Cambridge till about
the 1st of June & then have completely
done with the place. excepting some future
examination — I fully concur in
all that Hodgson said about St George's
and shall like very much to pursue
the medical practice there. I am
not quite sure what would be the
best thing to do immediately after
leaving Cambridge as the medical
lessons are not then going on. I
however rather think that London wd
be the best place in at all events

Figure 4.66: 1843-11. FG to STG page 2.

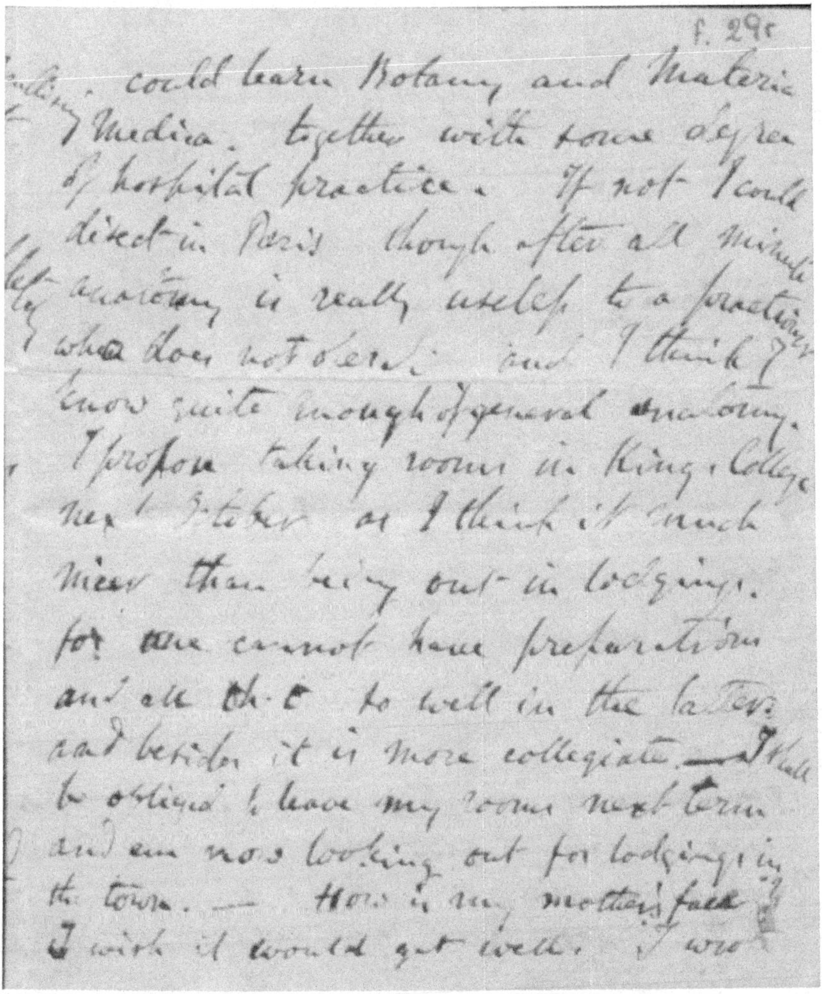

f. 29r

could learn Botany and Materia
Medica. together with some degree
of hospital practice. If not I could
dissect in Paris though after all minute
anatomy is really useless to a practitioner
who does not dissect; and I think I
know quite enough of general anatomy.
I propose taking rooms in Kings College
next term as I think it much
nicer than being out in lodgings.
for one cannot have preparations
and all that so well in the latter;
and besides it is more collegiate. — I shall
be obliged to leave my rooms next term
and am now looking out for lodgings in
the town. — How is my mother's face?
I wish it would get well. I wro

Figure 4.67: 1843-11. FG to STG page 3.

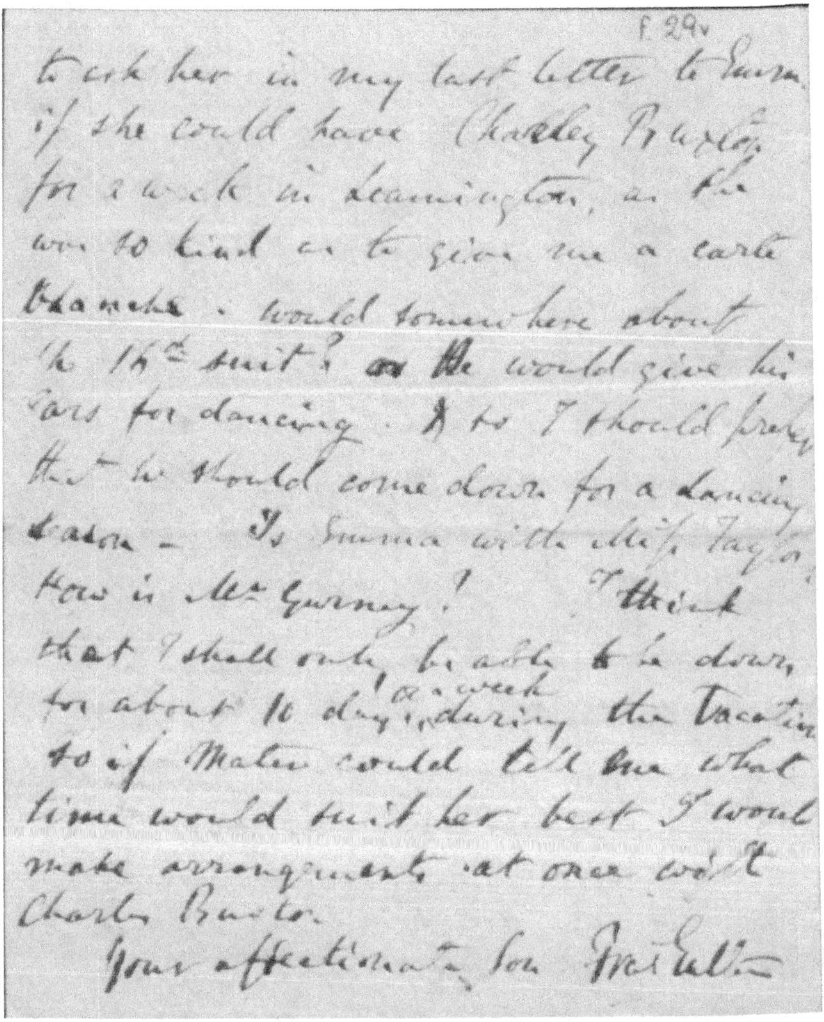
f. 29v

to ask her in my last letter to Emma
if she could have Charley Buxton
for a week in Leamington, as she
was so kind as to give me a carte
blanche. would somewhere about
the 15th suit? as he would give his
ears for dancing. & so I should prefer
that he should come down for a dancing
season. Is Emma with Miss Taylor?
How is Mr Gurney? I think
that I shall only be able to be down
for about 10 days or a week during the Vacation
so if Mater could tell me what
time would suit her best I would
make arrangements at once with
Charles Buxton.
Your affectionate Son Fras Galton

Figure 4.68: 1843-11. FG to STG page 4.

3. Sunday.
Breakfast A. Spring Rice[114]
Wine A. Stewart.

5. Tuesday
Evg party Mrs Aytoun[115] $8\frac{1}{2}$
Ball Mrs. Bond.

1843/12/05. Francis Galton to Samuel Tertius Galton

Tuesday Evening

My Dear Father

I take the earliest possible opportunity of writing to say that C Buxton is all sorrow & everything that is proper owing to his being quite unable to come to Leamington. Sir F. B.[116] is not well & as C.B.[117] can only spare about 10 days out of his vacation, & has had a letter expressing some anxiety that he sh^{d} spend them at home. Thank you & please thank my mother also very much for so kindly inviting him. I also received a letter from Miss Taylor inviting us to Moore Green[118] for a day before going to the concert, which I have answered full of gratitude. I propose not going down to you till a day or two after Xmas. Will Delly positively be in Leamington then, as she & I realise the favourite simile of Dr. Booth. [See Figure 4.69, p. 346.]

I have paid my subscription to the British Foreign Institute[119] and I really think I have been quite right in doing so. I think that I shall only been [sic] able to stay a week or so with you. I am all anxiety to begin Doctoring again putting on all steam. What is Rassy's[120] direction. I directed a letter to him some time since Cross near Bristol wh: I fear he has not received.

Your very affectionate Son
Fras. Galton

---

114 Aubrey Richard Spring Rice (1822-1897).
115 Aytoun, Roger Sinclair. See page 480.
116 Sir Fowell Buxton (1786-1845), brewer and M.P.
117 Charles Buxton (1823-1871), son of Sir Fowell.
118 Kings Norton, Worcestershire.
119 The British and Foreign Institute, which was officially founded that year by James Silk Buckingham, but only lasted just over four years.
120 Erasmus Galton.

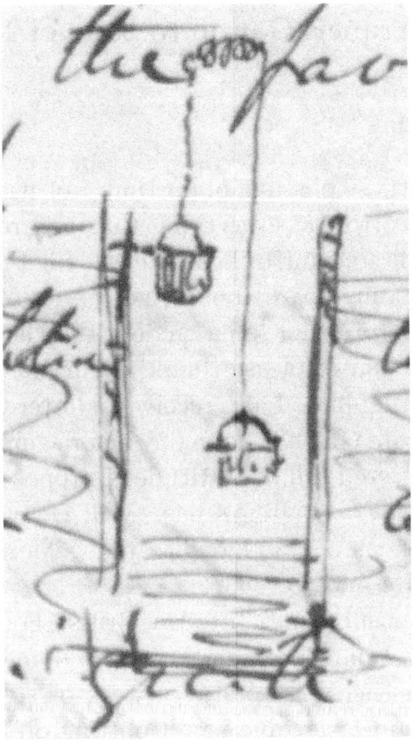

Figure 4.69: 1843-12-05 FG to STG: Dr. Booth's Favourite Simile.

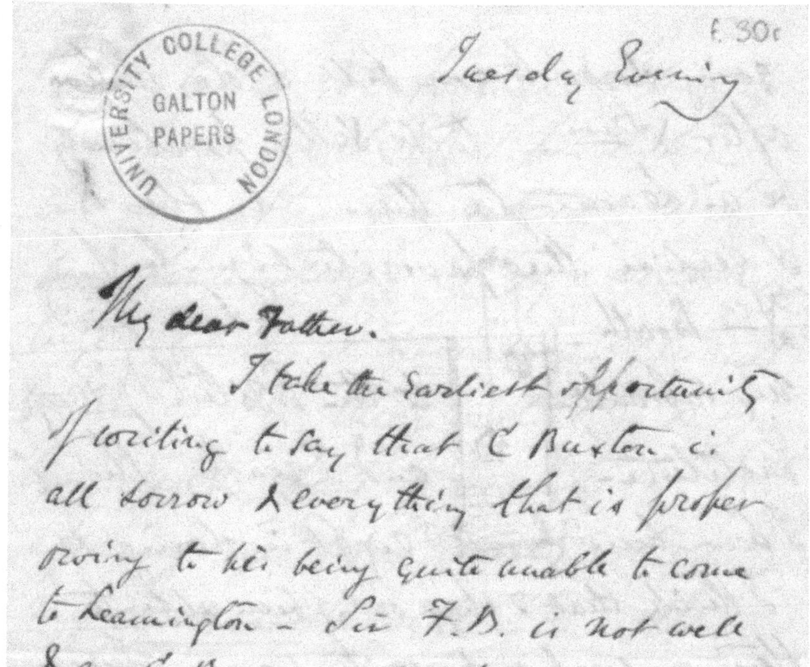

Tuesday Evening

My dear Father,

I take the earliest opportunity of writing to say that C Buxton is all sorrow & everything that is proper owing to his being quite unable to come to Leamington – Sir F.B. is not well & as C.B. can only spare about 10 days out of his vacation, ~~they~~ he has had a letter expressing some anxiety that he sh^d spend them at home. Thank you & please thank my Mother also very much for so kindly inviting him. I also received a letter from Mrs Taylor inviting us to Moore Green for a day before going to the Concert, which I have answered full of gratitude – I propose not

Figure 4.70: 1843-12-05. FG to STG page 1.

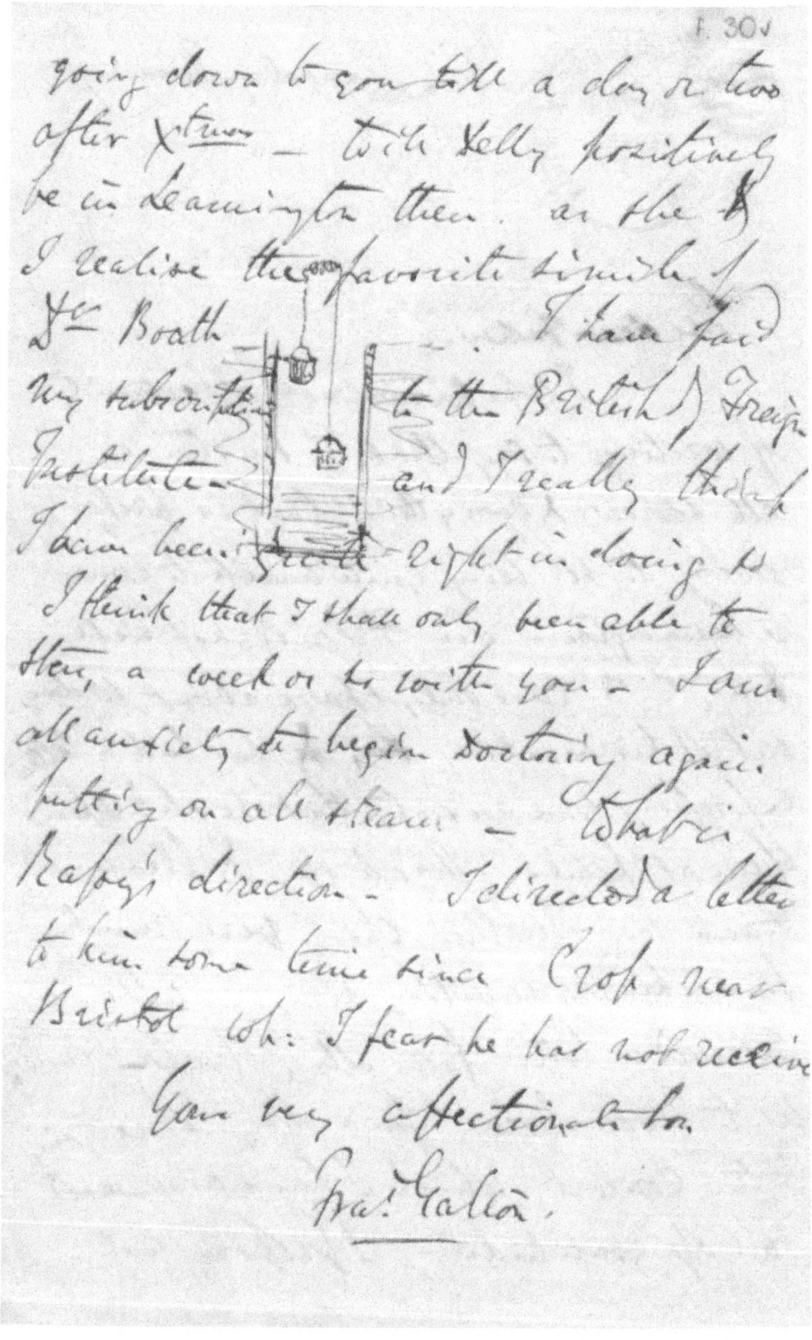
going down to you till a day or two
after Xtmas — will Kelly positively
be in Leamington then. as the [illegible]
I realise the favorite simile of
Dr Booth — I have paid
my subscription to the British & Foreign
Institute — and I really think
I have been right in doing so
I think that I shall only be able to
stay a week or so with you — I am
all anxiety to begin doctoring again
putting on all steam — What is
Rafey's direction — I directed a letter
to him some time since Crop[?] near
Bristol wh: I fear he has not receiv[ed]
Your very affectionate Son
Fras. Galton.

Figure 4.71: 1843-12-05. FG to STG page 2.

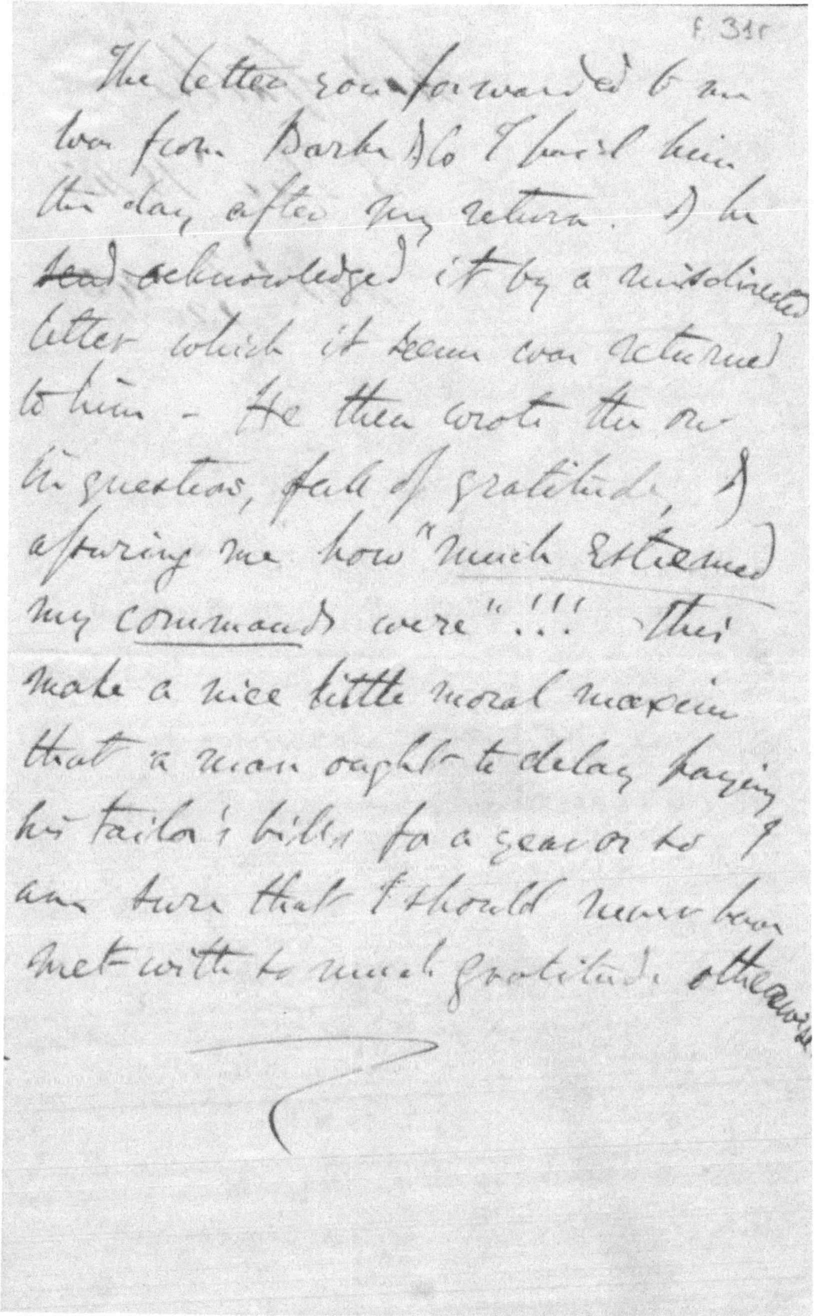

f. 315

The letter you forwarded to me was from Barker & Co I paid him the day after my return. & he ~~had~~ acknowledged it by a misdirected letter which it seems was returned to him – He then wrote the one in question, full of gratitude, & assuring me how "much esteemed my commands were"!!! This makes a nice little moral maxim that a man ought to delay paying his tailor's bills for a year or so I am sure that I should never have met with so much gratitude otherwise

Figure 4.72: 1843-12-05. FG to STG page 3.

Figure 4.73: 1843-12-05. FG to STG page 4.

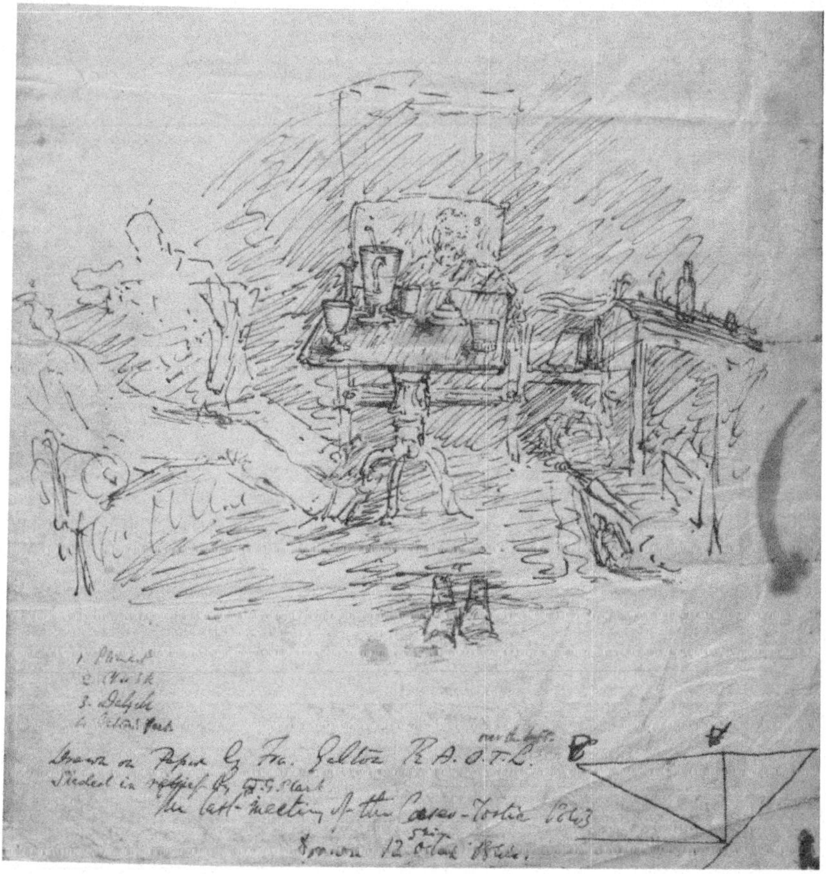

Figure 4.74: The Last Meeting of the Caseo-Tostic Society 1843. 1. Stewart 2. Clark 3. Dalyell 4. Galton's feet.

1843/12/07–27. FG Diary

7. Thursday.
Dine Newman[121]

27. Wednesday.
Subscription Ball Leamington.

[121] Possibly William Turner Newman (1820-1876).

Chapter 5

1844

1844/01 circa. Francis Galton to Samuel Tertius Galton

My dear Father

Your affectionate son is B.A., the ceremony having taken place this morning and I now wear a flowing gown with ribbons hung in front over either arm. [See Figure 5.1, p. 354.]

I send you a list. I am 44th at which I highly pride myself the Classics being below par & Medicine &c. only allowed me a month to get my subjects up. I was third in Mathematics but would have been first only for a misunderstanding in one question which lost me fifteen marks.

The place however in the Poll signifies nothing.

I see Hopkins[1] occasionally who often asks me out he has asked me to Dinner tonight and again on Tuesday night.

Thank you very much about the carriage to the Stratford ball. It will be most convenient and capital for me.

Please tell Emma that I fear my Schiller will not do for her on account of the type, besides I think that I had rather keep it. Tell her I was much obliged for her remembering my offer of selling it.

In great hurry

Your affectionate Son
Fras. Galton

I send you a Newspaper.

1 Hopkins, William. See page 484.

Figure 5.1: 1844-01 FG to STG: Your affectionate son is B.A.

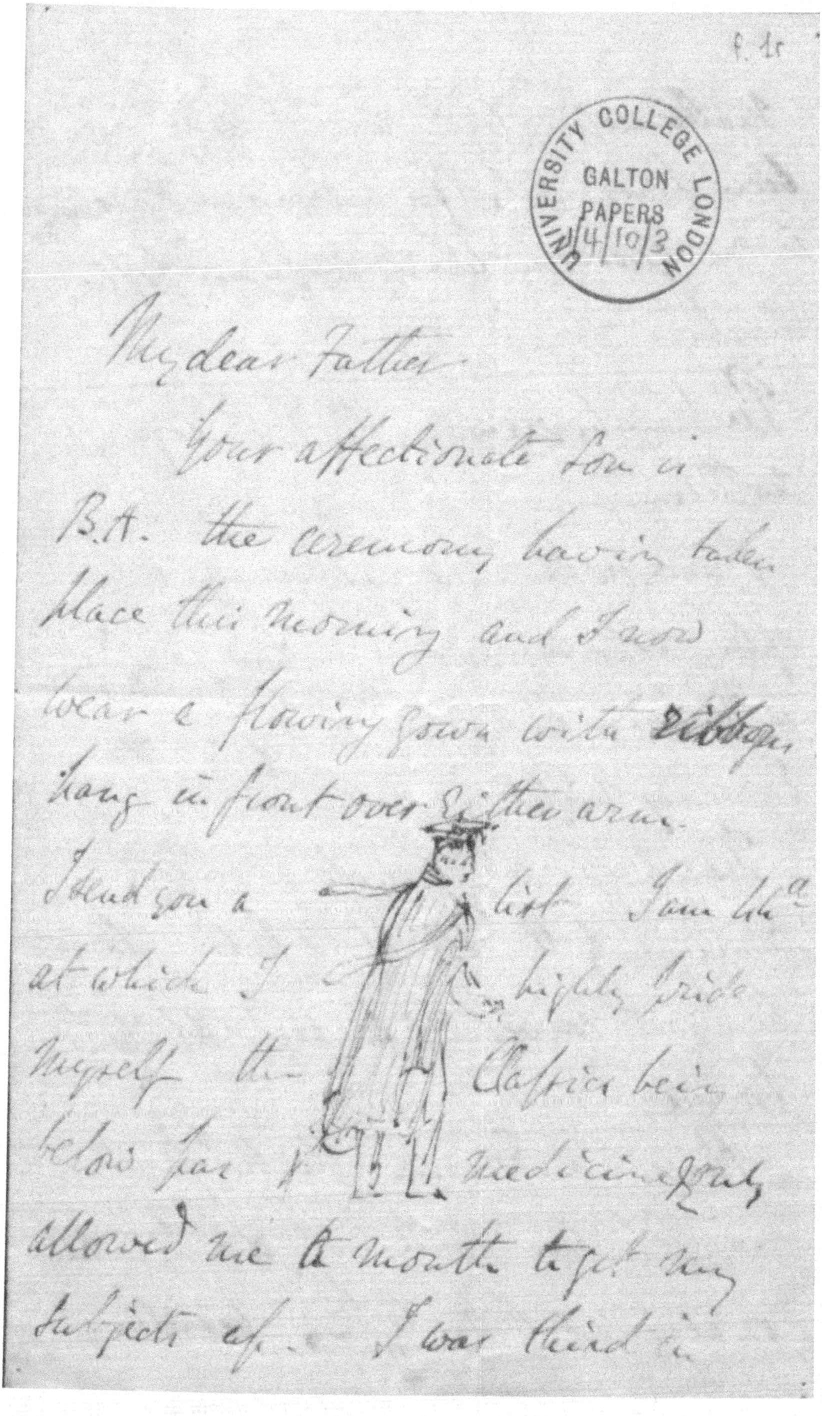

Figure 5.2: 1844-01. FG to STG page 1.

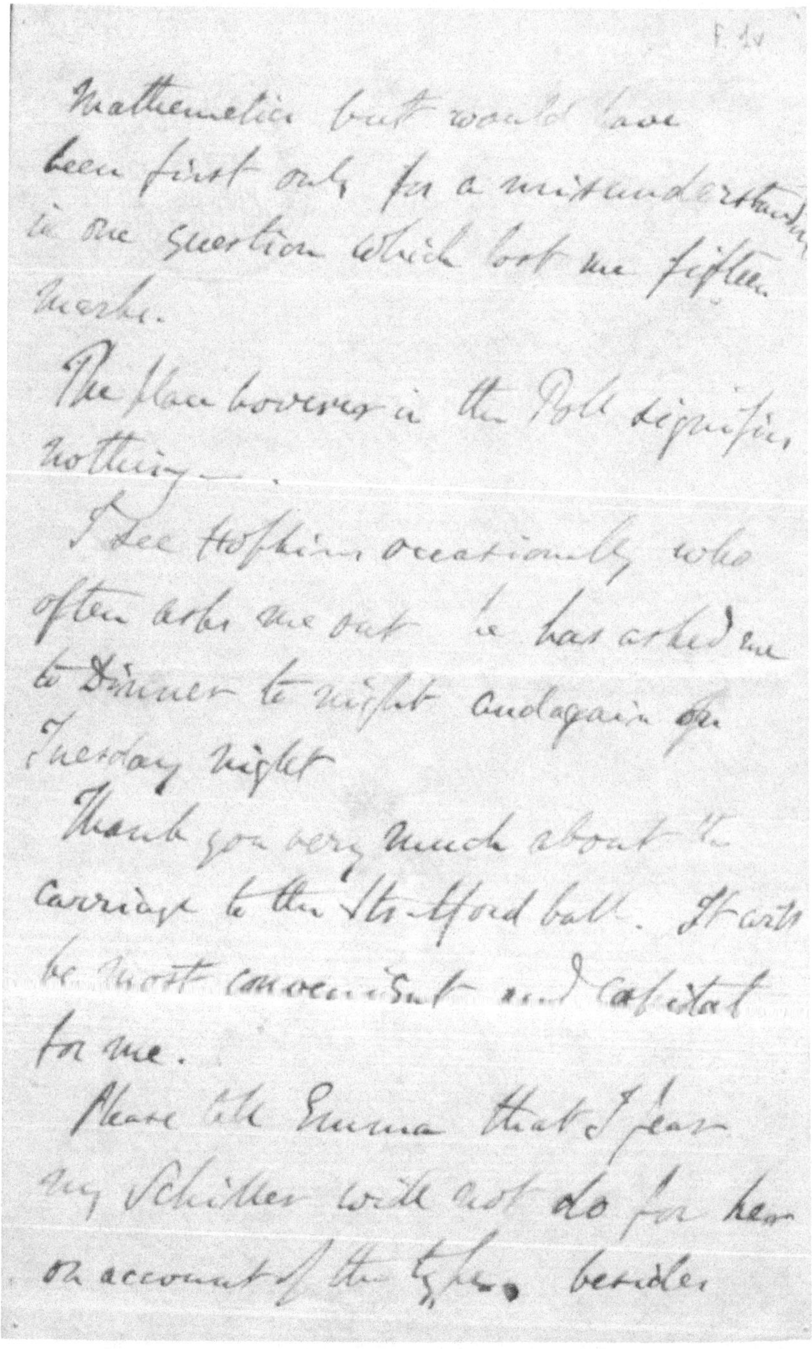
Mathematics but would have
been first only for a misunderstanding
in one question which lost me fifteen
marks.
The place however in the Poll signifies
nothing—.
I see Hopkins occasionally who
often asks me out he has asked me
to Dinner to night and again on
Tuesday night
Thank you very much about the
carriage to the Stratford ball. It will
be most convenient and capital
for me.
Please tell Emma that I fear
my Schiller will not do for her
on account of the type, besides

Figure 5.3: 1844-01. FG to STG page 2.

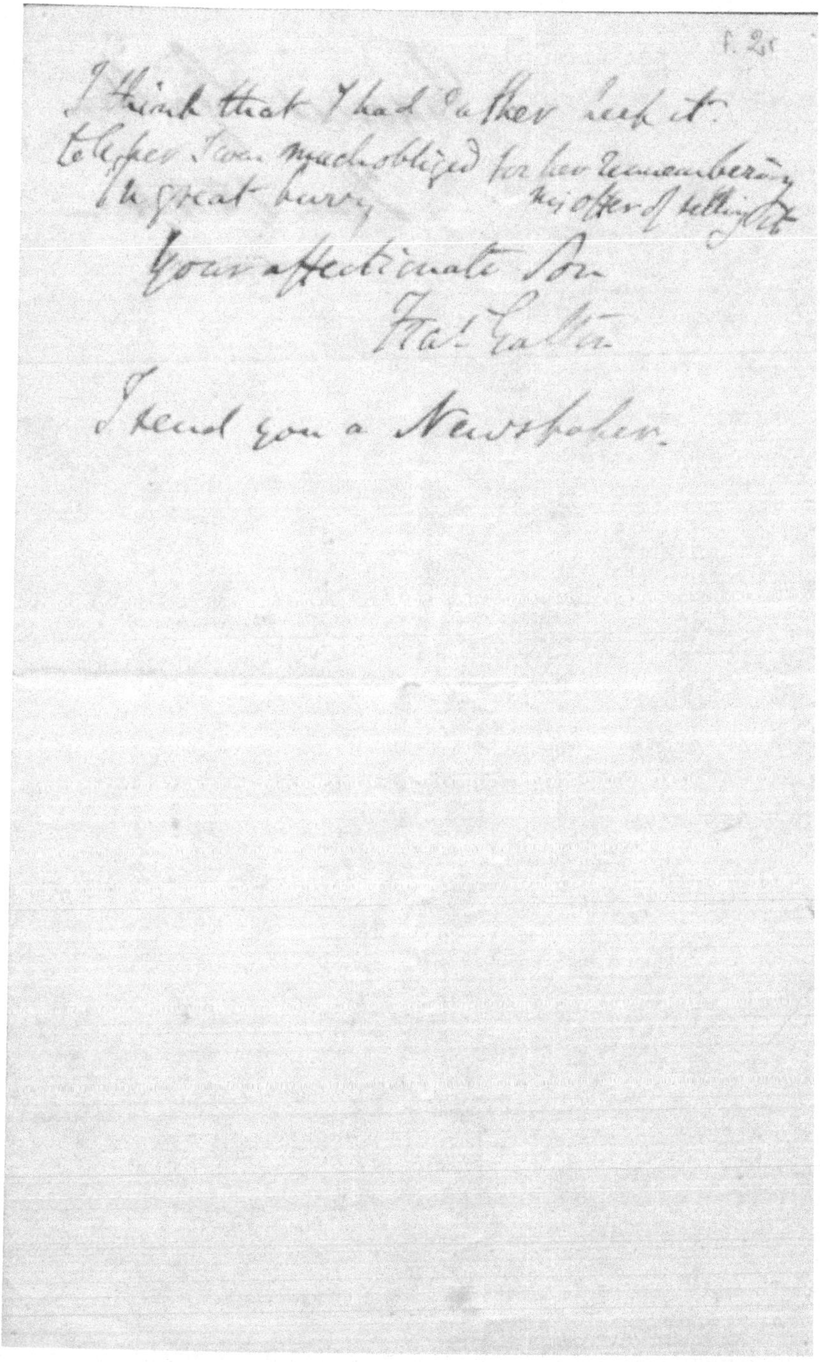

f. 25

I think that I had rather keep it.
tell her I am much obliged for her remembering
In great hurry my offer of selling it

Your affectionate Son
Fra^s. Galton

I send you a Newspaper.

Figure 5.4: 1844-01. FG to STG page 3.

f. 2v

Francis Galton
Jan 1844.

Figure 5.5: 1844-01. FG to STG page 4.

1844/01/24–27. Diary of Samuel Tertius Galton

January 24th. Wednesday. Leamington.

... Francis returned home from Cambridge and went to the Stratford Ball with Emma and Miss Hodgson.

January 27th. Saturday. Leamington.

Emma, Francis and myself drove to Claverdon and Edstone ...

.

1844/02/04. Samuel Tertius Galton to Francis Galton

23 Dover St. 4 Feb^{y}

My dear Francis,

Emma arrived in London on Friday quite recovered from the effects of the Russell's Ball. Prowett is making great progress and there is a probability of our being released by the latter end of the week, which I hope may be the case for we find the Metropolis most unamiable during this snowy, sloppy, foggy weather which is most uncongenial to the state of my lungs. Mr. Gurney gets gradually better. We have seen much of the Leonard Horners who tell us that the Hallams[2] are constantly speaking of your intimacy in Germany. We shall probably call there tomorrow with Emma, weather and lungs permitting. Theodore called yesterday and expressed his congratulations at your B.A. degree. He is studying law in good earnest with the intention of practicing upon the Midland Circuit which I was glad to hear, as occupation is always desirable for youth. He has good accounts from Howard who now suffers only from the debility consequent upon his late complaint. He tells me that Mr. J. Strutt had about 800 pictures—of these Howard is to have the choice of the first 100, and then to choose turn about with other members of the family whereby he will get about 70 more. He will also have some specimens of statuary of which there were several in Mr. Strutt's Gallery. As yet we have seen very few of the wonders of the Metropolis but shall be more industrious this week if the weather should improve. I observed the outside of <u>your</u> club as I passed thro' George St. Hanover Square. Your Mother desires her love. As Bessy has not doubt given you much salutary advice as to <u>exclusive</u> attention to Medicine I forbear repeating to

[2] Hallam, Henry Fitzmaurice. See page 483.

you all that Horner said to me on the importance of it to success in London practice as founded upon his own observation & the remarks of many leading Medical Men of his acquaintance. Believe me to remain ever, Your affectionate Father
S. Tertius Galton

If I can do anything for you whilst in London, let me know.

To Francis Galton Esq. B.A. Is the above your right address?

1844/02/11. Samuel Tertius Galton to Francis Galton

Leamington, 11 Feb. 1844

My dear Francis,

You will particularly oblige me by writing word, by return of post, how you are, for I find from your latest that you travelled outside with a bad cold & that you have not been heard from since your arrival at Cambridge—so pray relieve us at once from our anxiety.

Your Mother & myself returned yesterday from London, Prowett having done his work to perfection.[3] We left Emma in St. James's Square. Mr. Gurney[4] is not very materially better & Mrs. G. much worn down by attendance upon him. Theodore[5] dined with us one day took us to the Temple Church. He is studying Law in good earnest with the intention of attending the Midland business and making a business of it which I was glad to hear. The David Barclays[6] are still in Leamington & Mr. B. has expressed his hope of seeing you often when you reside in Town. Your Mother Bessy & Adèle are very well and send their love.

I remain
your affectionate Father
S. Tertius Galton

F. Galton Esq.

---

3 Possibly Robert Prowett, the West End dentist, who may have fitted Violetta with false teeth.

4 Hudson Gurney (1775-1864).

5 Theodore Galton, cousin to Francis.

6 David Barclay (1784-1861) of Eastwick Park, M.P. for Sunderland. His eldest son was Hedworth David Barclay (1820-1873), who would later accompany Francis down the Nile.

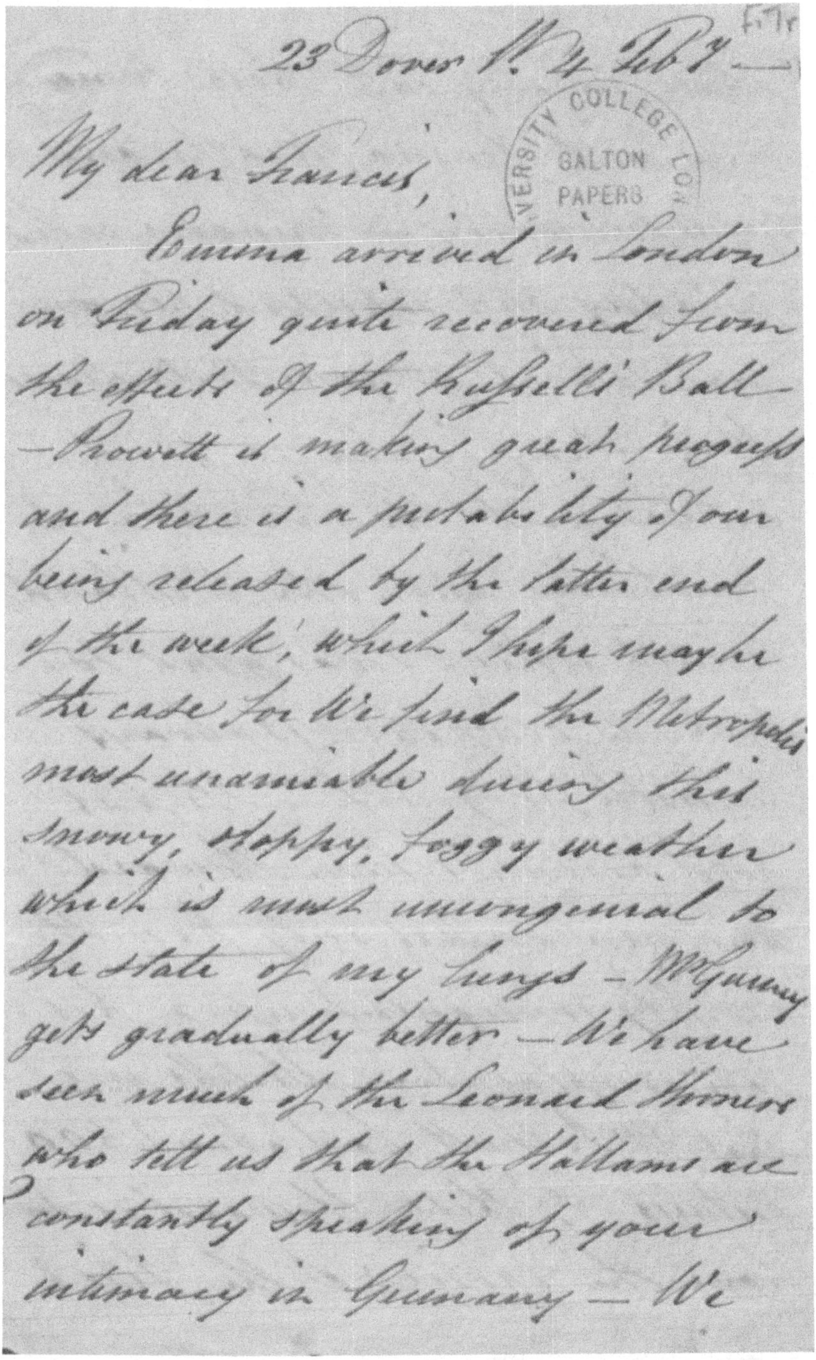

23 Dover St. 4 Feby —

My dear Francis,

Emma arrived in London on Friday quite recovered from the effects of the Russells' Ball — Prowett is making great progress and there is a probability of our being released by the latter end of the week, which I hope may be the case for we find the Metropolis most unamusable during this snowy, sloppy, foggy weather which is most uncongenial to the state of my lungs — Mr Gurney gets gradually better — We have seen much of the Leonard Horners who tell us that the Hallams are constantly speaking of your intimacy in Germany — We

Figure 5.6: 1844-02-04. STG to FG page 1.

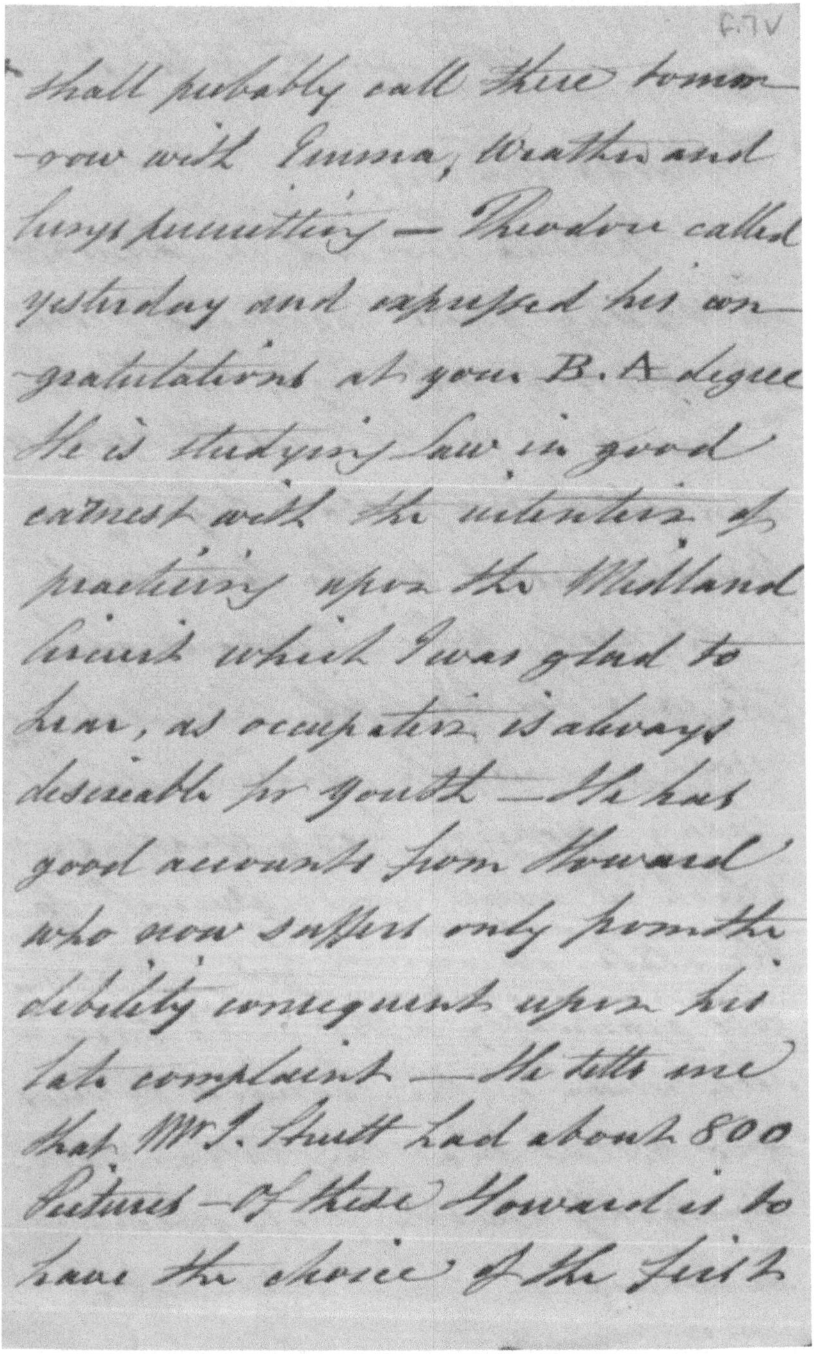

f.7v

shall probably call there tomor-
-row with Emma, weather and
hours permitting — Theodore called
yesterday and expressed his con-
-gratulations at your B.A degree
He is studying law in good
earnest with the intention of
practising upon the Midland
Circuit which I was glad to
hear, as occupation is always
desireable for youth — He has
good accounts from Howard
who now suffers only from the
debility consequent upon his
late complaint — He tells me
that Mr J. Strutt had about 800
Pictures — Of these Howard is to
have the choice of the first

Figure 5.7: 1844-02-04. STG to FG page 2.

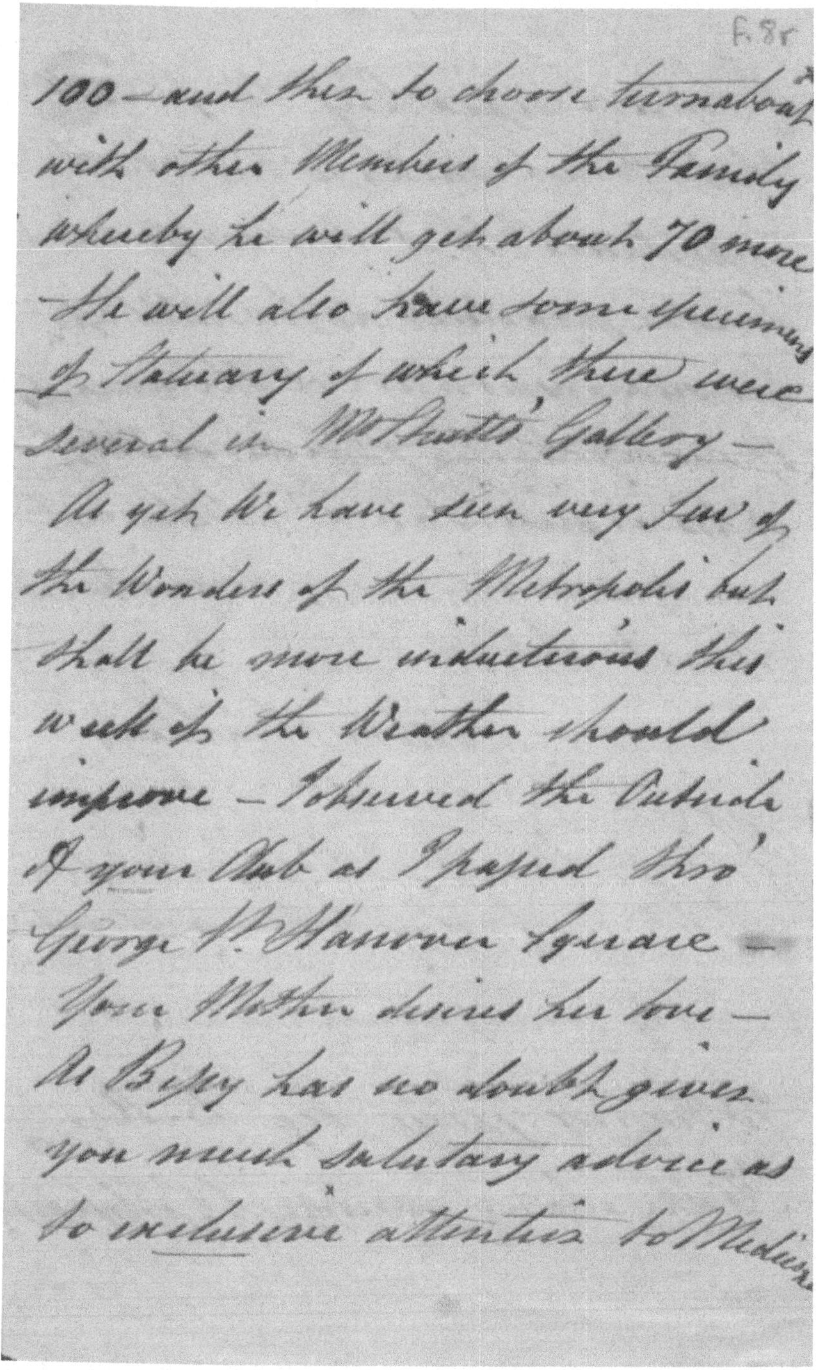

F.8r

100 — and then to choose turnabout
with other Members of the Family
whereby he will get about 70 more
— He will also have some specimens
of Statuary of which there were
several in Mr Scott's Gallery —
As yet We have seen very few of
the Wonders of the Metropolis but
shall be more industrious this
week if the Weather should
improve — I observed the Outside
of your Club as I passed thro'
George St. Hanover Square —
Your Mother desires her love —
As Bessy has no doubt given
you much salutary advice as
to exclusive attention to Medicine

Figure 5.8: 1844-02-04. STG to FG page 3.

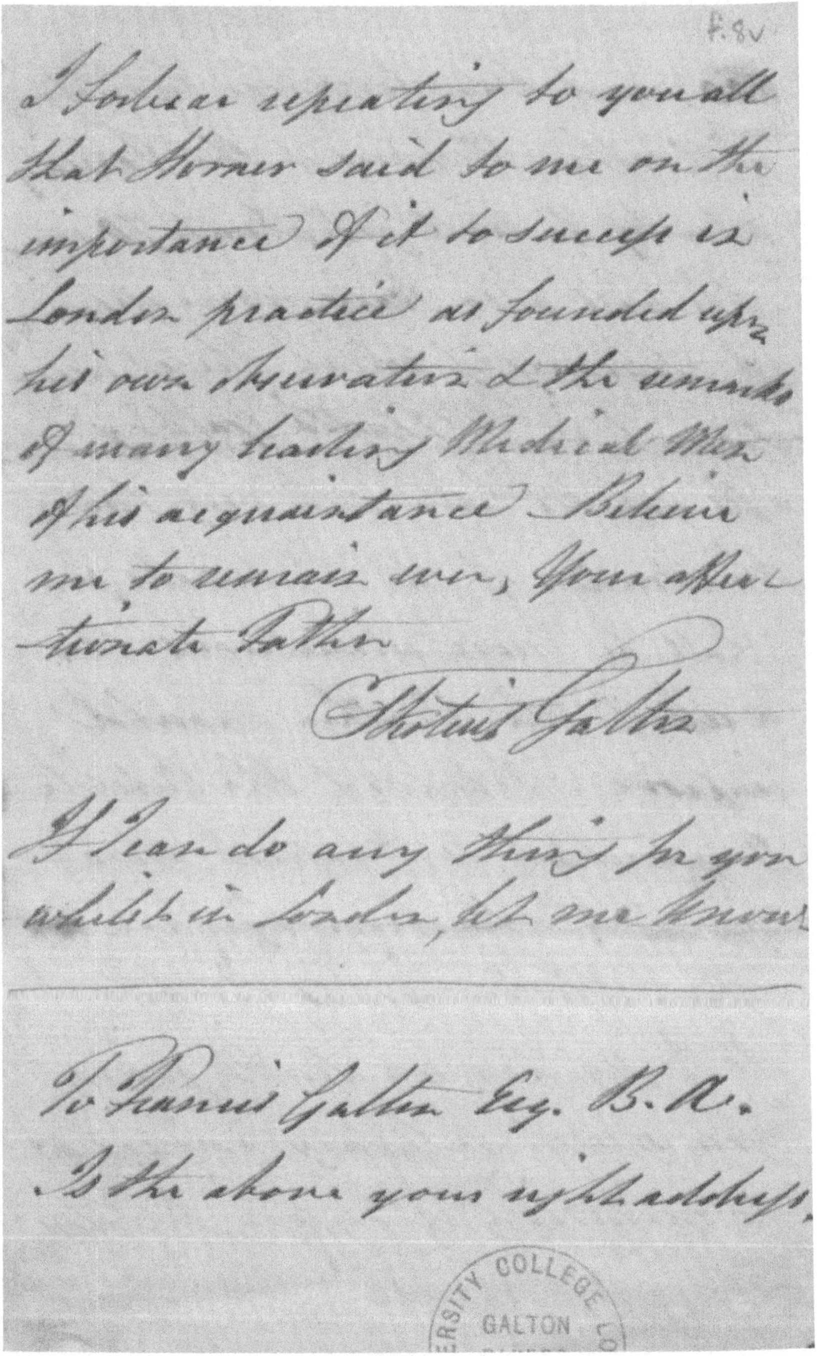
P.8v

I forbear repeating to you all
that Horner said to me on the
importance of it to success in
London practice as founded upon
his own observations & the remarks
of many leading Medical Men
of his acquaintance. Believe
me to remain ever, Your affec-
tionate Father

S. Tertius Galton

If I can do any thing for you
whilst in London, let me know.

To Francis Galton Esq. B.A.
Is the above your right address?

Figure 5.9: 1844-02-04. STG to FG page 4.

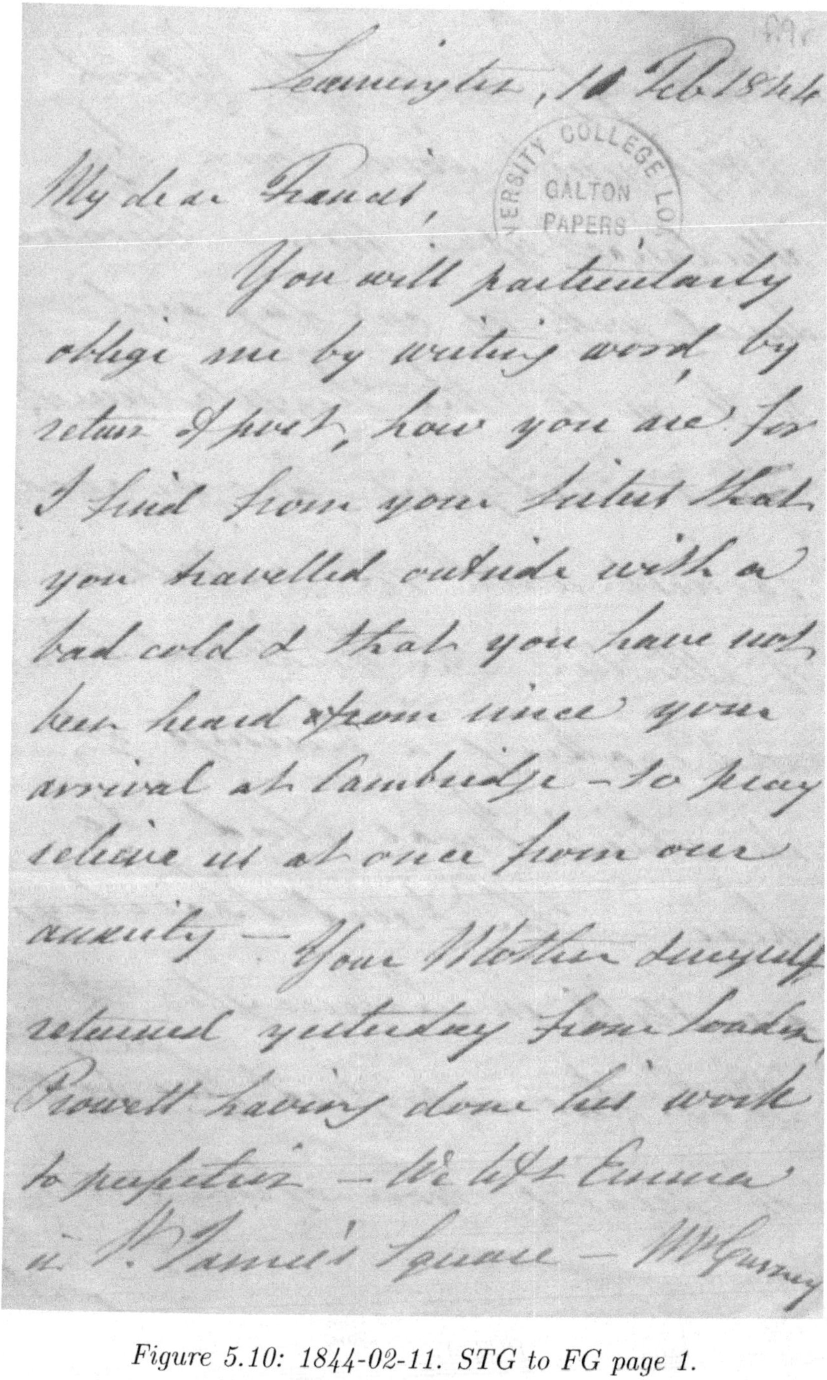

Leamington, 11 Feb 1844

VERSITY COLLEGE LO
GALTON
PAPERS

My dear Francis,

You will particularly oblige me by writing word, by return of post, how you are for I find from your Sister that you travelled outside with a bad cold & that you have not been heard from since your arrival at Cambridge – to pray relieve us at once from our anxiety – Your Mother & myself returned yesterday from London, Prowett having done his work to perfection – We left Emma in St James's Square – Mr Gurney

Figure 5.10: 1844-02-11. STG to FG page 1.

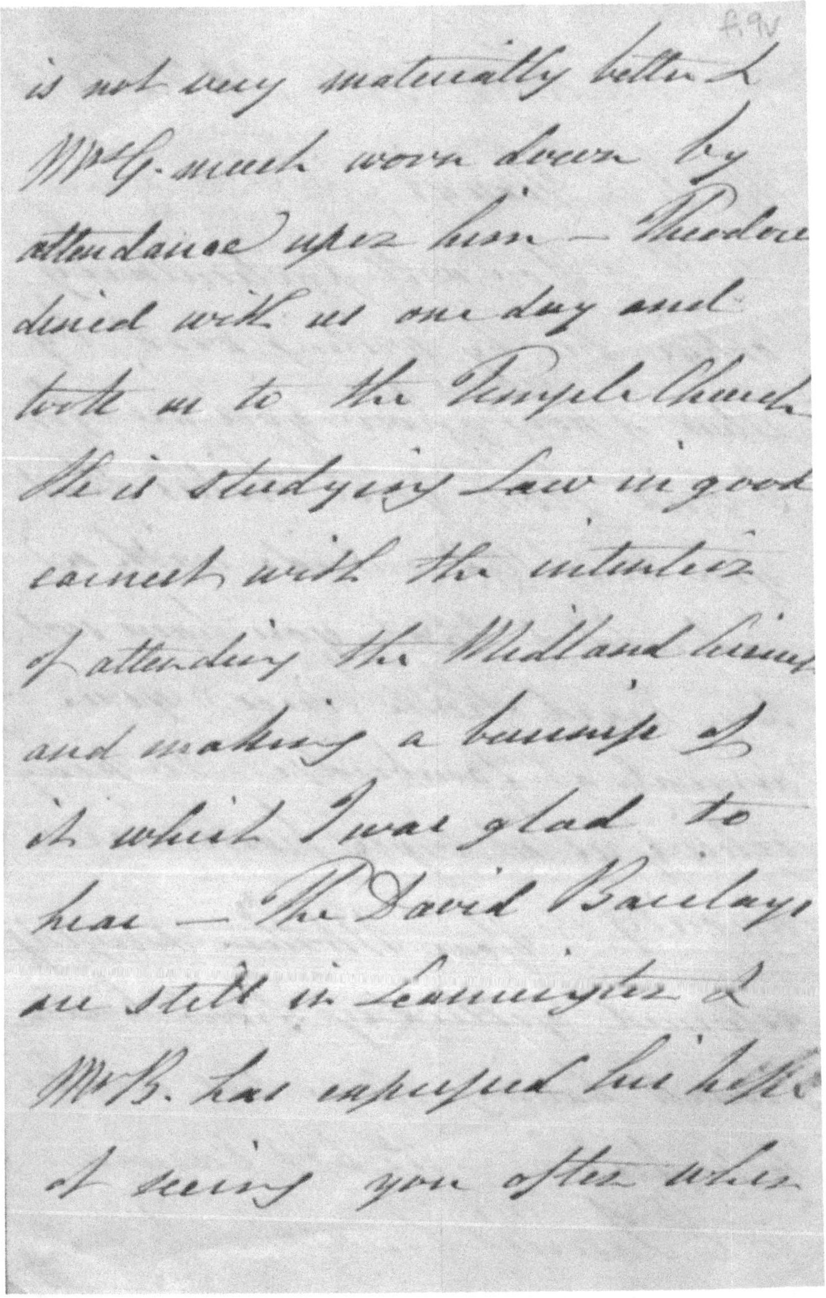
f9v

is not very materially better &
Mr G. much worn down by
attendance upon him — Theodore
dined with us one day and
took us to the Temple Church
He is studying Law in good
earnest with the intention
of attending the Midland circuit
and making a business of
it which I was glad to
hear — The David Barclays
are still in Leamington &
Mrs B. has expressed her hope
of seeing you often when

Figure 5.11: 1844-02-11. STG to FG page 2.

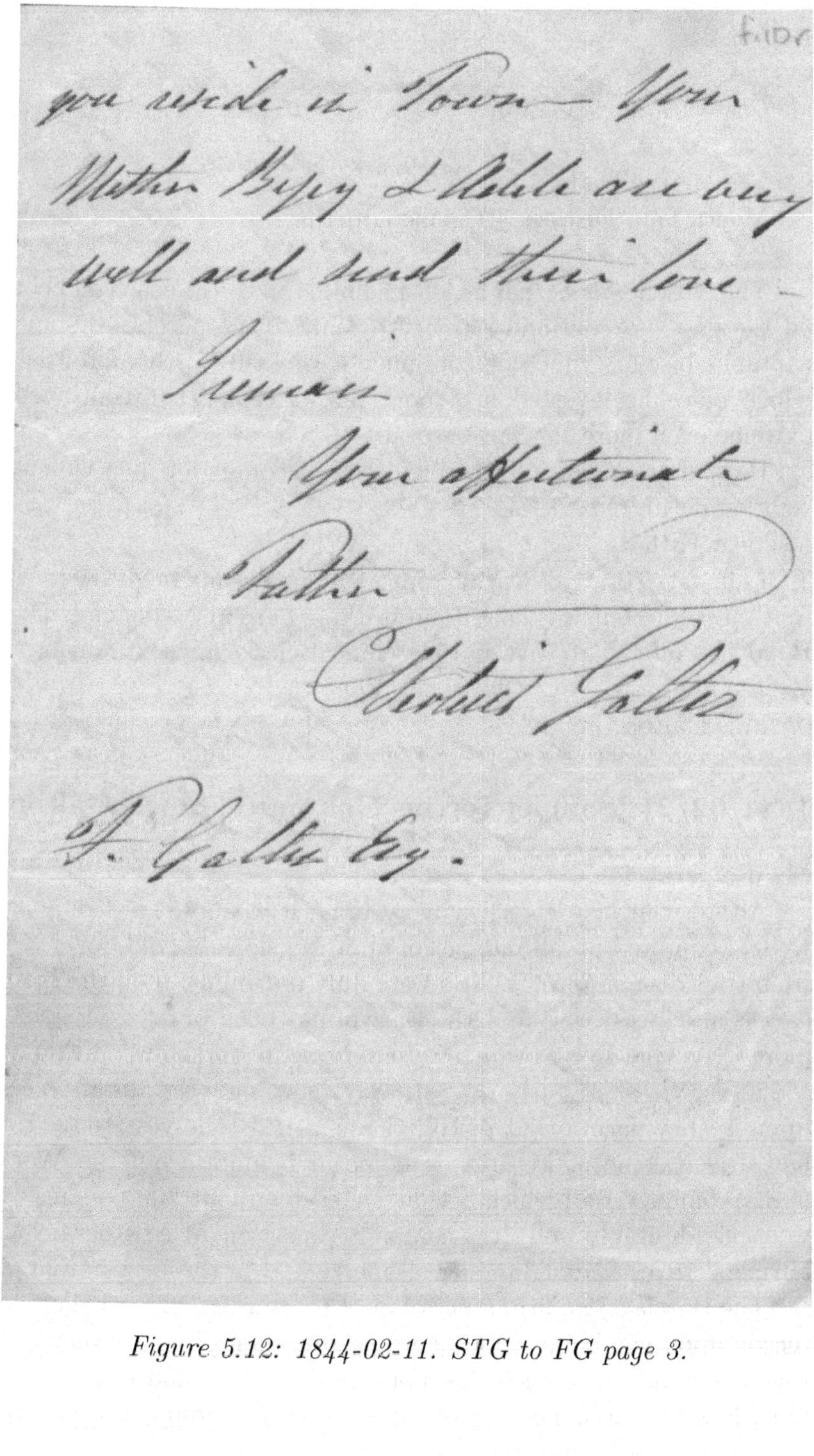

you reside in Town — Your
Mother Bessy & Adele are very
well and send their love —
I remain
Your affectionate
Father
S Tertius Galton

F. Galton Esq.

Figure 5.12: 1844-02-11. STG to FG page 3.

1844/02/13. Francis Galton to Samuel Tertius Galton

Bulletin

Case of Francis Galton.

Year 1844. Trade. Cantab.
Month Feb. Disease. Extreme appetite.

The patient states that he left Leamington by the coach on Feb. 3d the day was cold and rainy. At Southam he purchased some captain's biscuits which he continued eating till Northampton, at which place he invested in a pork pie. His appetite continued extreme even more so than natural.

Present state. Face flushed which he accounts for by a violent walk, appetite remarkable.

My dear Father

I am very sorry to hear of your asthma. I really ought to have written before but was lazy. I am working hard at Medicine & Dr. Bond has offered to give me his clinical clerkship next vacancy. Excuse my haste, your affectionate son

Fras Galton

1844/02/21. Samuel Tertius Galton to Francis Galton

My dear Francis

Altho' your last letter was very brief, it was most welcome as we had conjured up all manner of apprehensions of cold, fever &c from your dismal journey. You were duly remembered on the 10th as was also Bessy on the 19th. Darwin has been purchasing some more Cleveland Horses which were plowing in Yorkshire on Friday & in Warwickshire early on Saturday morning. He almost lives upon his new farm at [Cap] Hill. I was at Edstone yesterday. He had driven some of his friends home at 5 a.m. from the Warwick Yeomanry Ball which was attended by about 700 persons & went off admirably well—the company consisting of Aristocracy & Farmers' wives & daughters & all merry and in the best humor.

The Barclays are firm allies of ours & are with us most days & reckon upon you being a frequent guest at their house in London. Bassett Saunderson called this morning & tells me that he has seen Lord Rosse's telescope—6 feet diam^{r} & 50 feet long—the [...] of

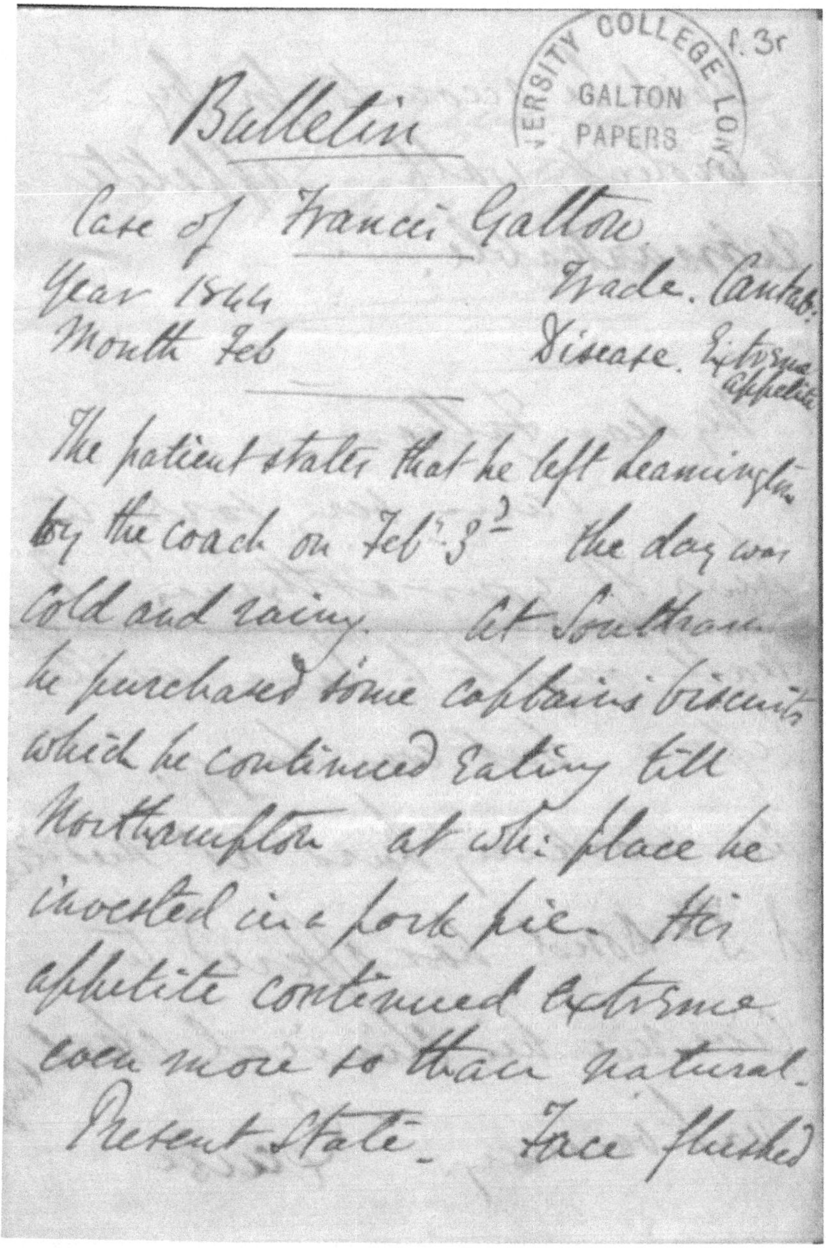

Bulletin

Case of Francis Galton
Year 1844 Trade. Cantab.
Month Feb Disease. Extreme appetite

The patient states that he left Leamington by the coach on Feb. 3d the day was cold and rainy. At Southam he purchased some captains biscuits which he continued eating till Northampton at whi: place he invested in a pork pie. His appetite continued extreme even more so than natural.
Present State. Face flushed

Figure 5.13: 1844-02-13. FG to STG page 1.

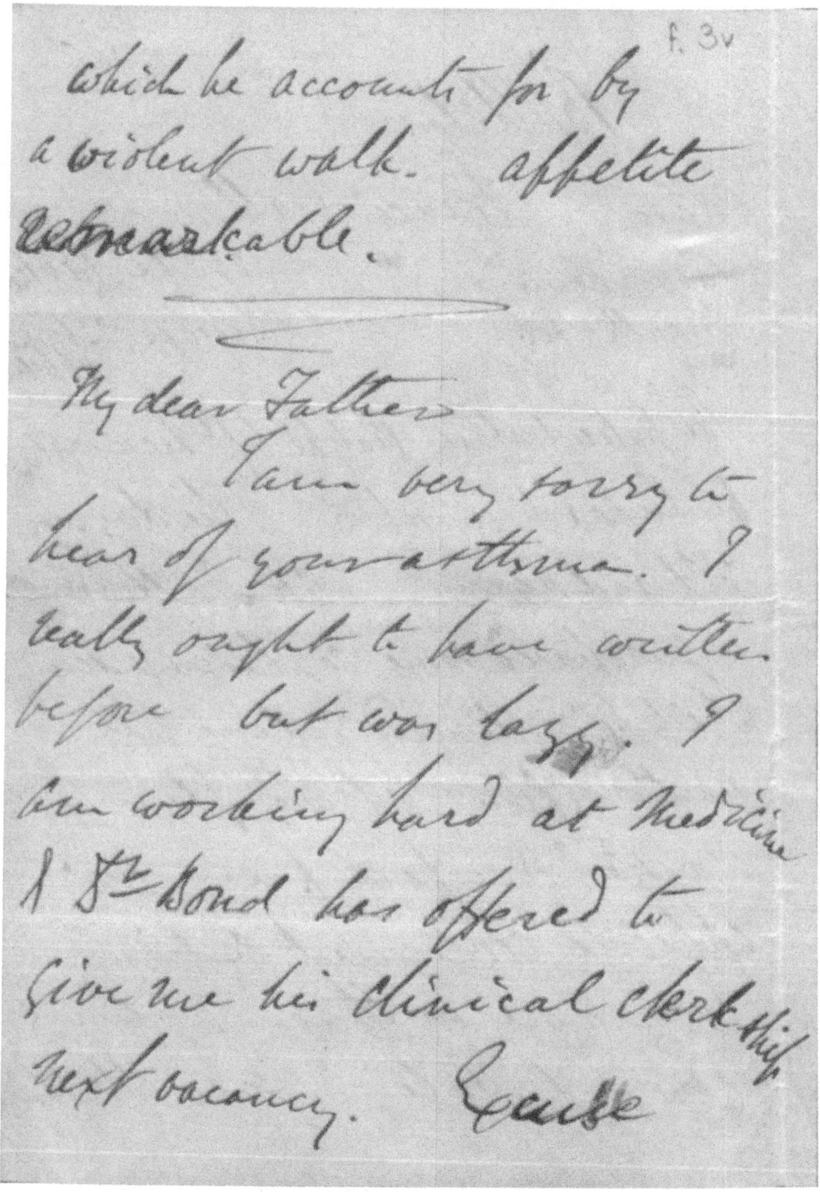

f. 3v

which he accounts for by
a violent walk. appetite
remarkable.

My dear Father
I am very sorry to
hear of your asthma. I
really ought to have written
before but was lazy. I
am working hard at Medicine
& Dr Bond has offered to
give me his clinical clerkship
next vacancy. Excuse

Figure 5.14: 1844-02-13. FG to STG page 2.

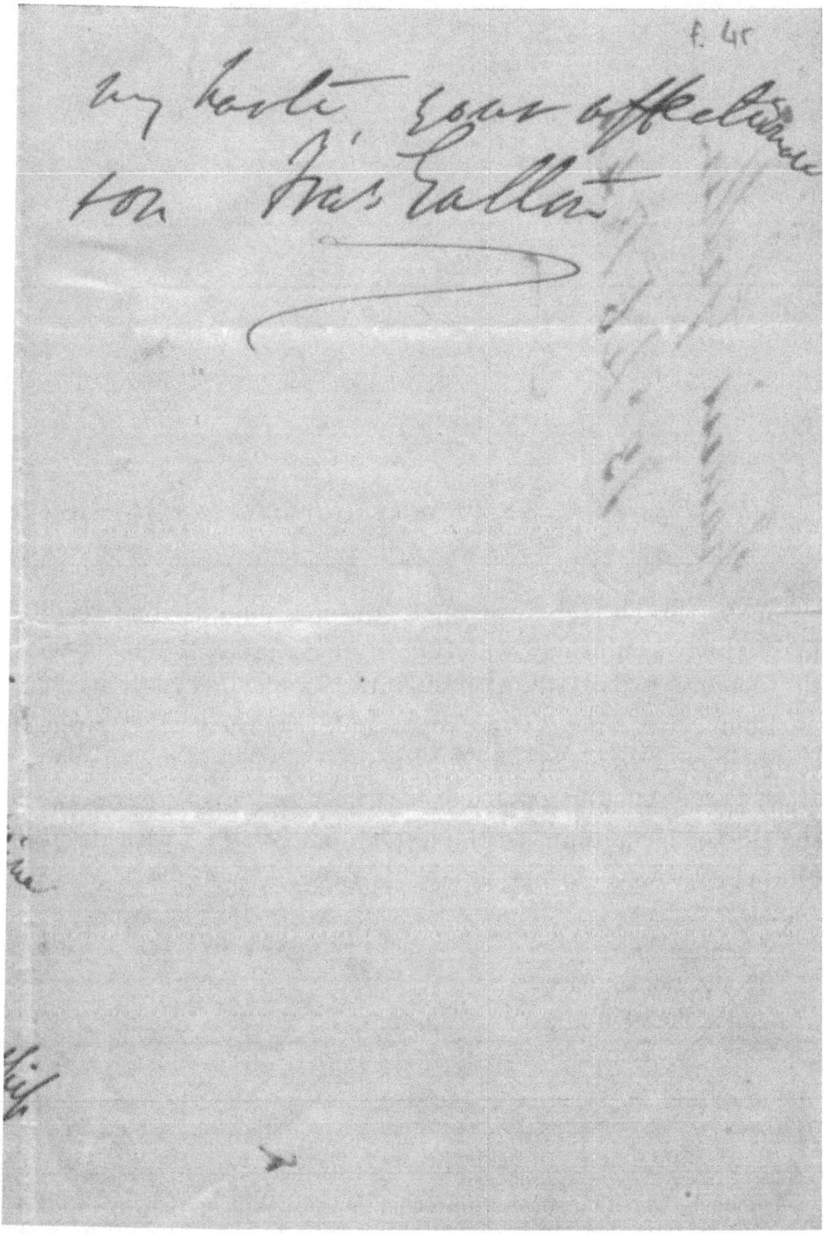

Figure 5.15: 1844-02-13. FG to STG page 3.

Figure 5.16: 1844-02-13. FG to STG page 4.

the moon thro' it is most distinct—the speculum $3\frac{1}{2}$ tons weight.[7] I go to Lucy's on Friday & to Hadzor on Saturday to remain there till Monday. I shall call upon $\mathrm{y^r}$ Aunt [Sophia[8]] on the way.

Emma gives a much worse report of Mr. Gurney whose strength is rapidly sinking. I am afraid he cannot last many days longer.

Bessy & Minnie Barclay have been poring over Hebrew but the Doctor has prohibited the latter from going on with it. Your Mother's Prowetts fit abominably.[9] All send their love.

I remain
Yr affectionate Father
S. Tertius Galton.

Leamington
21 Feb

[7] The Leviathan of Parsonstown, Ireland, built by William Parsons, 3rd Earl of Rosse (1800—1867), between 1842 and 1845. It was the largest telescope in the world for the remainder of the 19th century.

[8] Possibly Sophia Galton (1772–1863), a sister of Samuel Tertius. She was married to Charles Brewin in 1833.

[9] These may have been false teeth from the West End dentist Robert Prowett, of 16 Berkeley Square.

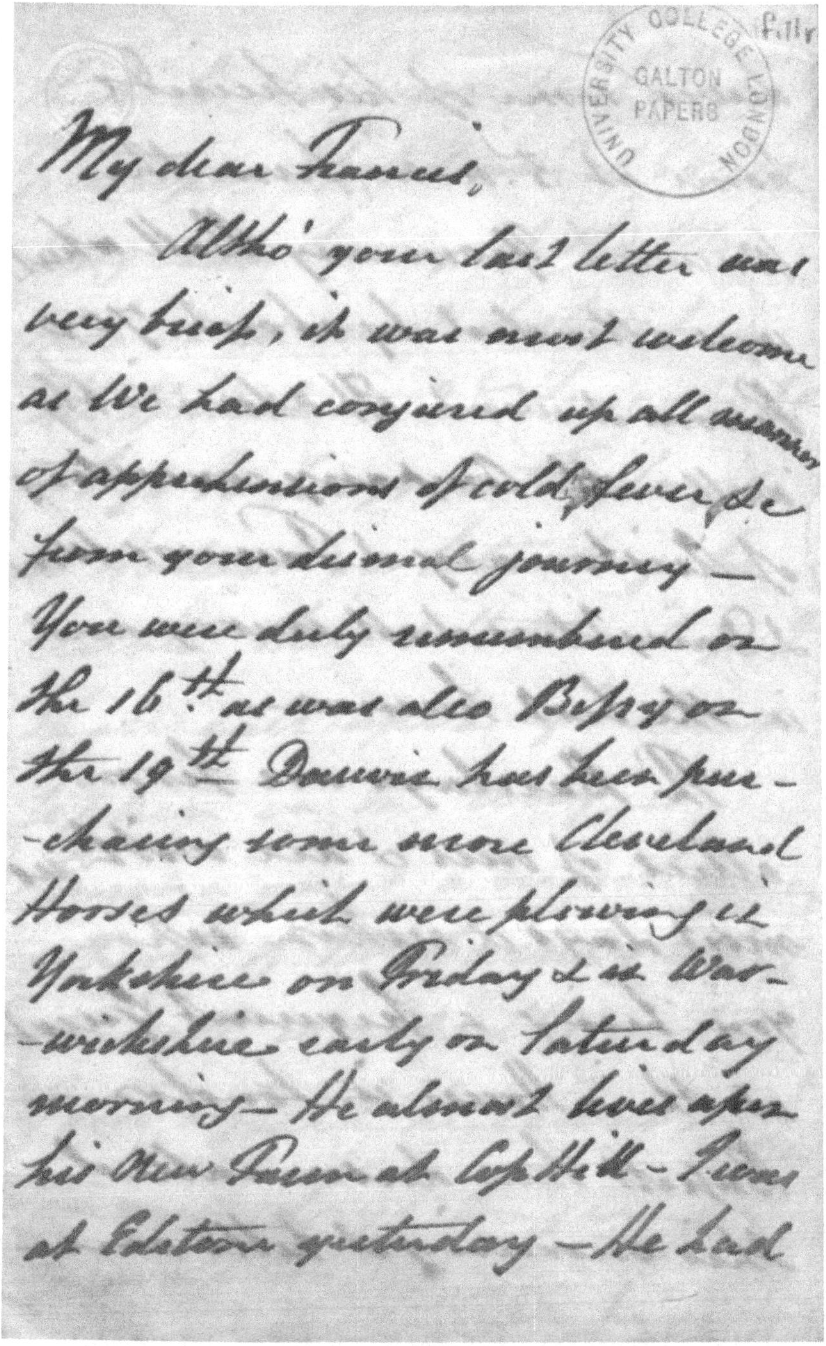

My dear Francis,

Altho' your last letter was very brief, it was most welcome as We had conjured up all manner of apprehensions of cold, fever &c from your dismal journey — You were duly remembered on the 16th as was also Bessy on the 19th Darwin has been pur-chasing some more Cleveland Horses which were plowing in Yorkshire on Friday & in War-wickshire early on Saturday morning — He almost lives upon his New Farm at Cop Hill — I dined at Edstone yesterday — He had

Figure 5.17: 1844-02-21. STG to FG page 1.

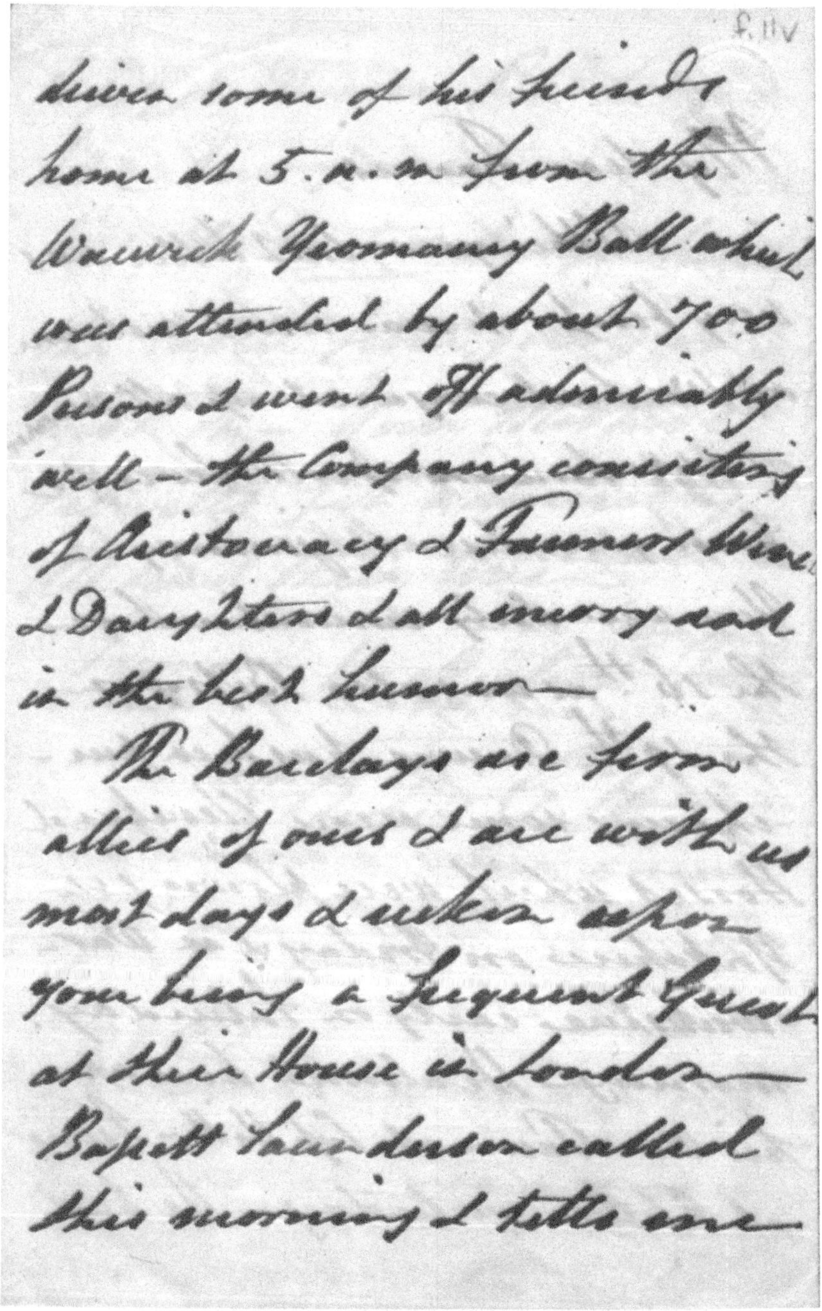

f. 11v

driven some of his friends
home at 5. a. m from the
Warwick Yeomanry Ball which
was attended by about 700
Persons & went off admirably
well — the Company consisting
of Aristocracy & Farmers Wives
& Daughters & all merry and
in the best humour —
The Barclays are firm
allies of ours & are with us
most days & reckon upon
your being a frequent Guest
at their House in London —
Bassett Saunderson called
this morning & tells me —

Figure 5.18: 1844-02-21. STG to FG page 2.

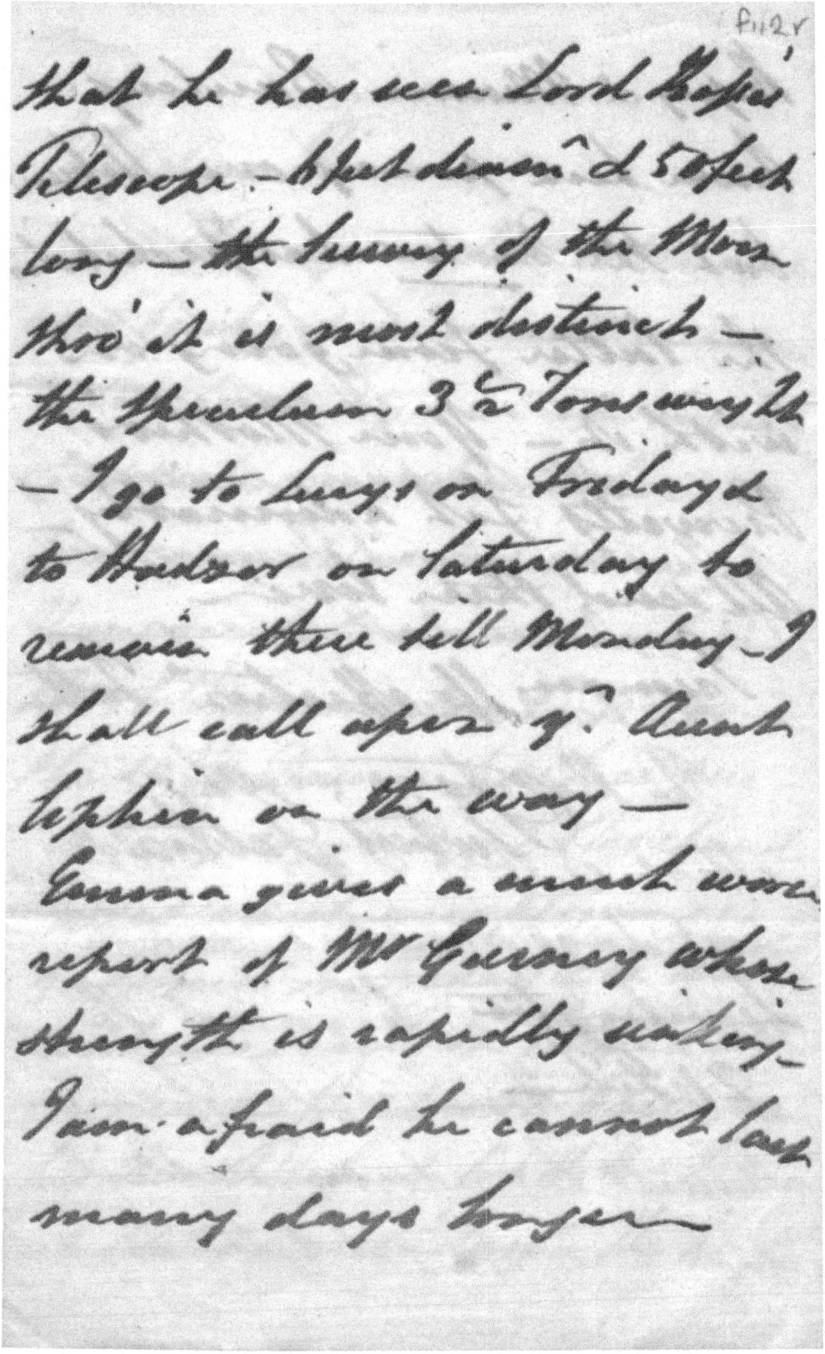
f112r

that he has seen Lord Rosse's
Telescope – 6 feet diam^n & 50 feet
long – the surface of the Moon
thro' it is most distinct –
the Speculum 3 1/2 Tons weight
– I go to Ways on Friday &
to Hudson on Saturday to
remain there till Monday – I
shall call upon y^r Aunt
Sophia on the way –
Emma gives a much worse
report of Mr Gurney whose
strength is rapidly sinking –
I am afraid he cannot last
many days longer –

Figure 5.19: 1844-02-21. STG to FG page 3.

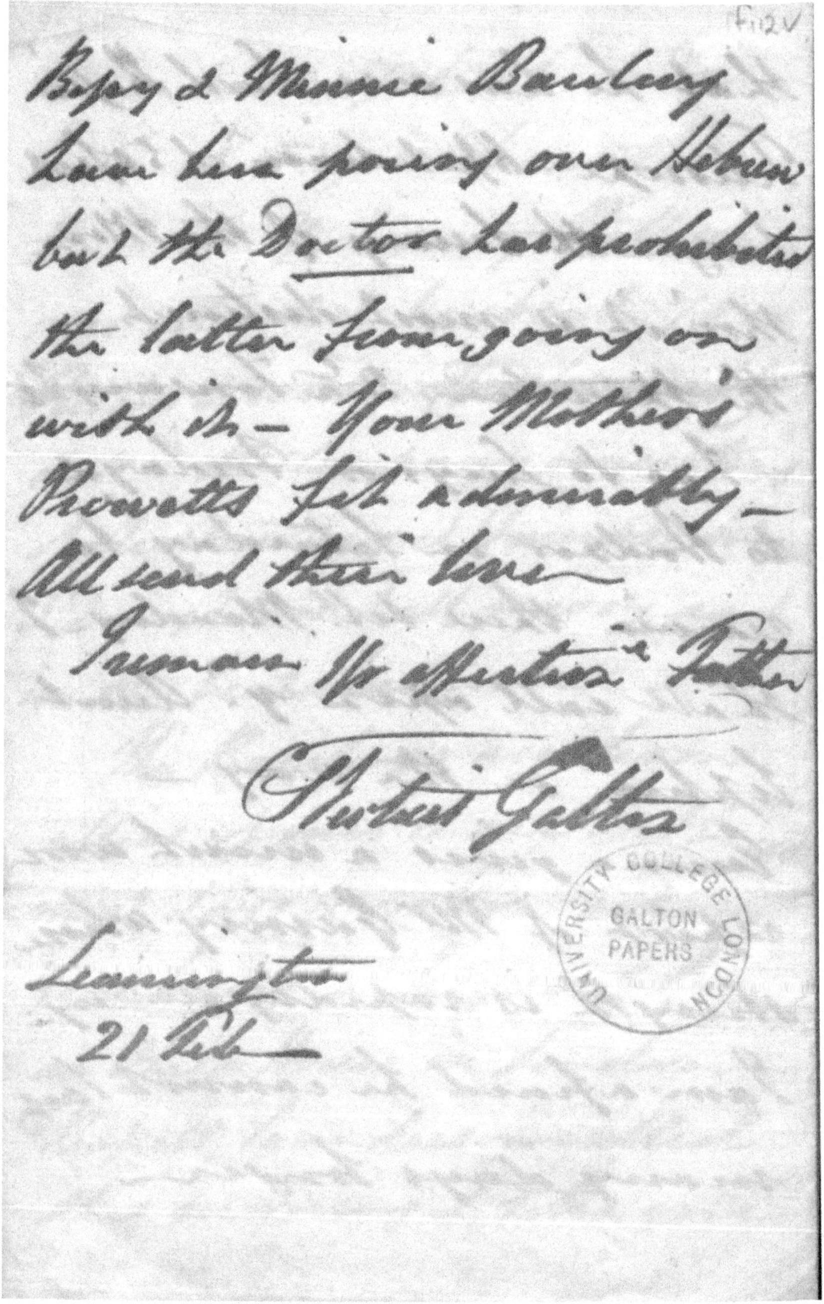
f.12v

Bessy & Minnie Barclay
have been poring over Hebrew
but the Doctor has prohibited
the latter from going on
with it — Your Mother's
Prowetts felt admirably —
All send their love —
I remain Yr affection^e Father
S Tertius Galton

Leamington
21 Feb

Figure 5.20: 1844-02-21. STG to FG page 4.

1844/03/03. Samuel Tertius Galton to Francis Galton

Leamington, 3 Mar. 1844

My dear Francis,

Your letter has given me much pleasure, both because it assures me that you are going on satisfactorily with your studies, and also because the sight of your hand-writing is always welcome to me.

I was last week at Hadzor and also at Selly Hall. Your Uncle Howard is recovering from a severe attack of Gout and comes here on Wednesday to deposit Cameron at Mrs. Hall's school at Warwick & then goes onto Derby, where he will be engaged in making selections from Mr. Strutt's numerous paintings, prints & books of which he is to have the first choice. You have probably heard of the tragical occurrence in Lucy's family by the self-destruction of their Butler, who was found suspended in the [Larders]. Poor Lucy has borne it with as much composure as could have been expected.

The David Barclays left us yesterday much pleased with their visit at Leamington and very anxious to show you every attention (except doctoring them which they decline until after the decease of Dr. Chambers) whilst you are in London. You will find them very agreeable and friendly. We have been most sorry to lose them. Mr. Barclay and Darwin are excellent friends being both very fond of Agriculture. Darwin is to visit him at Eastwick Park at the next British Agricultural Meeting. Darwin has received a letter from Erasmus detailing another invention of his for fastening field gates which is cheap, simple, and effectual. Adèle is busily engaged furnishing the Fry Cottage at Claverdon for her new school, which is to be managed by Mrs. Armitage. a daughter of Mrs Webb. Emma has been over the British Museum with the Hallams[10] where she met Miss Edgeworth. Your Mother & sisters send their kind love. I remain Your affectionate Father

S. Tertius Galton

F. Galton Esq.

[10] Hallam, Henry Fitzmaurice. See page 483.

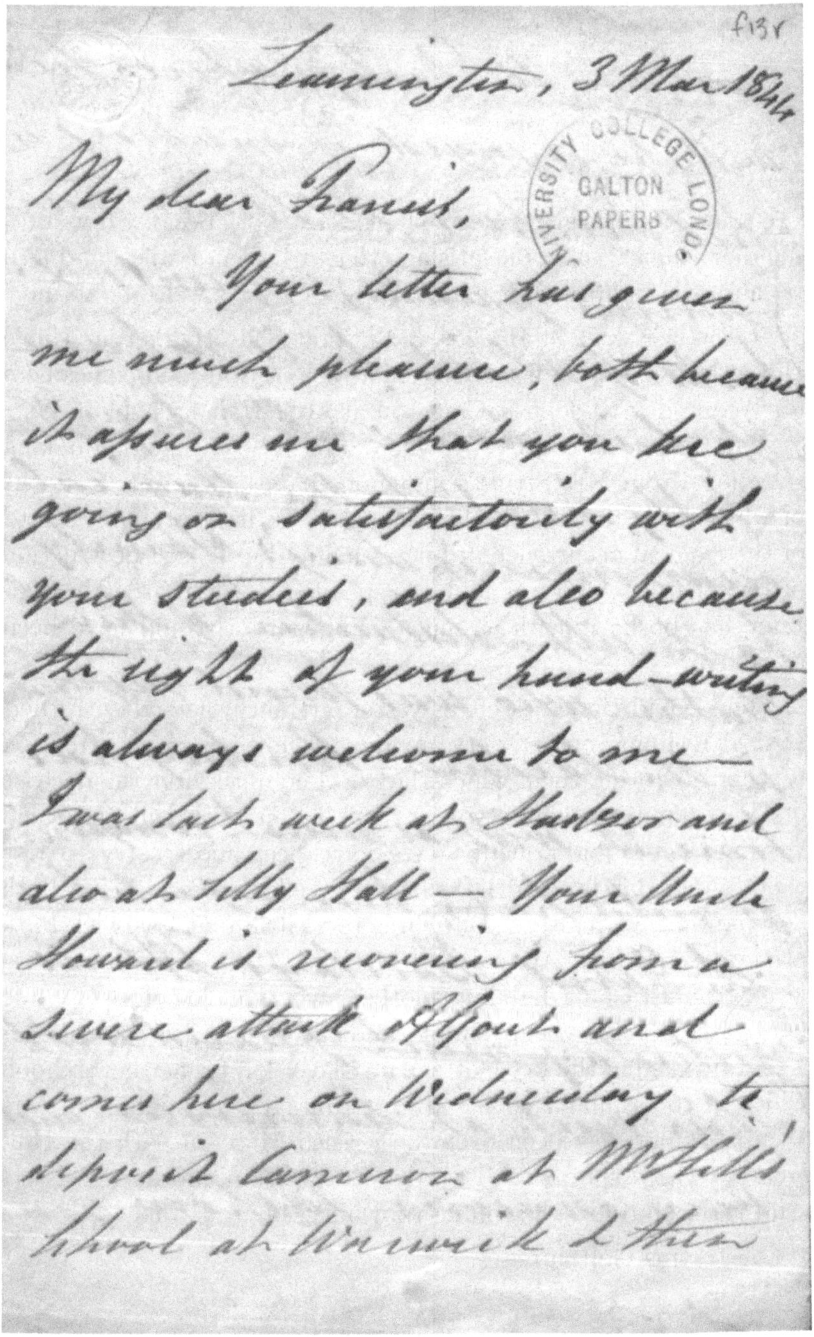

f13r

Leamington, 3 Mar 1844

My dear Francis,

Your letter has given me much pleasure, both because it assures me that you are going on satisfactorily with your studies, and also because the sight of your hand-writing is always welcome to me — I was last week at Hadzor and also at Tilly Hall — Your Uncle Howard is recovering from a severe attack of Gout and comes here on Wednesday to deposit Cameron at Mr Hill's School at Warwick & then

Figure 5.21: 1844-03-03. STG to FG page 1.

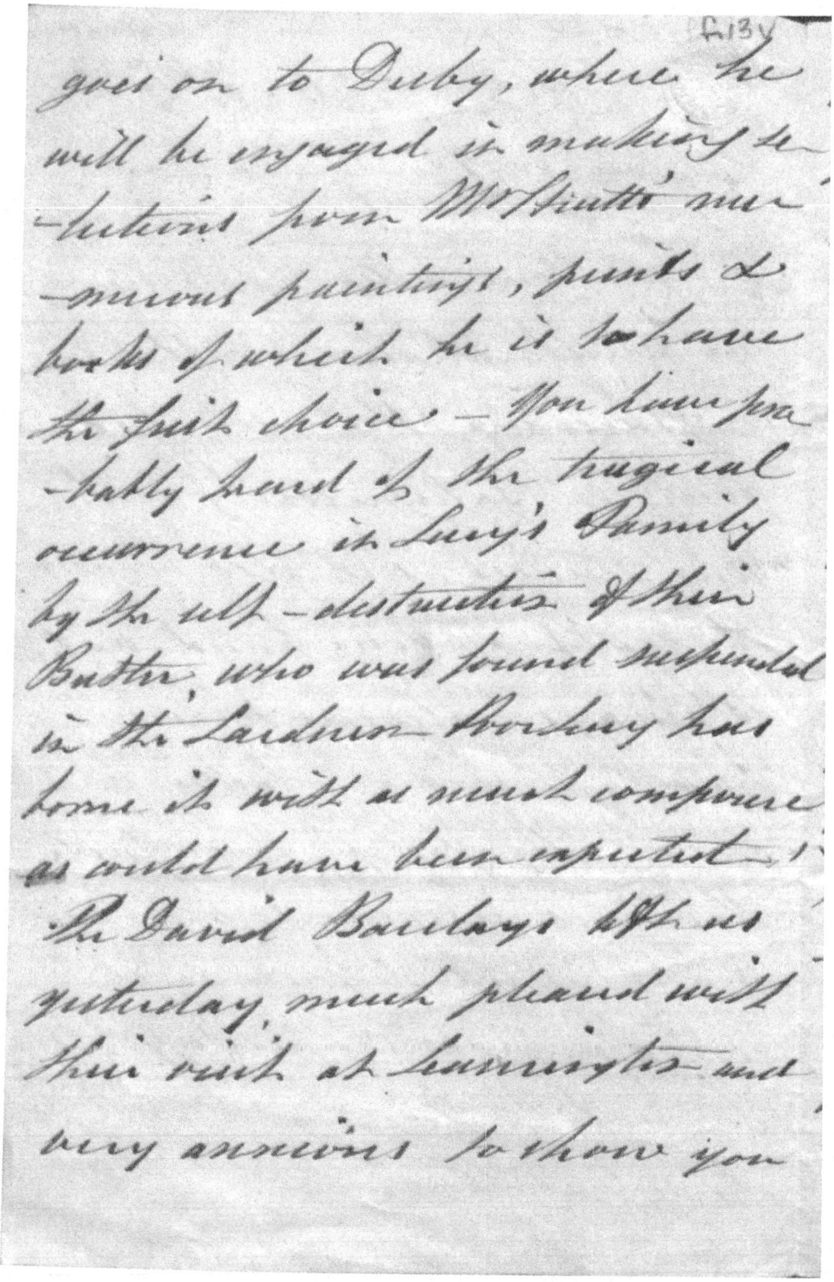

f13v

goes on to Derby, where he
will be engaged in making se-
-lections from Mr Strutt's nu-
-merous paintings, prints &
books of which he is to have
the first choice — You have pro-
-bably heard of the tragical
occurrence in Lucy's family
by the self-destruction of their
Brother, who was found suspended
in the Gardens — Woolsey has
borne it with as much composure
as could have been expected —
The David Barclays left us
yesterday, much pleased with
their visit at Leamington and
very anxious to show you

Figure 5.22: 1844-03-03. STG to FG page 2.

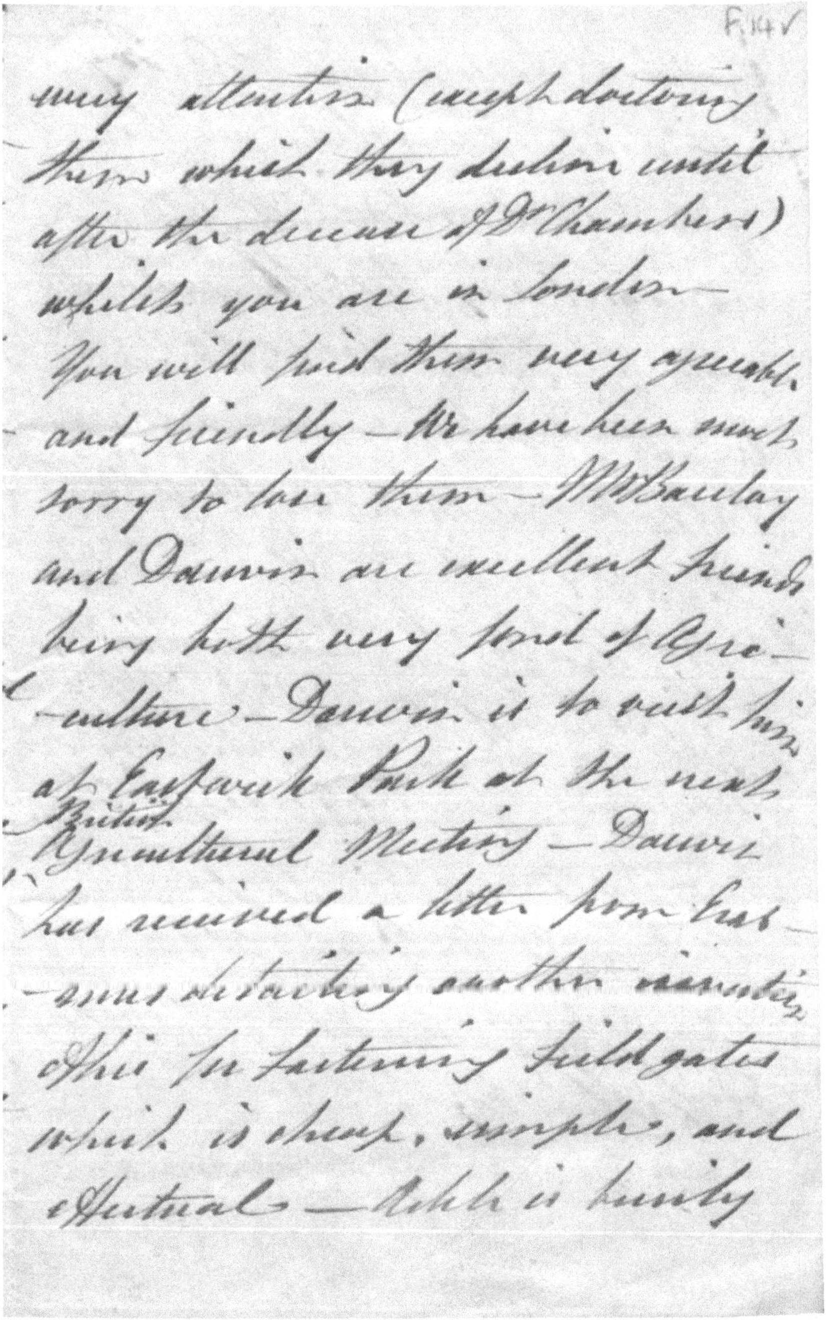
F.14 ✓

very attentive (except doctoring them which they decline until after the decease of Dr Chambers) whilst you are in London — You will find them very agreeable and friendly — We have been most sorry to lose them — Mr Barclay and Darwin are excellent friends being both very fond of Agri-culture — Darwin is to visit him at Eastwick Park at the next British Agricultural Meeting — Darwin has received a letter from Eras-mus detailing another invention of his for fastening field gates which is cheap, simple, and effectual — Adele is kindly

Figure 5.23: 1844-03-03. STG to FG page 3.

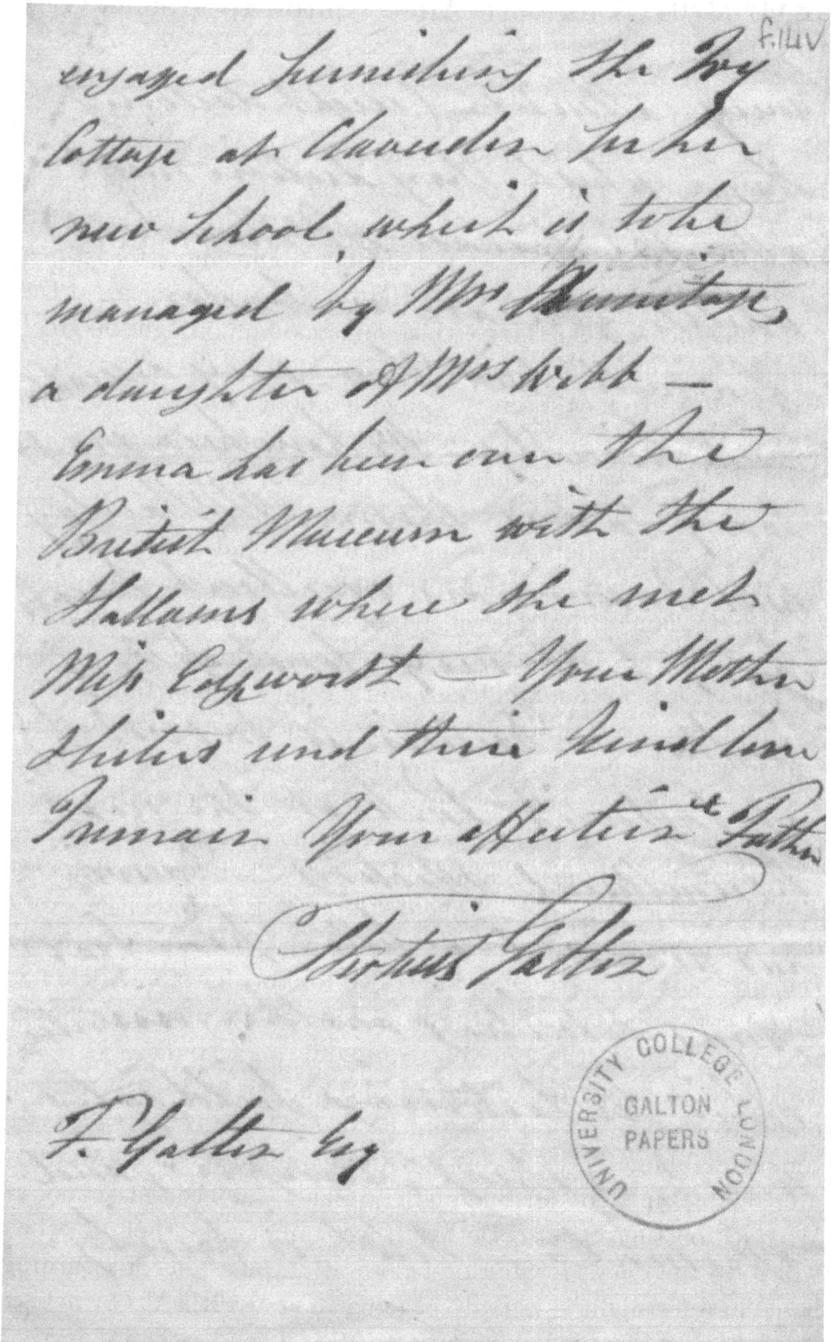
f.144v

engaged furnishing the Ivy
Cottage at Claverdon for her
new School, which is to be
managed by Mrs Armitage,
a daughter of Mrs Webb —
Emma has been over the
British Museum with the
Hallams where she met
Miss Coxworth — Your Mother
Sisters and their kind love
Remain Your affectionate Father

S Tertius Galton

F. Galton Esq

Figure 5.24: 1844-03-03. STG to FG page 4.

1844/03/04. Elizabeth Anne Galton to Francis Galton

My dearest Francis

Thank you much for your kind attention to my father. Your two letters, especially the one today, has done my father much good. He has had the one rec^{d} today read over & over again to him. He is very poorly, & we are anxious about him. His asthma left him suddenly when he was at Hadzor, & some messages at the canal meeting & other things which annoyed him brought on great depression & he returned home very low & ill, & has continued so ever since, & we have been very uneasy about him. This morning at 3 o'clock his nose began to bleed, & continued till 9 o'clock, when he had lost three quarters of a pint of blood, & was very feeble & faint. Jephson[11] & Dr. Prichard have both seen him, & say it was most fortunate that he had this nose bleeding as it has probably saved an attack of apoplexy. They say there is no immediate danger, but he will require care & to be occasionally bled with leeches. He has been very feeble & drowsy all day, but his mind [is], I am thankful to say, quite calm & comfortable again, & tonight he is certainly much better, & I do hope & trust now he will do well. I will write a line tomorrow to say how he is. In about a week, or before if you can, write to him again, just such a letter as you sent today. It will please him hearing from you, & knowing you are anxious about his health. I am very glad you are doing so well in the doctoring line, & like it so much. Darwin is here every day. Adèle is just come from Edstone, having been there with the Barclays. Mr. B. gave Dr. March £50 for the Jews, & Adèle has almost converted him to teetotalism. I am glad temperance agrees with you. I always thought you required it, on account of y^{r} blood shot eyes. I hope you will go to Shrewsbury at Easter, as you ought to see Uncle Bob[12] before he dies. My mother will write tomorrow about it, as she is too tired tonight. The Howards[13] are coming on Wed^{y} for one night to place Cameron at school at Warwick on their way to Derby. If we cannot take them in, they will go to the Landsdowne Hotel close by. Emma tells of returning home in a fortnight as a Lady is going to stay with A^{t} Gurney, & can take our place for a short time. We all send our best love to

---

[11] Jephson, Dr. Henry. See page 485.

[12] Dr. Robert Waring Darwin, father of the naturalist Charles Darwin.

[13] Uncle Howard Galton and his wife.

you, & my Father has said over & over again "give my affect. love to my dear Francis".

Believe me ever,
Dearest Fras.
Yr very affect. sister
EA Galton
Monday Even^{g} 4th March.
1844[14]

1844/03/06. Francis Galton to Samuel Tertius Galton

My dear Father

Will you tell Bessy that I received her letter just after I had put my last into the post and thank her much for it. I see young Barclay occasionally we have breakfasted at each other's room and are good friends when we meet, but I have now so little spare time at my disposal, being the whole morning in attendance on medical lectures &c. that I have not been able to go out much lately and consequently have rarely met him. I get more and more fond of medicine every day. I am trying some new ways of taking cases, or rather the outlines of cases by lines drawn under each particular symptom and varying according to its severity every day or every 2nd day as the case may be. In fact like the ordinary plan of statistical charts. It seems of great use for noting cases quickly, since you can do all you want by the bedside of the patient and when going round with the physician which w^{d} be quite out of the question in the ordinary way of proceeding and then many cases are noted wh: otherwise used [to] be neglected.

Do you remember my mother and myself talking about the con nection of gout and asthma? I asked several medical men whether they had ever observed any and they all said no, when curiously enough yesterday Dr. Haviland stated in his lectures that "from a wide observation he c^{d} not help thinking that gout and asthma had certain connections wh. have not yet been investigated." I shall certainly look out for cases that way, for it w^{d} be very curious if such apparently unlike diseases were after all related. Dr. Haviland spoke much of very strong coffee as often being of very great service in asthma—that or tea, which is much the same, for their active principles are identical, I know you have found [them] good only you don't take the former strong enough.

14 Pencil annotation.

Does my mother still adhere to her intention of accompanying me to Shrewsbury next Easter, will you ask her to write about it? I shall have I hope nearly a fortnight altogether but ~~should~~ must spend a week at home to talk over future plans and Bob Sawyer dodges for getting into practice & so on, with you.

You will probably have heard from Emma, who found it out through the Hallams, that I am a tea totaller of about a month standing. It suits with my constitution gloriously—but warm advocate as I am of the cause, whatever you do, my dear Father don't lower yourself, as wine is a most necessary medicine for you. I am very glad I have taken the pledge. I told Delly my reasons who will tell them you. It was not done without a term's previous consideration.

Your affectionate son
F. G.

1844/03/09. Francis Galton to Samuel Tertius Galton

Saturday morning

My dear Father

As I was not able myself to enter into learned consultation with Pritchard and Dr. Jephson,[15] I cannot altogether give up my privilege of "family doctor" and so will write this letter full of prescriptions. But first I must truly congratulate you on your convalescence which Delly tells me is in capital progress, and, as I presume, only wanting a little hospital-patient discipline to make it perfect. Now my prescriptions are

1st. That the Hospital Patient do on no occasion feel his pulse.

2nd. That the H. P. do never look in the glass to see whether his eyes are red.

3rd. That the H. P. do never examine his own health with a view to self-doctoring.

4th. That the H. P. do make improvements at Claverdon and commit prisoners at Leamington when so inclined, but that he never attend canal-meetings, nor put himself to inconvenience, or anxiety.

[15] Jephson, Dr. Henry. See page 485.

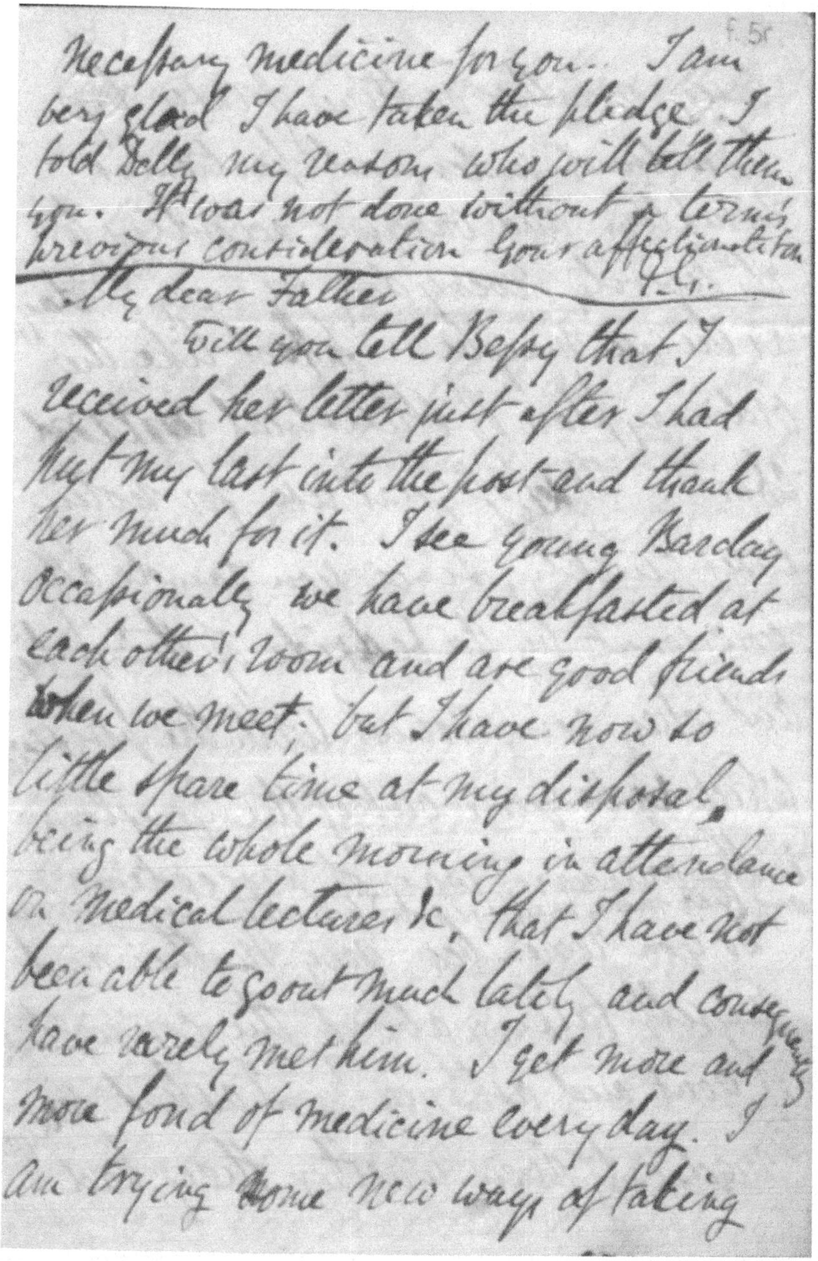

f. 5r

necessary medicine for you. I am
very glad I have taken the pledge. I
told Dolly my reasons who will tell them
you. It was not done without a very
previous consideration. Your affectionate son
F.G.

My dear Father

Will you tell Bessy that I
received her letter just after I had
sent my last into the post and thank
her much for it. I see young Barclay
occasionally we have breakfasted at
each other's room and are good friends
when we meet, but I have now so
little spare time at my disposal,
being the whole morning in attendance
on medical lectures &c, that I have not
been able to go out much lately and consequently
have rarely met him. I get more and
more fond of medicine every day. I
am trying some new ways of taking

Figure 5.25: 1844-03-06. FG to STG page 1.

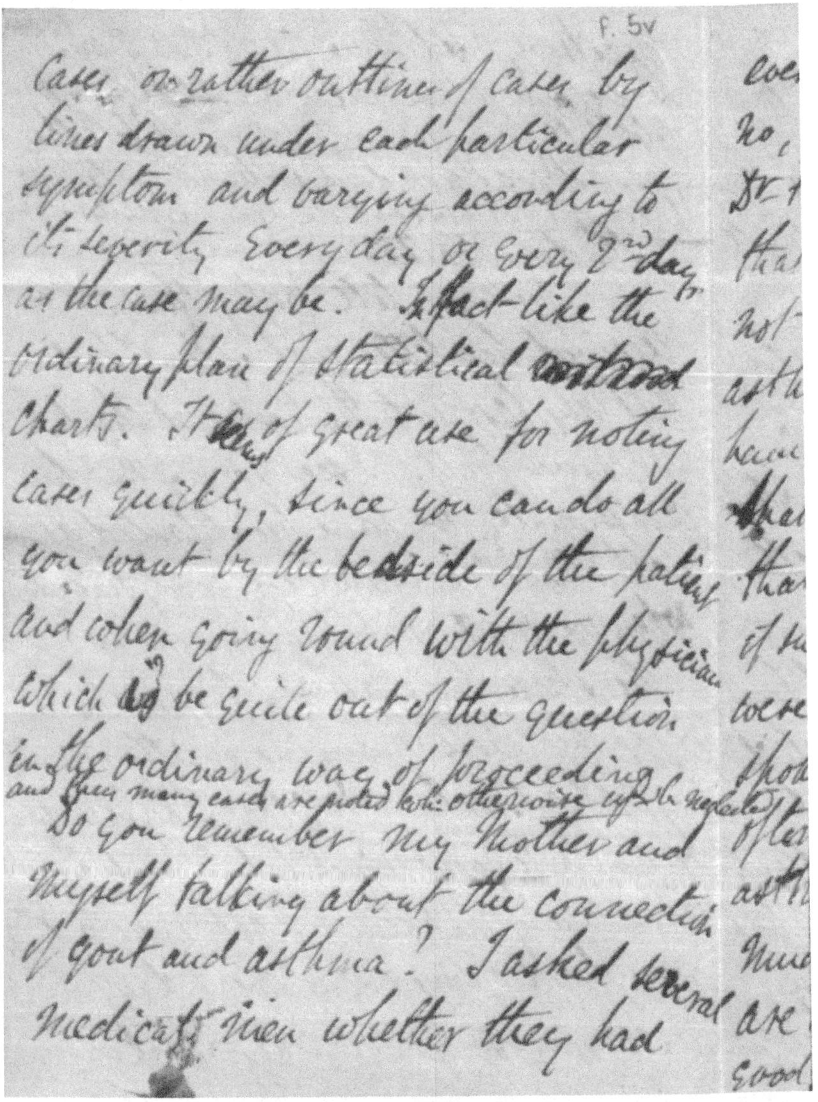
F. 5v

cases, or rather outlines of cases by
lines drawn under each particular
symptom and varying according to
its severity every day or every 2nd day
as the case may be. In fact like the
ordinary plan of statistical ~~[illegible]~~
charts. It is of great use for noting
cases quickly, since you can do all
you want by the bedside of the patient
and when going round with the physician
which is be quite out of the question
in the ordinary way of proceeding
and then many cases are noted which otherwise ought to be neglected
Do you remember my Mother and
myself talking about the connection
of gout and asthma? I asked several
medical men whether they had

Figure 5.26: 1844-03-06. FG to STG page 2.

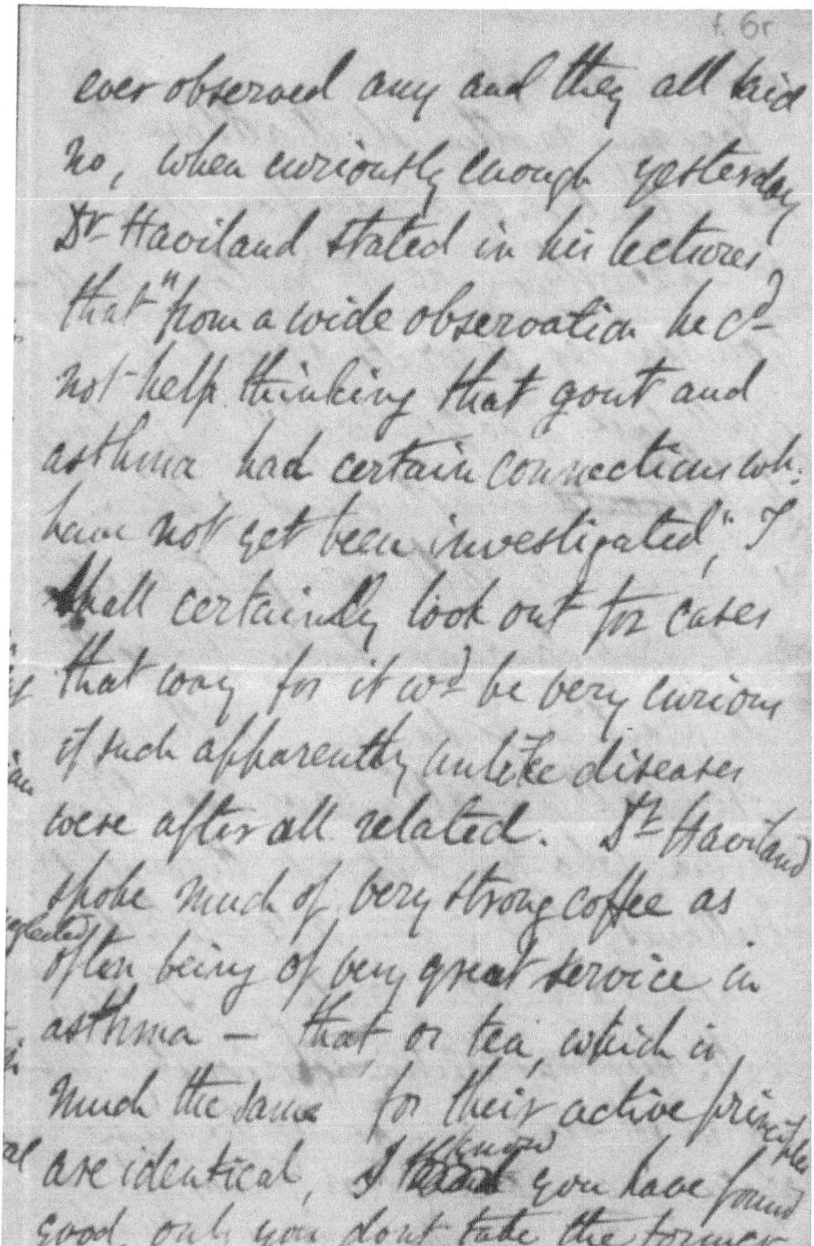
ever observed any and they all said
no, when curiously enough yesterday
Dr Haviland stated in his lectures
that "from a wide observation he cd-
not help thinking that gout and
asthma had certain connections wh.
have not yet been investigated". I
shall certainly look out for cases
that way for it wd be very curious
if such apparently unlike diseases
were after all related. Dr Haviland
spoke much of very strong coffee as
often being of very great service in
asthma — that or tea, which is
much the same for their active principle
are identical, I know you have found
good, only you dont take the former

Figure 5.27: 1844-03-06. FG to STG page 3.

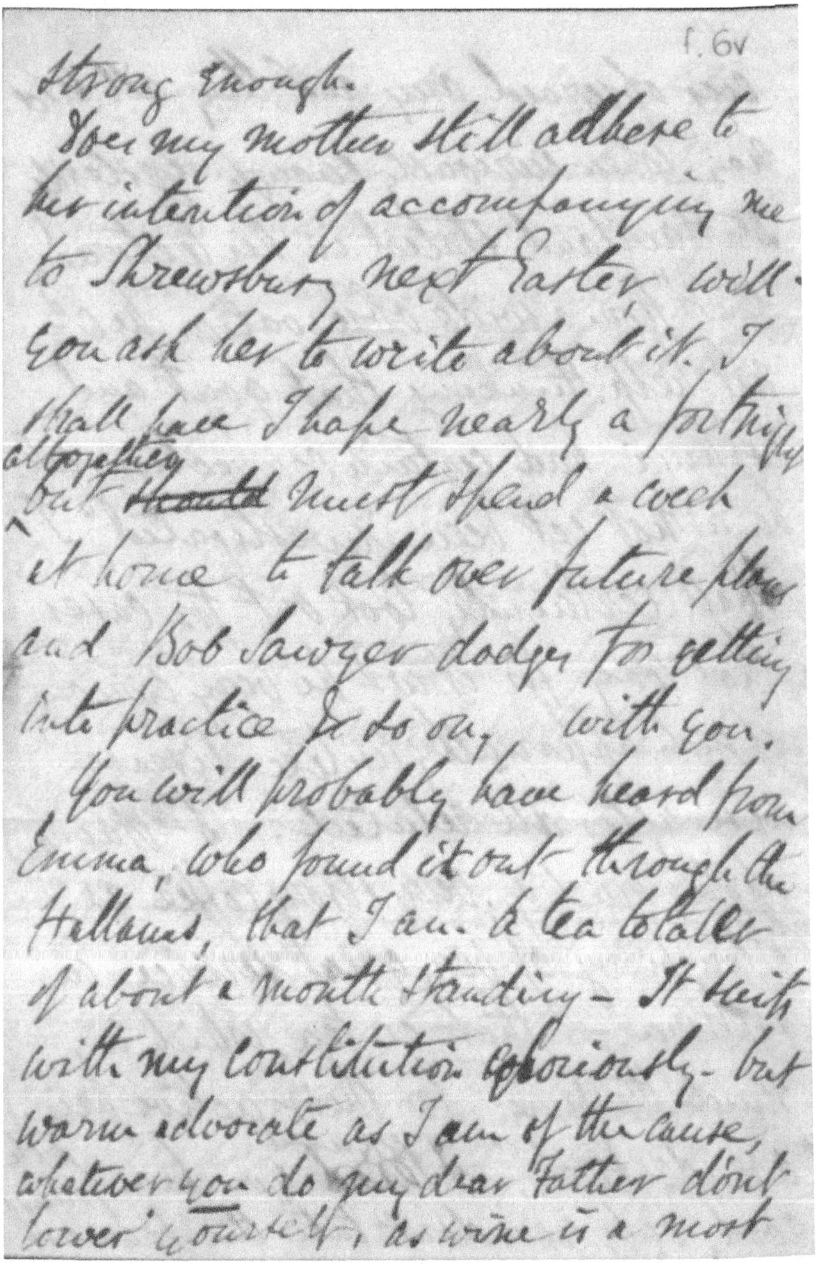
f. 6v

strong enough.
Does my mother still adhere to her intention of accompanying me to Shrewsbury next Easter will you ask her to write about it. I shall have I hope nearly a fortnight altogether but ~~I should~~ must spend a week at home to talk over future plans and Bob Sawyer dodges for getting into practice &c so on, with you.
You will probably have heard from Emma, who found it out through the Hallams, that I am a tea totaller of about a month standing— It suits with my constitution gloriously— but warm advocate as I am of the cause, whatever you do my dear Father don't lower yourself, as wine is a most

Figure 5.28: 1844-03-06. FG to STG page 4.

Figure 5.29: 1844-03-06. FG to STG page 5.

5th. That the H. P. do henceforth enjoy an "otium cum dignitate,"[16] and leave hard work to younger heads for whom it is a duty.

And now my dear Father I have finished doctoring for the present, but shall go on writing doctor's letters until I hear that you obey my rules, and treat your own constitution with the respect it deserves for having brought you through asthma, hard work at banking and anxieties of all sorts for so long. Indeed it is a highly meritorious constitution, and fairly deserves rest. I hope to be with you in about a fortnight, but the exact time is not yet fixed, however I shall know before another three days, when I will write. At present I still continue full work at medicine & am reading Hippocrates and Aretaeus in which we are examined for the MB degree.

It is now my lecture time and so not to lose a post I send my letter unfinished but will write again on Monday.

Your very affectionate Son
Fras. Galton

1844/03/09. Samuel Tertius Galton to Francis Galton

Leamington 9 Mar
1844

My dear Francis

Your letter cheered me much at a time I was again laid up with debility after a profuse bleeding of the nose from the effects of which I am gradually recovering, so as to be able to take a short ride out this morning for the first time—and as old age creeps upon one & these occasional relapses remind me that "non sum quales eram".[17] I shall take the hint and withdraw myself from all anxious business.

I am extremely glad that you take so fondly to your profession upon every account as an occupation useful to yourself and to others, and as a source of pecuniary independence which, after all, is among the number of our duties to promote. My own case favours the notion that there may be a relationship between asthma & gout—but I have found the former the first born by many years.

[16] Dignified leisure.
[17] I am not what I used to be.

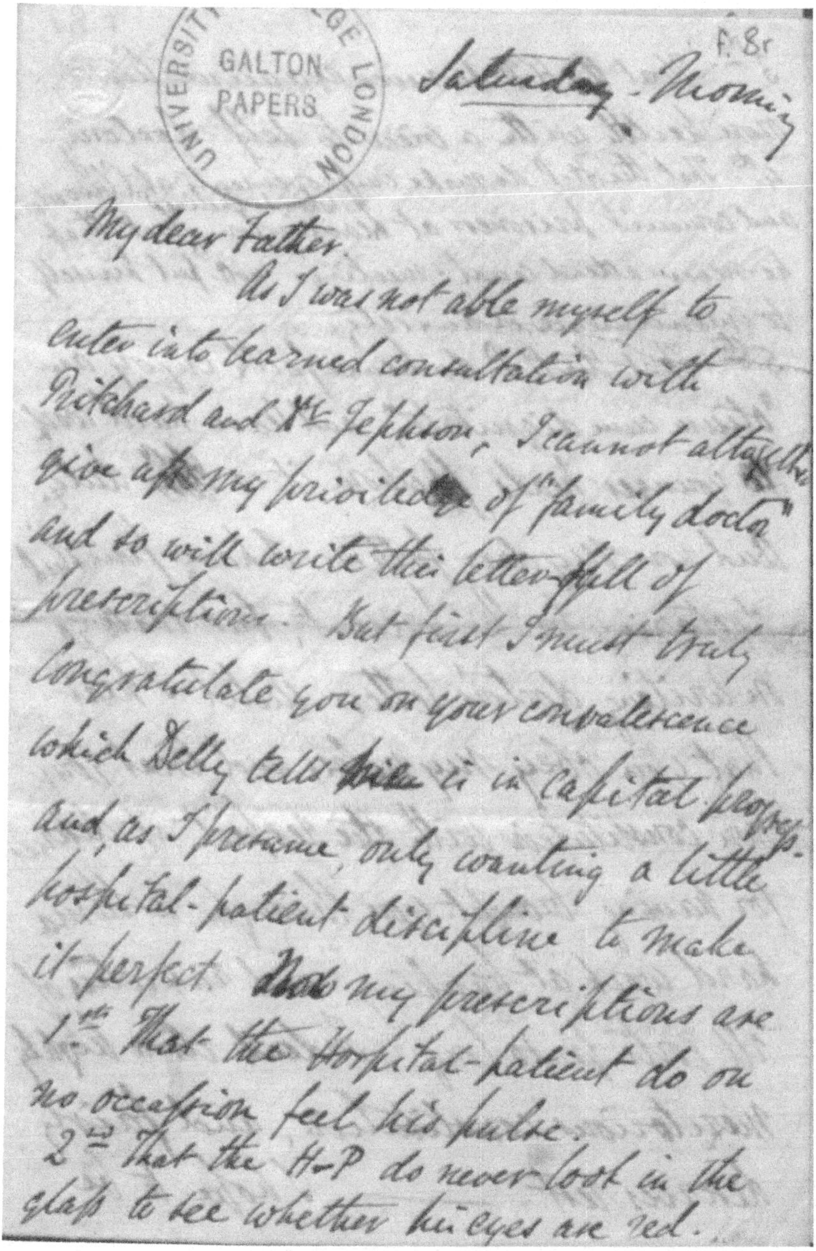

Saturday Morning

My dear Father.
As I was not able myself to enter into learned consultation with Pritchard and Mr Jephson, I cannot altogether give up my privilege of "family doctor" and so will write this letter full of prescriptions. But first I must truly congratulate you on your convalescence which Delly tells ~~him~~ is in capital progress. And, as I presume, only wanting a little hospital-patient discipline to make it perfect. ~~Now~~ my prescriptions are
1st That the Hospital-patient do on no occasion feel his pulse.
2nd That the H-P do never look in the glass to see whether his eyes are red.

Figure 5.30: 1844-03-09. FG to STG page 1.

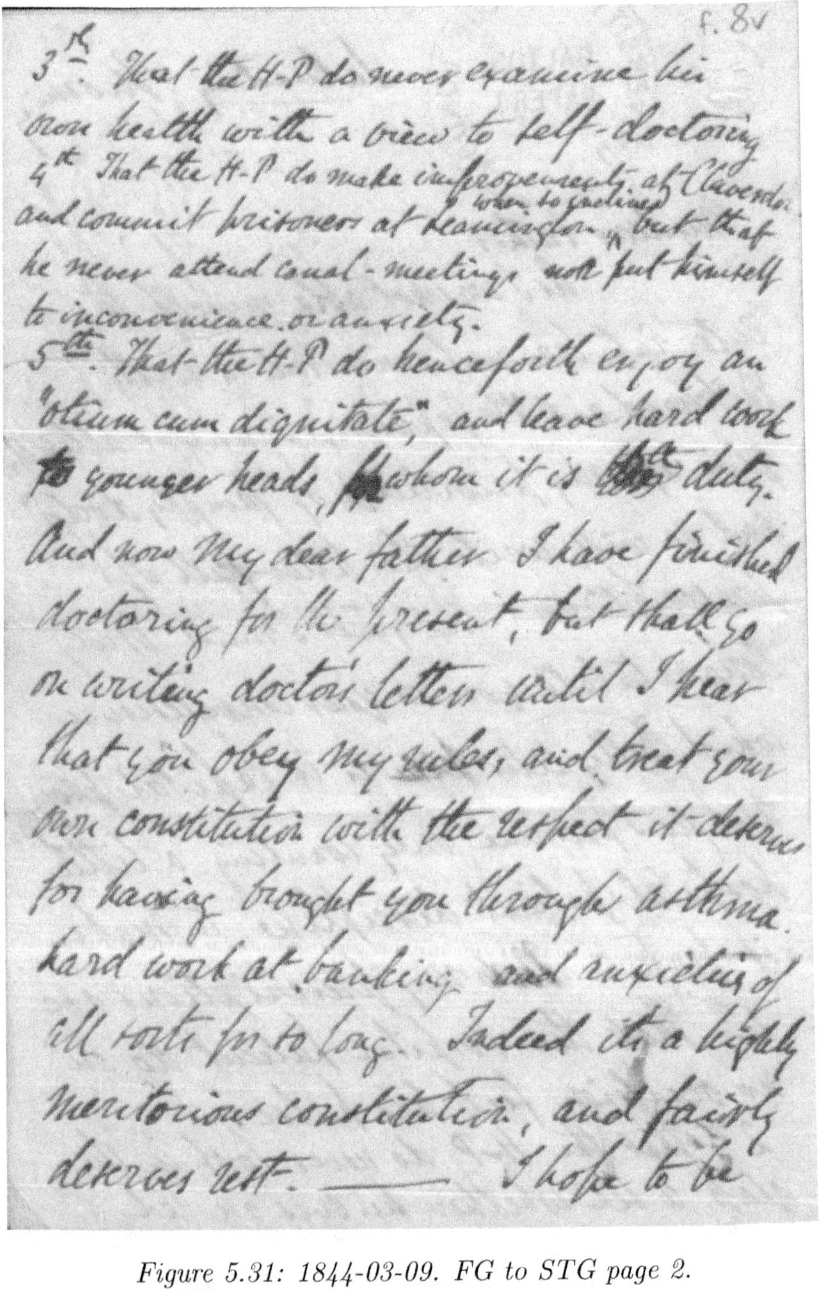

f. 8v

3rd. That the H-P do never examine his own health with a view to self-doctoring

4th That the H-P do make improvements at Claverdon and commit prisoners at Leamington, when so inclined, but that he never attend canal-meetings nor put himself to inconvenience or anxiety.

5th. That the H-P do henceforth enjoy an "otium cum dignitate," and leave hard work to younger heads, to whom it is the duty.

And now my dear father I have finished doctoring for the present, but shall go on writing doctor's letters until I hear that you obey my rules, and treat your own constitution with the respect it deserves for having brought you through asthma, hard work at banking, and anxieties of all sorts for so long. Indeed its a highly meritorious constitution, and fairly deserves rest. —— I hope to be

Figure 5.31: 1844-03-09. FG to STG page 2.

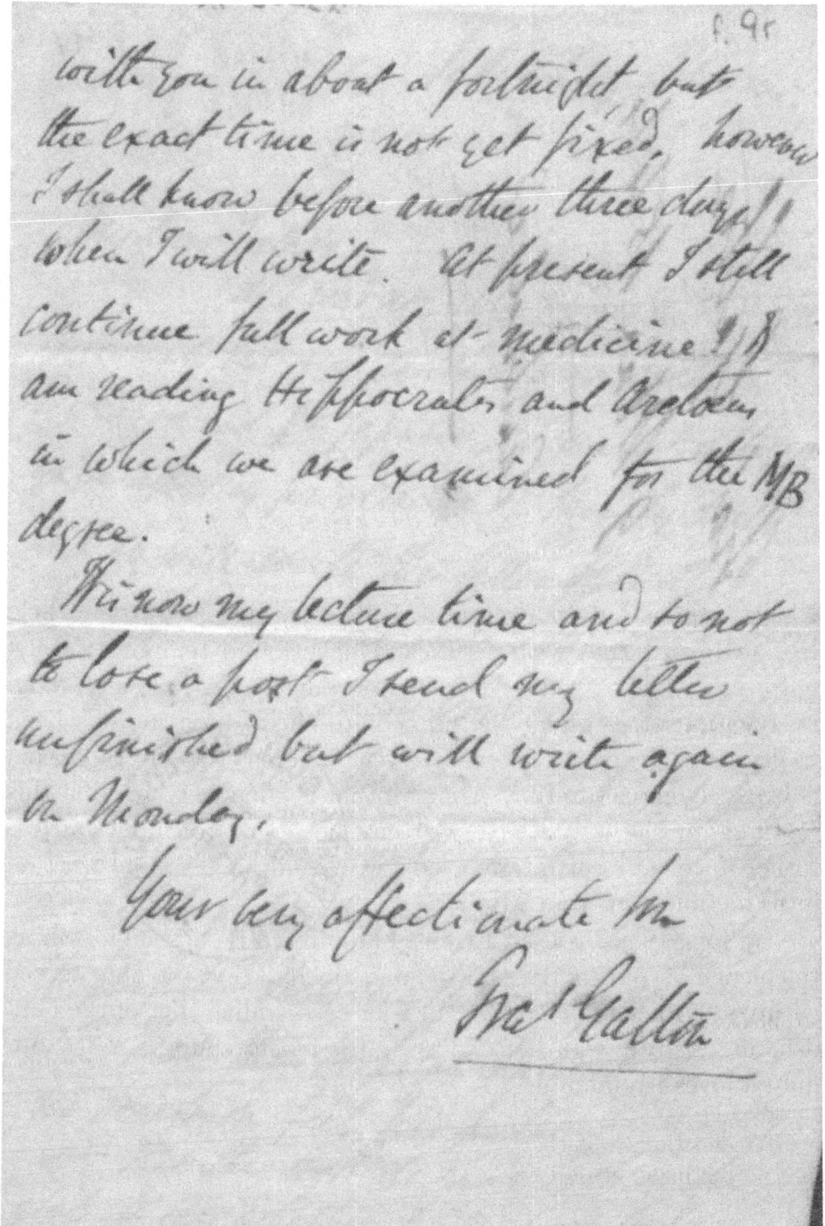

f. 95

with you in about a fortnight, but the exact time is not yet fixed, however I shall know before another three days when I will write. At present I still continue full work at medicine!! I am reading Hippocrates and Aretaeus in which we are examined for the MB degree.

It is now my lecture time and so not to lose a post I send my letter unfinished but will write again on Monday.

Your very affectionate son

Fra^s Galton

Figure 5.32: 1844-03-09. FG to STG page 3.

Figure 5.33: 1844-03-09. FG to STG page 4.

I have been asthmatic ever since 12—gouty since 45. Your Uncle Howard also began with asthma. I have found temporary relief only from Coffee & other stimulants, but always followed by an aggravation of asthma—the immediate effect upon me has been to dry up the expectorations by fever—whereas the effectual relief is by freely expectorating. Your Mother is gone to Claverdon this morning, or she would have written—her going to Shrewsbury is rather uncertain but in any event, we would have you go with an introduction from your Mother.

We long to see you at Easter. I admire your courage in taking the pledge & your motives for it—and am glad that the plan agreed with you. Adèle tells me that, in your case, unlike that of the Gin-drinking Lady, resolution was rewarded beforehand. With our united loves I remain ever

Yr affectionate Father
S. Tertius Galton

1844/03/10. Samuel Tertius Galton to Francis Galton

Leamington, 10 Mar
1844

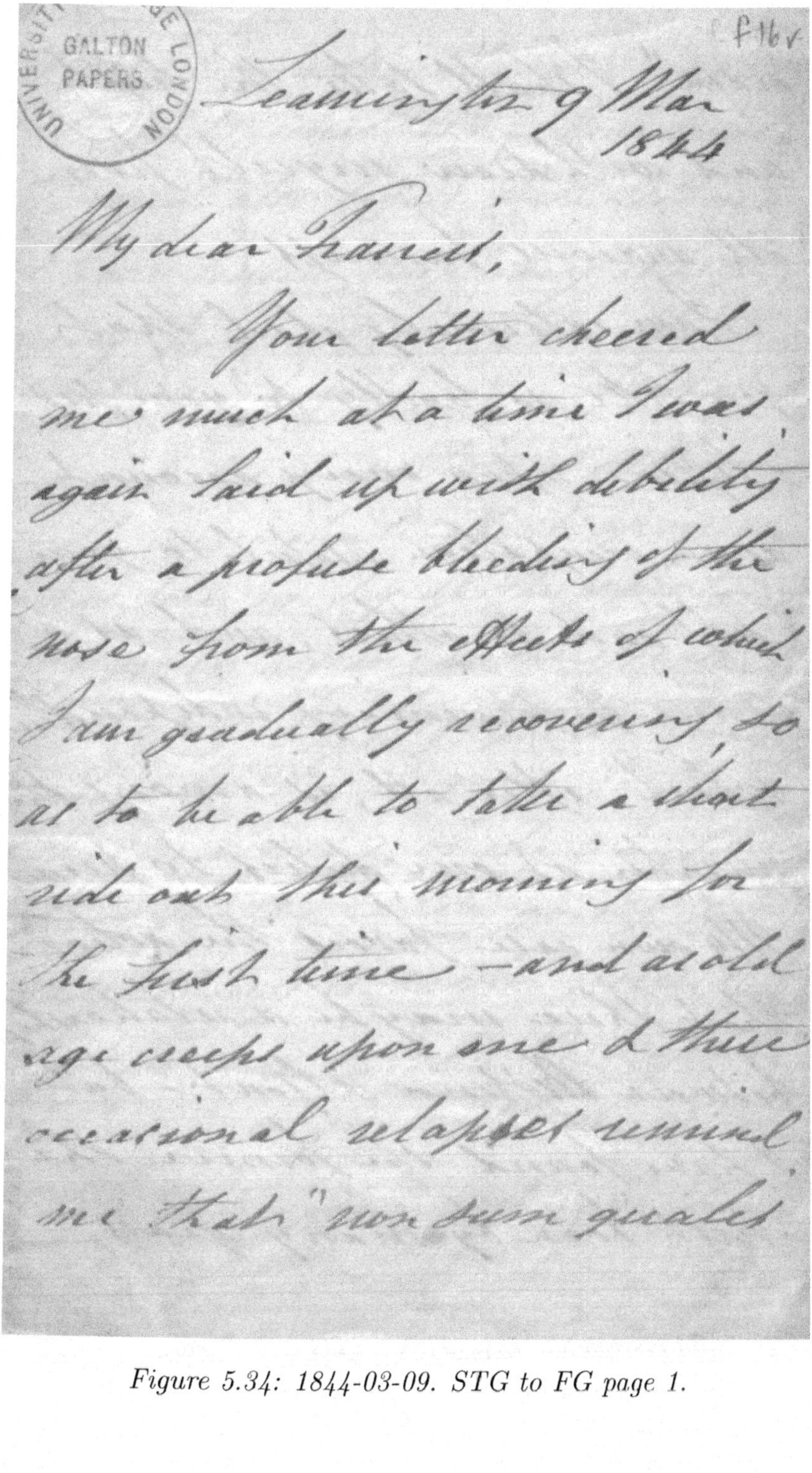

f16r

Leamington 9 Mar
1844

My dear Francis,

Your letter cheered me much at a time I was again laid up with debility after a profuse bleeding of the nose from the effects of which I am gradually recovering, so as to be able to take a short ride out this morning for the first time —and as old age creeps upon me & these occasional relapses remind me that "non sum quales

Figure 5.34: 1844-03-09. STG to FG page 1.

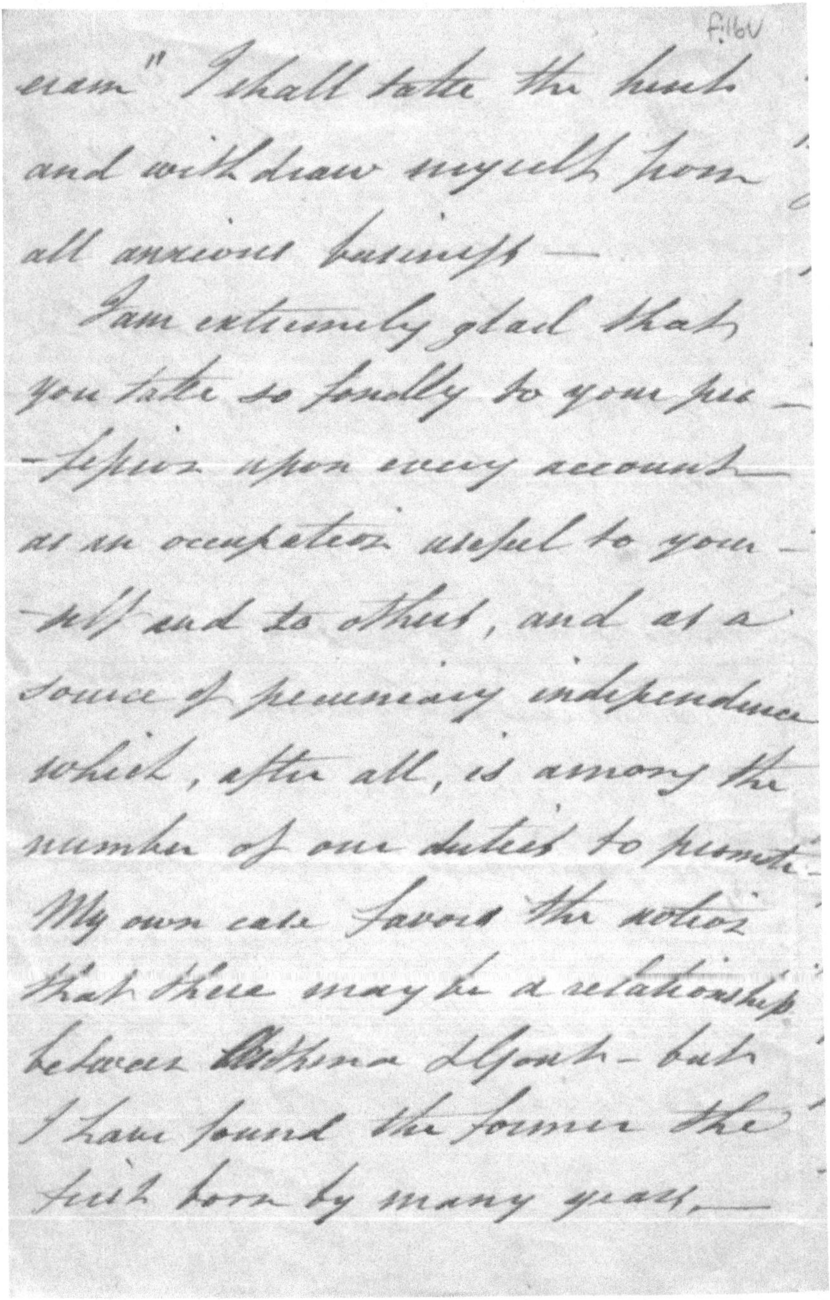

f.116v

ever" I shall take the hint
and withdraw myself from
all anxious business —
I am extremely glad that
you take so fondly to your pro-
-fession upon every account —
as an occupation useful to your-
-self and to others, and as a
source of pecuniary independence
which, after all, is among the
number of our duties to promote —
My own case favors the notion
that there may be a relationship
between Asthma & Gout — but
I have found the former the
first born by many years —

Figure 5.35: 1844-03-09. STG to FG page 2.

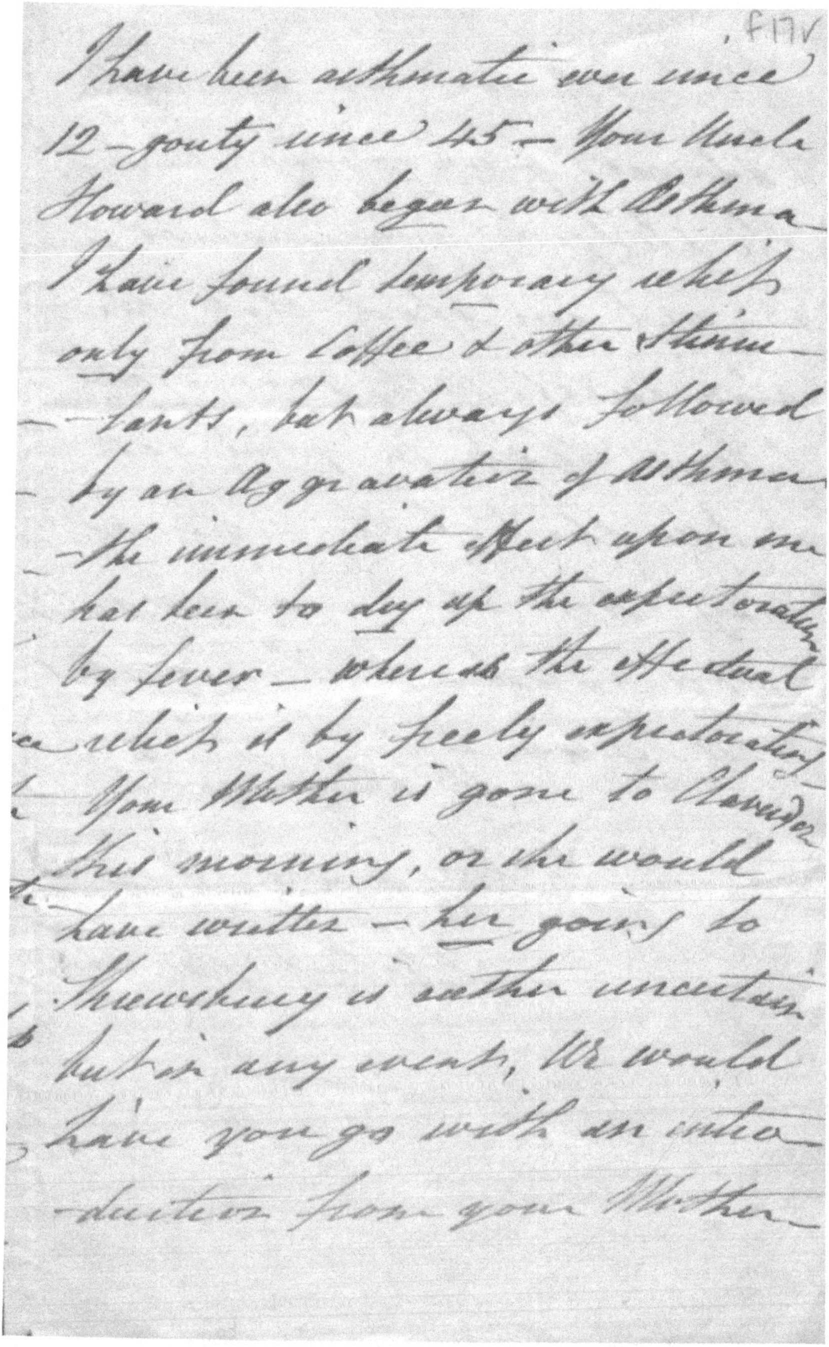
f17v

I have been asthmatic ever since
12 — gouty since 45 — Your Uncle
Howard also began with Asthma —
I have found temporary relief
only from Coffee & other Stimu-
-lants, but always followed
by an Aggravation of Asthma
— The immediate effect upon me
has been to dry up the expectoration
by fever — whereas the effectual
relief is by freely expectorating
Your Mother is gone to Clevedon
this morning, or she would
have written — her going to
Shrewsbury is rather uncertain
but in any event, We would
have you go with an intro-
-duction from your Mother —

Figure 5.36: 1844-03-09. STG to FG page 3.

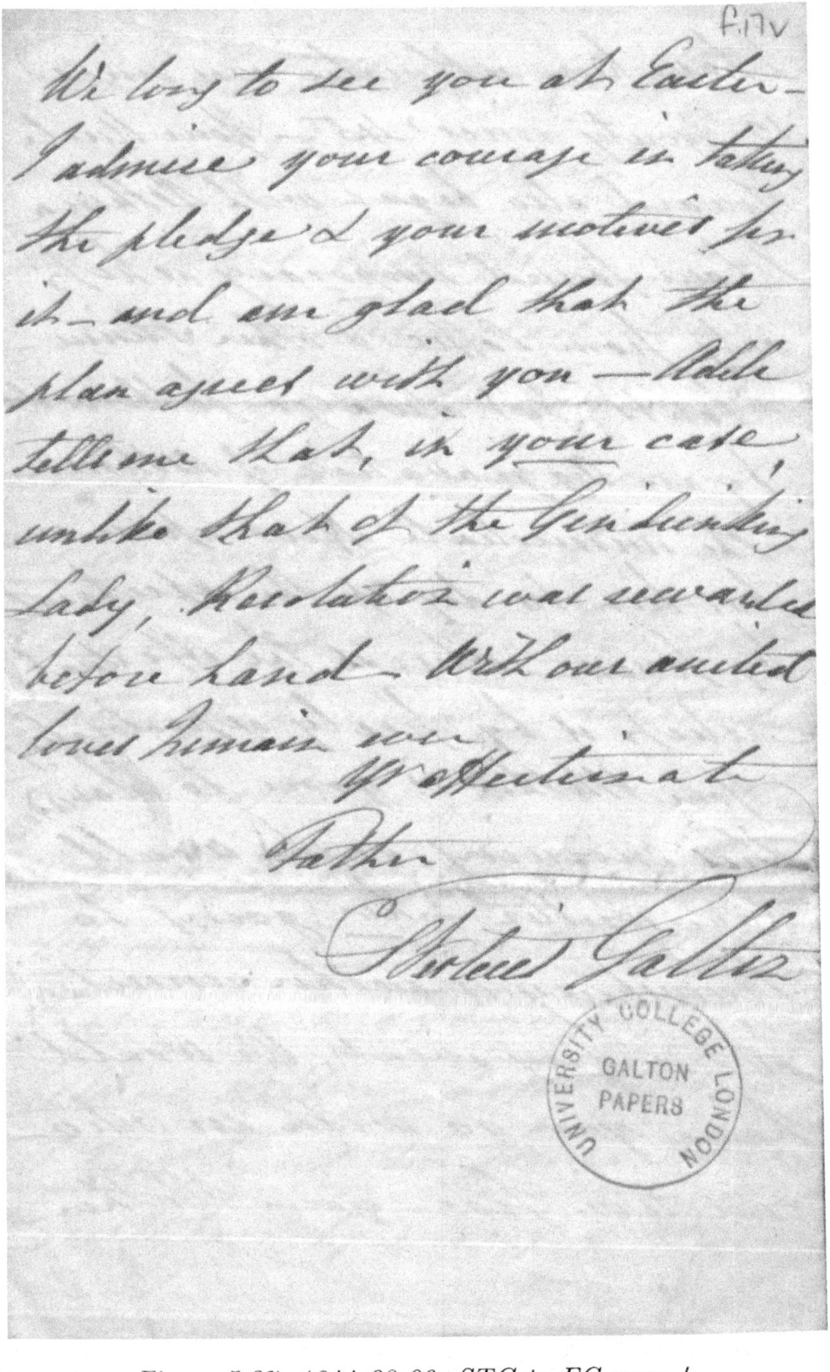
f.17v

We long to see you at Easter—
I admire your courage in taking
the pledge & your motives for
it—and am glad that the
plan agrees with you—Adele
tells me that, in your case,
unlike that of the Penitentiary
Lady, Reformation was rewarded
before hand—With our united
loves I remain ever
Yr affectionate
Father
S Tertius Galton

Figure 5.37: 1844-03-09. STG to FG page 4.

My dear Francis

I much feel your kindness in writing to me & assure you that your letters, tho' unfee-ed, are full as efficacious as the prescriptions of Jephson[18] & Prichard[19] and am able to say that between you all I am rapidly progressing in spirits altho' a degree of burning on my very toe whispers the probability of a degree of forthcoming gout. I shall certainly act upon your suggestion of giving up all anxious business and am gradually sending in resignations. Prichard who called this morning mentions his opinion that the exertions of Liebig would constitute a very important era in medicine when I told him that you had made the same observation to me in a late letter. I was not aware till you mentioned it that Aretaeus[20] was one of the great Fathers of Medicine. I am sorry to observe that the Birm. doctors are again at their quarrels. John Ledsam has printed a furious address to the Medical Profession on the subject of undue influence at Hospital elections, and dragged into it the old affair with Hodgson,[21] and Gutteridge has been at work again. Altho' I do not admire the spirit and temper of the interminable appeals of the latter, it is not improbable that much good may arise out of it. Prichard told me that medical practice has much declined in Leamington from the greatly diminished number of visitors and that a considerable number of houses are likely to be unfurnished this spring.

Your Mother was at Claverdon all yesterday, and her Farmyard Journal records that one of the geese has been set upon eleven eggs. I hope you will approve the changes we have been making there. With our united loves I remain ever

Your affectionate Father
S. Tertius Galton

1844/03/24. Samuel Tertius Galton to Francis Galton

My dear Francis,

Your Mother has written by this day's post to Shrewsbury to propose to your Uncle to appoint some day for receiving you

---

18 Jephson, Dr. Henry. See page 485.
19 John Prichard, a surgeon at Westrock House, Leamington.
20 Aretaeus of Cappadocia.
21 Hodgson, Joseph. See page 483.

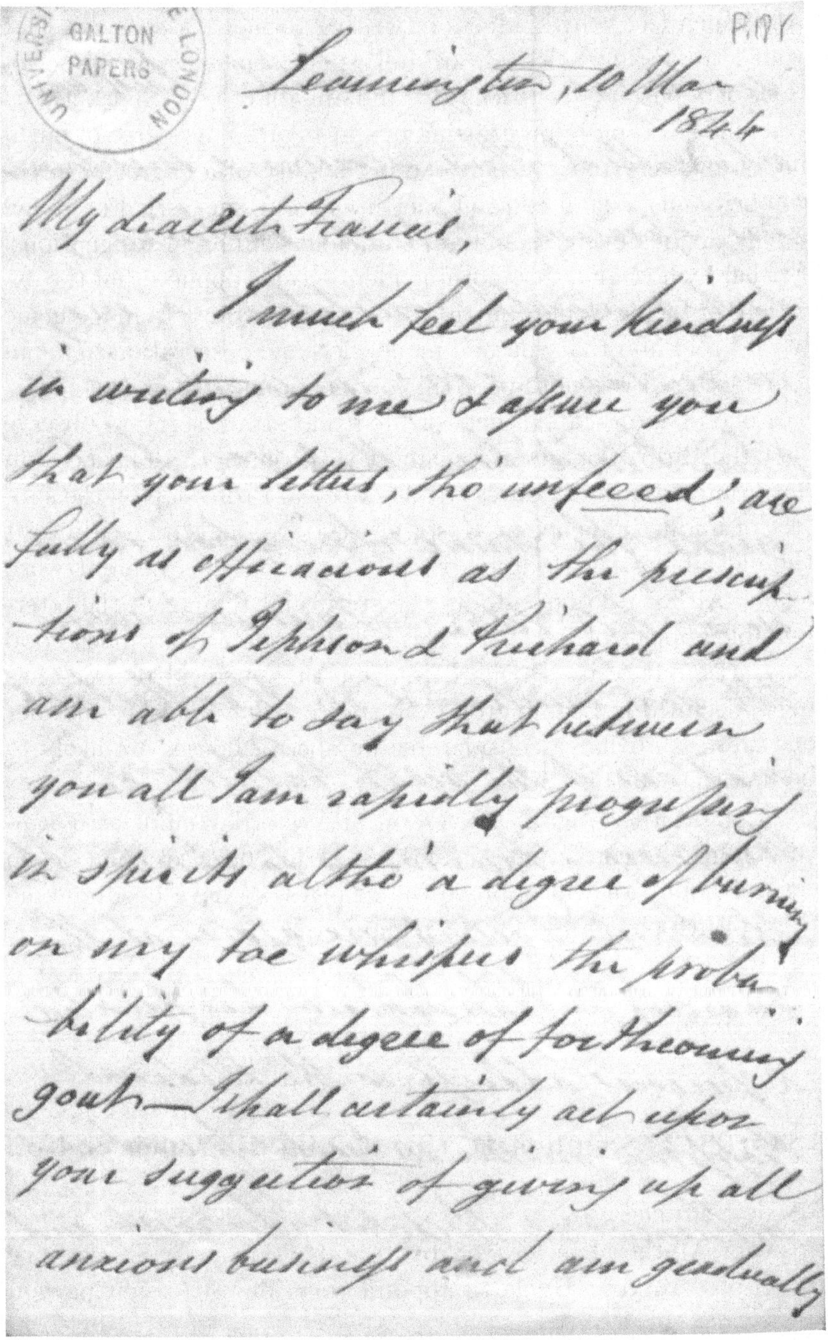

Leamington, 10 Mar
1844

My dearest Francis,

I much feel your kindness in writing to me & assure you that your letters, tho unfeed, are fully as efficacious as the prescriptions of Jephson & Prichard and am able to say that between you all I am rapidly progressing in spirits altho' a degree of burning on my toe whispers the probability of a degree of forthcoming gout — I shall certainly act upon your suggestion of giving up all anxious business and am gradually

Figure 5.38: 1844-03-10. STG to FG page 1.

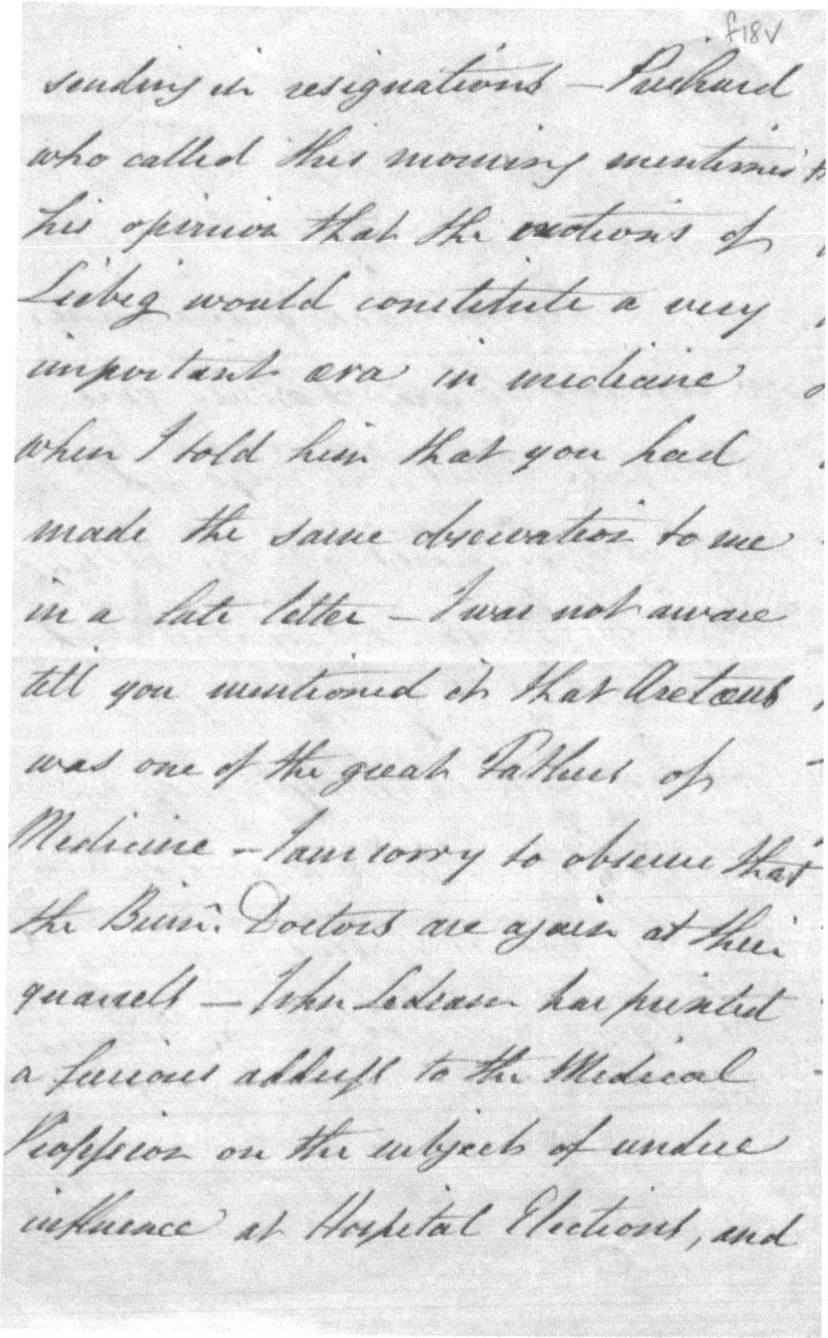

f18v

sending in resignations — Prichard
who called this morning mentioned
his opinion that the motions of
Liebig would constitute a very
important era in medicine
when I told him that you had
made the same observation to me
in a late letter — I was not aware
till you mentioned it that Aretaeus
was one of the great Fathers of
Medicine — I am sorry to observe that
the Birm: Doctors are again at their
quarrels — John Ledsam has printed
a furious address to the Medical
Profession on the subject of undue
influence at Hospital Elections, and

Figure 5.39: 1844-03-10. STG to FG page 2.

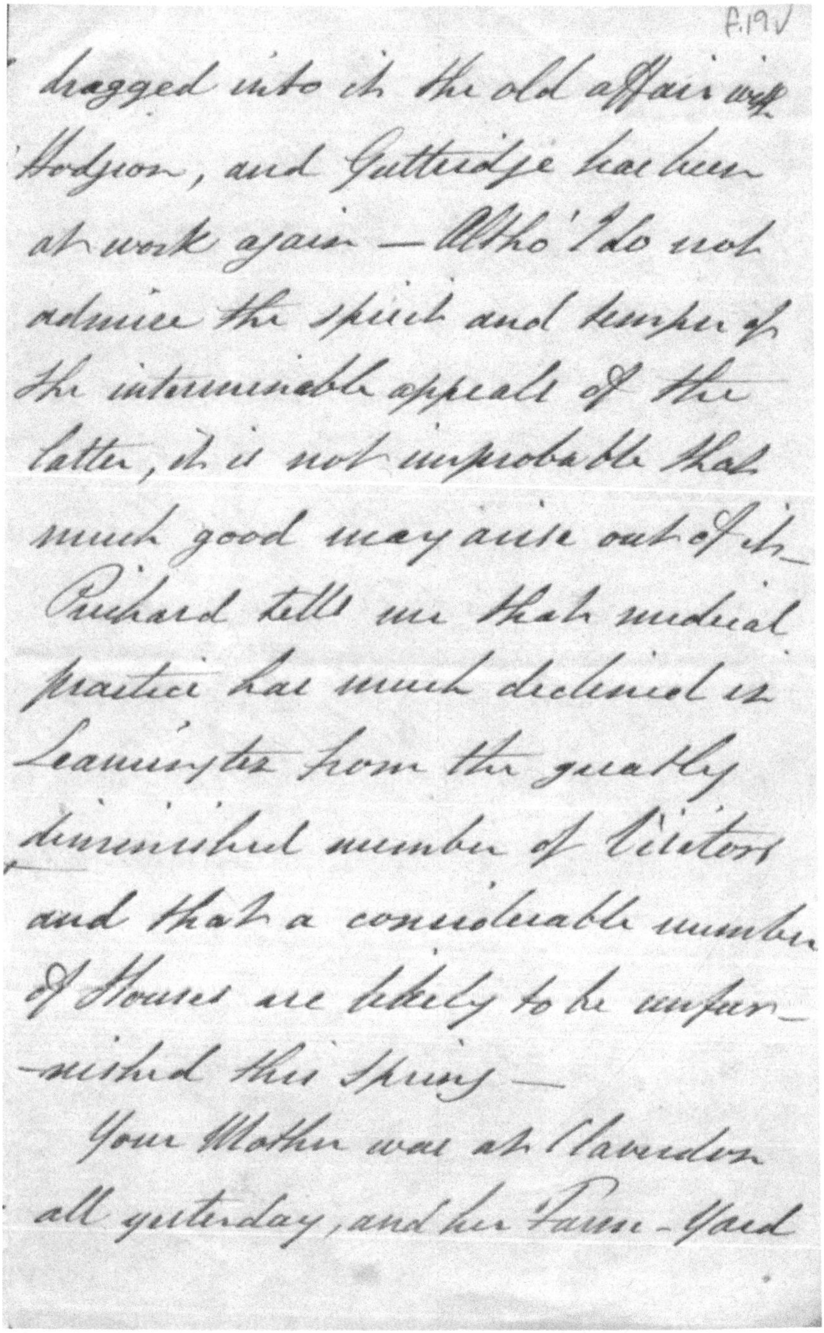
f.19v

dragged into it the old affair with
Hodgson, and Gutteridge has been
at work again — Altho' I do not
admire the spirit and temper of
the interminable appeals of the
latter, it is not improbable that
much good may arise out of it —
Prichard tells me that medical
practice has much declined in
Leamington from the greatly
diminished number of Visitors
and that a considerable number
of Houses are likely to be unfur-
-nished this Spring —
Your Mother was at Clarendon
all yesterday, and her Farm-Yard

Figure 5.40: 1844-03-10. STG to FG page 3.

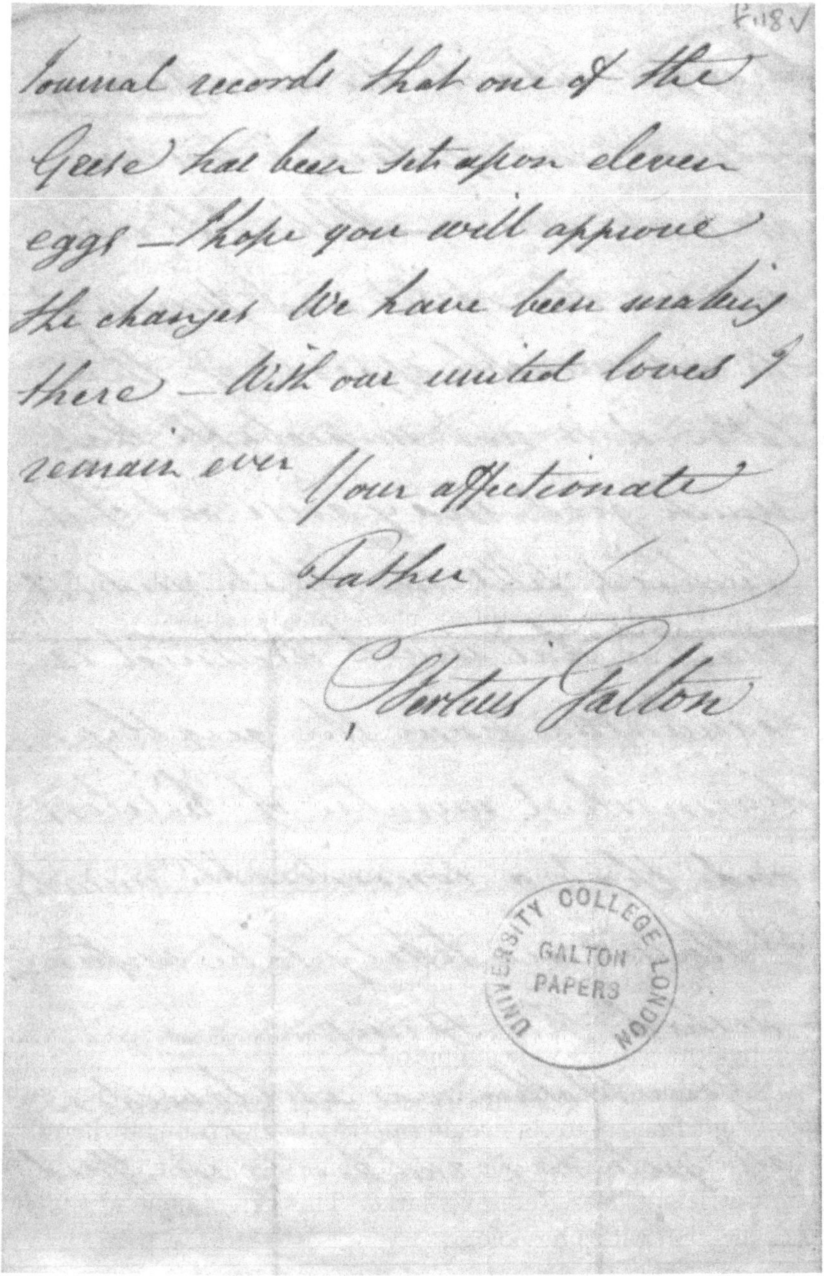

F.118 V

Journal records that one of the
Geese has been set upon eleven
eggs — I hope you will approve
the changes we have been making
there — With our united loves I
remain ever Your affectionate
Father

S Tertius Galton

Figure 5.41: 1844-03-10. STG to FG page 4.

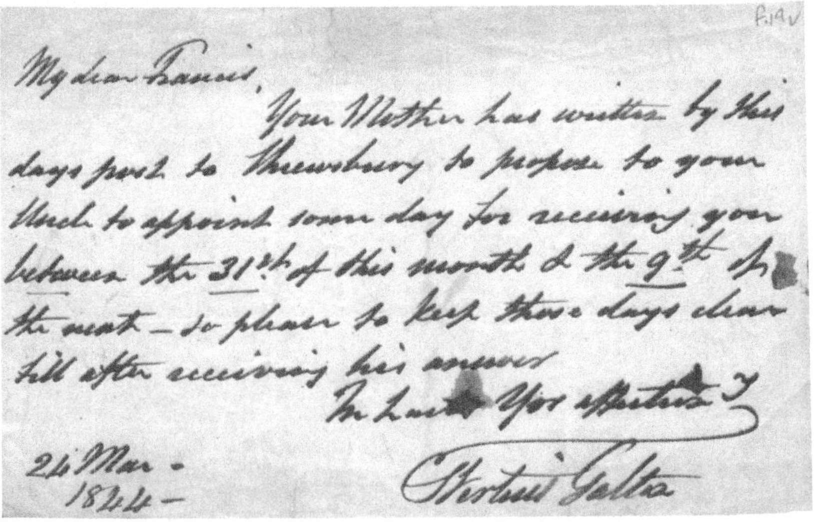

P.19v

My dear Francis,

Your Mother has written by this days post to Shrewsbury to propose to your Uncle to appoint some day for receiving you between the 31st of this month & the 9th of the next—so please to keep those days clear till after receiving his answer

In haste Your affectionate F

24 Mar. 1844—

S Tertius Galton

Figure 5.42: 1844-03-24. STG to FG page 1.

between the 31st of this month & the 9th of the month—so please do keep those days clear till after receiving his answer.

In haste Your affectionate F.

S. Tertius Galton
24 Mar. 1844.

1844/03/29–30. Diary of Samuel Tertius Galton

March 29th. Thursday. Leamington.

. . . Francis came from Cambridge.

March 30th. Saturday. Leamington.

. . . Captain Barclay arrived to dinner and I took him with Bessy and Francis to an evening party at Mrs Youngs where we witnessed from extraordinary performances of mesmerism by a Mr. Cattle and Mr. Ramsay Clarke. This experiment was made in connection with phrenology.

1844/04/01–08. Diary of Samuel Tertius Galton

April 1st. Monday. Leamington.

I rode with Bessy and went with Francis to see the gelatine works at Emscote and the disc engine. Serene day.

April 2nd. Tuesday. Leamington.
Violetta, Francis and myself went to Claverdon where Darwin brought Captain Barclay. ...

April 3rd. Wednesday. Leamington.
I took a ride with Francis and crossed over the new Milverton bridge for the first time since its erection. ...

April 4th. Thursday. Leamington.
Violetta took Francis to Claverdon on his way to Edstone. ...

April 6th. Saturday. Leamington.
I went with Violetta to Claverdon where the Edstone party came with Francis who returned home with us. ..

April 8th. Monday. Leamington.
Easter Monday. We breakfasted at 7:30 a.m. Francis started at 8 o'clock for Birmingham on his way to Shrewsbury. ...

1844/04/10. Francis Galton to Samuel Tertius Galton

Shrewsbury
Wednesday[22]

My dear Father

I shall not be with you until Saturday evening, but I will write tomorrow and let you know the exact coach by which I propose to come. I return to Birm^{n} to morrow and stay with Lucy who hangs out dinner for me to meet James [Russell].[23]

At 3 p.m. yesterday I arrived at my Uncle's gates, the palms of my hands were decidedly moist—the courage was oozing. The fly drove up to the door and I was heartily welcomed by my cousin Susan. I made many apologies which were directly stopped short as everything was made up & excused & then I was taken into the

[22] A visit to his mother's half-brother, Dr. Robert Waring Darwin, The Mount, Shrewsbury.

[23] Possibly Russell, James, see page 490.

dining room to eat luncheon and then in came my uncle who welcomed me if possible 5 times more heartily and who also stopped short all apologies having however first shewn me the delinquent letter, which was wonderful free from all dates. Not a word have I heard of my iniquities since then up to the present time and they have all been as goodnatured and as warm-hearted as possible. They wanted me much to stay, but I thought I had better not, lest my uncle should feel the excitement too much, and also because they wanted me to have some amusements all day and Shrewsbury does not afford any and so I fear they may be afraid that they are not as hospitable as they might be, which is far from the case, but still they might think it.

My uncle is very much better & stronger than I had expected—and I have enjoyed my visit extremely.

Your affectionate son F. Galton.

1844/04/14–20. Diary of Samuel Tertius Galton

April 14th. Sunday. Claverdon.

Emma and Francis walked to Hatton Church in the morning. Bessy, Emma, Francis and myself went to Claverdon Church in the afternoon. ...

April 15th. Monday. Claverdon.

... Violetta and Francis drove to Henley. In the evening Lion, the dog, was almost suddenly seized with violent palpitations and foaming at the mouth attended with great debility. Francis gave him some croton oil which made him excessively thirsty and occasioned free vomiting. We locked him up in the loose box at night. ..

April 16th. Tuesday. Claverdon.

Beautiful day. The dog, Lion, much better and nearly as well as usual towards evening. ... I took measurements in Paul's Piece with Francis and we made the foot of the lower gate post about 55 feet lower than the terrace walk step. ...

April 19th. Friday. Claverdon.

... Emma, Francis and myself dined with the Cartwrights at Preston Bagot ...

April 20th. Saturday. Claverdon.

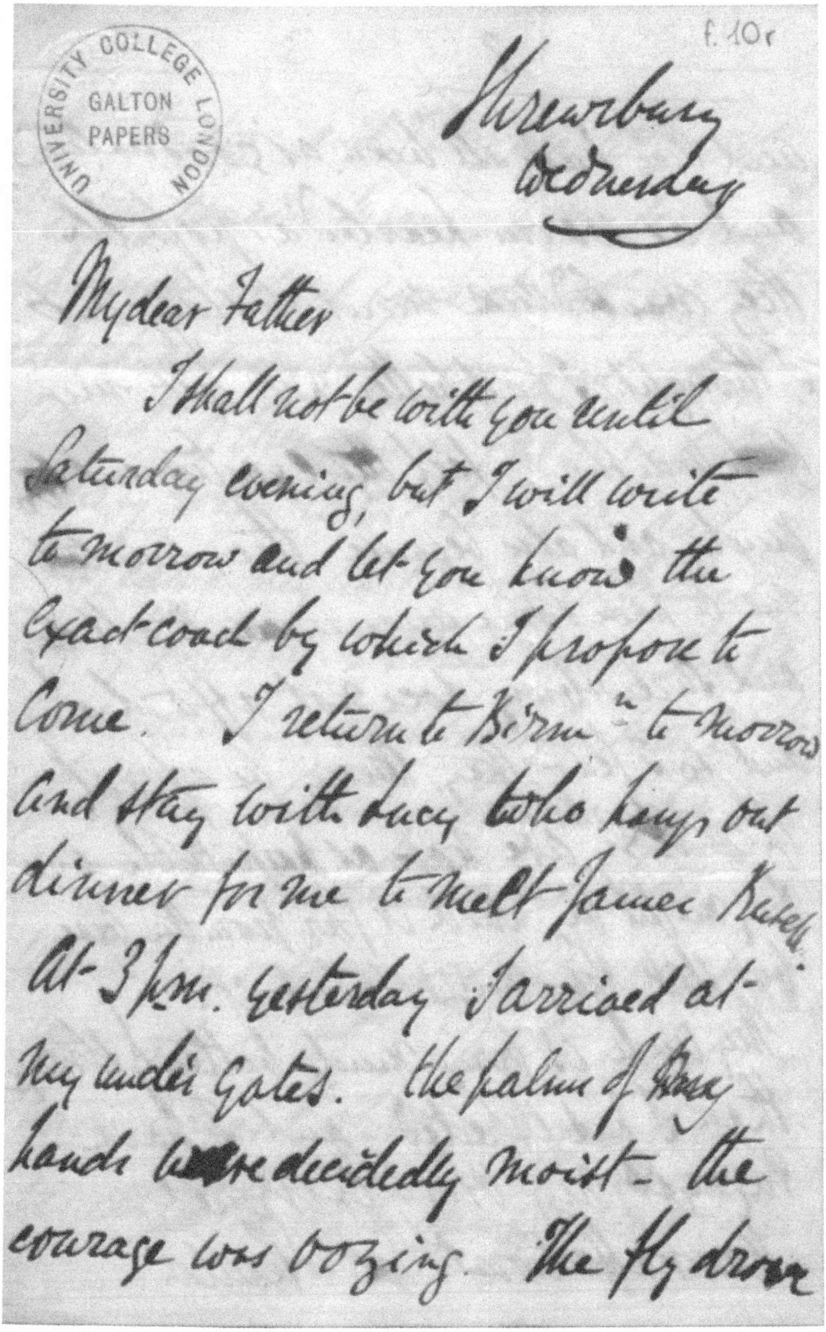

f. 10r

Shrewsbury
Wednesday

My dear Father
I shall not be with you until Saturday evening, but I will write to morrow and let you know the exact coach by which I propose to come. I return to Birm^{m} to morrow and stay with Lucy who keeps our dinner for me to meet James Russell at 3 p.m. Yesterday I arrived at my uncle's Gates. the palm of my hands were decidedly moist – the courage was oozing. The fly drove

Figure 5.43: 1844-04-10. FG to STG page 1.

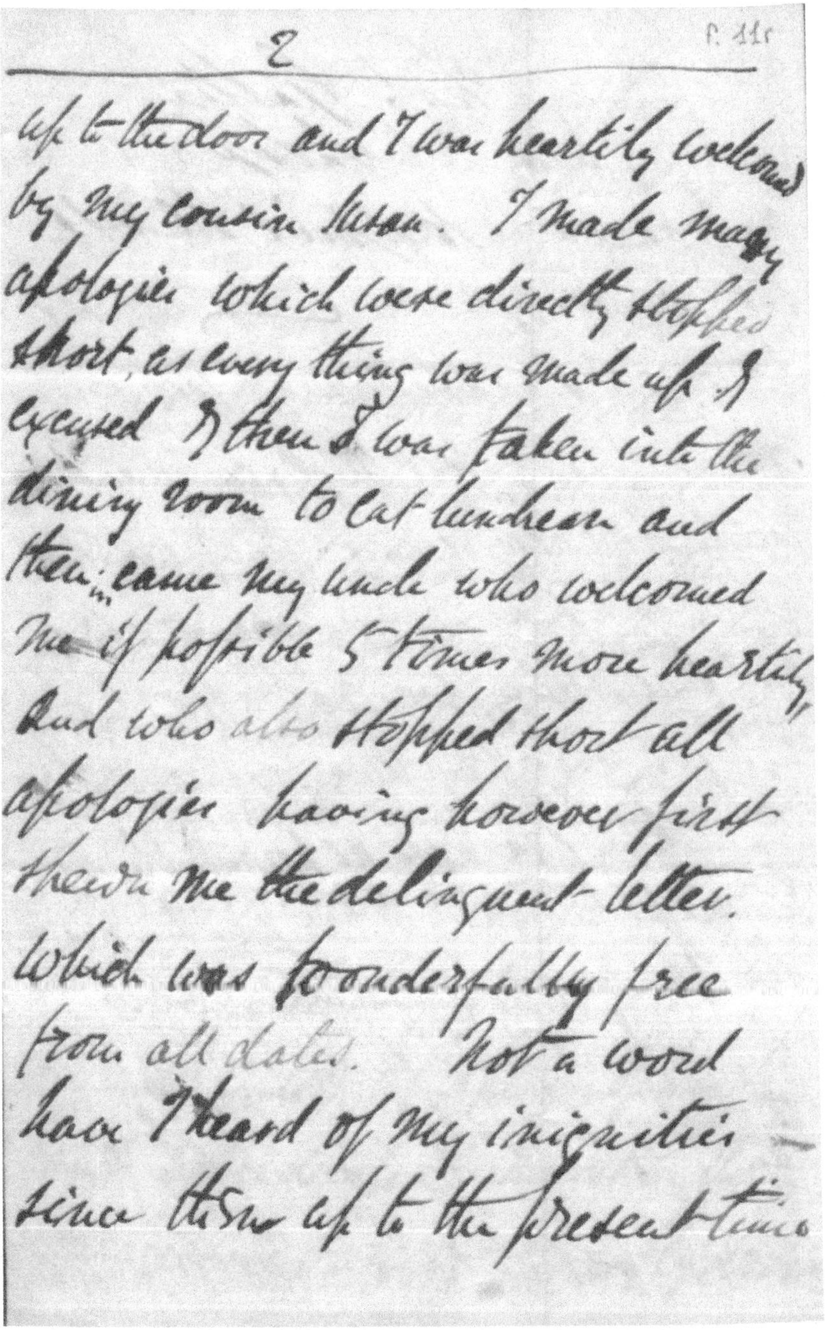
2

f. 11r

up to the door and I was heartily welcomed
by my cousin Susan. I made many
apologies which were directly stopped
short, as every thing was made up &
excused & then I was taken into the
dining room to eat luncheon and
then in came my uncle who welcomed
me if possible 5 times more heartily
and who also stopped short all
apologies having however first
shewn me the delinquent letter.
which was wonderfully free
from all dates. Not a word
have I heard of my iniquities
since then up to the present time

Figure 5.44: 1844-04-10. FG to STG page 2.

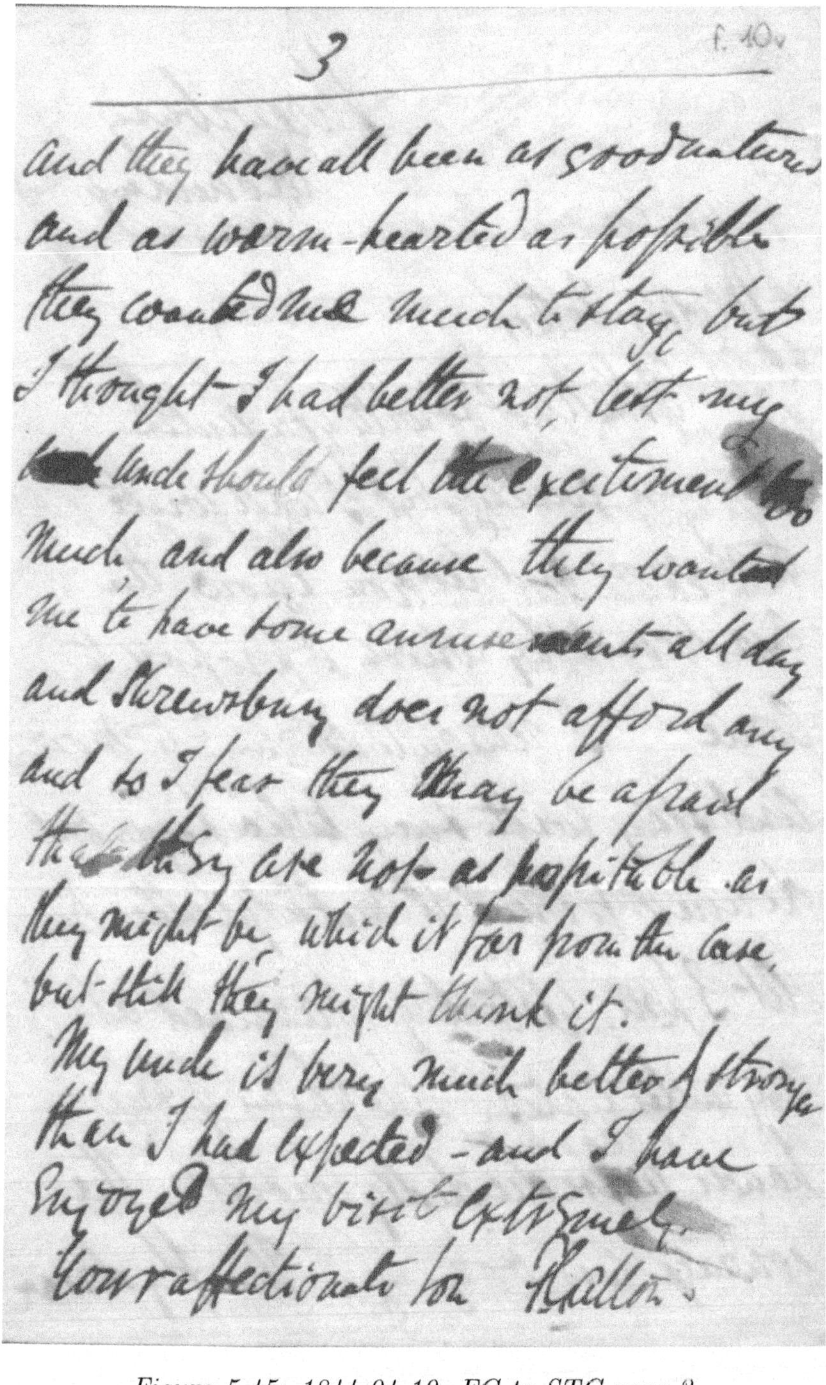

3 f. 10v

and they have all been as goodnatured
and as warm-hearted as possible
they wanted me much to stay, but
I thought I had better not, lest my
uncle should feel the excitement too
much, and also because they wanted
me to have some amusements all day
and Shrewsbury does not afford any
and so I fear they may be afraid
that they are not as hospitable as
they might be which is far from the case,
but still they might think it.
My uncle is very much better & stronger
than I had expected - and I have
enjoyed my visit extremely.
Your affectionate son Galton.

Figure 5.45: 1844-04-10. FG to STG page 3.

f. 11v

Fra! Galton
Salop
10 April 1844

Figure 5.46: 1844-04-10. FG to STG page 4.

Francis left Claverdon this morning for Cambridge being taken to Leamington by Violetta and Adèle. ...

1844/05/01. Francis Galton to Samuel Tertius Galton

Trin Coll:

My dear Father

According to promise and upon a respectably sized sheet I begin to write you an account of my proceedings. Medicine flourishes, though Cambridge certainly affords but very little to do, and I shall not be sorry to begin working in London. It appears that a Cambridge degree is almost necessary for a man who wishes in any degree to get into London practice as a physician. Robert Frere who was my senior at Partridge's,[24] and who has indeed taken a surgeon's degree has had this fact so much urged upon him by different London Doctors that he intends coming to Cambridge as a freshman next term. So I was right after all, notwithstanding Hodgson's forebodings,[25] in wishing for a Cantab: education The reason of this general preference in favour of university degrees I do not know but I suppose that Doctors act very much "en clique" and make a habit of decrying those who have not been so fully educated as themselves.

Tell Delly that as soon as I came up, and through the medium of Tooke I served a writ for total abstinence on Selwyn. He professes himself not quite decided as yet, though undoubtedly in favour of the cause. Tooke introduced me to him the day after. The Epigram Society flourishes in great vigour, we meet next time to write epitaphs on the various dons now in authority. Selwyn I hear desires to join us. The Kays come up next week to take degrees in Freemasonry, and then they are to tell me about any lodgings near them they have found out for me in town. How does Cob Hill get on & how is Mrs [Barlow], what an extraordinary affair that was. Is Lion convalescent?

The answer to the riddle that I sent yesterday is two pigs. I humbly beg everybody's pardon for having sent it to them too soon & in particular those who tried at it the hardest. "Macaulay's Riddle" is said not to be Macaulay's, for he disowned having written it. Bessy is of course with Aunt Gurney.

24 Partridge, Richard. See page 490.

25 Hodgson, Joseph. See page 483.

Good bye, your affectionate son
Francis Galton.

1844/05/06. Francis Galton to Samuel Tertius Galton

Monday Morning

My dear Father

Thank you very much for your two letters & I have just received Emma's also with the account of Miss E. I should think Mr T. was not a person of a very sanguine temperament. What is the correct thing for a lady to do under those circumstances? I always thought that the bridegroom was made to breakfast with the bridal party before the ceremony, and never lost sight of till after it was over, lest he might bolt. It seems so odd to make an appointment to meet & be married at ~~such an~~ a given hour at a church. Had Mr T. only been a disciple of my Father's he would have been shivering at least half an hour before his time waiting for the church-door to ~~be~~ open, & not be so grievously late. I am glad Mrs Onslow is getting better, did you say the ball went right through her? My old friend Joe Kay[26] is in Cambridge and he tells me of several lodgings, about where I should like to be in London. I have not actually fixed upon the street but certainly sh^{d} prefer one of those running out of Park Lane for then I shall be near St George's Hospital & the Park and close by the Kay's Campbell's C. Buxton & not far from Hallam[27] & indeed some others. I fancy that I shall get from Kay's description quite as good rooms as I shall want for 20 shillings a week

Pray be sure to remember me to Charlotte Batt when you see her, I am glad her children are such fine ones.

We had such a glorious May day here. [See Figure 5.51, p. 417.] I determined to ~~manage~~ improve upon my last year's one and together with two other men raised a shilling or a sixpenny subscription to £3 with this we got 80 buns 240 oranges 600 small biscuit cakes & materials for 8 gallons of tea (to be made ready sugared &c. in cans) then we got a maypole 9 feet above the ground with a great wreath on the top & 2 flags one on either side fastened to sticks and held by boys, then an arch through which

[26] Kay, Joseph. See page 486.
[27] Hallam, Henry Fitzmaurice. See page 483.

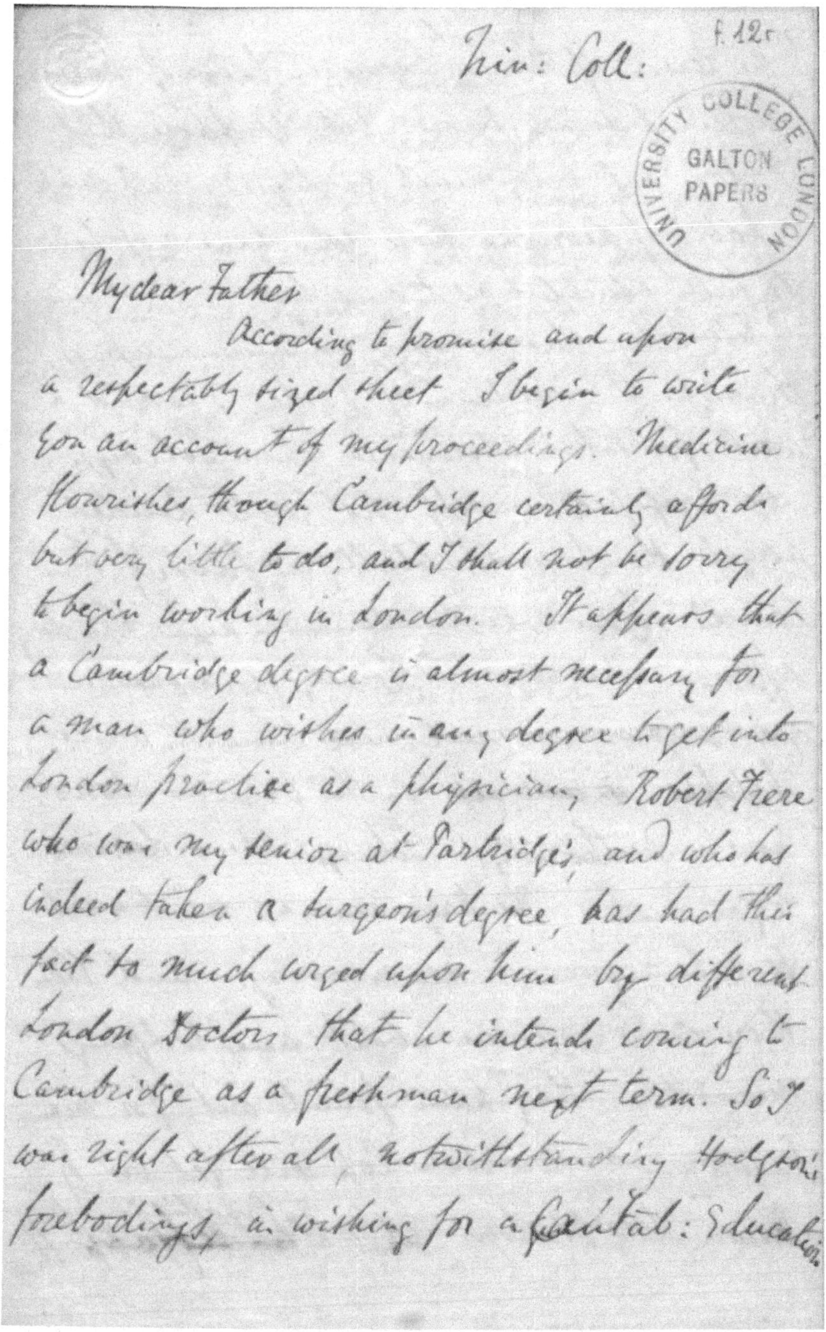
Trin: Coll:

f. 12r

My dear Father

According to promise and upon
a respectably sized sheet I begin to write
you an account of my proceedings. Medicine
flourishes, though Cambridge certainly affords
but very little to do, and I shall not be sorry
to begin working in London. It appears that
a Cambridge degree is almost necessary for
a man who wishes in any degree to get into
London practice as a physician, Robert Frere
who was my senior at Partridge's, and who has
indeed taken a surgeon's degree, has had this
fact so much urged upon him by different
London Doctors that he intends coming to
Cambridge as a freshman next term. So I
was right after all, notwithstanding Hodgson's
forebodings, in wishing for a Cantab: Education

Figure 5.47: 1844-05-01. FG to STG page 1.

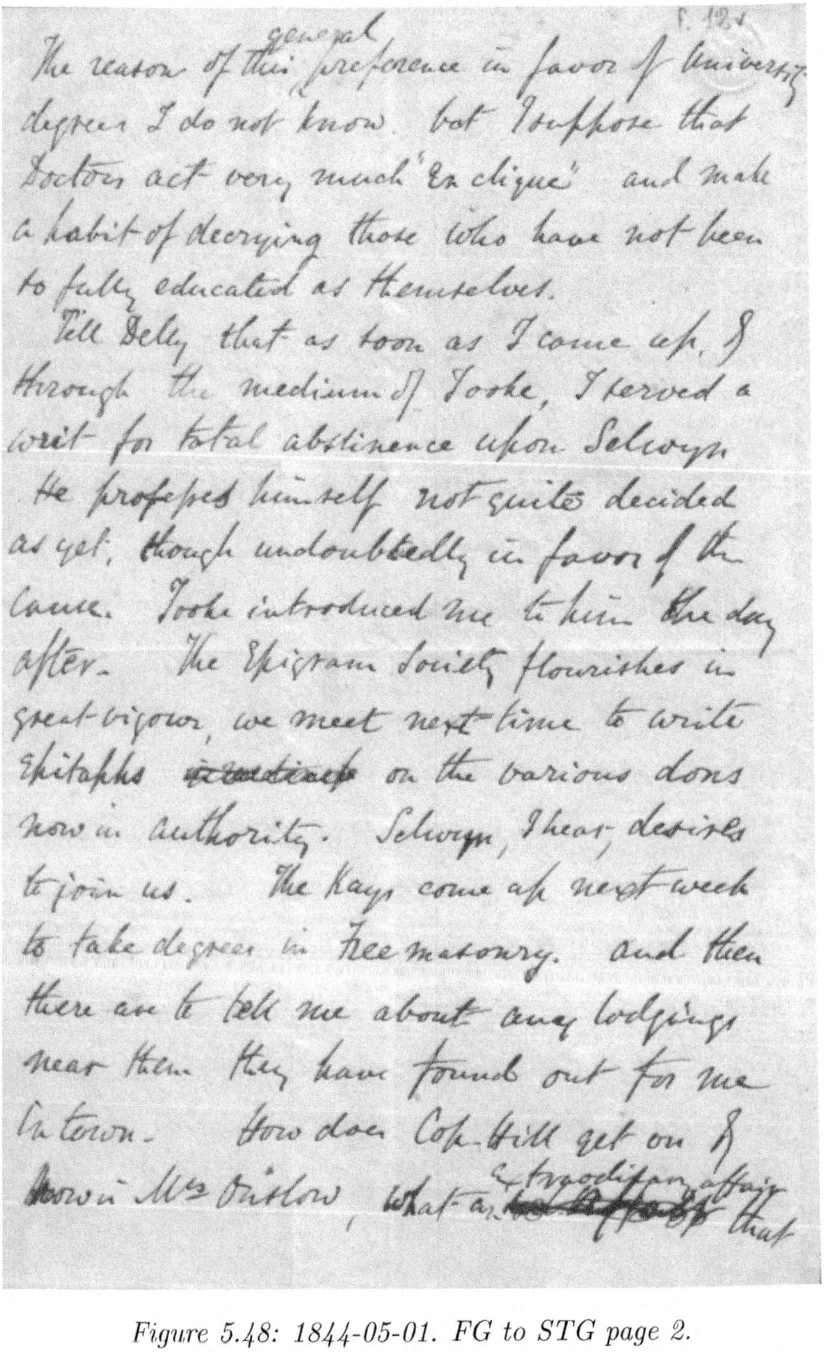
The reason of this general preference in favor of University
degrees I do not know. but I suppose that
Doctors act very much "en clique" and make
a habit of decrying those who have not been
so fully educated as themselves.
Tell Delly that as soon as I come up, &
through the medium of Tooke, I served a
writ for total abstinence upon Selwyn
He professed himself not quite decided
as yet, though undoubtedly in favor of the
cause. Tooke introduced me to him the day
after. The Epigram Society flourishes in
great vigour, we meet next time to write
Epitaphs ~~[illegible]~~ on the various dons
now in authority. Selwyn, I hear, desires
to join us. The Kays come up next week
to take degrees in Free masonry. and then
there are to tell me about any lodgings
near them they have found out for me
in town. How does Cop-Hill get on &
how is Mrs Onslow, what an ~~[illegible]~~ extraordinary affair that

Figure 5.48: 1844-05-01. FG to STG page 2.

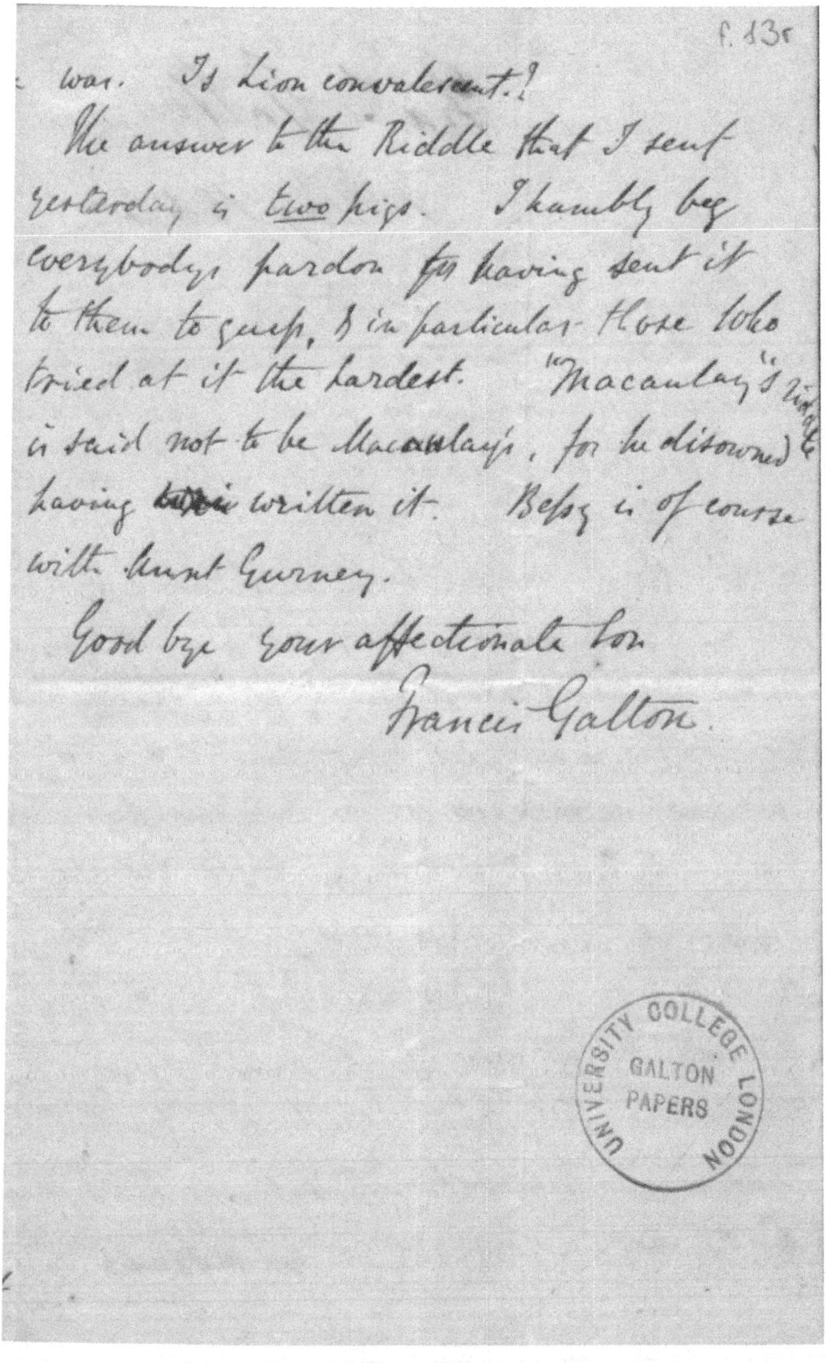

f. 13r

war. Is Lion convalescent.?

The answer to the Riddle that I sent yesterday is <u>two</u> pigs. I humbly beg everybody's pardon for having sent it to them to guess, & in particular those who tried at it the hardest. "Macaulay's" riddle is said not to be Macaulay's, for he disowned having ~~[illegible]~~ written it. Bessy is of course with Aunt Gurney.

Good bye your affectionate son

Francis Galton.

Figure 5.49: 1844-05-01. FG to STG page 3.

Fras. Galton
1 May 1844?

Figure 5.50: 1844-05-01. FG to STG page 4.

only 2 could pass at a time, receiving each of them a bun or 2 or three biscuits and an orange (in transitu). Then we chose a Queen of the May, the prettiest little girl I ever remember to have seen & Mrs. Hoppit took her in charge & washed her & attired her in a royal diadem & then the undergrads present, about 60, shoved their way in to the May pole & took hands and spread out leaving a large vacant ring in the centre of which was the May pole & the Queen of the May. [See Figure 5.52, p. 418.]

She chose her partner & with 5 other girls & their partners danced the College hornpipe, then we let in a number of other smaller girls who took hands and danced in a wide ring round them & the May pole, & after that another ring, concentric, that danced the opposite way round, so all three were going at once. We had one of those street hurdy gurdy things for the music, so the plan of the proceedings was thus. [See Figure 5.53, p. 419.]

There were more than 200 children & the undergraduates' arms were at full stretch. The maypole was put up in the college green.

Riddle. If a man wants to obtain a vegetable time piece at what hour should he rise ?

Answer. He must get "up at eight o'clock" (must get a potato clock).

Goodbye your affectionate Son
Fras. Galton.

P.S. I will always write on Mondays as on this & last week.

Figure 5.51: 1844-05-06 FG to STG: May Pole.

1844/05/13. Francis Galton to Samuel Tertius Galton

Trin: Coll

Figure 5.52: 1844-05-06 FG to STG: May Queen.

Tuesday (instead of Monday).

My dear Father

Pray excuse the small sheet of paper, for I have so much of it on hand that I can find no way for its disposal. A number of my old college friends have come up during the past week, and most of them gone down again. Mathew Boulton[28] just appeared for two days & a night to take his degree. He asked me much to come to Tew,[29] and I have accepted his invite gladly for some far future time as I am afraid that I shall have no time for holidays in London. Every body is making up long vacation parties & I have had earnest entreaties to travel with different allies to St. Petersburg to Madrid & I don't know where else, but after all it is satisfactory to have something better to do than to join them. Fowell Buxton has also been here, he is working at the brewery in London but unfortunately ~~will~~ keeps a long way off from where I shall. The 2 Kays came up, one to be made a Freemason, the others to be passed to the 2d degree & I was raised to the third on the same night. Frederick Bristowe has been in Cambridge the last week to see his brother, who takes his degree next year. At the Epigram meeting last time we had the most amusing collection that as yet we have been favoured with. One short one was on Griffin[30] a Johnian Senior Wrangler who has written the most stale abstract & uninteresting books on optics it is possible to

---

[28] Boulton, Matthew Piers Watt. See page 480.

[29] The Boulton country estate in Oxfordshire.

[30] William Nathaniel Griffin (1815–1892).

Figure 5.53: 1844-05-06 FG to STG: Dancing Circle.

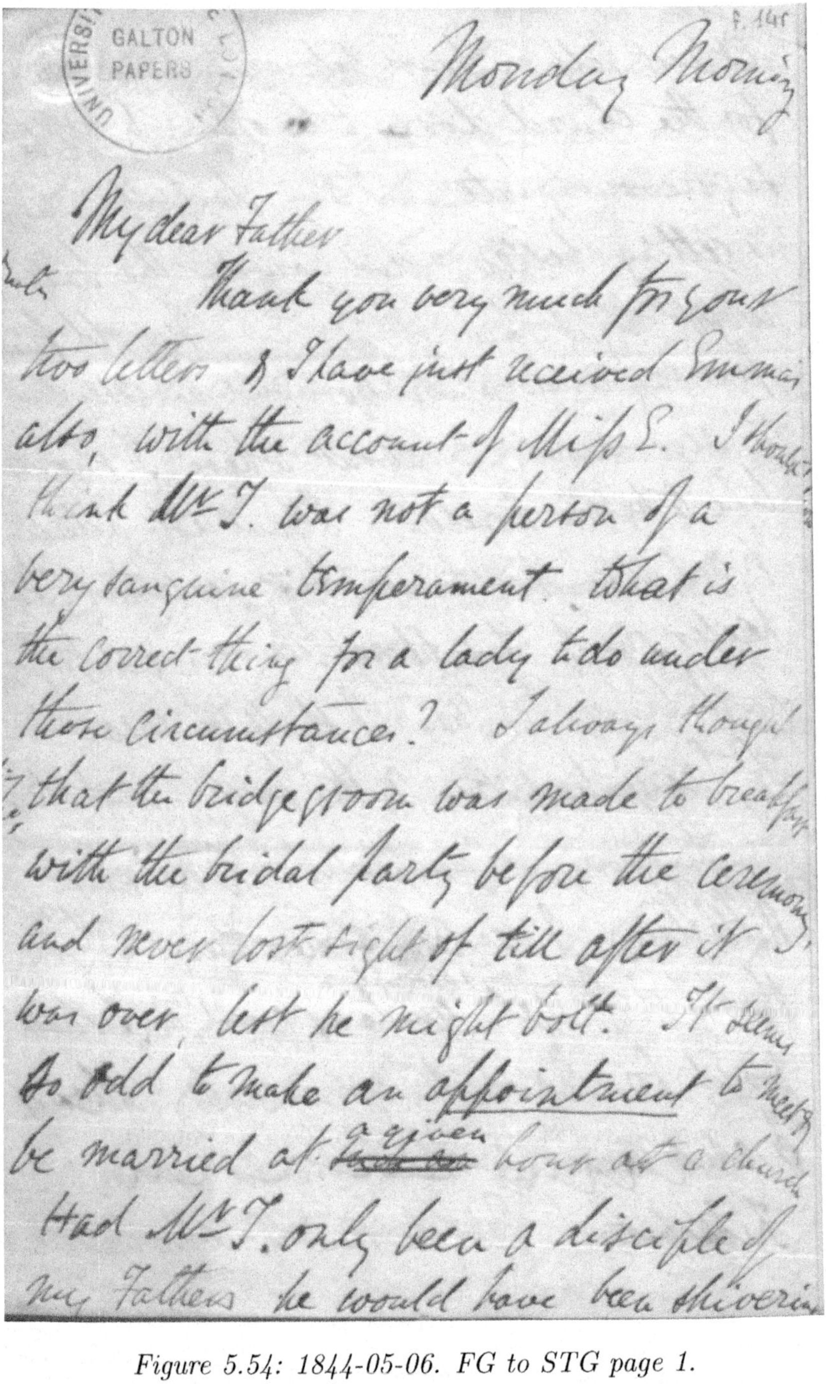

f. 145

Monday Morning

My dear Father

Thank you very much for your two letters & I have just received Emma's also, with the account of Miss E. I should think Mr T. was not a person of a very sanguine temperament. What is the correct thing for a lady to do under those circumstances? I always thought that the bridegroom was made to breakfast with the bridal party before the ceremony and never lost sight of till after it was over lest he might bolt. It seems so odd to make an appointment to meet & be married at a given hour at a church. Had Mr T. only been a disciple of my Fathers he would have been shivering

Figure 5.54: 1844-05-06. FG to STG page 1.

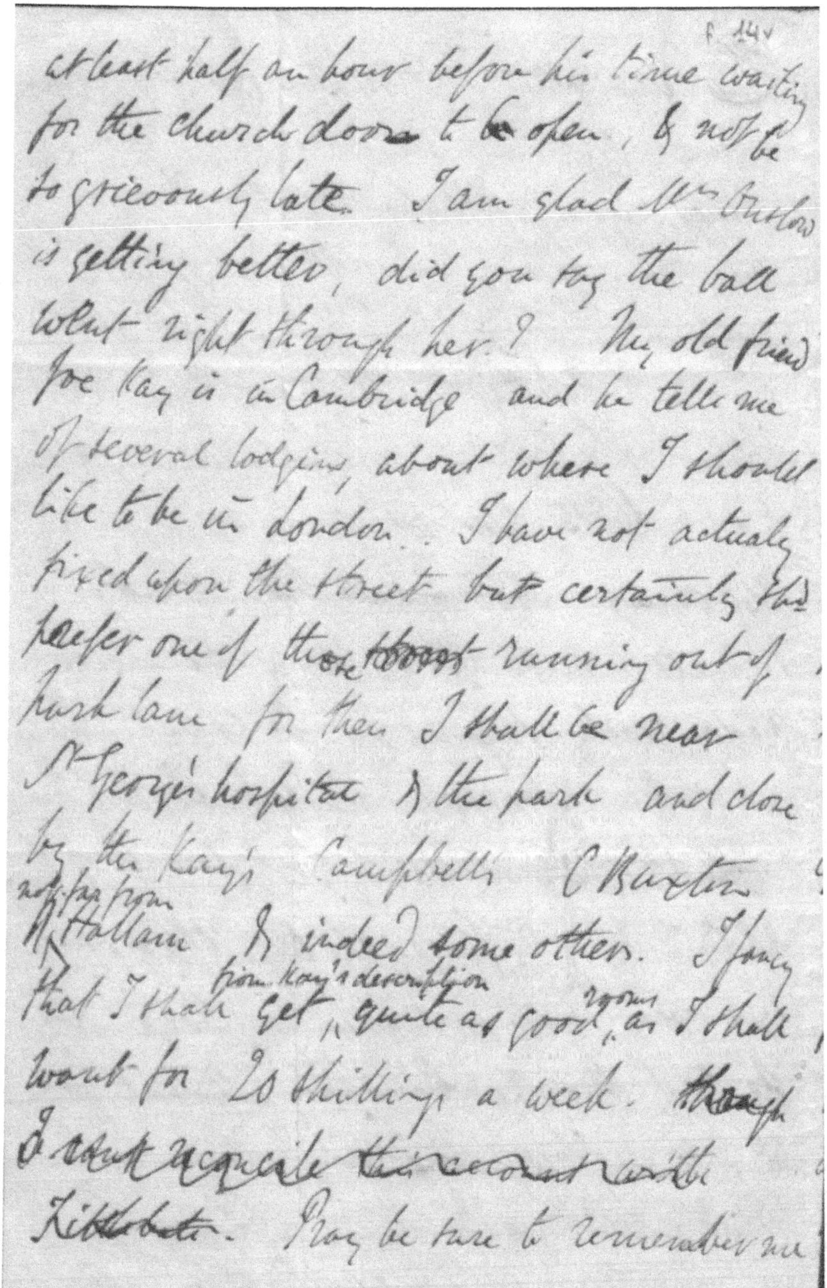
f. 14v

at least half an hour before his time waiting for the church door to be open, & not be so grievously late. I am glad Mrs Onslow is getting better, did you try the ball what right through her? My old friend Joe Kay is in Cambridge and he tells me of several lodgings about where I should like to be in London. I have not actually fixed upon the street but certainly I'd prefer one of those ~~streets~~ running out of Park lane for then I shall be near St George's hospital & the park and close by the Kays Campbells C Buxton not far from H Hallam & indeed some others. I fancy that I shall get from Kay's description quite as good rooms as I shall want for 20 shillings a week. ~~though I cant reconcile this account with Kittoboto~~. Pray be sure to remember me

Figure 5.55: 1844-05-06. FG to STG page 2.

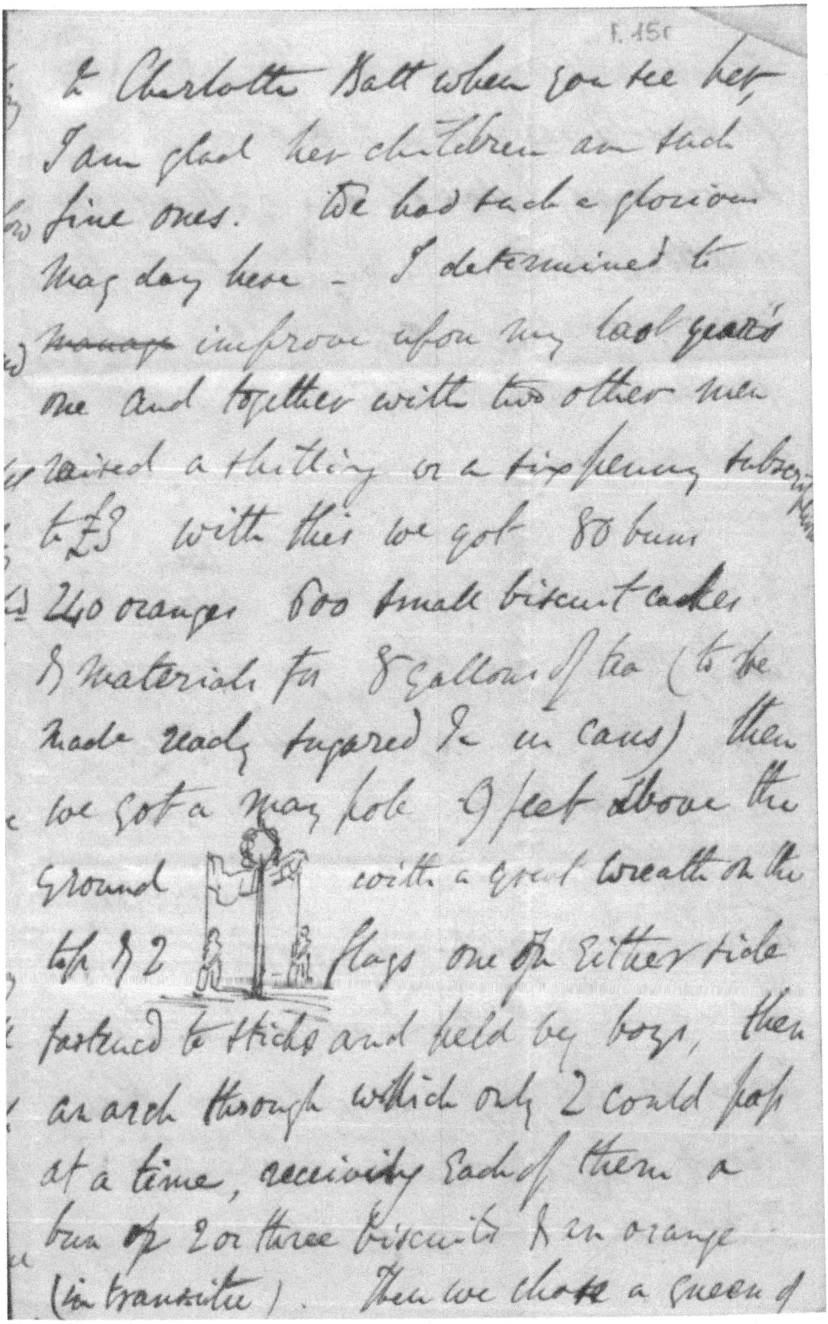

f. 151

to Charlotte Batt when you see her,
I am glad her children are such
fine ones. We had such a glorious
May day here – I determined to
~~manage~~ improve upon my last year's
one and together with two other men
raised a shilling or a sixpenny subscrip-
to £3 with this we got 80 buns
240 oranges 600 small biscuit cakes
& materials for 8 gallons of tea (to be
made ready sugared &c in cans) then
we got a may pole 9 feet above the
ground with a great wreath on the
top & 2 flags one on either side
fastened to sticks and held by boys, then
an arch through which only 2 could pass
at a time, receiving each of them a
bun or 2 or three biscuits & an orange
(in transitu). Then we chose a queen of

Figure 5.56: 1844-05-06. FG to STG page 3.

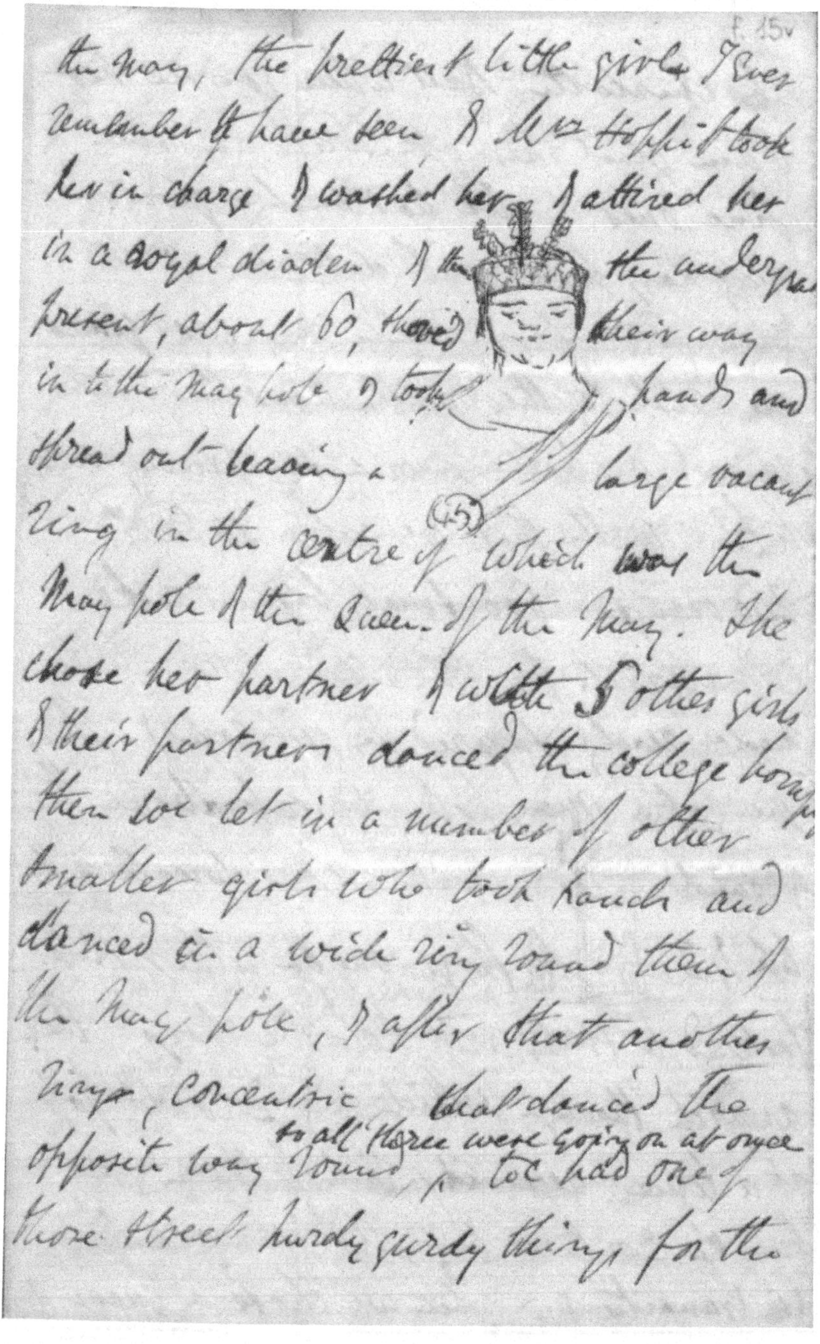
the May, the prettiest little girl I ever
remember to have seen & Mrs Hoppit took
her in charge & washed her & attired her
in a royal diadem & the the undergar
present, about 60, moved their way
in to the May pole & took hands and
spread out leaving a large vacant
ring in the centre of which was the
May pole & the Queen of the May. She
chose her partner & with 5 other girls
& their partners danced the college hornpipe
then we let in a number of other
smaller girls who took hands and
danced in a wide ring round them &
the May pole, & after that another
rings, concentric that danced the
opposite way round, to all three were going on at once & we had one of
those street hurdy gurdy things for the

Figure 5.57: 1844-05-06. FG to STG page 4.

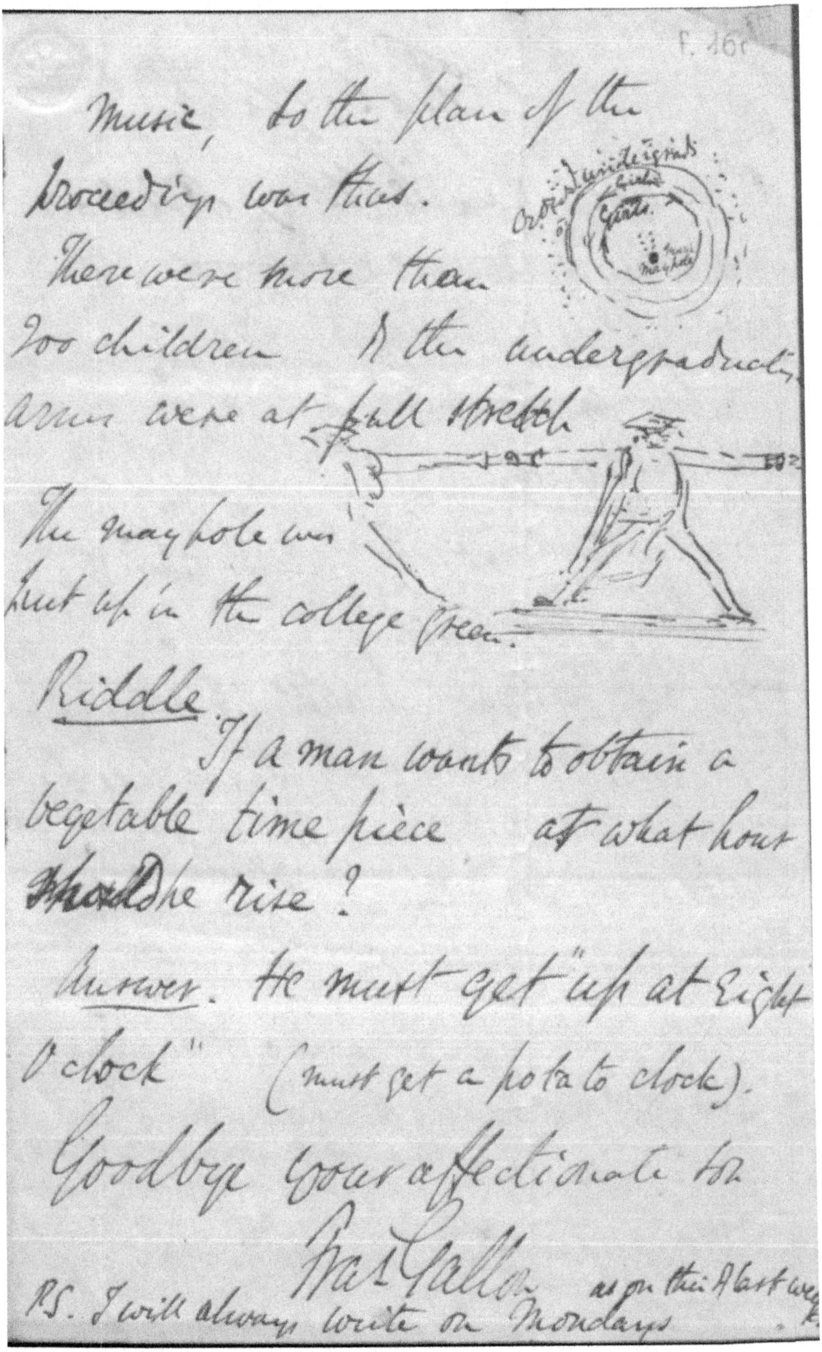

f. 16r

music, to the plan of the proceedings was thus.

There were more than 200 children & the undergraduates arms were at full stretch

The maypole was put up in the college green.

<u>Riddle</u>

If a man wants to obtain a vegetable time piece at what hour should he rise?

<u>Answer</u>. He must get "up at Eight O clock" (must get a potato clock).

Goodbye your affectionate son

Frank Galton

P.S. I will always write on Mondays as on this & last week. F.

Figure 5.58: 1844-05-06. FG to STG page 5.

Figure 5.59: 1844-05-06. FG to STG page 6.

conceive & quite spoilt the beauty of the science.[31] I ought to say the subject given was "An Epitaph." It was:

Who'll weep for Griffin?
Not I said the eye,
He has made me so dry,
I cannot weep for Griffin.

Bessy wrote to me the other day, it was principally on good advice.

The Kays tell me that they are going to build a splendid street in London longer than any at present existing & closed at either end with large metal doors and archways it is to be by Kensington Gardens. Westmacott has nearly finished his bas reliefs for the basement of the new Royal Exchange they are said to be splendidly executed. I do not know how the figures are grouped, but they form an allegory relating to Commerce and the figures are in modern not ancient dress, I believe not unlike those in the old penny postage envelope. The British & Foreign Institute is going to build extensively there are now about 1250 members the prices for dining are the same as those of the Athenaeum, which are high. A very fair library has suddenly sprung up by all the principal publishers giving very handsome presents of books to Buckingham as a return for his exertions in that part of the copyright bill by which the number of copies of each publication that must be sent to different libraries has been diminished, and these books he has made over to the Institute. Are Lucy's kinchins[32] still with you? give my love to them, if they are and also to Lucy.

Your very affectionate Son
Francis Galton.

How is my mother's health? & do you still teach Adèle's school-children. Your chess-board is invaluable, we lie lazily on the banks of the river in the sun playing chess after hall which is luxurious to a degree.

I did not read the speech of Sir Robert that you mentioned but should have been very glad to have been able to have cheered him for the passage in question.

[31] Griffin, Senior Wrangler in 1837, wrote *A Treatise on Optics* (Cambridge University Press, 1842) and many other textbooks on Algebra and Trigonometry.

[32] Children.

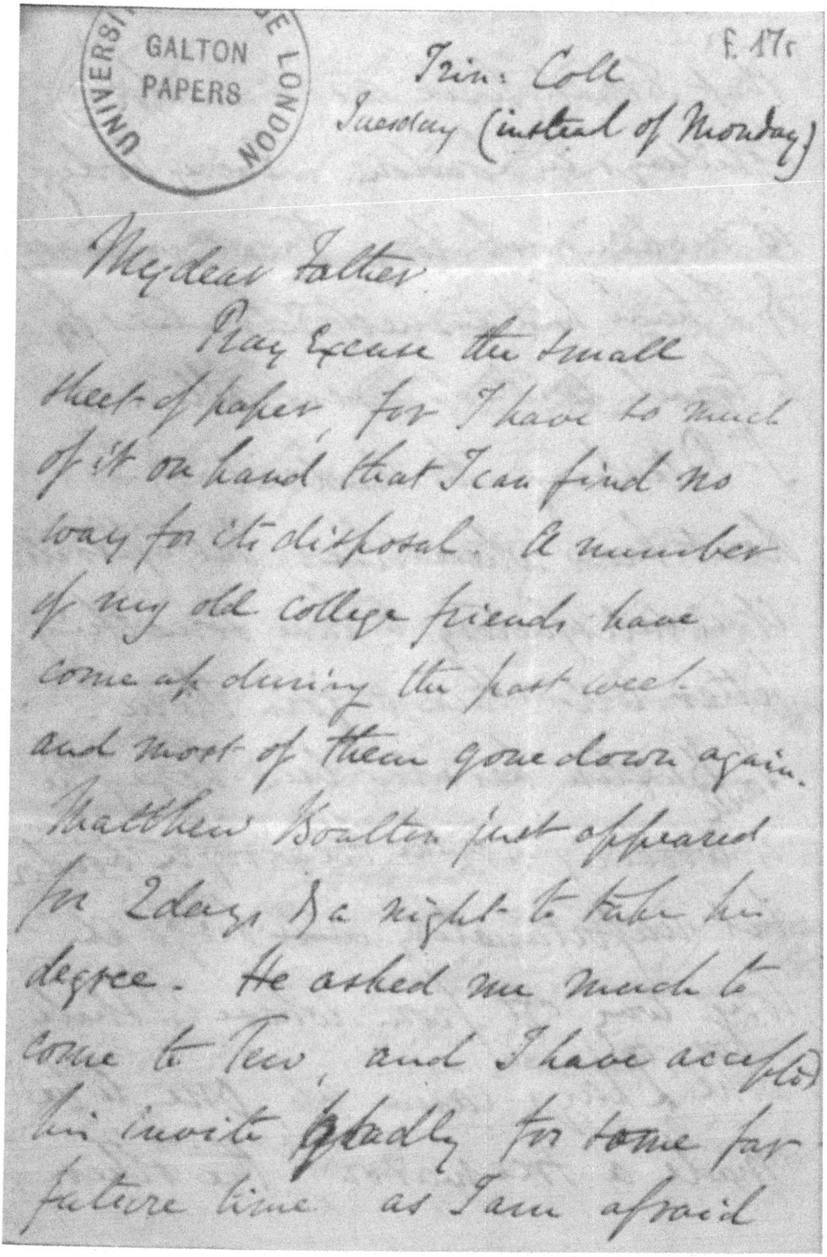

f. 17r

Trin: Coll
Tuesday (instead of Monday)

My dear Father

Pray excuse the small sheet of paper, for I have so much of it on hand that I can find no way for its disposal. A number of my old college friends have come up during the past week and most of them gone down again. Matthew Boulton just appeared for 2 days & a night to take his degree. He asked me much to come to Tew, and I have accepted his invite gladly for some fur- future time as I am afraid

Figure 5.60: 1844-05-13. FG to STG page 1.

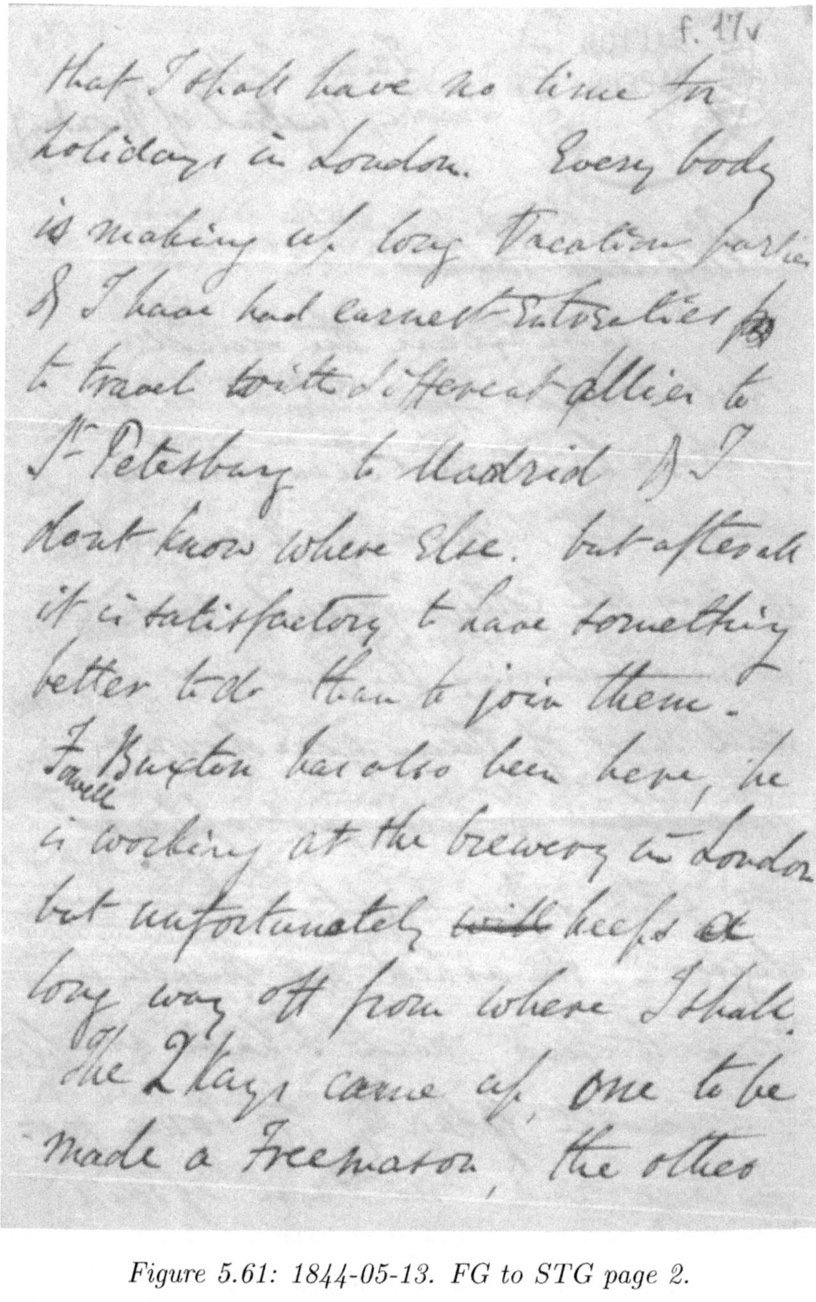
f. 17v

that I shall have no time for
holidays in London. Every body
is making up long vacation parties
& I have had earnest entreaties
to travel with different fellows to
St Petersburg to Madrid & I
dont know where else. but after all
it is satisfactory to have something
better to do than to join them.
Fowell Buxton has also been here, he
is working at the brewery in London
but unfortunately ~~will~~ keeps a
long way off from where I shall.
The 2 Days came up, one to be
made a Freemason, the other

Figure 5.61: 1844-05-13. FG to STG page 2.

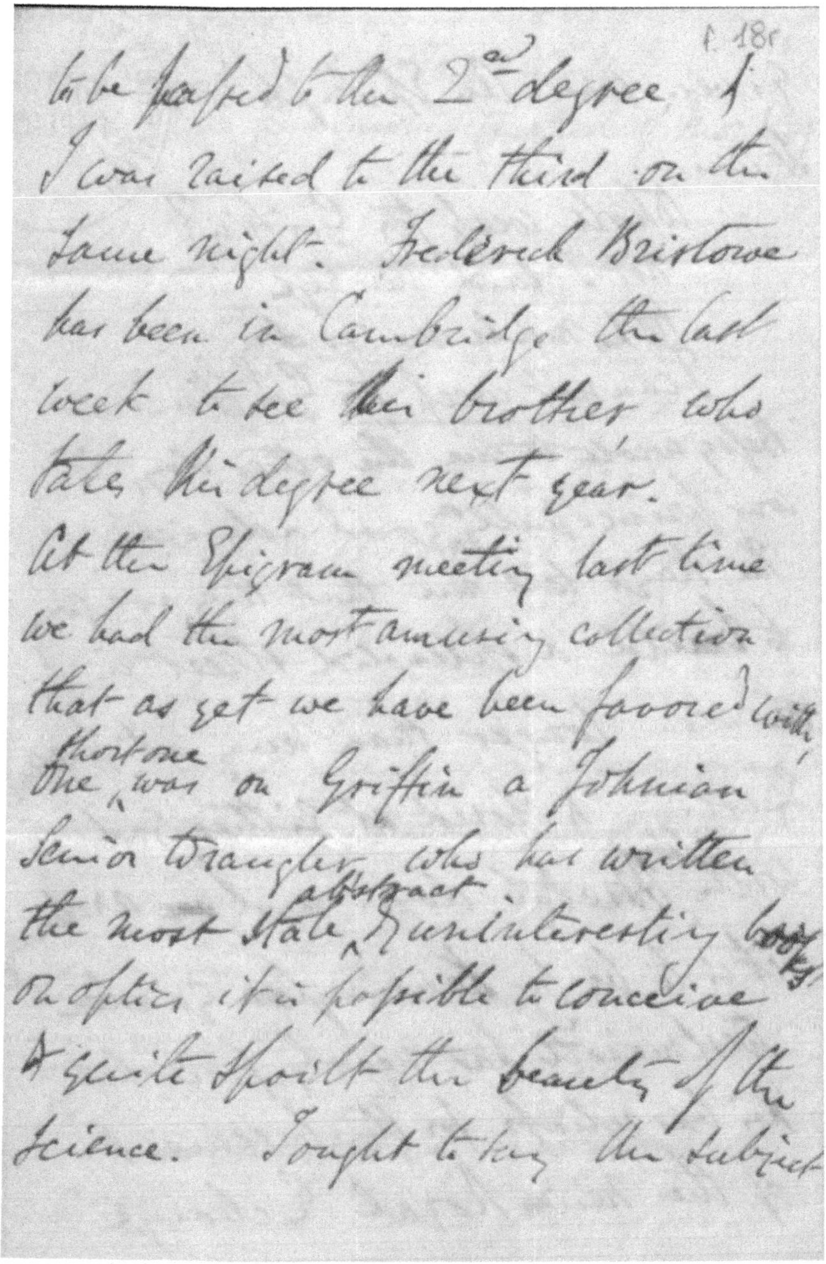

f. 18r

to be passed to the 2nd degree &
I was raised to the third on the
same night. Frederick Bristowe
has been in Cambridge the last
week to see his brother who
takes his degree next year.
At the Epigram meeting last time
we had the most amusing collection
that as yet we have been favored with;
one short one was on Griffin a Johnian
Senior Wrangler who has written
the most stale abstract & uninteresting books
on Optics it is possible to conceive
& quite spoilt the beauty of the
science. I ought to keep the subject

Figure 5.62: 1844-05-13. FG to STG page 3.

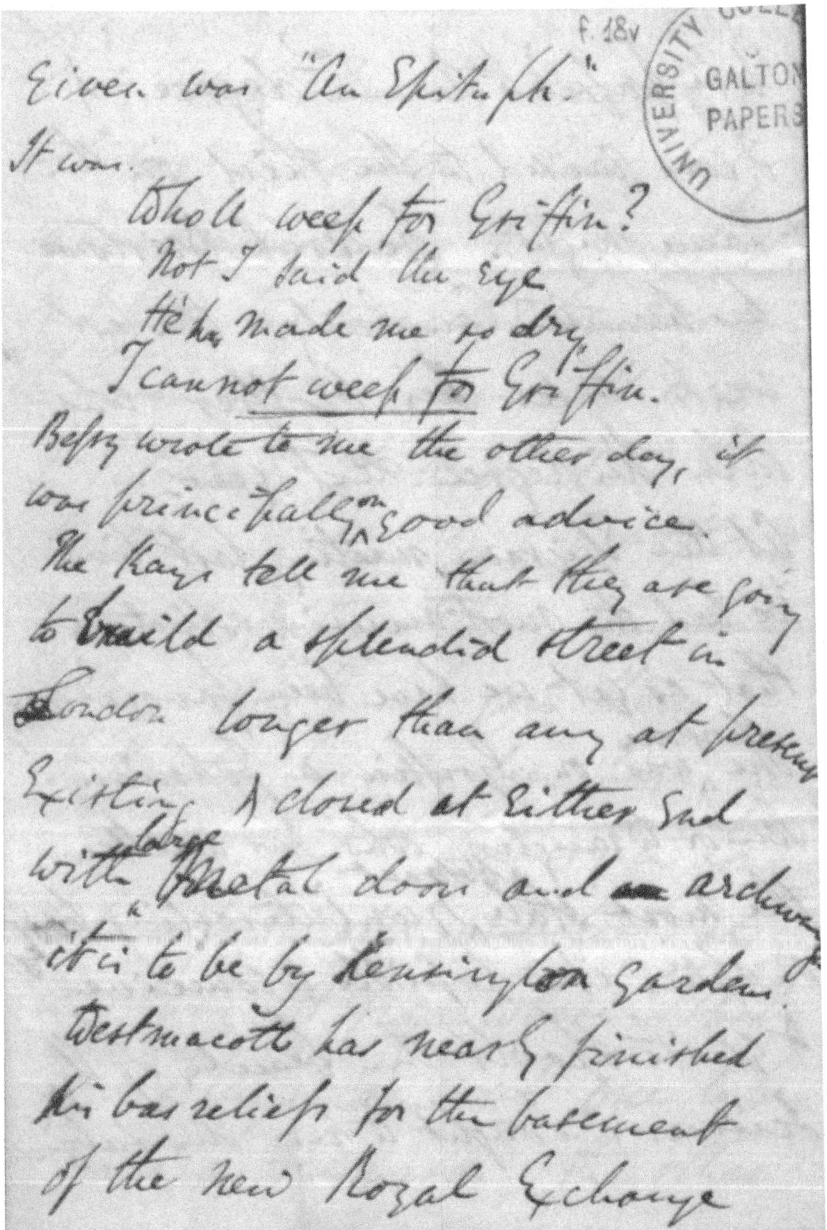

Given was "An Epitaph."
It was.
Who'll weep for Griffin?
Not I said the eye
He has made me so dry,
I cannot weep for Griffin.
Betsy wrote to me the other day, it was principally on good advice.
The Kays tell me that they are going to build a splendid street in London longer than any at present existing & closed at either end with large metal doors and archway it is to be by Kensington gardens.
Westmacott has nearly finished his bas reliefs for the basement of the new Royal Exchange

Figure 5.63: 1844-05-13. FG to STG page 4.

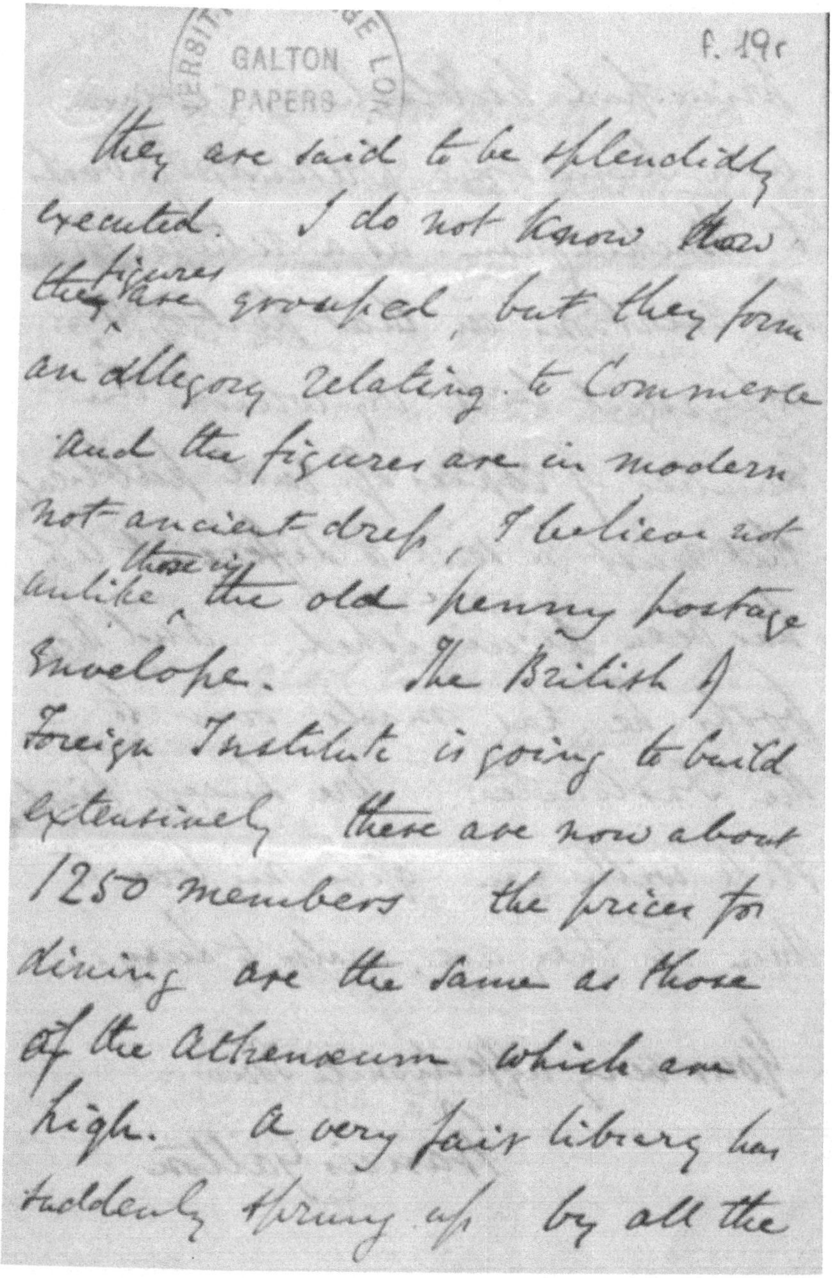

f. 19r

they are said to be splendidly executed. I do not know how the figures are grouped, but they form an allegory relating to Commerce and the figures are in modern not ancient dress, I believe not unlike those in the old penny postage Envelope. The British & Foreign Institute is going to build extensively there are now about 1250 members the prices for dining are the same as those of the Athenaeum which are high. A very fair library has suddenly sprung up by all the

Figure 5.64: 1844-05-13. FG to STG page 5.

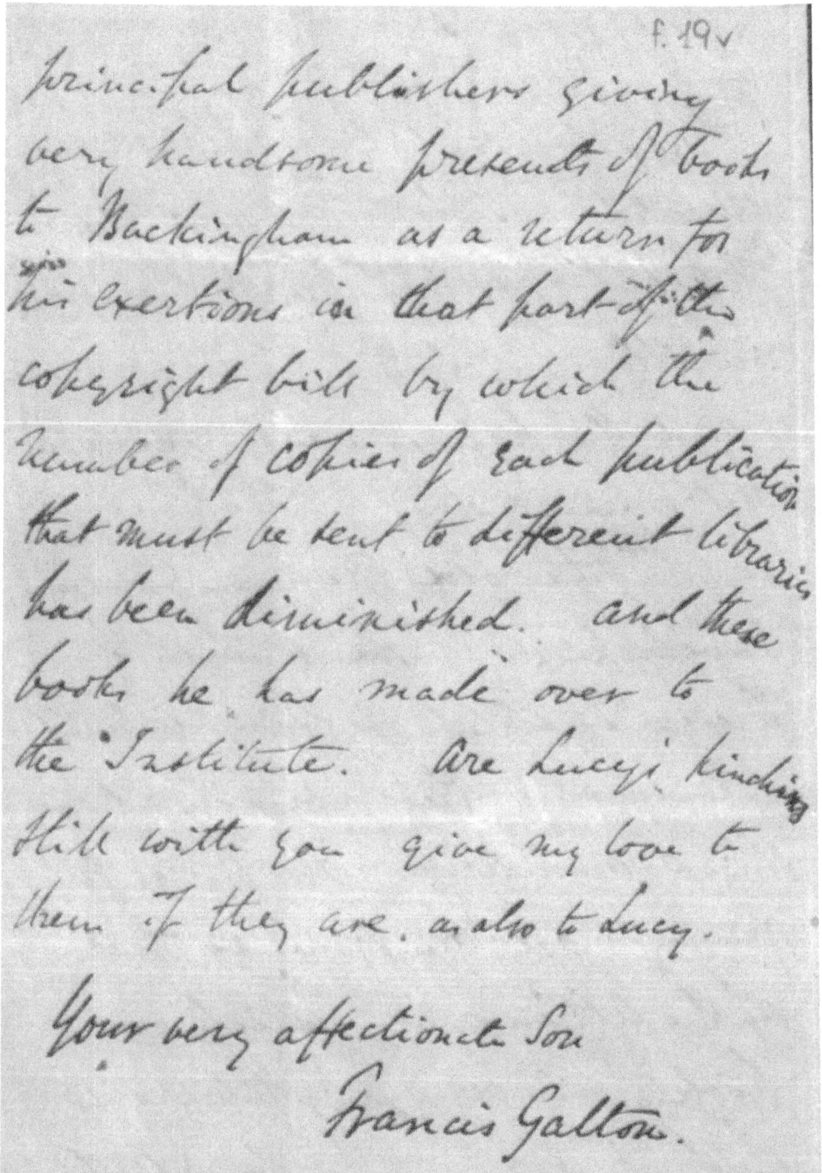
f. 19v

principal publishers giving very handsome presents of books to Buckingham as a return for his exertions in that part of the copyright bill by which the number of copies of each publication that must be sent to different libraries has been diminished. and these books he has made over to the Institute. Are Lucy's kindred still with you give my love to them if they are, and also to Lucy.

Your very affectionate Son

Francis Galton.

Figure 5.65: 1844-05-13. FG to STG page 6.

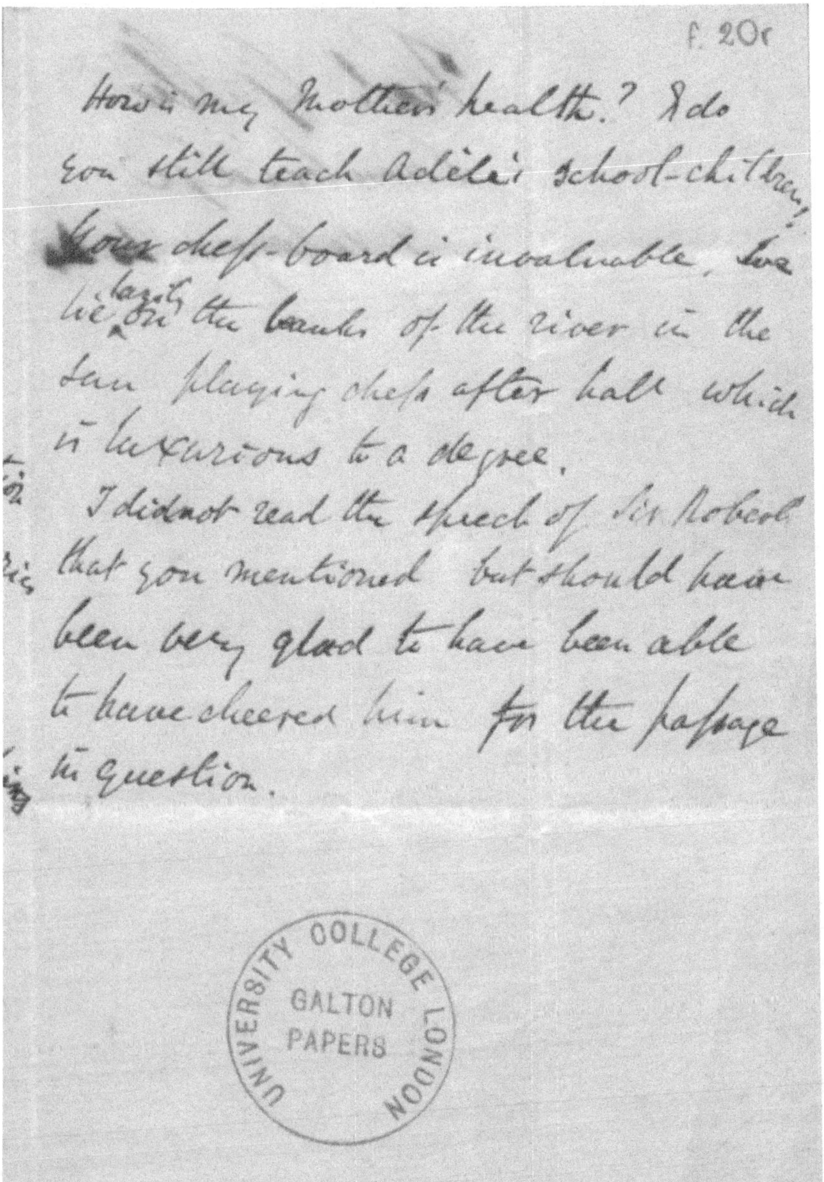

f. 20r

How is my Mother's health? & do
you still teach Adela's school-children?
Your chess-board is invaluable, we
lie lazily on the banks of the river in the
sun playing chess after hall which
is luxurious to a degree.

I did not read the speech of Sir Robert
that you mentioned but should have
been very glad to have been able
to have cheered him for the passage
in question.

Figure 5.66: 1844-05-13. FG to STG page 7.

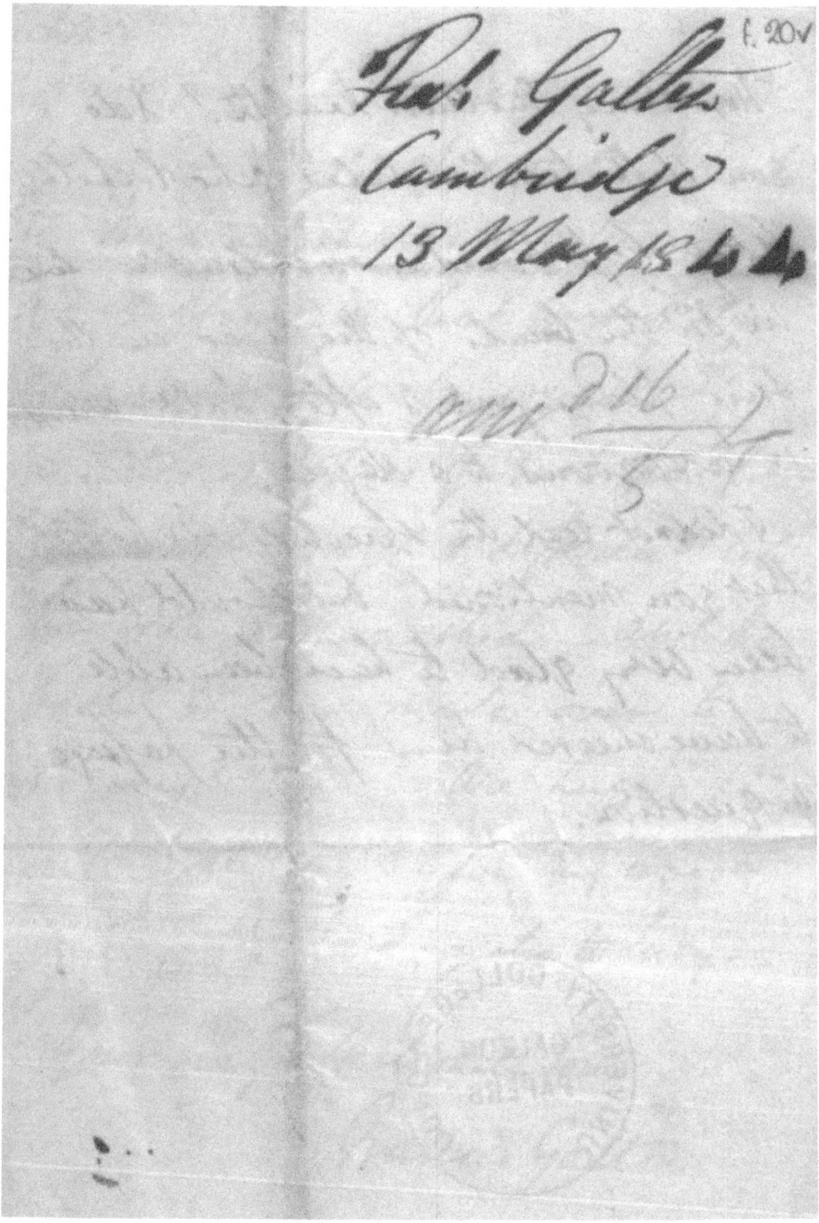

Figure 5.67: 1844-05-13. FG to STG page 8.

1844/06/09. Samuel Tertius Galton to Francis Galton

Claverdon 9 June 1844

My dear Francis

A feverish attack connected, I suppose, with asthma season, has rather weakened me & must be accepted as my apology for not having earlier acknowledged your last letter to me. I am at present under the care of Jephson[33] & Prichard by whose advice I am going to Leamington for a few days that they may the better observe me.

Bessy's letters to us contain very satisfactory accounts of your proceedings, & I shall hope to hear from you again when your plans have been matured. Let me know how your letters are to be addressed to you. I am glad you have been staying with Darwin and Mary and that you have been introduced to the Byrons, who are very kind-hearted and to be relied upon as steady friends.

Darwin has sent me a wonderful account of his observations of the electrical telegraph at Slough.

We are sadly dried up here having had only an occasional shower of short duration. We all return to Leamington tomorrow for a few days. With our united loves to yourself [& co.] I remain ever Your affectionate Father

STG

1844/06/12. Francis Galton to Samuel Tertius Galton

16 King St
Covent Garden, London
Wednesday.

My dear Father

Mr. Partridge[34] and Dr. Budd have been for the last week making many enquiries from the different physicians whom they thought best to consult as to that hospital to which I had better attach myself; the result is that the opinions are decidedly in favour of either the Middlesex or St George's, but some prefer one, some the other. St George's if I could get appointed to it as physician is undeniably the best but it appears quite blocked up,

---

[33] Jephson, Dr. Henry. See page 485.

[34] Partridge, Richard. See page 490.

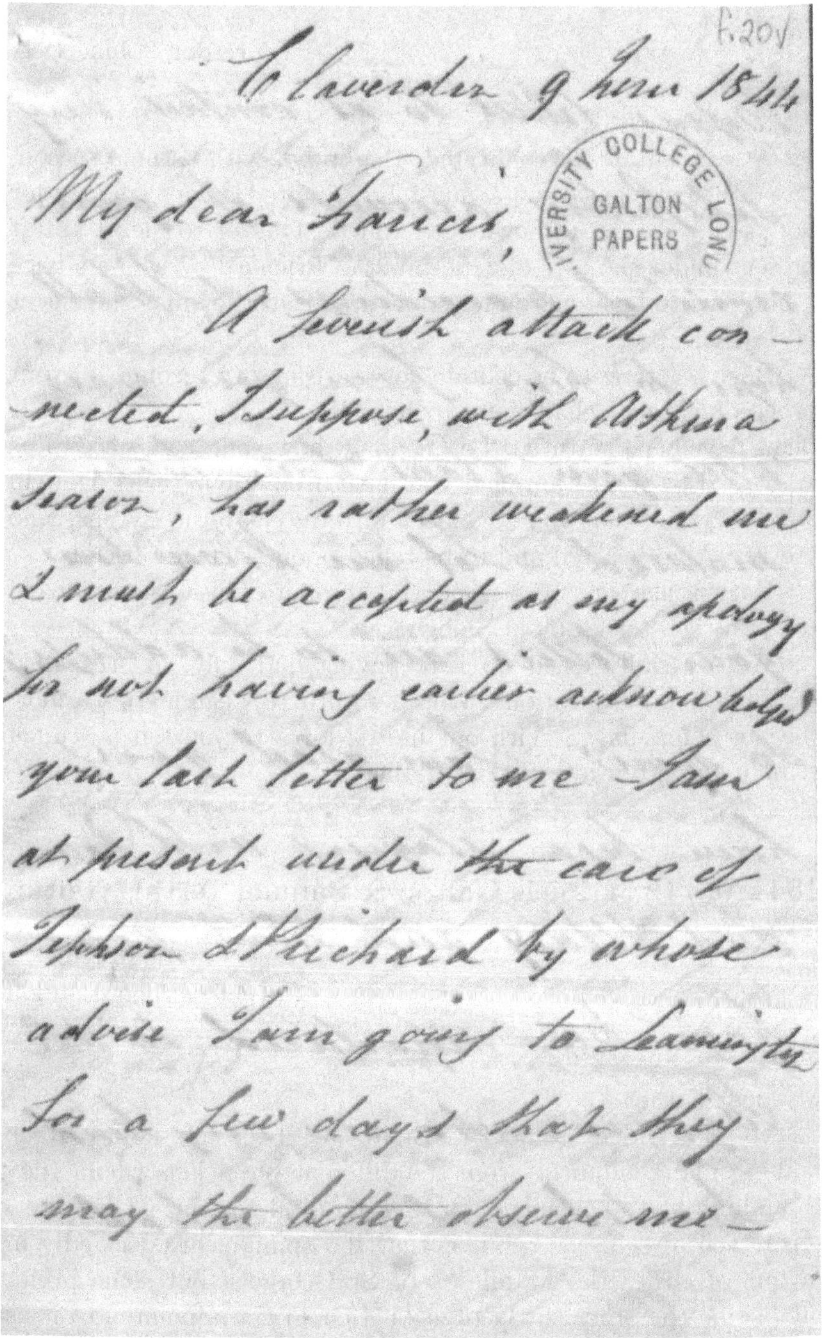

f.20v

Claverdon 9 June 1844

My dear Francis,

A feverish attack connected, I suppose, with Asthma season, has rather weakened me & must be accepted as my apology for not having earlier acknowledged your last letter to me — I am at present under the care of Jephson & Prichard by whose advice I am going to Leamington for a few days that they may the better observe me —

Figure 5.68: 1844-06-09. STG to FG page 1.

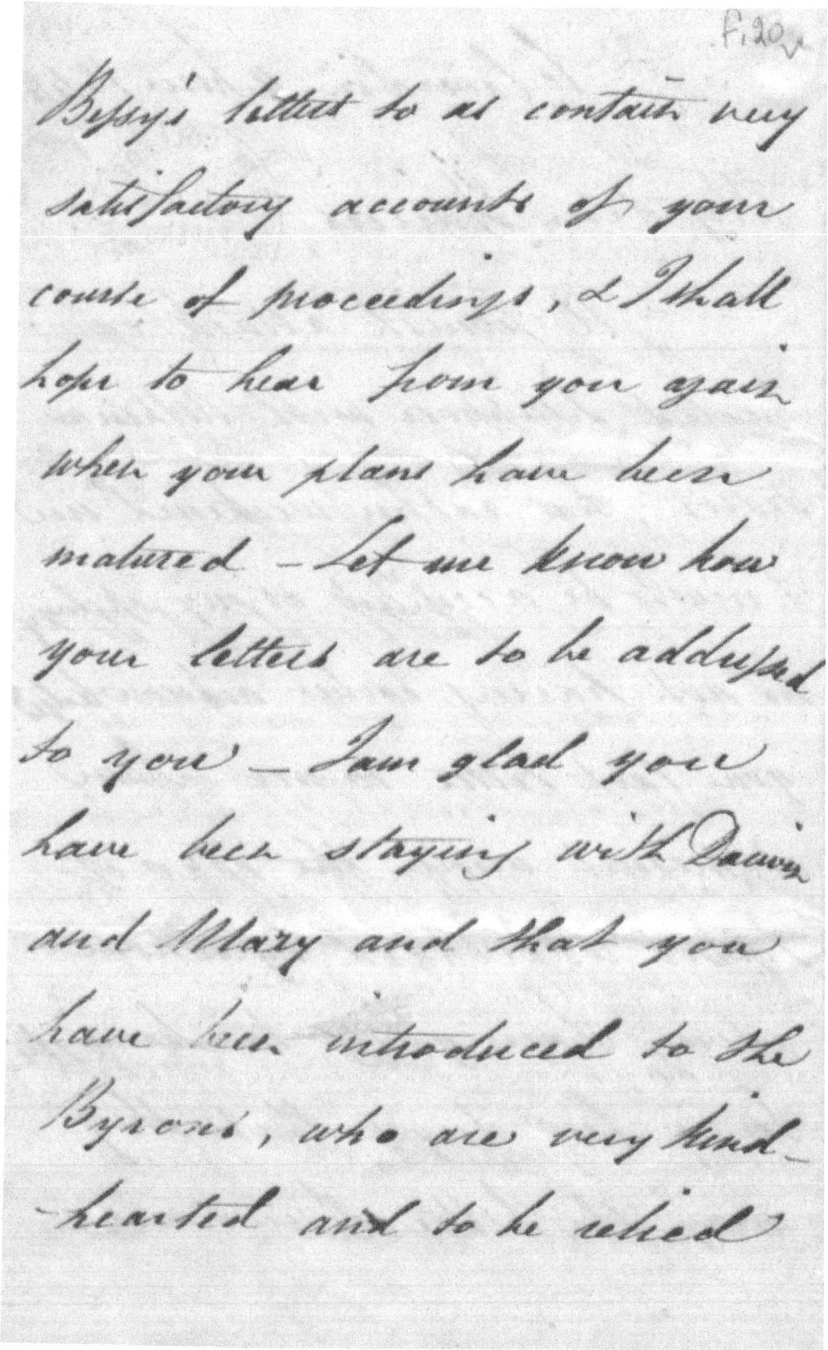

f.20v

Bessy's letter to us contains very
satisfactory accounts of your
course of proceedings, & I shall
hope to hear from you again
when your plans have been
matured — Let me know how
your letters are to be addressed
to you — I am glad you
have been staying with Darwin
and Mary and that you
have been introduced to the
Byrons, who are very kind-
-hearted and to be relied

Figure 5.69: 1844-06-09. STG to FG page 2.

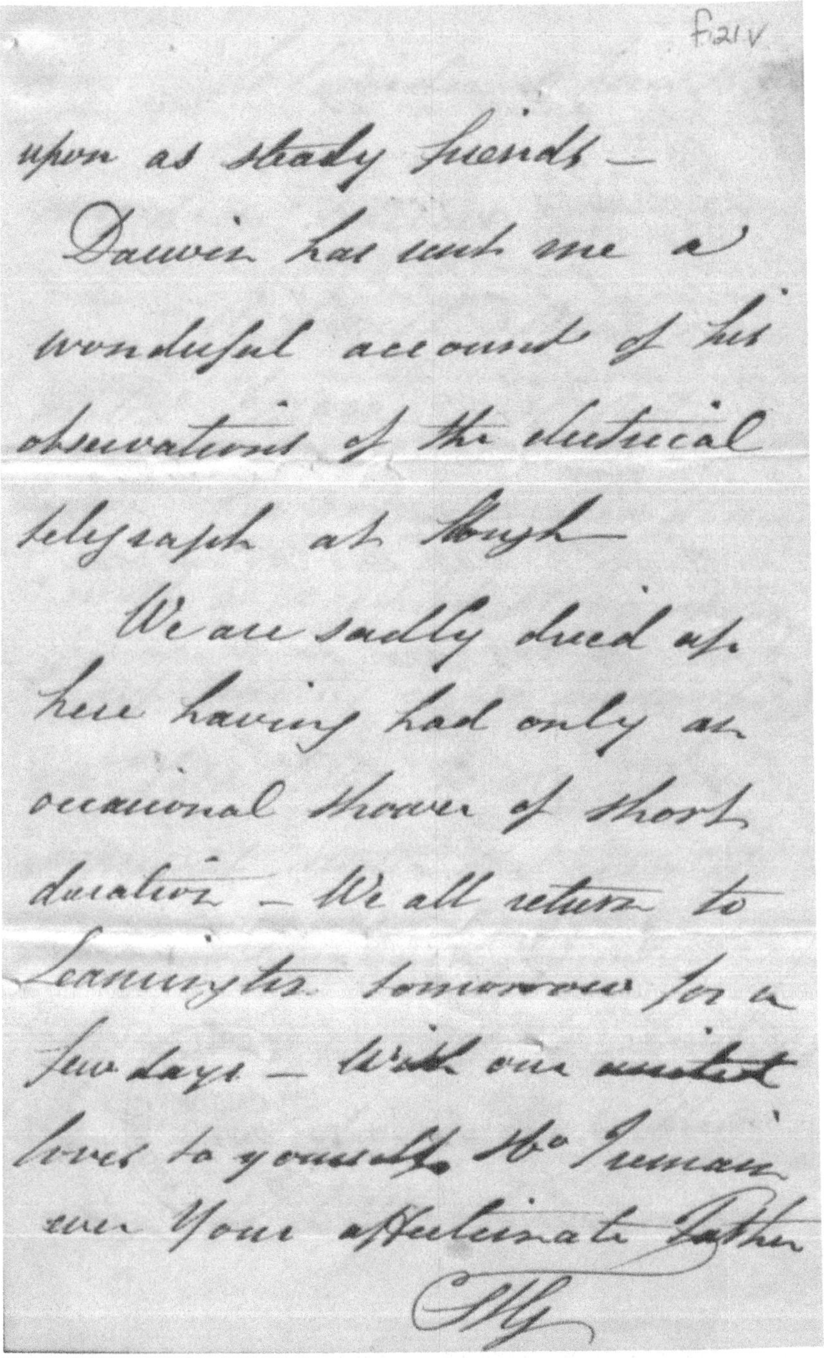
f.21v

upon as steady friends —

Darwin has sent me a wonderful account of his observations of the electrical telegraph at Slough —

We are sadly dried up here having had only an occasional shower of short duration — We all return to Leamington tomorrow for a few days — With our united loves to yourself & I remain ever Your affectionate Father

STG

Figure 5.70: 1844-06-09. STG to FG page 3.

whilst the Middlesex is at present perfectly open, but then the appointment there wd have no influence whatever (by its connection) upon my practice. Notwithstanding Dr Watson & Dr Latham recommend it, though not strongly; and under these circumstances I have written a full account of what the various doctors have said to Mr Hodgson[35] and said that as I had always felt that he had taken a kind interest in me I was unwilling to take any step in my medical plan without his sanction and therefore hoped he wd be so kind as to give me his opinion by which I sh^{d} abide. I forget the exact words but I put them as sweetly as I could, very glad that the case presented itself in such a manner that the asking of Hodgson's opinion in that manner w^{d} be a real advantage to myself as well as a compliment quite due to him. That letter I have just put in the post, and as soon as I receive an answer I will write, that you may be informed as soon as possible about my plans. I have comfortable lodgings for 15 shillings a week as 16 King St Covent Garden, into which I have removed this morning. Will you have the kindness to transfer the money I have in your hands together with some of my next quarter's allowance say on the whole £60 to Barclay's as I shall have to pay at once for my hospital practice as soon as I know my destination & I have not enough money to do so at present.

I am sincerely sorry that your continue feverish but without doubt Jephson[36] & Leamington air will set you up. With love to all and acknowledgments to Pemmy for the parcel per Bouquet

Believe me your affectionate son
F. G.

1844/06/13. Samuel Tertius Galton to Francis Galton

Leamington 13 June 1844

My dear Francis

Your report of your proceedings is very satisfactory & I trust your plans will be decided upon on receiving Hodgson's reply.[37] I have this day given you a credit £60 with Barclay & Co., as you desire.

---

[35] Hodgson, Joseph. See page 483.
[36] Jephson, Dr. Henry. See page 485.
[37] Hodgson, Joseph. See page 483.

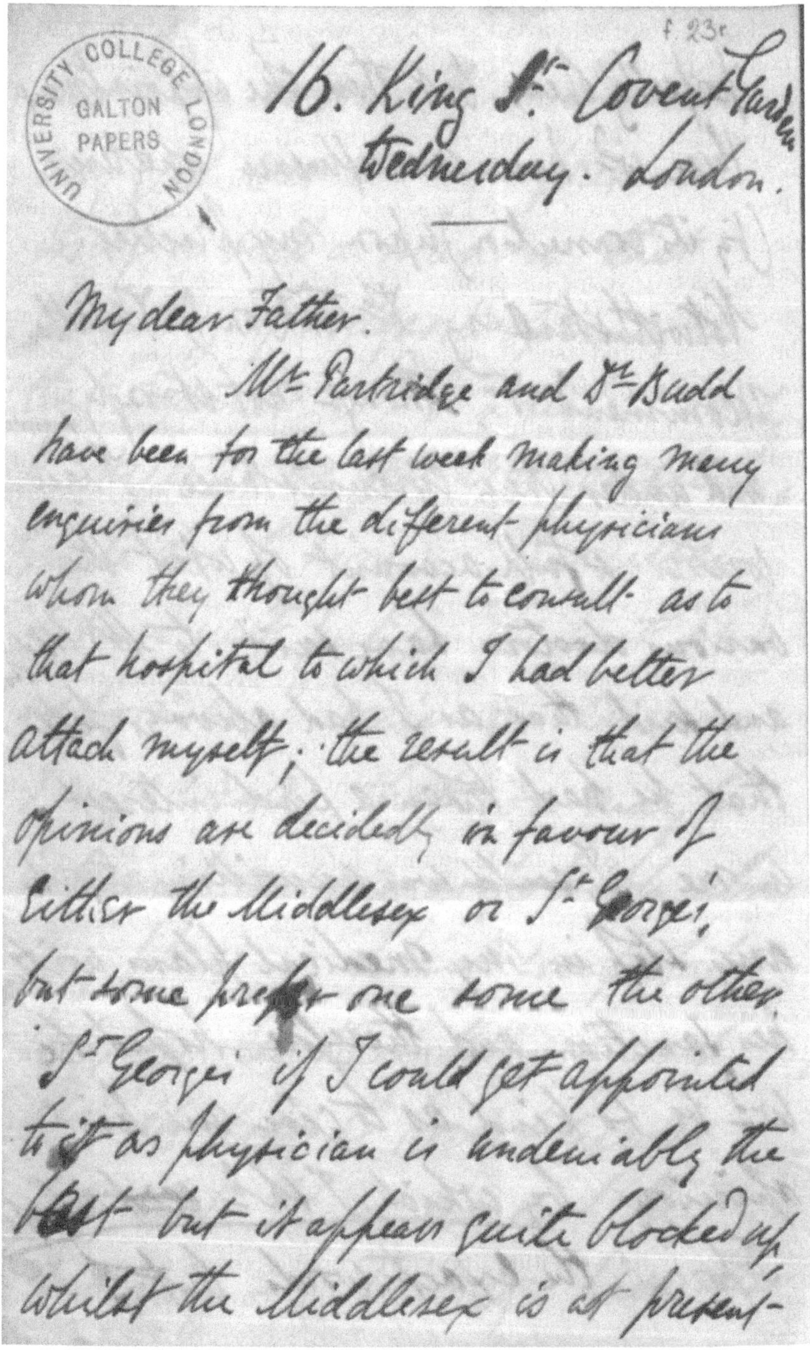

16. King St. Covent Garden
Wednesday. London.

My dear Father.

Mr Partridge and Dr Budd have been for the last week making many enquiries from the different physicians whom they thought best to consult as to that hospital to which I had better attach myself; the result is that the opinions are decidedly in favour of either the Middlesex or St George's, but some prefer one some the other. St George's if I could get appointed to it as physician is undeniably the best but it appears quite blocked up whilst the Middlesex is at present-

Figure 5.71: 1844-06-12. FG to STG page 1.

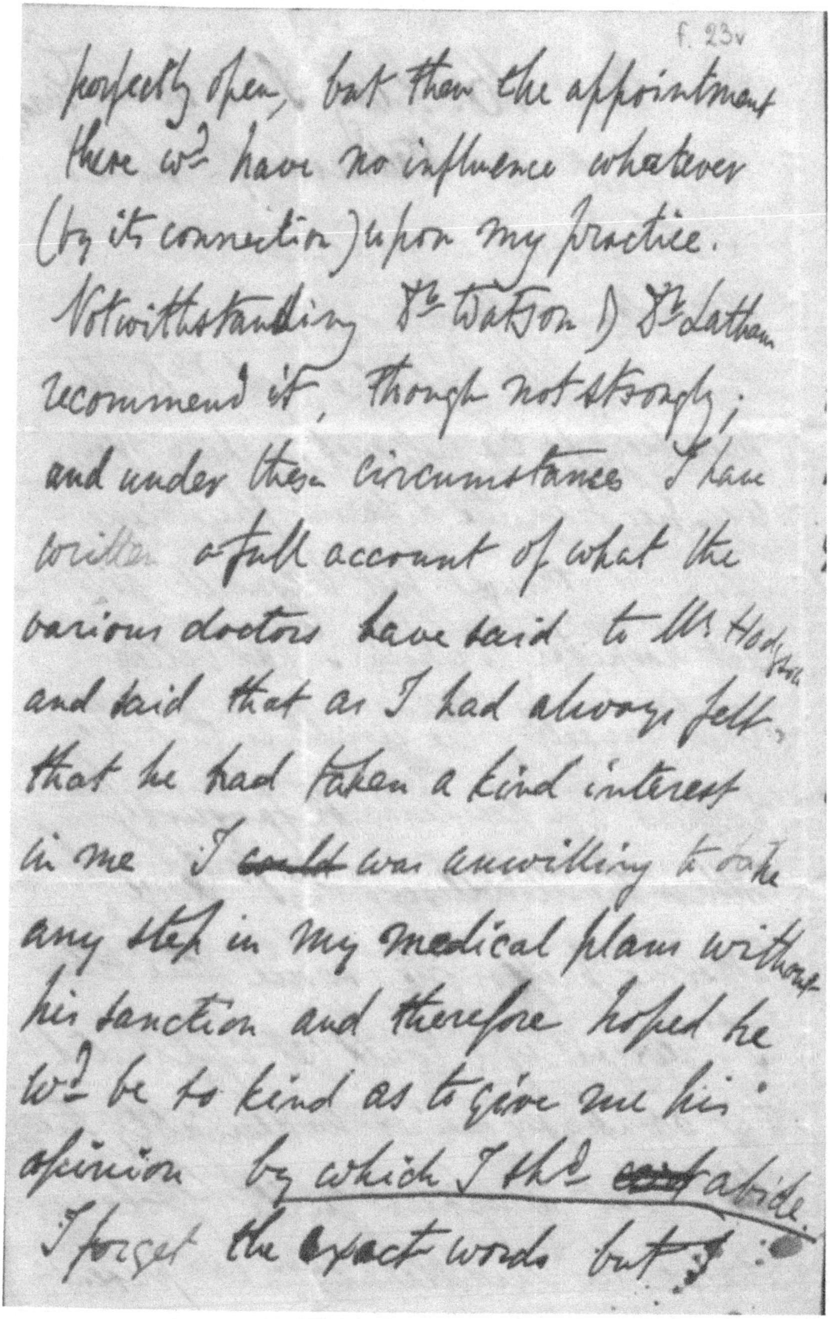

f. 23v

perfectly open, but then the appointment
there wd have no influence whatever
(by its connection) upon my practice.
Notwithstanding Dr Watson & Dr Latham
recommend it, though not strongly;
and under these circumstances I have
written a full account of what the
various doctors have said to Mr Hodgson
and said that as I had always felt
that he had taken a kind interest
in me I ~~could~~ was unwilling to take
any step in my medical plans without
his sanction and therefore hoped he
wd be so kind as to give me his
opinion by which I shd ~~cont~~ abide.
I forget the exact words but

Figure 5.72: 1844-06-12. FG to STG page 2.

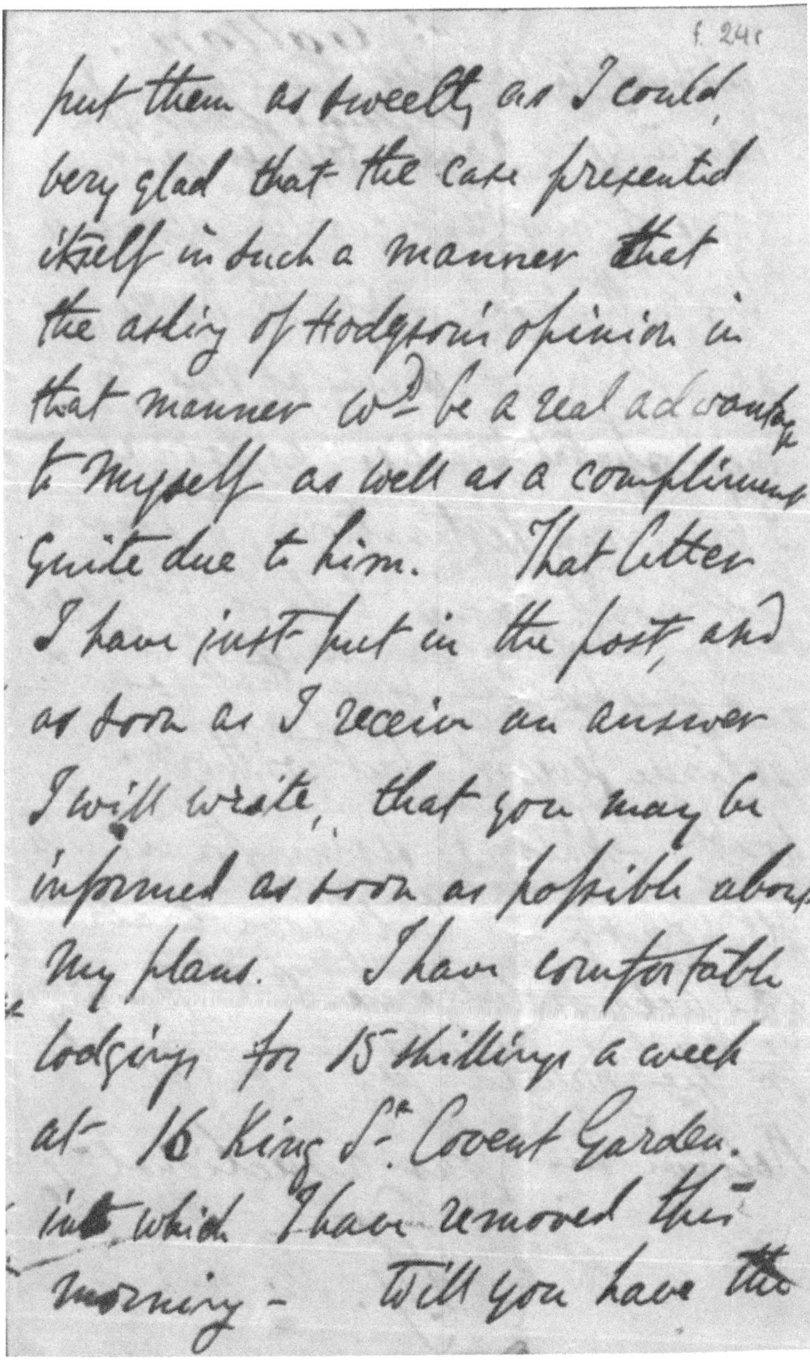
f. 24r

put them as sweetly as I could,
very glad that the case presented
itself in such a manner that
the asking of Hodgson's opinion in
that manner w^{d} be a real advantage
to myself as well as a compliment
quite due to him. That letter
I have just put in the post, and
as soon as I receive an answer
I will write, that you may be
informed as soon as possible about
my plans. I have comfortable
lodgings for 15 shillings a week
at 16 King S^{t}. Covent Garden.
into which I have removed this
morning – Will you have the

Figure 5.73: 1844-06-12. FG to STG page 3.

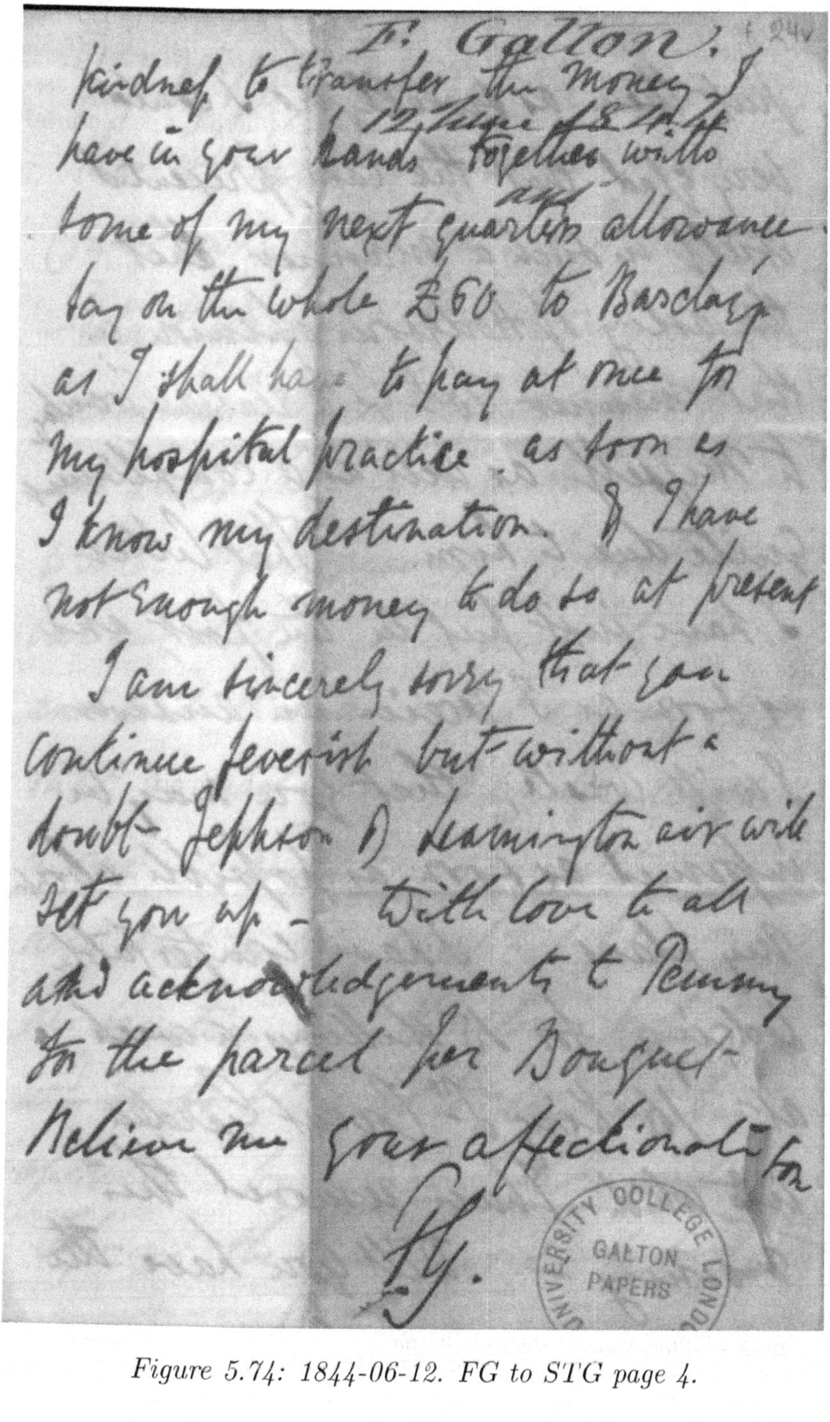

F. Galton.

kindness to transfer the money I
have in your hands [12 Shares of £8 [illegible]] together with
some of my next quarters allowance [and]
pay on the whole £60 to Barclays
as I shall have to pay at once for
my hospital practice as soon as
I know my destination. & I have
not enough money to do so at present
I am sincerely sorry that you
continue feverish but without a
doubt Jephson & Leamington air will
set you up — With love to all
and acknowledgements to Emmy
for the parcel for Bouquet
Believe me your affectionate son
FG.

Figure 5.74: 1844-06-12. FG to STG page 4.

Altho' you may not get an appointment at St. George's Hospital I cannot but think there must be some advantage in being in contact with those who hold the appointments—but this is a matter upon which Hodgson is a much better judge than I can be. I am very glad you have consulted him as he has taken a kind interest in you from the beginning. I am glad you are comfortably lodged.

My fever is better but my tongue very white, as usual in Asthma season, but as yet I am quite free from Asthma. The surface of the fields all around is very much in the same state as my tongue but with a browner tinge. The Leamington Railway Station Houses are growing rapidly. Leamington is very empty at present, there being placards in 5 or 6 Houses in Lansdowne Place. Give our united loves to Bessy & Darwin & Mary[38] when you see them.

I remain
Your affectionate Father
S. Tertius Galton

Give my best compliments to Partridge[39] & Bowman[40] when you see them.

1844/06/18. Francis Galton to Samuel Tertius Galton

My Dear Father

Mr Hodgson[41] wrote me a very kind reply saying that he would be in London on Tuesday (today) and w^{d} then talk with me about my medical plans. This he has done most kindly and most fully, strongly advising St George's on account of its clique & name both in helping practice & in aiding candidates to appointments generally, whether hospital or other. He advises me to continue at Kings' College for the next 6 months or so, going occasionally only to St George's, since the physicians at the former are enthusiastic and of higher repute than perhaps any elsewhere & therefore more desirable as masters at first. He also said that if, after belonging [5] years or so as the case might be at St George's, an undoubted opening shd appear at the Middlesex or elsewhere that I might

[38] Mary Phillips, wide of Darwin Galton.
[39] Partridge, Richard. See page 490.
[40] Bowman, William. See page 481.
[41] Hodgson, Joseph. See page 483.

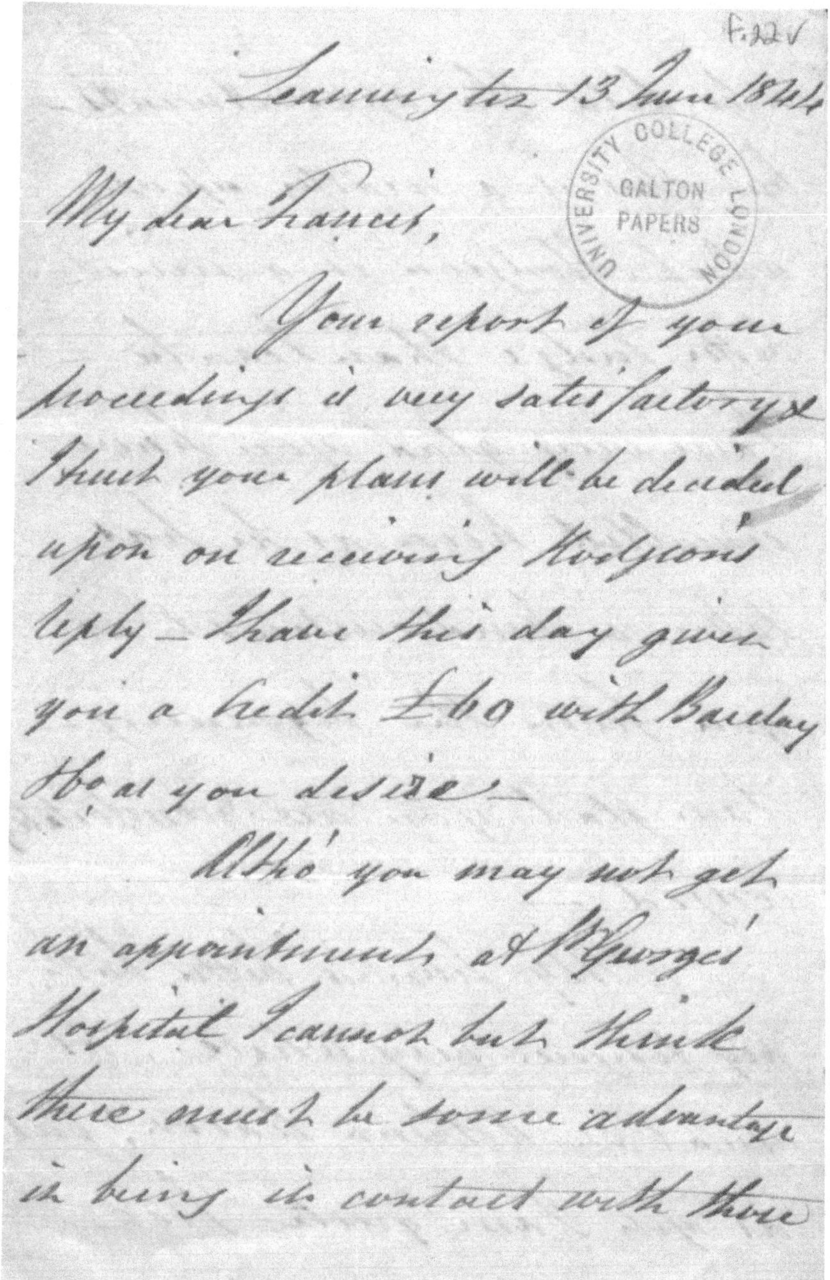

F.22v

Leamington 13 June 1844

My dear Francis,

Your report of your proceedings is very satisfactory & I think your plans will be decided upon on receiving Hodgson's reply — I have this day given you a credit £60 with Barclay & Co. as you desire —

Altho' you may not get an appointment at St George's Hospital I cannot but think there must be some advantage in being in contact with those

Figure 5.75: 1844-06-13. STG to FG page 1.

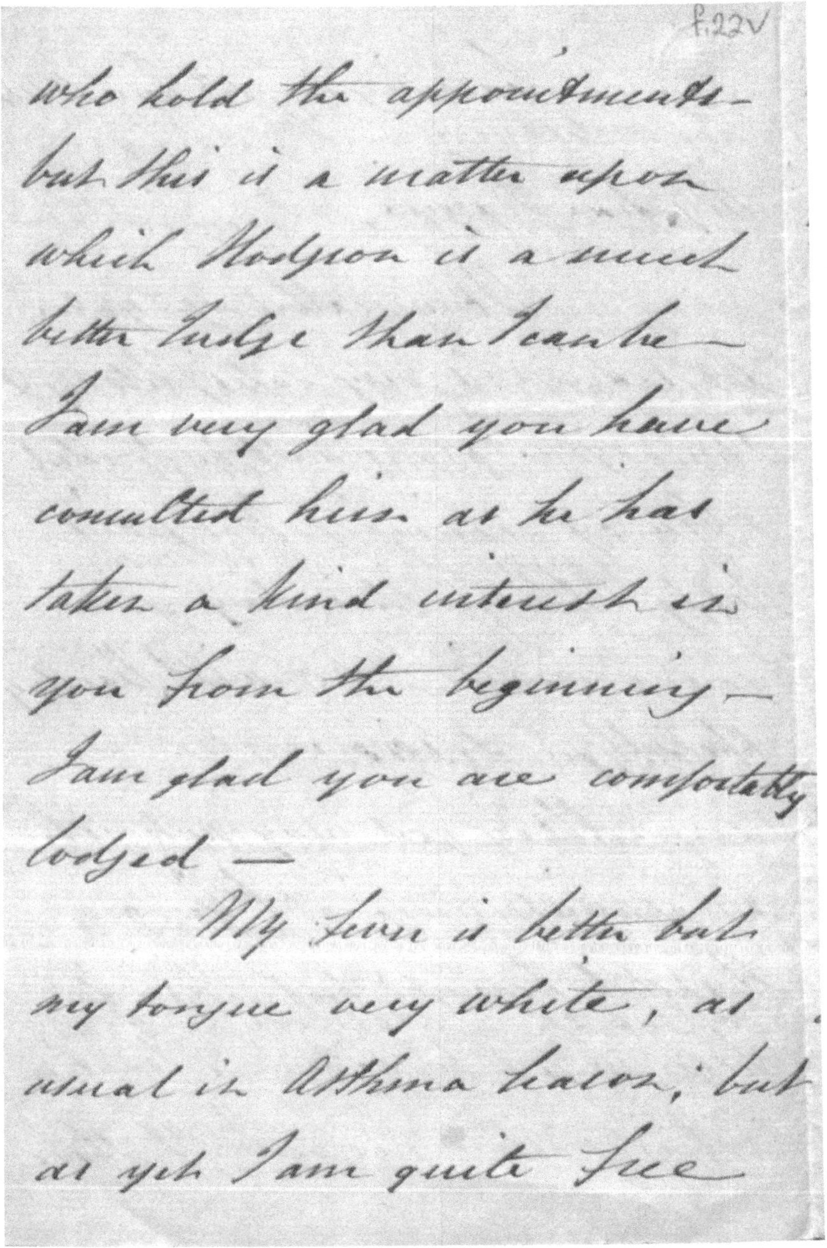
f.22v

who hold the appointments—
but this is a matter upon
which Hodgson is a much
better Judge than I can be—
I am very glad you have
consulted him as he has
taken a kind interest in
you from the beginning—
I am glad you are comfortably
lodged—

My fever is better but
my tongue very white, as
usual in Asthma season; but
as yet I am quite free

Figure 5.76: 1844-06-13. STG to FG page 2.

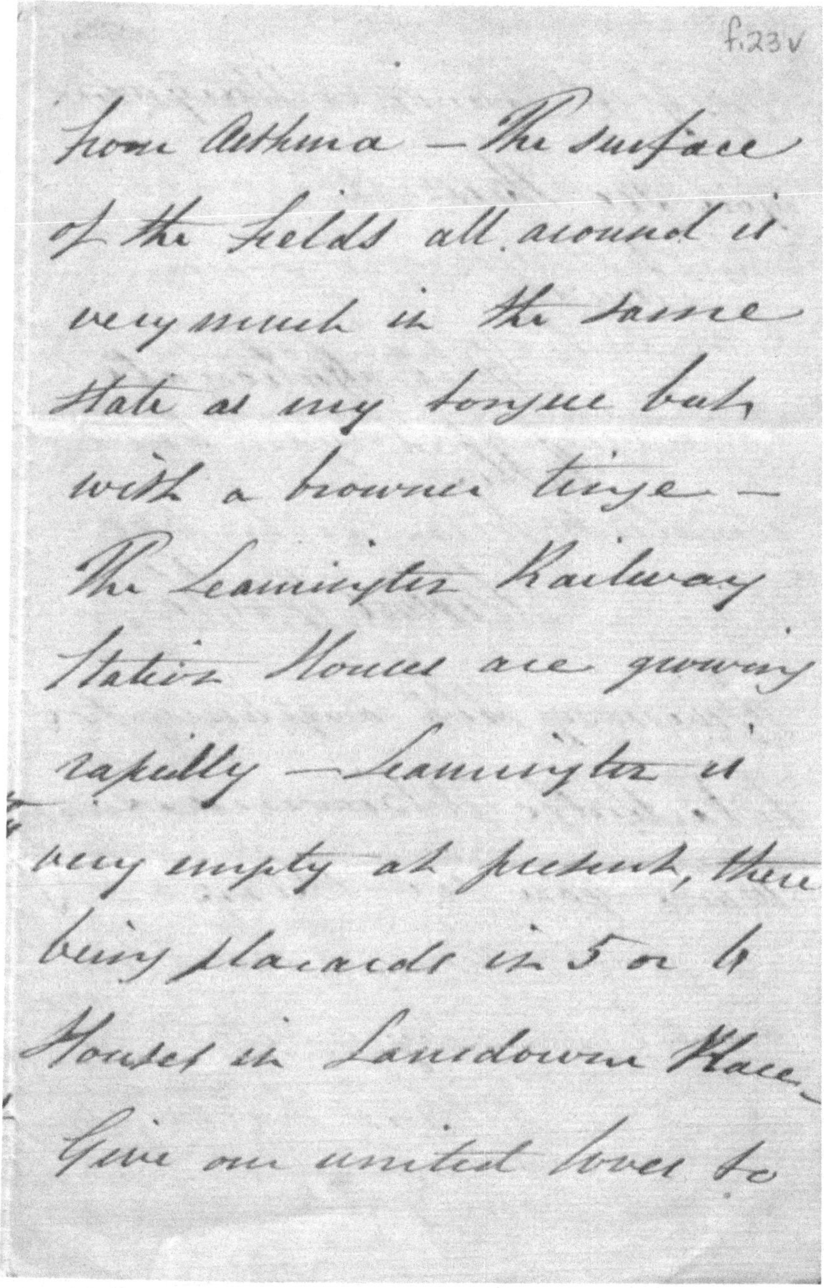
f.23v

from Asthma — The surface
of the fields all around is
very much in the same
state as my tongue but
with a browner tinge —
The Leamington Railway
Station Houses are growing
rapidly — Leamington is
very empty at present, there
being placards in 5 or 6
Houses in Lansdowne Place —
Give our united loves to

Figure 5.77: 1844-06-13. STG to FG page 3.

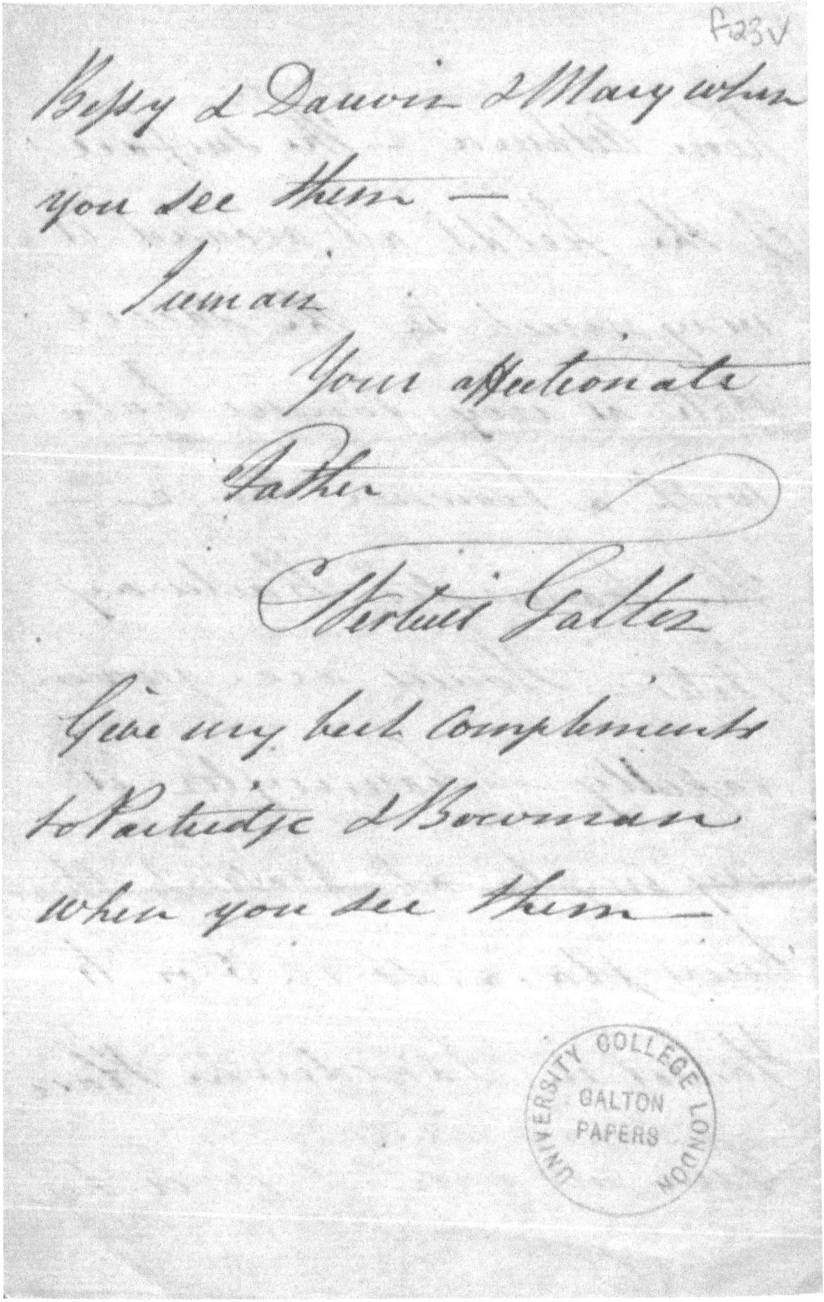

f23v

Bessy & Darwin & Mary when
you see them —
Tumair
Your affectionate
Father
S. Tertius Galton

Give my best compliments
to Partridge & Bowman
when you see them —

Figure 5.78: 1844-06-13. STG to FG page 4.

again change but that is quite a future matter. I left him at 11 on his way to Oxford and then went straight to St George's with a letter of introduction in my pocket. Have already paid my fees and entered as a perpetual pupil and have been introduced to all the authorities there, telling them that my attendance w^{d} not be regular as I had not yet left K. College, that they might not construe rare attendance into negligence. I attend K. College tomorrow – and all the more readily as it was the plan I had myself wished for. I will write to you very shortly again to make up for this sadly curtailed letter, but I wished to save the post & I am very sleepy. I sincerely hope the hay fever is not still annoying.

I dine with the Barclays at Leighton on Friday.

Your affectionate son
F Galton.

NOTICE

A supplement containing a picture of my room will be forwarded with the next letter.

Tuesday
16 King St. Covent Garden.

1844/06/20. Samuel Tertius Galton to Francis Galton

Leamington, 20 June
1844

My dear Francis

I have not for a long time received any letter that has afforded me greater satisfaction than the one you sent me yesterday. You have done well in consulting your earliest, steadiest, & best medical friend and acting promptly upon his advice. You have begun your London medical career most judiciously and if you pursue it properly, I do not doubt of your ultimate success. In the great Metropolis, talent and good sense can command success—but you must always consider yourself medical, & play no tricks. In London there are thousands of competitors, & as the same people do not often come into contact, & any peculiarity or eccentricity once indulged in stands recorded against a person (often unjustly) as being a fixed habit. If any of your relations should once smell a cigar breath, you will be stamped by the whole clique as being

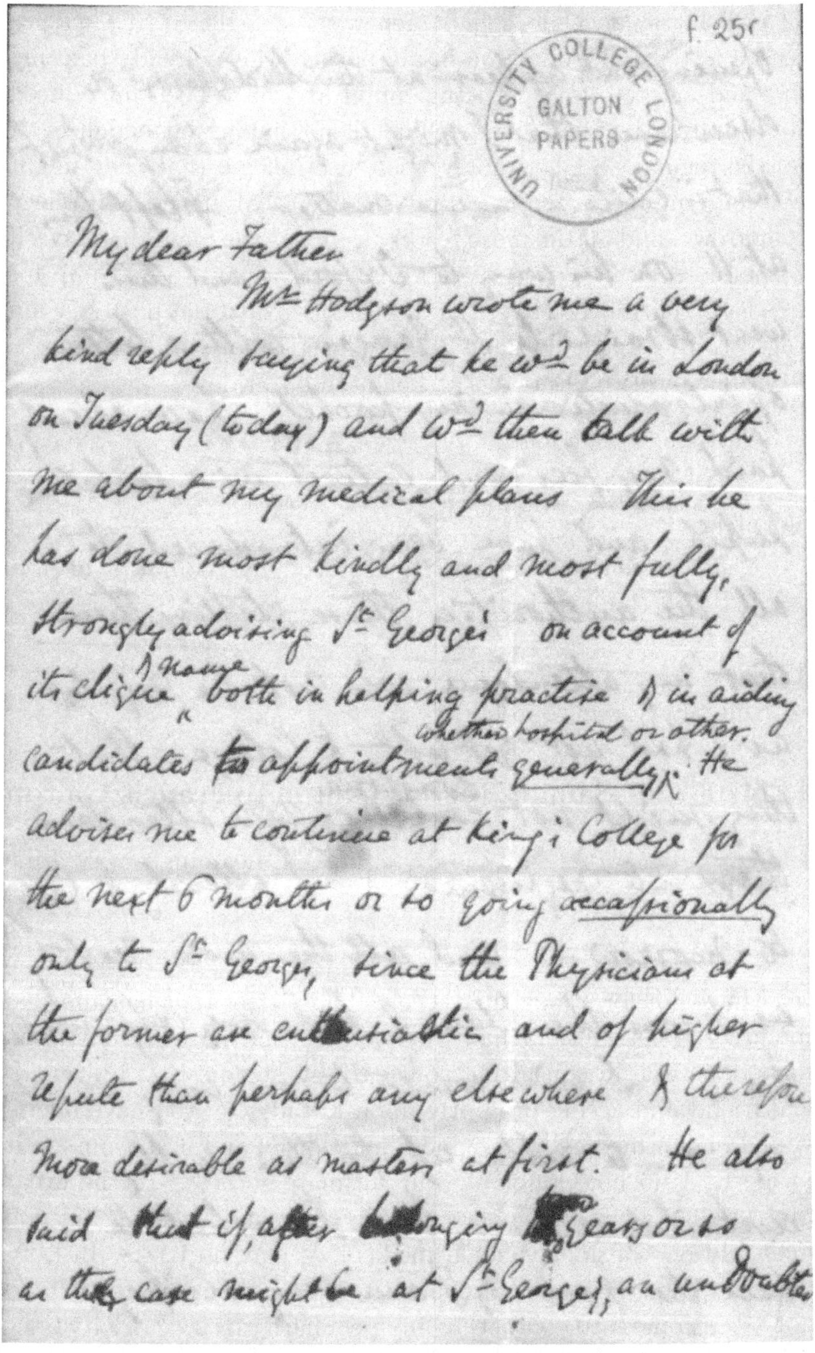
f. 250

My dear Father
Mr Hodgson wrote me a very
kind reply saying that he wd be in London
on Tuesday (today) and wd then talk with
me about my medical plans. This he
has done most kindly and most fully,
strongly advising St Georges on account of
its clique name both in helping practise & in aiding
candidates to appointments generally whether hospital or other. He
advises me to continue at King's College for
the next 6 months or so going occasionally
only to St Georges, since the Physicians at
the former are enthusiastic and of higher
repute than perhaps any elsewhere & therefore
more desirable as masters at first. He also
said that if, after [illegible] years or so
as the case might be at St Georges, an undoubte[d]

Figure 5.79: 1844-06-18. FG to STG page 1.

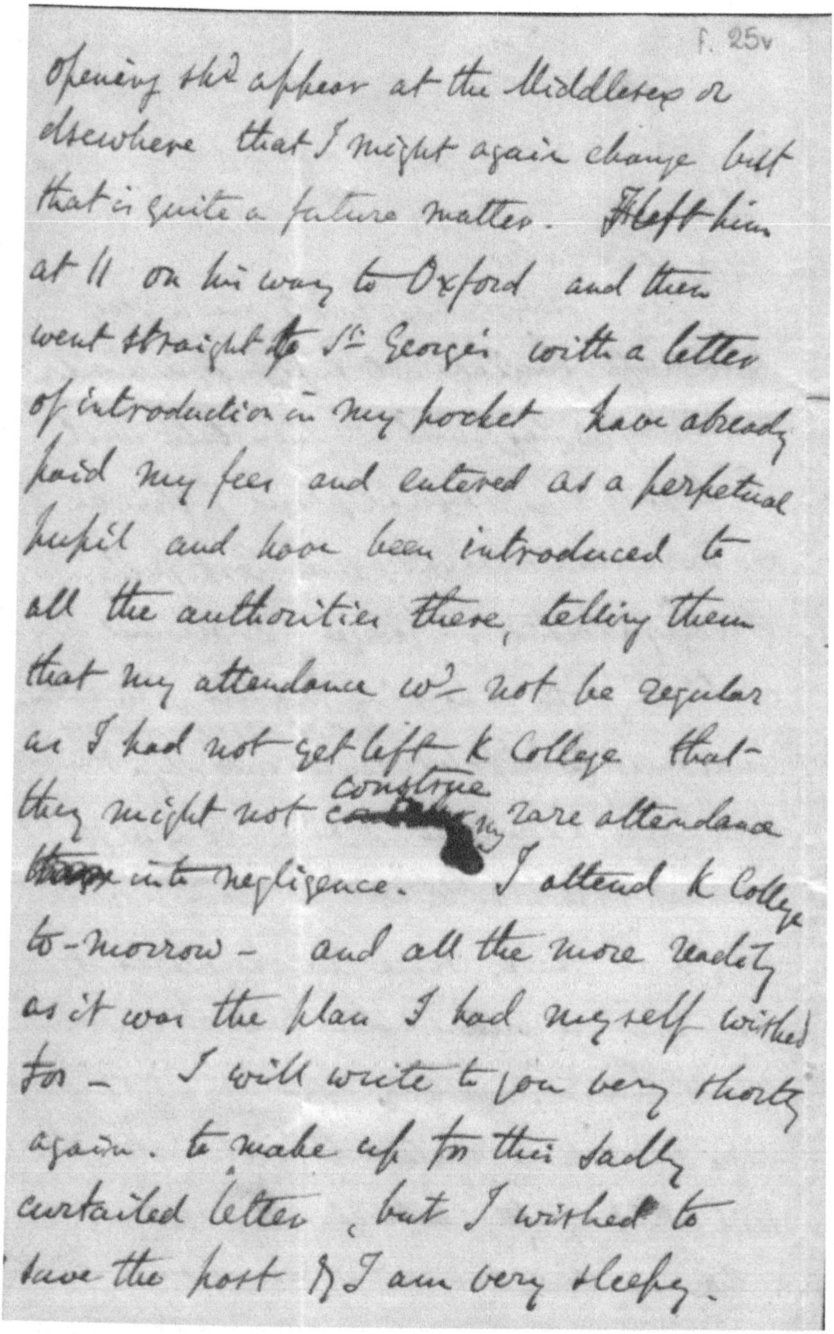
f. 25v

Opening shd appear at the Middlesex or
elsewhere that I might again change but
that is quite a future matter. ~~I~~ left him
at 11 on his way to Oxford and then
went straight to St George's with a letter
of introduction in my pocket have already
paid my fees and entered as a perpetual
pupil and have been introduced to
all the authorities there, telling them
that my attendance wd not be regular
as I had not yet left K College that
they might not ~~contribute~~ construe my rare attendance
~~there~~ into negligence. I attend K College
to-morrow – and all the more readily
as it was the plan I had myself wished
for – I will write to you very shortly
again. to make up for this sadly
curtailed letter, but I wished to
save the post & I am very sleepy.

Figure 5.80: 1844-06-18. FG to STG page 2.

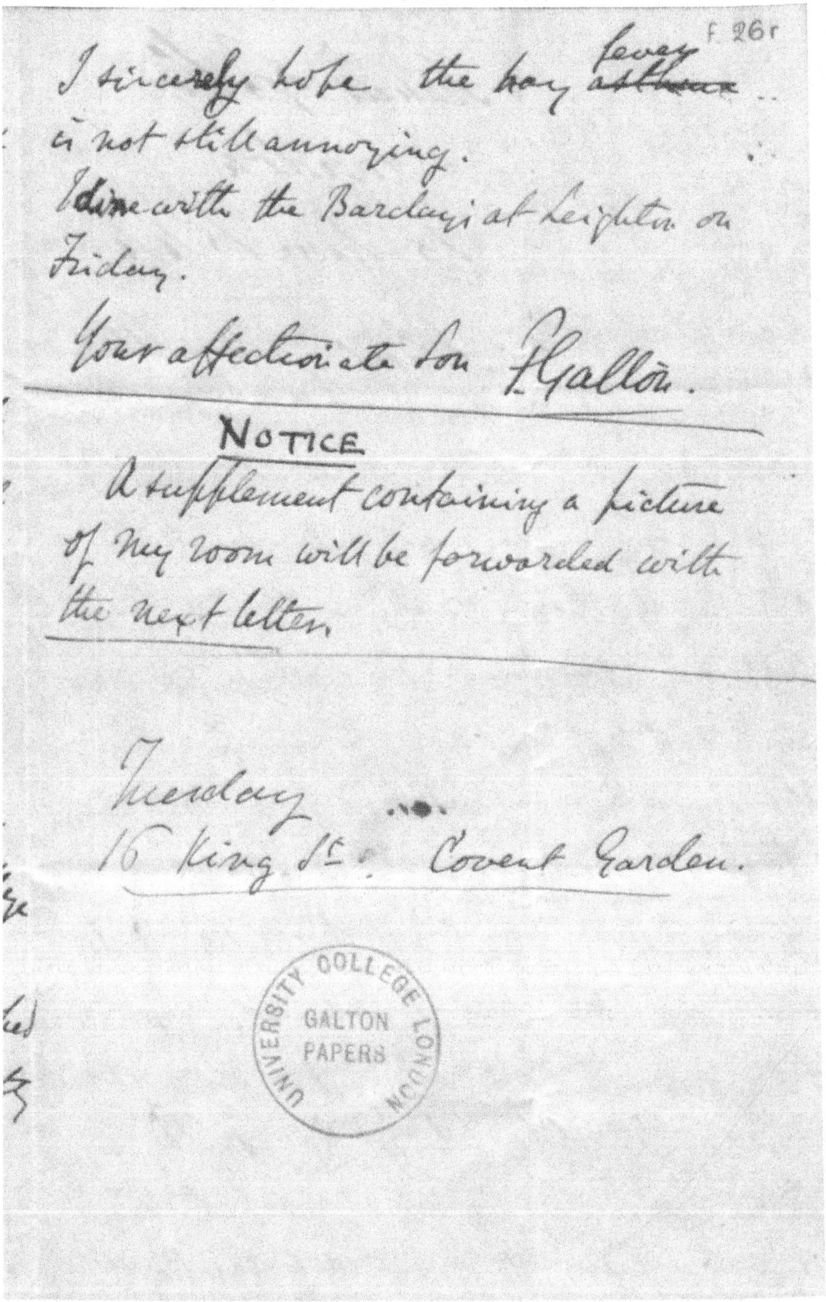

f 26r

I sincerely hope the hay ~~asthma~~ fever is not still annoying.
I dine with the Barclays at Leighton on Friday.

Your affectionate son F. Galton.

NOTICE

A supplement containing a picture of my room will be forwarded with the next letter.

Tuesday
16 King St. Covent Garden.

Figure 5.81: 1844-06-18. FG to STG page 3.

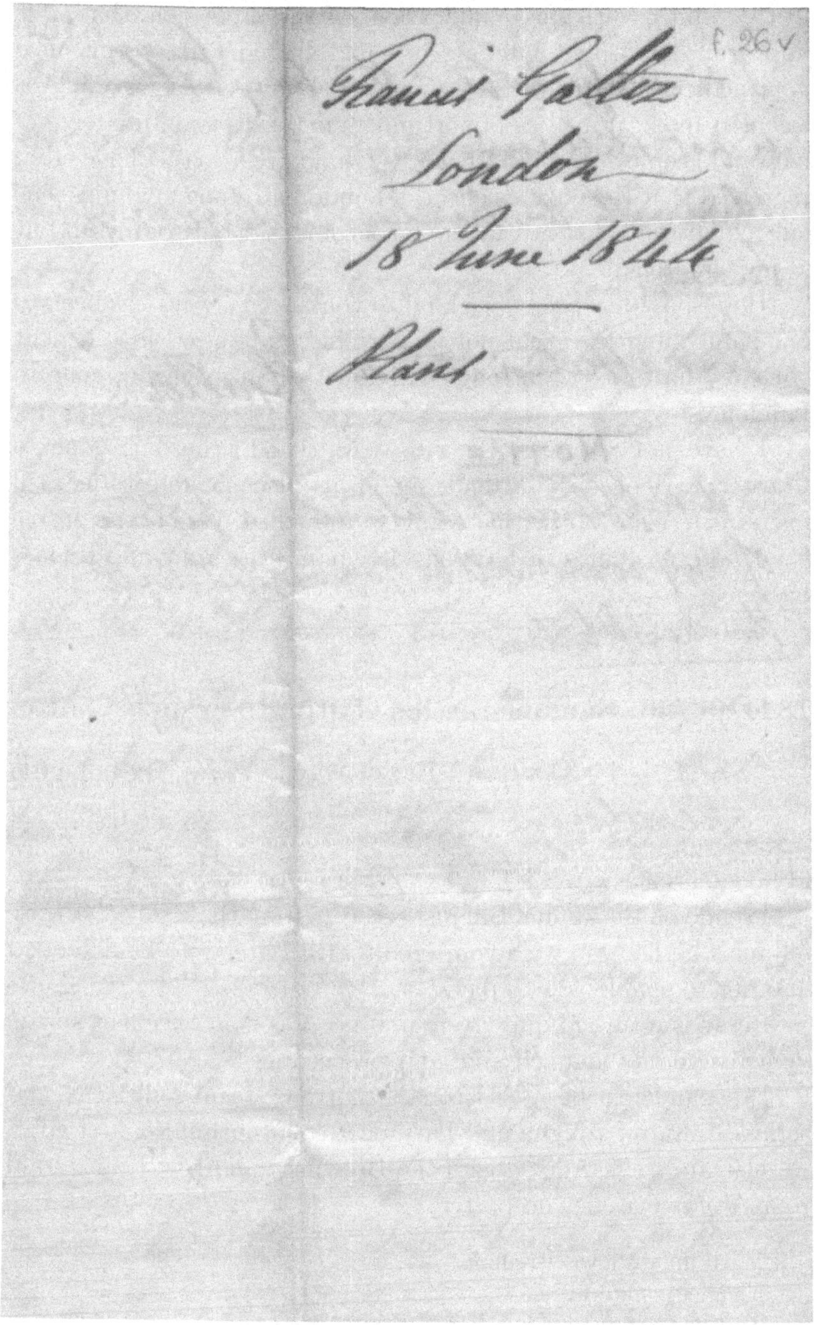

f. 26v

Francis Galton
London
18 June 1844

Plant

Figure 5.82: 1844-06-18. FG to STG page 4.

an inveterate smoker—and I would recommend you to be so perfectly imbued with the qualifications of the profession, that your appearance (without humbug or cant) should rather pronounce you medical. Cultivate the acquaintance & society of leading or talented medical men as opportunities naturally & intimately occur, and keep up your old friends & London relations. I have often regretted that the noxious effect of London air upon my lungs, has cut me off from that advantage on account not only of myself but of my children.

Hodgson[42] has been very kind to you, as he always has been to our family, and I think it not improbable that the wish to benefit you may have been a strong motive with him in having gone to London.

I have not yet seen Darwin & Mary. Prichard is gone to Chester, & Jephson[43] attends me in his place, & refuses fees till his return. Your Mother has a kind of Influenza. I am better & free from Asthma, but still feverish. I remain ever Your affectionate Father

S. Tertius Galton

1844/06/30. Samuel Tertius Galton to Francis Galton

Leamington
30 June 1844

My dear Francis

I send you the annual Bill just received from Blakesley.[44] Please tell me whether you wish your credit with Barclay to be extended in order to enable you to pay it.

Bessy is returned quite well and has put us in excellent spirits by her report of yourself and your proceedings.

Erasmus is to be at Edstone tomorrow. I am still weak and listless from dry tongue and have occasional intimations of gout. All else are well. Adèle visits the Brewins tomorrow I am afraid your Aunt's eyes get no better.

With united loves I remain
Yr affectionate father
Tertius Galton.

---

42 Hodgson, Joseph. See page 483.
43 Jephson, Dr. Henry. See page 485.
44 Galton's classical tutor, Blakesley, Joseph Williams, see page 480.

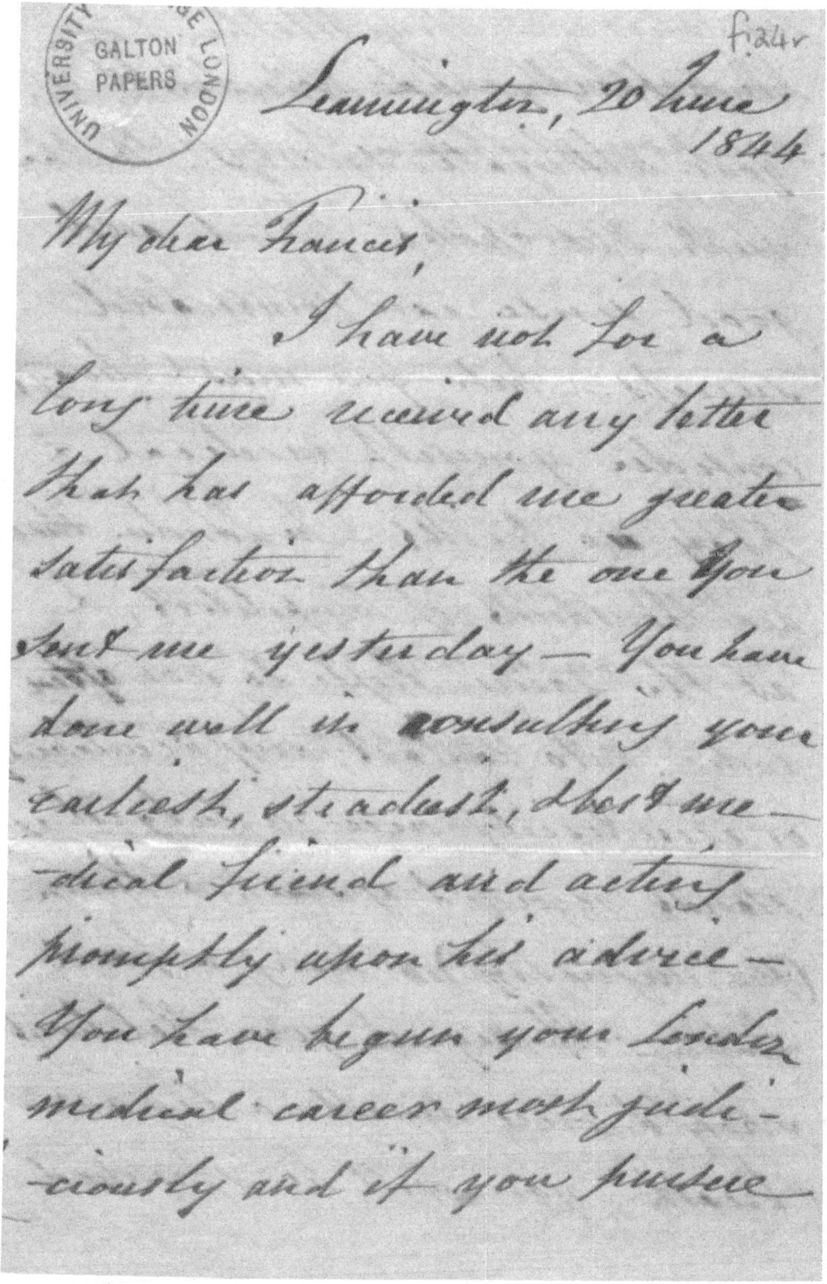

f.24r

Leamington, 20 June
1844

My dear Francis,

I have not for a long time received any letter that has afforded me greater satisfaction than the one you sent me yesterday — You have done well in consulting your earliest, steadiest, & best medical friend and acting promptly upon his advice — You have begun your London medical career most judiciously and if you pursue

Figure 5.83: 1844-06-20. STG to FG page 1.

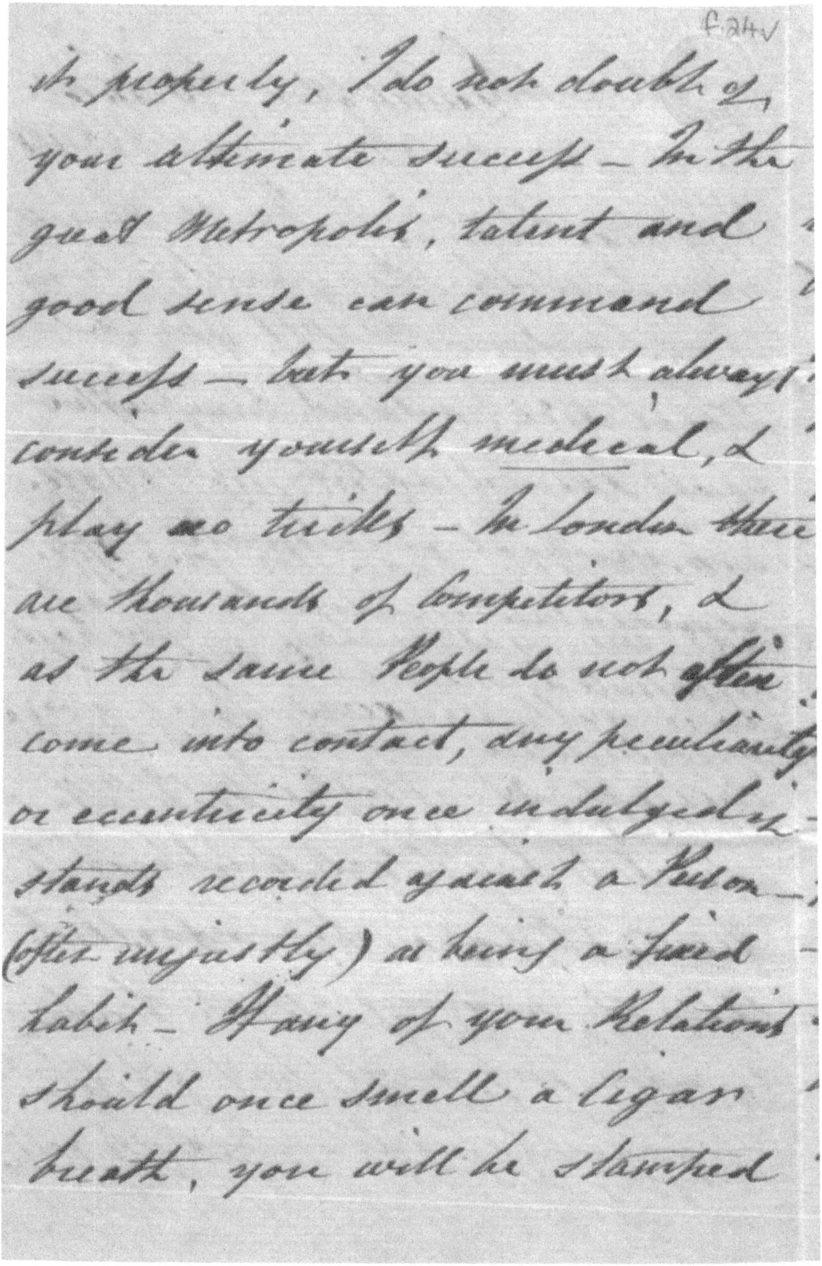
f.24v

it properly, I do not doubt of
your ultimate success – In the
great Metropolis, talent and
good sense can command
success – but you must always
consider yourself medical, &
play no tricks – In London there
are thousands of Competitors, &
as the same People do not often
come into contact, any peculiarity
or eccentricity once indulged in
stands recorded against a Person –
(often unjustly) as being a fixed
habit – If any of your Relations
should once smell a Cigar
breath, you will be stamped

Figure 5.84: 1844-06-20. STG to FG page 2.

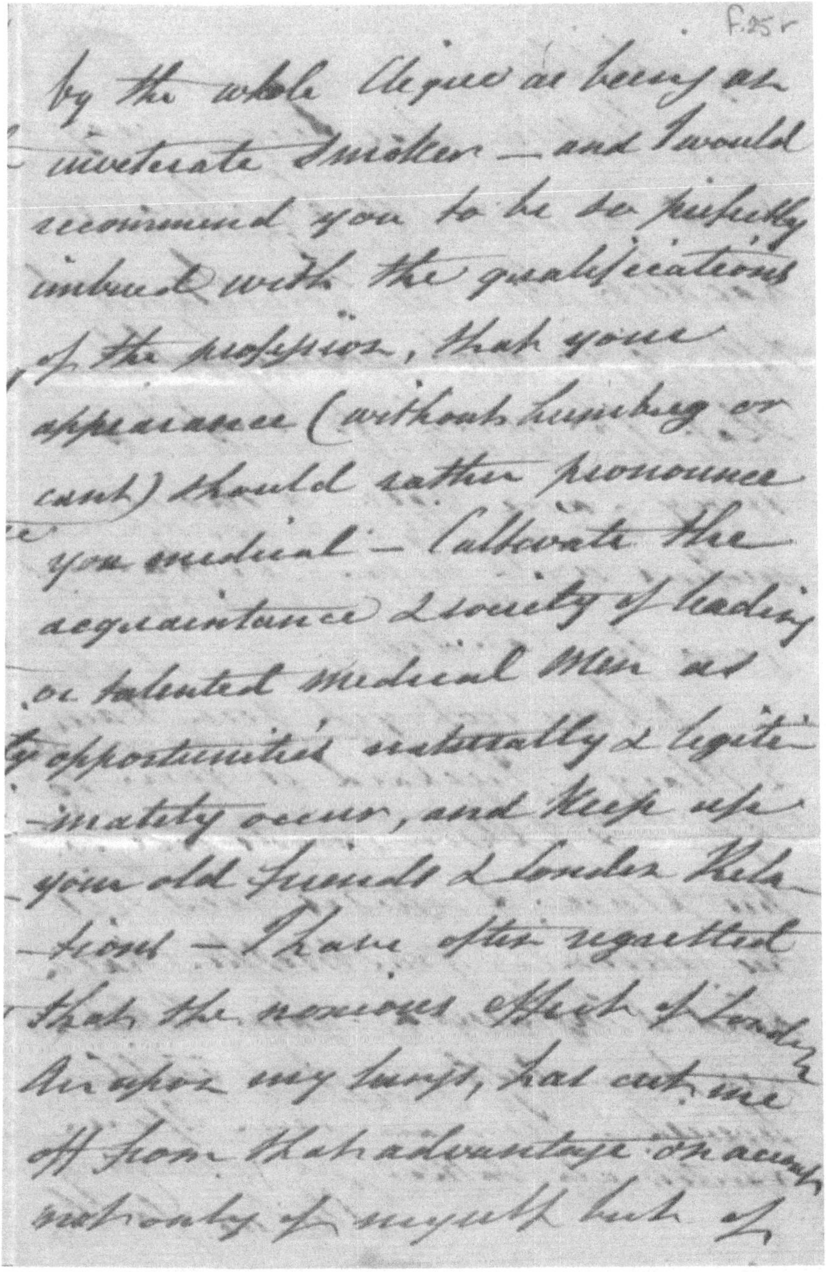
f.25r

by the whole Clique as being an
inveterate Smoker — and I would
recommend you to be so perfectly
imbued with the qualifications
of the profession, that your
appearance (without humbug or
cant) should rather pronounce
you medical — Cultivate the
acquaintance & society of leading
or talented Medical Men as
opportunities naturally & legiti-
-mately occur, and keep up
your old Friends & London Rela-
-tions — I have often regretted
that the noxious effect of London
Air upon my lungs, has cut me
off from that advantage on accou[nt]
not only of myself but of

Figure 5.85: 1844-06-20. STG to FG page 3.

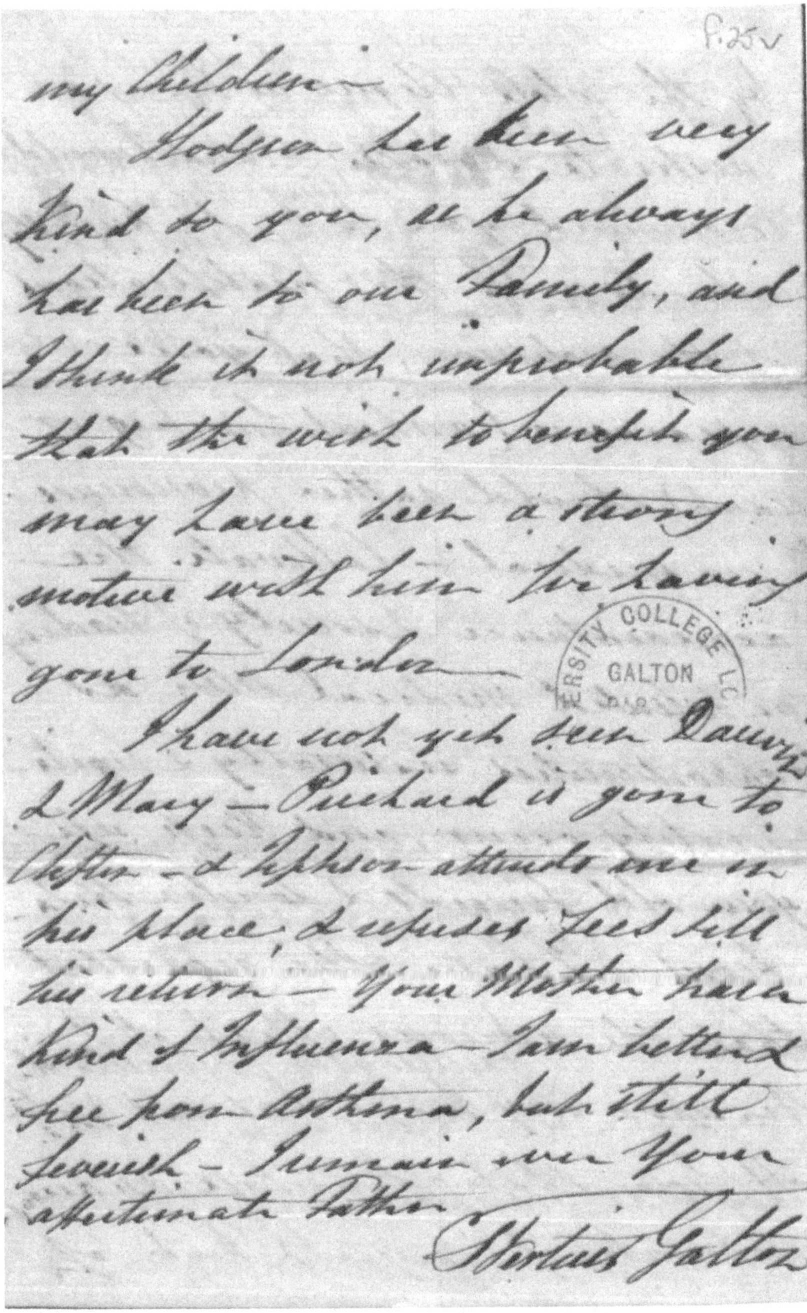

Figure 5.86: 1844-06-20. STG to FG page 4.

1844/07. Francis Galton to Samuel Tertius Galton

Monday

My dear Father

I am afraid that I have two unanswered letters on the score against me, and thank you much for the letter of introduction to Mr. Walker contained in the last. I will tell you the result after I have taken it. I meet with numbers of my Cambridge friends so that I am pretty sure of a call every day and this with working till about 4 at medicine, and again before going to bed makes my existence about as jolly and as cozy as I ever expect to aspire to. Mr. Hallam[45] gave me a ticket for the private view of the fresco-paintings now in Westminster Hall (on Saturday). They are said to be decidedly inferior to what was expected and nearly half of them were rejected as unworthy of exhibition. The two best (and I had come to the same conclusion without hearing it before) are considered to be "Rachel & Jacob" by Cope[46] & a study by Armitage, a female figure looking something like Britannia.[47] Armitage & Cope were two of the three that got £300 prizes for their cartoons last year.

Dr Todd is very good natured to me.[48] He has invited me to spend next Sunday with him at a country cottage of his near Streatham Common.

I enclose you two scrawls on one piece of paper intended respectively to represent different views of my room. Tell Bessy that since she was in it my landlord has allowed me a glorious damask green little bit of a sofa which fits as snugly as possible into the room. [See Figure 5.87, p. 460.]

Sir Arthur Brooke said he w^{d} take me to see Alexis the mesmeriser about whom so much has been written in the Times Chronicle & Herald by & in reply to Colonel Gurwood.[49] I thought I had better go as he is said to be by far the most successful clairvoyant, he won't exhibit publicly and this time Mr. Ramsay Clarke had him in his rooms. It was entirely a failure, he certainly played

---

45 Hallam, Henry Fitzmaurice. See page 483.

46 Charles West Cope.

47 "Ophelia" by Edward Armitage.

48 Robert Bentley Todd (1809-1860).

49 Alexis Didier (1826-1886) the clairvoyant. He was mesmerized by Marcillet into the (supposedly) clairvoyant state. Colonel John Gurwood (1790–1845), the Peninsular Wars veteran, had been persuaded by him.

at cards with his eyes blindfolded but that is not conclusive; but in not one instance could he read words written aside & put into boxes, which he professed to be able to do.

Will you be so kind as to let me know on what days Warwick Castle is visible as Mr. Hallam[50] spends a day next week over Warwick Coventry & Kenilworth, not Leamington. If you have any thin guide book to those places which might be sent easily by post I should be much obliged if you would send it to me, as I want to do everything obliging for the "Antik Vogel." Emma will explain.

Your affectionate son F. G.

Dear Emma—She is sweeter than ever. F. G.

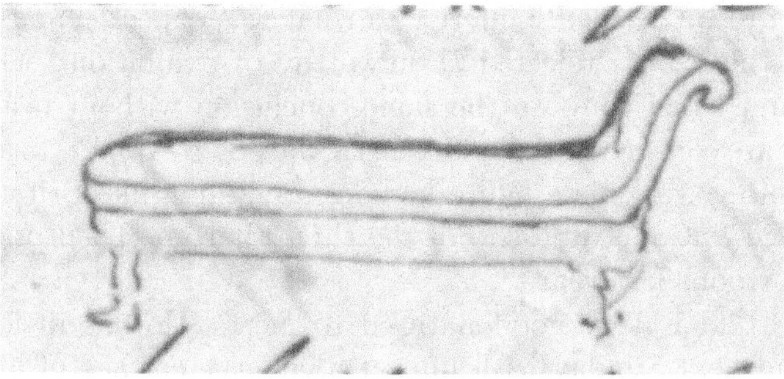

Figure 5.87: 1844-07 FG to STG: Sofa.

1844/07/08. Diary of Samuel Tertius Galton

July 8th. Monday. Leamington.

... Francis came to attend me from London at the latter end of this month. I have been too weak to continue my diary during the latter part of July and beginning of August.

1844/08/19–26. Diary of Samuel Tertius Galton

August 19th. Monday. Leamington.

... Francis and Emma dined at Dr. Jephson's.[51] ...

[50] Hallam, Henry Fitzmaurice. See page 483.
[51] Jephson, Dr. Henry. See page 485.

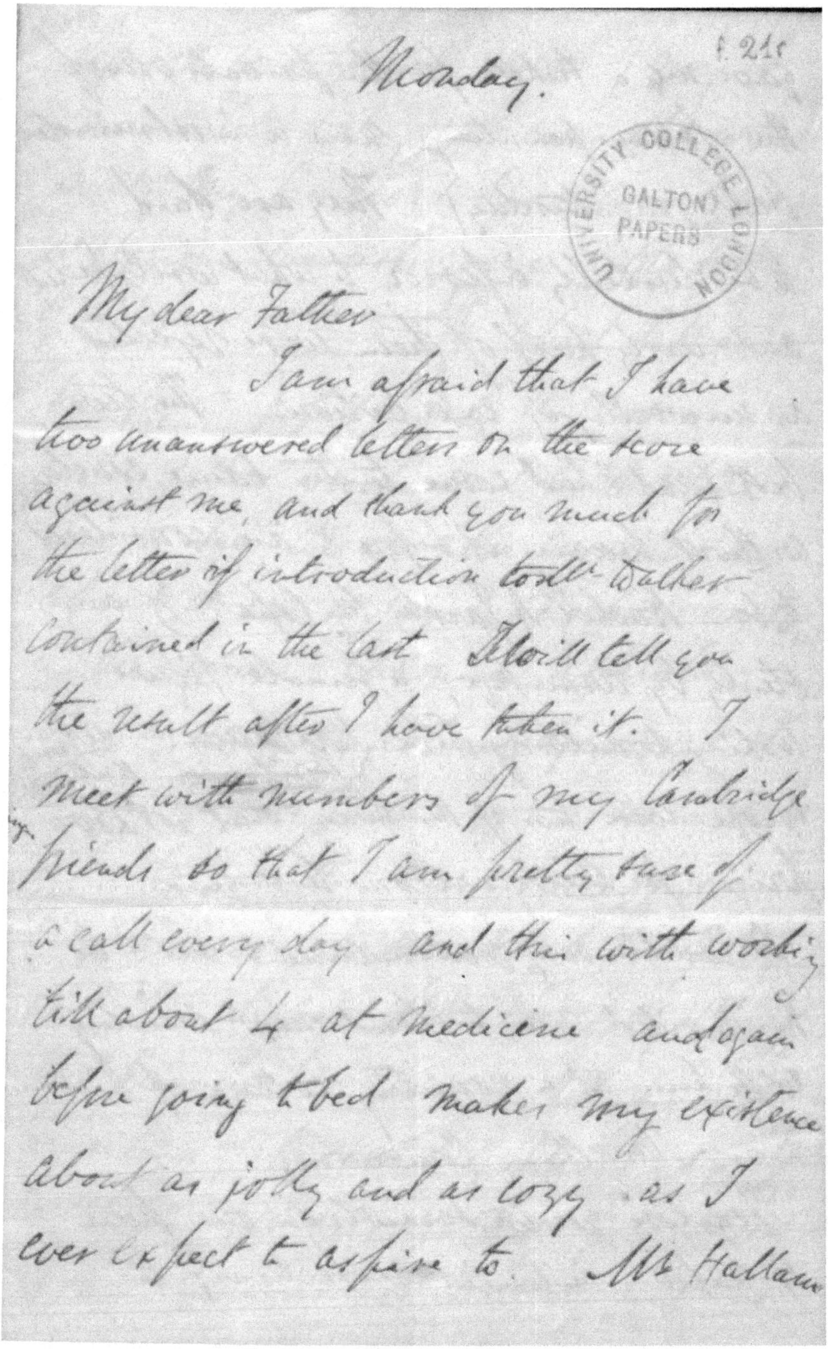

Monday.

f. 215

My dear Father
I am afraid that I have
two unanswered letters on the score
against me, and thank you much for
the letter of introduction to Mr Walker
contained in the last. I will tell you
the result after I have taken it. I
meet with numbers of my Cambridge
friends so that I am pretty sure of
a call every day and this with working
till about 4 at medicine and again
before going to bed makes my existence
about as jolly and as cozy as I
ever expect to aspire to. Mr Hallam

Figure 5.88: 1844-07. FG to STG page 1.

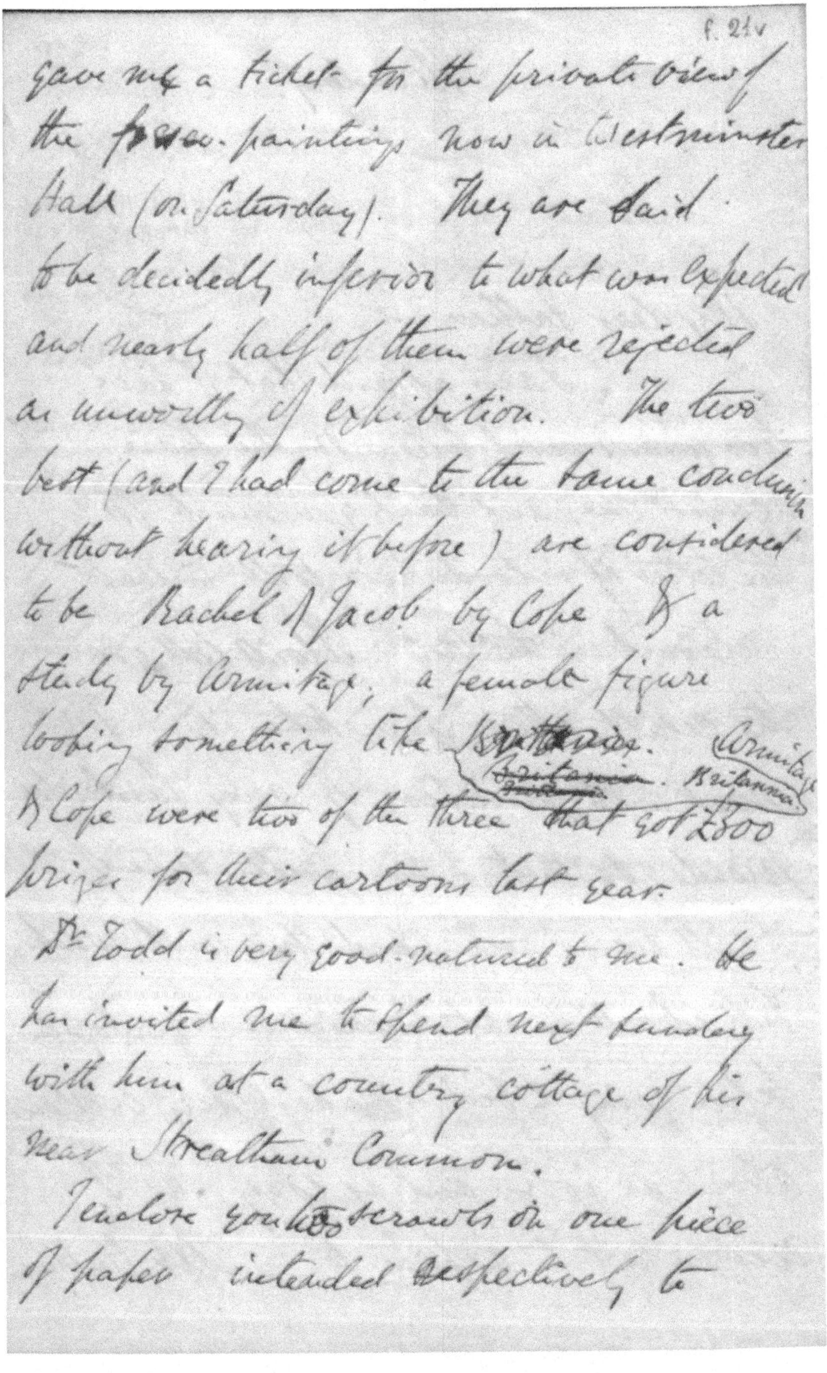
f. 21v

gave me a ticket for the private view of
the fresco-paintings now in Westminster
Hall (on Saturday). They are said
to be decidedly inferior to what was expected
and nearly half of them were rejected
as unworthy of exhibition. The two
best (and I had come to the same conclusion
without hearing it before) are considered
to be Rachel & Jacob by Cope & a
study by Armitage, a female figure
looking something like ~~Britannia~~. Britannia. Armitage
& Cope were two of the three that got £300
prizes for their cartoons last year.
Dr Todd is very good-natured to me. He
has invited me to spend next Sunday
with him at a country cottage of his
near Streatham Common.
I enclose you two scrawls on one piece
of paper intended respectively to

Figure 5.89: 1844-07. FG to STG page 2.

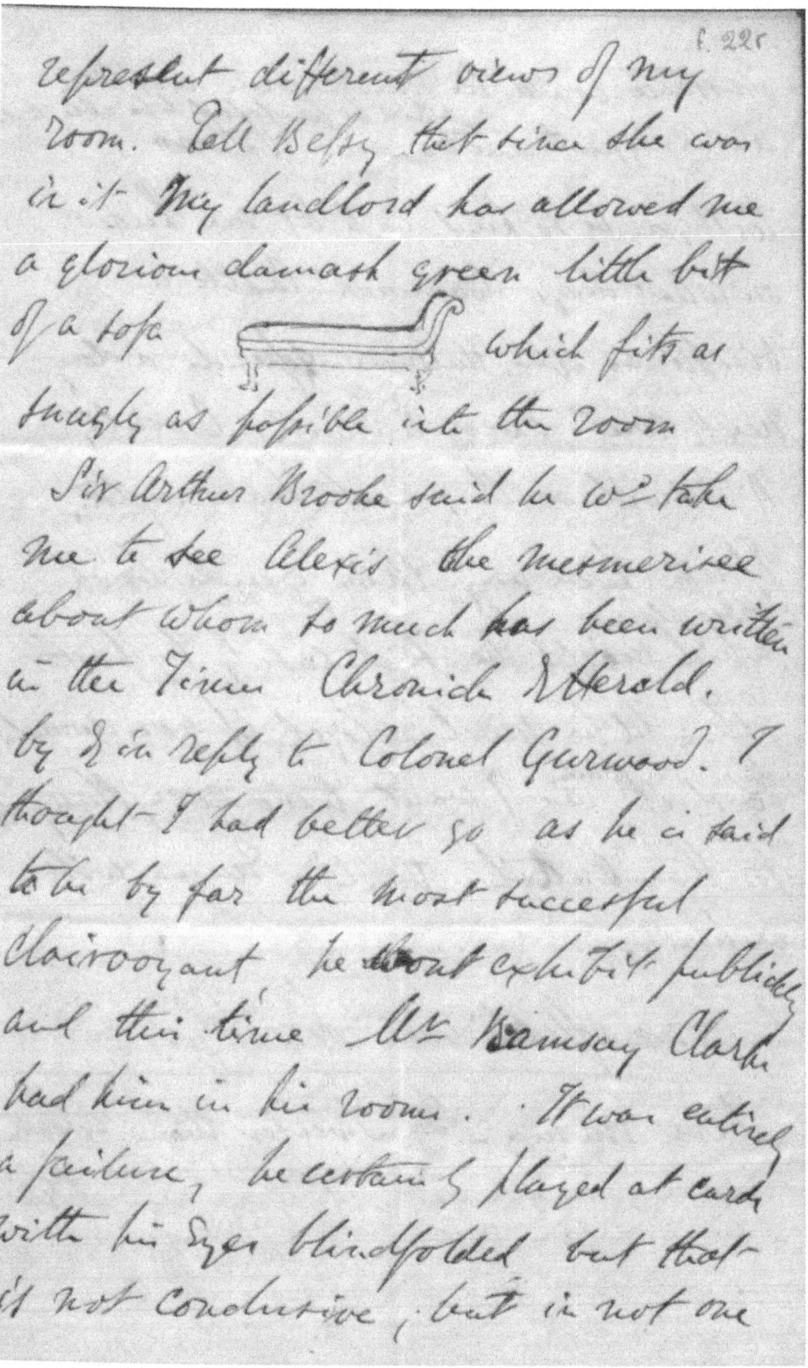

f. 225

represent different views of my
room. Tell Betsy that since she was
in it my landlord has allowed me
a glorious damask green little bit
of a sofa which fits as
snugly as possible into the room
Sir Arthur Brooke said he wd take
me to see Alexis the mesmerisee
about whom so much has been written
in the Times Chronicle & Herald.
by & in reply to Colonel Gurwood. I
thought I had better go as he is said
to be by far the most successful
clairvoyant he exhibits publicly
and this time Mr Ramsay Clarke
had him in his rooms. It was entirely
a failure, he certainly played at cards
with his eyes blindfolded but that
is not conclusive, but in not one

Figure 5.90: 1844-07. FG to STG page 3.

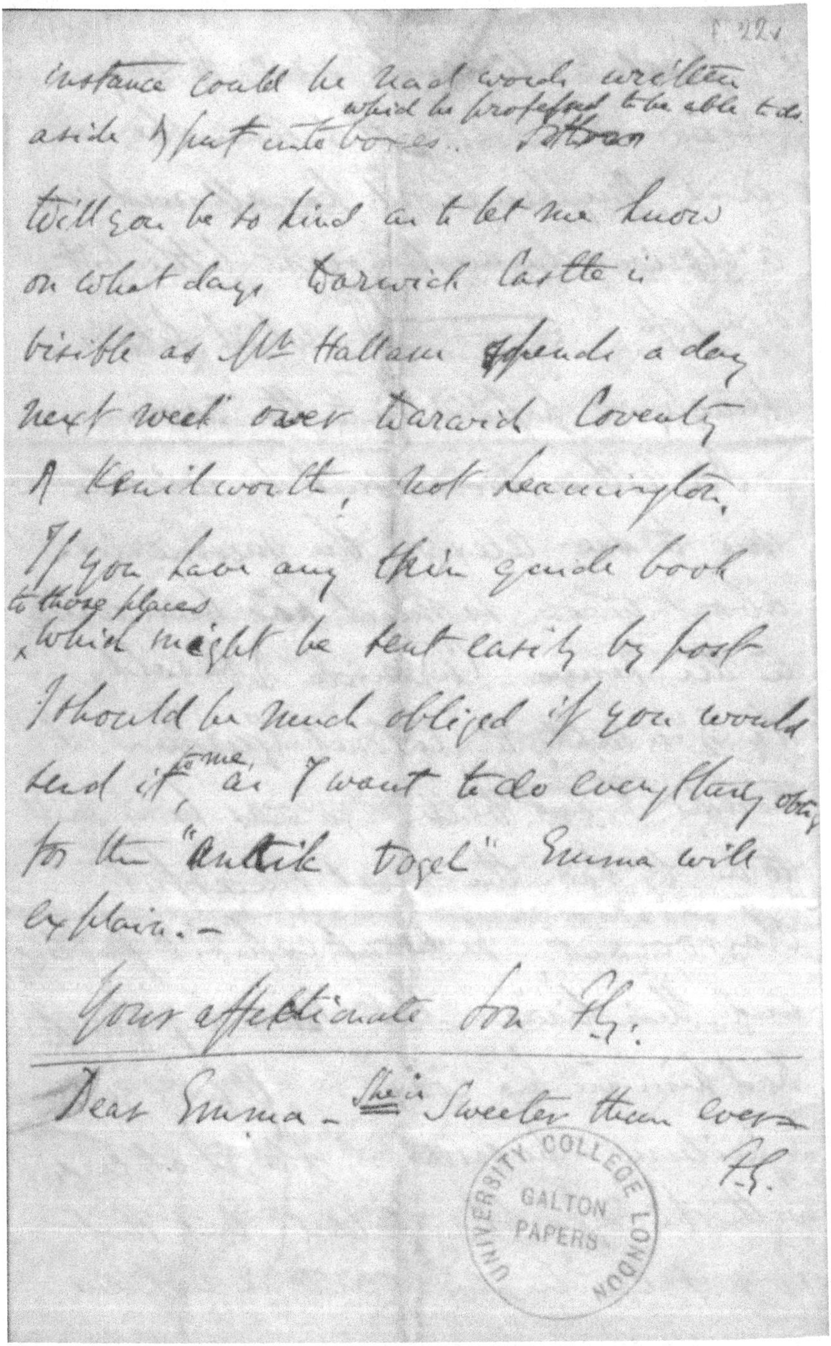

instance could be read words written aside & put into boxes.. which he professed to be able to do [illegible]

Will you be so kind as to let me know on what days Warwick Castle is visible as Mr Hallam spends a day next week over Warwick Coventry & Kenilworth, not Leamington. If you have any thin guide book to those places which might be sent easily by post I should be much obliged if you would send it to me, as I want to do everything cheaply for the "Antik Vogel" Emma will explain.—

Your affectionate son Fy.

Dear Emma — She is sweeter than ever —

F.G.

Figure 5.91: 1844-07. FG to STG page 4.

August 21st. Wednesday. Leamington.

I took a drive with Violetta and Francis by the Milverton railway. ...

August 22nd. Thursday. Leamington.

... I took a drive to Radford with Emma and Francis. Violetta, Emma, Francis and myself played at whist in the evening.

August 26th. Monday. Leamington.

.. Francis occupied himself with putting up a telegraph at Cop Hill. ...

1844/09/09–26. Diary of Samuel Tertius Galton

September 9th. Monday. Leamington.

Emma, Erasmus, Francis and myself drove to Coventry. ... Francis and myself put up at the Euston where Edward Hurst joined us in the coffee room. Francis went in a cab to execute some commissions.

September 10th. Tuesday. London.

Francis and myself breakfasted with Jos. Yates, his wife and daughter and started by the 11:30 a.m. train for Tunbridge and proceeded in a car to the Wells and put up at the Kentish Hotel. We drove to the Rocks, etc.

September 11th. Wednesday. Tunbridge Wells.

Francis called upon Mrs Hubert Galton at Tavistock House abd at 10 a.m. we took a carriage to St. Leonards ...

September 12th. Thursday. St. Leonards

Fine day. I went with Francis to the castle at Hastings. ... I managed my local complaint this day for the first time without assistance. ...

September 13th. Friday. St. Leonards.

I walked with Francis to the nearest Martello tower on the Archery ground and afterwards to Hastings where we called upon Mr. Winthrop whom we found better and met the Priors who have bee staying here some time. We returned in a car. ... I managed my local complaint this day for the second time without assistance. ...

September 17th. Tuesday. St. Leonards.
I walked to Hastings and Francis took a much longer walk in the same direction. ...

September 18th. Wednesday. St. Leonards.
I was taken ill in the night with Cholera and we have therefore determined on remaining here for the present. ...

September 20th. Friday. St. Leonards.
Bright hot summer's day. I took Francis and W Grice a sail of two hours in a boat under the castle hill, Hastings. ...

September 25th. Wednesday. St. Leonards.
I felt very shiverish this morning and did not get up till nearly 11 a.m. From this time I am endeavouring to revert to my practice of having my water drawn three times only in each day. Francis started soon after 7 a.m. for Hastings for the purpose of going in a fishing boat. ...

September 26th. Thursday. St. Leonards.
I was feverish and shivery in the night but much better towards morning.[52]

Medical Lectures Attended

In 1844 Galton collected a number of certificates attesting that he had attended courses of medical lectures at Cambridge over the previous two years.[53]

1. Lectures in Botany, 1843, J S Henslow.

2. Lectures in Medical Jurisprudence, 1844, Dr W. Fisher.[54]

3. Public lectures in Chemistry, 1844, Trinity College Cambridge.

[52] Samuel Tertius Galton died at Hastings on October 23rd 1844, at half-past one.
[53] GALTON/1/2/6/1/2.
[54] William Webster Fisher.

4. Lectures in Material Medicine and Pharmacy, 1844, Dr. W. Fisher.

5. Lectures in Anatomy & Physiology of the Human Body, 1844, W. Clark.

List of Figures

Register

Atwood, H. A. S.

1800.01.13–1877.06.22. Reverend Atwood ran a school in Kenilworth, where he was briefly a curate (not the Vicar, contrary to Karl Pearson 1914 and the many who have echoed that mistake). Galton was educated at his school, along with the Boulton brothers, between August 1832 and December 1834. The following obituary appeared in the *Monthly Notices of the Royal Astronomical Society* 38 (1878), 141. 'The Rev. Henry Adams Sergison Atwood was born on January 13, 1800, in the parish of St. Margaret's, Westminster, where his father, who died young, was curate. His father was a grandson of the Rev. George Atwood, Archdeacon of Taunton, and nephew of George Atwood, Fellow and Tutor of Trinity College, Cambridge, who invented the well-known contrivance called Atwood's machine, which is still used for lecture-room experiments to demonstrate the laws of uniformly accelerated motion. Henry Adams Sergison Atwood was entered at Queen's College, Oxford. He took his M. A. degree in 1824, and shortly afterwards, having obtained orders, he became curate first of Banbury and then of Kenilworth. In 1839 he was collated to the rectory of Ashelworth, in the county of Gloucester, where, in addition to the duties of his parish, he devoted himself most successfully to tuition. Several of his pupils have since taken distinguished positions in the political world. In 1827 he was elected a Fellow of the Astronomical Society, and, consequently, at the time of his death he was one of the oldest Fellows of the Society. He was always devotedly attached to the study of astronomy, though his many duties prevented him from indulging his tastes, as he no doubt would have done if he had been less occupied with his parish and his pupils. In 1860 he accompanied the expedition which was sent out in *H.M.S. Himalaya* to observe the total eclipse of the Sun in Spain. At the time of totality Mr. Atwood was stationed, together with Mr. Francis Galton and Mr. Gray, on a tower outside the little town of La Guardia, near Logroño, from which they had an excellent view of the surrounding country. It had been arranged that Mr. Atwood should devote himself to meteorological observations, but he also found time to observe the changes of tint upon the clouds and sky as the shadow advanced. During totality he observed the positions of several of the red prominences through a small telescope, and scrutinised the structure of the corona. Immediately after totality was over he was fortunate in observing the rays from the cusps of the thin solar crescent which have only been seen and described by a few other observers. An account of his observation is given in vol. xli. of the *Memoirs*. Mr. Atwood died on June 22, 1877'. 73

Aytoun, Roger Sinclair

1823—1904. A contemporary of Galton at Trinity College. He was the son of John Aytoun from Inchdairnie in Scotland. Galton visited the family there in 1842, after accompanying a Cambridge reading party to Aberfeldy. 212, 234, 249, 286, 337, 345

Blakesley, Joseph Williams

1808–1885. Galton's classical tutor at Cambridge, 'an accomplished classical scholar, contemporary with Tennyson and his set, and subsequently Dean of Lincoln' (Galton 1908, 58). He was an early member of the Cambridge Apostles Club, and 21st Wrangler and 3rd Classic in 1830. 2, 121, 454

Boulton, Matthew Piers Watt

1820–1894. The Boultons were old Birmingham friends of the Galton family. Matthew Piers Watt Boulton was at Rev. Atwood's school in Kenilworth with Francis, and also a contemporary at Trinity College in the early 1840s, where they formed a close friendship. He was the grandson of the Birmingham 'toy' manufacturer and steam power pioneer Matthew Boulton (1728–1809), who had been a member of the Lunar Society along with Galton's grandfather Samuel Galton Junior. In later years, after inheriting the family fortune, Boulton became reclusive, though he published various works on philosophy and translations from the classics, and filed a number of mechanical patents, including an early aeronautical component, the aileron.

'Matthew was then [at school], subsequently at Cambridge, and again for some years afterwards, an object of reverence to me. I have known few or any who seemed to me his natural superiors in breadth and penetration of intellect, but he was cursed with a fortune far in excess of his simple though cultured needs, which exacted duties from him that he hated. His large fortune also removed the stimulus which necessity gives for getting through work and having done with it, instead of lingering indefinitely. He consequently grew amateurish, wasting thought on ingenious paradoxes and literary trifles, and failed to check a natural tendency towards recluseness and some other oddities of disposition. He gained the University prizes for Greek and Latin Epigrams at Cambridge in 1841, but did not care to compete for other honours. His artistic sense was of a high and classical order. His ideal, like that of Goethe, was a uniform culture of all the higher faculties. There was nothing ignoble in his nature. Whenever I talked with him about my own occasional annoyances, they seemed to become petty through his broad way of looking at things, I may almost say under the mere influence of his presence. His photograph, which is near me as I write, testifies to a personality that accords with the grandeur of his character. I owe much to his influence, and still remain conscious of the void in my friendships caused by his death very many years ago' (Galton 1908, 19–20). 15, 41, 73, 223, 249, 272–274, 418

Bowman, William

1816–1892. Bowman was from the Birmingham area and had trained as a surgeon under Joseph Hodgson. He accompanied Galton, along with James Russell, on a medical tour to the continent in 1938—'I greatly enjoyed the tour and the companionship of Bowman, from whom I doubtless imbibed and assimilated more than I can now distinguish' (Galton 1908, 25). Later Bowman became a distinguished eye specialist. 444

Bristed, Charles Astor

1820–1874. An American contemporary of Galton's at Trinity College, who moved in the same circles and left an important memoir of the period in his *Five Years in an English University* (1852). Bristed was from a wealthy family in New York City, and was educated at Yale before going to Cambridge. He appears several times in Galtons' letters and diaries from the period, though in his *Memories* Galton dismisses him as 'an American who came to Cambridge for a couple of years or so, and whose racy ways made him everywhere an acceptable guest' (Galton 1908, 77). In fact, Bristed was at Cambridge longer than Galton, and a Prizeman in 1845. 223, 250, 294, 330

Broadley, Anne 'Fanny'

1819-x-02–? A family friend of the Galtons. She helped Francis with the drawings for the *Telotype* in 1849 or 1850. Her diaries survive and are in the New York Public Library. She was the daughter of John and Anne Elizabeth Broadley of Kirk Ella Yorkshire, and married Henry Nicholson Burrows (1809–1859) on June 22, 1845. Burrows was educated at Trinity College, Cambridge, and was a schoolmaster. 21

Dalyell, Robert Alexander Osborne

1821–1886. A Cambridge undergraduate friend of Galton's. 'I received a kindly welcome from time to time after leaving Cambridge, in the homes of not a few of my fellow-undergraduates. One was that of Robert, afterwards Sir Robert Dalyell. His father, Captain Sir William Dalyell, was a naval veteran with a scar across his face left by a severe gash, who had quarters in Greenwich Hospital as one of the Captains in command, the constitution of Greenwich Hospital being then totally different from what it is now. The family consisted of himself, Lady Dalyell, and their two daughters. Numerous friends appeared every Sunday. We visitors walked and had tea, spending healthful and delightful summer afternoons, usually returning to London by river. The life of a young bachelor in not over elegant lodgings is vastly cheered by such occasional outings' (Galton 1908, 78). 249, 273, 337, 338

Eddis, Arthur Shelly

1817–1893. A classics coach at Trinity. He was on the reading party to the Lake District that Galton accompanied in 1841. Later he practiced at the Bar. 122, 136, 162

Hallam, Henry Fitzmaurice

1824–1850. A close friend of Galton from Cambridge, where they were in the Historical Society together. They were part of a circle including Henry Sumner Maine, Charles Astor Bristed and Franklin Lushington. 'Harry' was a son of Henry Hallam the historian, and a younger brother of the poet Arthur Henry Hallam (1811–1833), the lamented subject of Tennyson's *In Memoriam* (1850). He shared his brother's knack for inspiring worshipful devotion, and was a brilliant but uneven and idiosyncratic scholar, gaining a Chancellor's Medal. After failing to be selected as a Fellow at Trinity, he read for the bar instead, joining the Midlands circuit. He died on October 25th 1850 of fever while travelling with his family in Italy, at Siena. The autopsy suggested that Hallam, who was always delicate, could not have expected to live much longer. He was buried inside St Andrew's Church ('obscure and solitary') in Clevedon, North Somerset, along with his siblings. 223, 273, 286, 311, 337, 339, 359, 377, 412, 459, 460

Hallam, Julia Maria Frances

1818–1880. Sister of Henry Fitzmaurice Hallam, daughter of Henry Hallam. She married Sir John Farnaby Lennard in 1852. She toured Germany with her brother Harry, Francis Galton and Emma Galton in 1843. 312

Hodgson, Joseph

1788–1869. A notable surgeon who practiced in Birmingham, mainly at the Birmingham General Hospital. He attended Galton's birth at the Larches, Duddeston, in 1822 and was physician to his father Samuel Tertius. In later years he frequently offered career advice regarding Frank to the Galton family. 22, 28, 49, 97, 114, 157, 170, 193, 194, 224, 244, 274, 288, 311, 318, 340, 399, 411, 439, 444, 454

Hopkins, William

1793–1866. Galton's (second) mathematics coach at Cambridge. Hopkins was one of the two most illustrious freelance coaches in Victorian Cambridge, approached only (in later years) by his own pupil Edward J. Routh. He was Seventh Wrangler himself in 1827, having entered college at the advanced age of 30 after an earlier career as a farmer. He only took pupils in their second year; these included Stokes, Cayley, Sylvester and Thomson. His (second) marriage as a student meant that he could not be elected a fellow, but he earned considerably more as the leading private tutor than most full professors. 'I never had a tutor whom I reverenced and loved so entirely as Hopkins' (Galton 1908, 65). He was part of the committee that recommended Galton as the second General Secretary to the British Association in 1863; Hopkins was the other secretary. 'The office is a very pleasant and gentlemanly one', he wrote to his former pupil, 'I should rejoice if you could be installed as my partner' (Hopkins to Galton, 1863-08-03, GALTON/3/2/2/23). Galton accepted the position, but noticed sad changes, as Hopkins' 'health had by then declined and his fine intellect begun to fail'. Galton's diary for February 1865 already noted 'letters about Hopkins failure of intellect'. Hopkins died in Northumberland House Lunatic Asylum, Green Lanes, Stoke Newington (Craik 2008, 107). 1, 102, 152, 157, 162, 169, 176, 177, 223, 243, 245, 330, 338, 340, 353

Horner, Leonard

1785–1864. A family friend and confidant of the Galtons. He was a linen merchant with interests in geology and educational reform, and a Fellow of the Royal Society. He met Samuel Tertius Galton at Edinburgh University, where Galton was sent for a term in 1803 by his father. 'Leamington, 9th September, 1840. I arrived at Coventry at half past one and got into the Warwick mail, and was soon in Galton's house. They are all very well, Mrs. Galton better, and looking better than she has done for some years. The last account they had from Francis' was from Pesth, and he was to set out next day for Constantinople. They read me his letter, which is an admirable one, evincing great powers of observation. They also read me another letter from him giving an account of an accident which very nearly cost him his life. You remember at the Oxford and Cambridge rowing match, your seeing the collision of a steamer, and three boats run down, he was on board the steamer, and by the shock was thrown headlong overboard, under the paddle, and had he not been a very skilful swimmer and diver, and possessed of rare courage and presence of mind, he must have perished, from the description he gives of the situation in which he was. He does so, not at all boastingly, but playfully. We have seen him more than once since, and he never said a word about it, that I remember. They are a little uneasy at his travelling in those countries in such hot weather, all alone' (Horner 1890, Vol. 2, 19). 6, 10, 32, 53, 305

Jephson, Dr. Henry

1798–1878. An immensely successful doctor who practiced in Leamington, tending to the well-heeled, including Samuel Tertius Galton. He had risen rapidly, starting as an apothecary's apprentice in 1823, and taking time off to complete a medical degree at Glasgow in 1826-8, before returning to his burgeoning practice at Leamington in 1828. At times he earned over £20,000 a year, a very large sum for the times, allowing him to build a mansion, 'Beech Lawn', in Leamington (since demolished). In 1848 he went blind and was forced to retire to the consolations of well-connected philanthropy—though he continued to treat some patients after this, including Francis Galton himself in 1866-7. His success was founded on the inspiration of his personality and his command of patients. His treatments consisted mostly of 'plain' diets—especially avoiding stimulants, but not excluding essentials like sherry and game—outdoor exercise and taking the sulphurous waters of the Spa, internally and externally. One of his stand-by medicines was an emetic preparation called 'the broom'. 247, 287, 288, 321, 382, 384, 399, 435, 439, 454, 460

Kay, Edwin Ebenezer

1822–1897. A fellow student of Galton's at Trinity, the half-brother of Joseph Kay. The Kays were from Lancashire. Eben later became a Judge of the High Court, Chancery. Galton remembered that Kay 'had rooms on the same staircase as myself, and we wasted a great deal of time together, both in term and in my second summer vacation' (Galton 1908, 68). 207, 234, 248, 249, 272, 273, 286, 339

Kay, Joseph

1821–1878. A fellow student of Galton's at Trinity, the half-brother of Eben Kay. He became a judge, like his half-brother, with additional interests in economics and social reform. After graduating in 1844 he immediately took up, for a few years, the previously-neglected position of the official 'Travelling Bachelor of Cambridge', tasked by the Senate to travel through the continent to report back on social conditions there. He exploited this role to advocate domestic social, educational and economic reforms, in several published tracts, to the annoyance of many of his colleagues. *The Spectator* was moved to comment that 'Wherever he has been on the Continent, (excepting perhaps the Austrian dominions,) he finds comparatively a paradise among the poor, and in this country a hell upon earth'; Kay had, they said, all the veracity of an auctioneer (*Spectator*, Vol. 23, 1850, 352). 223, 294, 330, 337, 412

Maine, Henry James Sumner

1822–1888. A close friend of Galton at Cambridge, though he was at Pembroke College. They were both members of the Epigram Society, and associated with H.F. Hallam. Maine went on to become a distinguished jurist and historian. According to Galton, he 'was one of the few non-Trinity men who became thoroughly at home in Trinity itself. In later years, when he had become an eminent jurist and had filled with distinction the highest legal post in India, I used to enjoy long talks with him at the Athenaeum Club, mostly on topics connected with Primitive Culture' (Galton 1908, 66). 294

Mathison, William Collings

1818–1870. The Fifth Wrangler in 1839. He was a mathematics coach at Cambridge when he led the reading party to the Lake District that Galton accompanied in 1841. He went on to be Vice-Master of Trinity. 41, 102, 105, 108, 121, 135, 136, 152, 162

O'Brien, Matthew

1814–1855.viii.23. Galton's first private tutor at Cambridge. The son of a successful doctor, from Ennis in County Clare, he was educated at Ennis, Trinity College Dublin, then Caius College, Cambridge. A pupil of William Hopkins, he attained Third Wrangler in 1838. After a brief junior fellowship at Caius in 1840-1, O'Brien was Professor of Natural Philosophy and Astronomy at King's College, London, from 1844 to 1854. From 1849 he was also a lecturer in astronomy at the Royal Military Academy in Woolwich. He published treatises on calculus, geometry and mathematical geography, and anticipated many of the methods of vector analysis and differential geometry. He resigned from Kings College in 1854 when he was made Professor of Mathematics at Woolwich. O'Brien became ill in 1855 and travelled to Petit Ménage in Jersey in search of health, but died that August, his 40th year. He was succeeded at Woolwich by J. J. Sylvester. 6, 10, 21, 41, 73, 102

Ossolinski, Vladimir Boris

1808?-1859. The Polish Count Vladimir Boris Ossolinski, of Chestnut Hill, Ambleside, Westmorland (his name is sometimes given as 'Ossalinsky'). He married the wealthy heiress Mary Jackson (1820–1902) in 1839, at Crosthwaite Church in Keswick. He was 31 at the time of his marriage (*Transactions of the Cumberland & Westmorland Antiquarian & Archaeological Society*, Volume 98, 1998, 268). Galton met him in 1841 while on a Cambridge summer 'reading party' to the Lake District. 'My story is that of a Polish Count, O., who appeared at Keswick with scant introductions, took a house, and made himself most agreeable. I fell at once under his influence, for he seemed to me extraordinarily accomplished. He had all sorts of books and instruments, and even a tame monkey! So the Count throve and prospered for a while. But a lady resident in the neighbourhood who had been connected in her youth with one of the German Courts, and who studied the Almanach de Gotha and the like, insisted that the Count's claims to the title were totally unfounded. So a small warfare raged' (Galton 1908, 62-63).

Ossolinksi seems to have arrived in the Lake District some

time in the mid to late 1830s. On March 3, 1838, *The Westmorland Gazette* reported that he was one of the notables promenading on an iced-over Bassenthwaite Lake. George Ticknor met him in Kendall on May 8, 1838, through the poet Robert Southey. 'Southey received us as usual, in his nice and somewhat peculiar library, but seemed more sad, and abstracted even, than he did when we last saw him. One of his daughters only was at home, Bertha, a very pleasing person; and there was, besides, Mrs. Lovell, the sister of his late wife, and a Polish Count, a very intelligent man, who seemed to have travelled everywhere' (George Ticknor, *Life, Letters, and Journals* (Sampson Low, 1876), Volume 2).

Ossolinksi's wife Mary was a daughter of the Keswick attorney Edward Washington Jackson. The family had originally made money in the silk trade in London, and owned a large estate on the Armboth Fells in the Thirlmere valley. In May 1839 the *Cumberland Pacquet* was pleased to 'announce the marriage of Count Ossalinsky, of Chestnut Hill, Keswick, and Miss Jackson, the amiable and accomplished grand-daughter of Wilson Jackson, of Armboth House, Esq. The Count has resided at Keswick several years, is of a noble polish family, and was obliged to leave his native country owing to his devoted patriotism, and consequent resistance of the tyranny of the Northern Autocrat. Ever since his arrival in England, the Count has conducted himself as became his station and high character, and has always moved in the best Society. The lovely and highly accomplished lady, whom the Count has chosen for his bride, has not yet attained her twentieth year. The marriage caused great rejoicings at Keswick, and those joyous instruments on such occasions, the parish bells, rung many merry peal at intervals throughout the day, whilst a splendid flag waved triumphantly from the steeple of the Church' (*Cumberland Pacquet, and Ware's Whitehaven Advertiser*, Tuesday 21 May 1839).

On January 11, 1840, the *Carlisle Journal* reported that 'On Christmas-eve, Count Ossalinsky entertained upwards of thirty mechanics, whom he has at present in employment about his premises, at Keswick, with a good substantial supper: the Count has also given various sums of money. &c. to a number of poor people in the neighbourhood of his residence'. On March 1 that year, the Ossalinskis had a son, Vladimir Boris Jackson Ossalinsky, who later served in the Royal Cumberland Regiment of Militia. In April 1841 they had a daughter, Nathalia.

Ossolinski was a member of the British Association for the Advancement of Science, and was listed as such between 1843 and 1846. He seems to have been active as an amateur scientist. In August 1841, he was lecturing local audiences about a floating island which had suddenly appeared in Derwent Lake on the 19th of July. 'Monday the 2nd instant, this phenomenon was made the subject of one of Count Ossalinsky's lectures, delivered at Miss Marshall's School-room, at Keswick, where a numerous and highly respectable audience were entertained by his experiments upon the gas which he had obtained upon the spot' (*Kendal Mercury*, Saturday 21 August 1841, 3).

There is a further sighting of Ossolinksi on July 15, 1845, in the *Cork Commercial Courier*. 'The Count Ossalinsky, arrived Valentia Hotel, Kerry, on the 7th inst, where he intends remaining some time, having taken one of Mr. Blennerhassett's new Bathing Lodges'. Again, on Saturday 30 October 1847, the *Western Times* locates him at the Royal York Hotel in Sidmouth.

According to Galton, the Count fleeced and then deserted his wife: 'the Count won the affections of a simple girl, the orphan child of a somewhat wealthy 'statesman,' that is what we should call a yeoman farmer. He married her, and afterwards ran away with as much of her money as he could get hold of, leaving her with the questionable title of Countess as her only consolation. This finale occurred after I had left' (Galton 1908, 62-63).

After his death in 1859 his widow was well-known in the Lake District as 'Countess Ossalinsky', and gained some notoriety by holding out against plans to submerge her land under a reservoir for the city of Manchester, demanding a large sum in compensation (she was initially offered £20,000, but after arbitration obtained £70,000). His son used the surname 'Jackson rather than 'Ossolinski' or 'Ossalinsky'. 122

Partridge, Richard

1805.i.19–1873.iii.25. Professor of Anatomy at King's College in Clare Market, London. Galton lodged with him while he was a medical student there. Partridge had gained fame in 1831 for turning in the London 'Burkers', who had tried to sell him the freshly killed corpse of a boy for dissection. In 1837 he was admitted to the Royal Society. He was also nominated in 1862 to travel to Italy to attend to Garibaldi, but failed to spot a bullet in the politician's ankle. 32, 49, 195, 305, 411, 435, 444

Perry, Charles

1807–1891. Senior Wrangler and Smith's Prize winner in 1828. He later became the Bishop of Melbourne. Galton attended his mathematics lectures at Cambridge. 6, 41, 102

Russell, James

1818–1885. A physician, originally trained at Kings College, who was attached to the Birmingham General Dispensary, and later to the General Hospital there. He accompanied Galton, along with William Bowman, on a medical tour of Europe in 1838. (See *Photographs of Eminent Medical Men, of All Countries* Volumes 1-2 edited by Thomas Herbert Barker (London: Churchill 1868): 108.) Karl Pearson refers to him only as 'Russell' (Pearson 1914, 92). He later lectured at Queen's and then Sydenham College, and practiced at 91 New Hall Street, Birmingham. The Russells were an old Quaker family, long familiar to the Galtons. His father James was also a surgeon, and known as a philanthropist. See the Royal College of Surgeons *Lives of the Fellows* Munk's Roll IV, where he is memorialized as 'a nonconformist and a Liberal'. 53, 243, 405

Strickland, Frederick

1820.ii.18–1849.ix.18. A friend of Galton from his Cambridge days. He was the second son of Sir George Strickland, M.P., from Bridlington. Galton and Strickland were at Trinity College and went on a reading party to Keswick together. 'Strickland was the son of a well-known Yorkshire baronet. He joined me in many pleasant walks from London after my college days, of which I especially recollect one in the then rural Isle of Wight, when there was little more than a single house at Shanklin, and that was its pretty, rustic hotel. The times of travel from London so fitted in, that the walk from Ryde about Easter-time began well before twilight, and we reached Shanklin not too late to be taken in, and to thoroughly enjoy the moonlit evening. Strickland was a strong swimmer, but he got into some difficulty next morning owing to the surf and undercurrents at the place where he entered the sea. He returned safely to shore, to my great relief, but much tired from long battling with the water. His end was tragic. It occurred in North America, when winter had just set in, near some well-known watering-place whose name I forget, separated by a low range of hills from another watering place about sixteen miles off. The road between the two was perfectly simple and easy in summer, but not so in the snowdrifts and darkness of winter. Strickland would attempt it, though much was said to dissuade him: he never reached his destination. A relief party tracked his wanderings. He seemed to have acted as one demented by the hardship, for he had stripped off his clothes and thrown them away, one after the other, even his boots, so that his dead body was almost wholly undressed. That was the story I heard from two persons' (Galton 1908, 63-64). Strickland was educated at Rugby, and was admitted as pensioner at Trinity in 1838, and graduated BA in 1843. His brothers Charles W. and Henry were also at Cambridge. He died in the White Mountains in northern New Hampshire, on October 19th 1849, attempting to descend Mount Washington, in the Ammonoosuc ravine, near Notch House. An Englishman, a Mr. Carpenter, accompanied him but turned back with their guide. Strickland is buried in Mount Auburn Cemetery, Cambridge, Mass. His brother Henry had been travelling with him in the United States but had left for home before Frederick made for Mount Washington. Other accounts confirm that he stripped off his trousers and drawers. (See Frank H Burt 'First to die on Mount Washington' (*Appalachia*, 1938) reprinted in *Mount Washington: Narratives and Perspectives* edited by Mike Dickerman, History Press, 2017) Also, *Annual Register*, January 1850, 17.) 135

Bibliography

Bristed, Charles Astor (1852). *Five Years in an English University.* Reprinted as An American Victorian in Cambridge edited by Christopher Stray, 2008. New York: Putnam, 1852.

Craik, A. D. D. (2008). *Mr. Hopkins' Men.* London: Springer Verlag, 2008.

Galton, Francis (1908). *Memories of My Life.* London: Methuen, 1908.

Horner, Leonard (1890). *Memoir of Leonard Horner. Consisting of Letters to His Family and from Some of His Friends.* Ed. by K. M. Lyell. 2 volumes. London: Women's Printing Society, 1890.

Pearson, Karl (1914). *The Life, Letters and Labours of Francis Galton. Volume I: Birth 1822 to Marriage 1853.* Cambridge: Cambridge University Press, 1914.

Made in United States
North Haven, CT
05 November 2026

21760261R00278